Computational Design of Chemicals for the Control of Mosquitoes and Their Diseases

QSAR in Environmental and Health Sciences

Series Editor

James Devillers
CTIS-Centre de Traitement de l'Information Scientifique
Rillieux La Pape, France

Aims & Scope

The aim of the book series is to publish cutting-edge research and the latest developments in QSAR modeling applied to environmental and health issues. Its aim is also to publish routinely used QSAR methodologies to provide newcomers to the field with a basic grounding in the correct use of these computer tools. The series is of primary interest to those whose research or professional activity is directly concerned with the development and application of SAR and QSAR models in toxicology and ecotoxicology. It is also intended to provide the graduate and postgraduate students with clear and accessible books covering the different aspects of QSARs.

Published Titles

Computational Design of Chemicals for the Control of Mosquitoes and Their Diseases, *James Devillers,* 2018

Computational Methods for Reproductive and Developmental Toxicology, *Donald R. Mattison,* 2015

Computational Approaches for the Prediction of pKa Values, *George C. Shields and Paul G. Seybold,* 2014

Juvenile Hormones and Juvenoids: Modeling Biological Effects and Environmental Fate, *James Devillers,* 2013

Three Dimensional QSAR: Applications in Pharmacology and Toxicology, *Jean Pierre Doucet and Annick Panaye,* 2010

Endocrine Disruption Modeling, *James Devillers,* 2009

Computational Design of Chemicals for the Control of Mosquitoes and Their Diseases

Edited by
James Devillers

CRC Press is an imprint of the
Taylor & Francis Group, an informa business

CRC Press
Taylor & Francis Group
6000 Broken Sound Parkway NW, Suite 300
Boca Raton, FL 33487-2742

Printed on acid-free paper

International Standard Book Number-13: 978-1-4987-4180-4 (Hardback)
International Standard Book Number-13: 978-1-315-15165-6 (eBook)

Library of Congress Cataloging-in-Publication Data

Names: Devillers, James, 1956-
Title: Computational design of chemicals for the control of mosquitoes and their diseases / [edited by] James Devillers.
Description: Boca Raton : CRC Press, [2018] | Includes bibliographical references and index.
Identifiers: LCCN 2017016932 | ISBN 9781498741804 (hardback : alk. paper) | ISBN 9781315151656 (ebook)
Subjects: LCSH: Mosquitoes as carriers of disease. | Mosquitoes--Control. | Insecticides. | Insecticides--Health aspects.
Classification: LCC RA640 .C56 2017 | DDC 571.9/86--dc23
LC record available at https://lccn.loc.gov/2017016932

Visit the Taylor & Francis Web site at
http://www.taylorandfrancis.com

and the CRC Press Web site at
http://www.crcpress.com

Contents

Series Introduction

The correlation between the toxicity of molecules and their physicochemical properties can be traced to the nineteenth century. Indeed, in a French thesis entitled *Action de l'alcool amylique sur l'organisme* (*Action of Amyl Alcohol on the Body*), which was presented in 1863 by A. Cros before the Faculty of Medicine at the University of Strasbourg, an empirical relationship was made between the toxicity of alcohols and their number of carbon atoms, as well as their solubility. In 1875, Dujardin-Beaumetz and Audigé were the first to stress the mathematical character of the relationship between the toxicity of alcohols and their chain length and molecular weight. In 1899, Hans Horst Meyer and Fritz Baum, at the University of Marburg, showed that narcosis or hypnotic activity was in fact linked to the affinity of substances to water and lipid sites within the organism. At the same time, at the University of Zurich, Ernest Overton came to the same conclusion, providing the foundation of the lipoid theory of narcosis. The next important step was made in the 1930s by Lazarev in St. Petersburg who first demonstrated that different physiological and toxicological effects of molecules were correlated with their oil–water partition coefficient through formal mathematical equations in the form: $\log C = a \log P_{oil/water} + b$. Thus, the quantitative structure–activity relationship (QSAR) discipline was born. Its foundations were definitively fixed in the early 1960s by the seminal works contributed by C. Hansch and T. Fujita. Since that period, the discipline has gained tremendous interest and now QSAR models represent key tools in the development of drugs as well as in the hazard assessment of chemicals. The rather new REACH (registration, evaluation, authorization, and restriction of chemicals) legislation on substances, which recommends the use of QSARs and other alternative approaches instead of laboratory tests on vertebrates, clearly reveals that this discipline is now well established and is an accepted practice in regulatory systems.

In 1993, the journal *SAR and QSAR in Environmental Research* was launched by Gordon and Breach to focus on all the important works published in the field and to provide an international forum for the rapid publication of SAR (structure–activity relationship) and QSAR models in (eco)toxicology, agrochemistry, and pharmacology. Today, the journal, which is now owned by Taylor & Francis and publishes three times more issues per year, continues to promote research in the QSAR field by favoring the publication of new molecular descriptors, statistical techniques, and original SAR and QSAR models. This field continues to grow very rapidly, but many subject areas that require larger developments are unsuitable for publication in a journal due to space limitation.

This prompted us to develop a series of books entitled *QSAR in Environmental and Health Sciences* to act in synergy with the journal. I am extremely grateful to Colin Bulpitt and Fiona Macdonald for their enthusiasm and invaluable help making the project become a reality.

This book is the sixth of the series and the very first dedicated to the use of QSAR and other *in silico* techniques to find new molecules acting as repellents against mosquitoes or killing their larvae and adults. The design of new drugs against the numerous diseases transmitted by mosquitoes is also addressed. Last, exposure modeling in mosquito control is also discussed through different scenarios of indoor and outdoor treatments.

I gratefully acknowledge Hilary Lafoe for her willingness to assist me in the development of this series.

James Devillers

Acknowledgments

I am extremely grateful to the authors of the chapters for their agreeing to participate in this book and for preparing valuable contributions. To ensure the scientific quality and clarity of the book, each chapter was sent to two referees for review. I would like to thank all the referees for their useful comments. A special thank to Jean-Baptiste Ferré (EID Méditerranée) to offer me the possibility to use one of his mosquito pictures for the cover of the book. Finally, I would like to thank all the publication team at CRC Press for making the publication of this book possible.

Contributors

Natasha M. Agramonte
USDA Agricultural Research Service
Center for Medical, Agricultural, and Veterinary Entomology
Gainesville, Florida

Rahul B. Aher
Drug Theoretics and Cheminformatics Laboratory
Department of Pharmaceutical Technology
Jadavpur University
Kolkata, India

Chanan Angsuthanasombat
Institute of Molecular Biosciences
Mahidol University, Salaya Campus
Nakhon Pathom, Thailand

and

Laboratory of Molecular Biophysics and Structural Biochemistry
Biophysics Institute for Research and Development
Bangkok, Thailand

Stephen J. Barigye
Department of Chemistry
Federal University of Lavras
Lavras, MG, Brazil

and

Facultad de Medicina
Universidad de Las Américas
Quito, Pichincha, Ecuador

Ulrich R. Bernier
USDA Agricultural Research Service
Center for Medical, Agricultural, and Veterinary Entomology
Gainesville, Florida

Apurba K. Bhattacharjee
Department of Microbiology and Immunology
Georgetown University School of Medicine
Washington, DC

Marie-Michelle Clémente
Centre de Démoustication et de Recherches Entomologiques-Lutte Anti Vectorielle
Collectivité Territoriale de Martinique
Fort-de-France, France

James Devillers
Centre de Traitement de l'Information Scientifique
Rillieux La Pape, France

Jean-Pierre Doucet
Interfaces, Traitements, Organisation et Dynamique des Systèmes
Université Paris-Diderot
Paris, France

Annick Doucet-Panaye
Interfaces, Traitements, Organisation et Dynamique des Systèmes
Université Paris-Diderot
Paris, France

Mariene H. Duarte
Department of Chemistry
Federal University of Lavras
Lavras, MG, Brazil

Elias Eliopoulos
Department of Biotechnology
Agricultural University of Athens
Athens, Greece

Sébastien Estaran
Entente Interdépartementale pour la Démoustication du Littoral Méditerranéen
Montpellier, France

Matheus P. Freitas
Department of Chemistry
Federal University of Lavras
Lavras, MG, Brazil

Jorge Galvez
Molecular Topology and Drug Design Research Unit
Department of Physical Chemistry
Faculty of Pharmacy
University of Valencia
Valencia, Spain

María Galvez-Llompart
Molecular Topology and Drug Design Research Unit
Department of Physical Chemistry
Faculty of Pharmacy
University of Valencia
Valencia, Spain

Ramón García-Domenech
Molecular Topology and Drug Design Research Unit
Department of Physical Chemistry
Faculty of Pharmacy
University of Valencia
Valencia, Spain

Chalermpol Kanchanawarin
Theoretical and Computational Biophysics Laboratory
Department of Physics
Faculty of Science
Kasetsart University
Bangkok, Thailand

Christophe Lagneau
Entente Interdépartementale pour la Démoustication du Littoral Méditerranéen
Montpellier, France

Armand Lattes
Laboratoire des Interactions Moléculaires et Réactivités Chimiques et Photochimiques
Université Paul Sabatier
Toulouse, France

Kenneth J. Linthicum
USDA Agricultural Research Service
Center for Medical, Agricultural, and Veterinary Entomology
Gainesville, Florida

Hubert Matondo
Laboratoire des Interactions Moléculaires et Réactivités Chimiques et Photochimiques
Université Paul Sabatier
Toulouse, France

Alexander A. Oliferenko
EigenChem Technologies Inc.
Alachua, Florida

Polina V. Oliferenko
EigenChem Technologies Inc.
Alachua, Florida

Girinath G. Pillai
Department of Chemistry
University of Florida
Gainesville, Florida

and

Department of Chemistry
University of Tartu
Tartu, Estonia

Gennady Poda
Drug Discovery Program
Ontario Institute for Cancer Research
and
Leslie Dan Faculty of Pharmacy
University of Toronto
Toronto, Ontario, Canada

Phisit Pouyfung
Department of Biochemistry
Faculty of Science
Mahidol University
Ratchathewi, Bangkok, Thailand

Aruna Prasopthum
Department of Biochemistry
Faculty of Science
Mahidol University
Ratchathewi, Bangkok, Thailand

Pornpimol Rongnoparut
Department of Biochemistry
Faculty of Science
Mahidol University
Ratchathewi, Bangkok, Thailand

Kunal Roy
Drug Theoretics and Cheminformatics Laboratory
Department of Pharmaceutical Technology
Jadavpur University
Kolkata, India

Trias Thireou
Department of Biotechnology
Agricultural University of Athens
Athens, Greece

Katerina E. Tsitsanou
Institute of Biology, Medicinal Chemistry and Biotechnology
National Hellenic Research Foundation
Athens, Greece

André Yébakima
Centre de Démoustication et de Recherches Entomologiques-Lutte Anti Vectorielle
Collectivité Territoriale de Martinique
Fort-de-France, France

Riccardo Zanni
Molecular Topology and Drug Design Research Unit
Department of Physical Chemistry
Faculty of Pharmacy
University of Valencia
Valencia, Spain

Spyros E. Zographos
Institute of Biology, Medicinal Chemistry and Biotechnology
National Hellenic Research Foundation
Athens, Greece

1 Repurposing Insecticides and Drugs for the Control of Mosquitoes and Their Diseases

James Devillers

CONTENTS

ABSTRACT

Neglected diseases transmitted by mosquitoes mainly affect the poorest people in developing countries, where they are responsible for important morbidity, mortality, and economic losses. Since the development of new insecticides and drugs is lengthy and expensive, repurposing strategies have been proposed as attractive fast-track approaches to speed up the identification process of new molecules. In this chapter, the repurposing studies aiming at finding new potential larvicides and adulticides to control mosquitoes are reviewed. In the same way, drug repurposing examples for malaria, dengue, West Nile fever, Zika, Japanese encephalitis, chikungunya, and Rift Valley fever are analyzed. The pros and cons of the methods applied to neglected diseases are discussed.

KEYWORDS

- Insecticides
- Xenobiotics
- Drug
- Repurposing
- Malaria
- *Flaviviridae*
- *Togaviridae*
- *Bunyaviridae*

1.1 INTRODUCTION

Unfortunately, mosquito-borne diseases (MBDs) are flourishing worldwide. MBDs include malaria, dengue, West Nile fever, chikungunya, yellow fever, Japanese encephalitis, Saint Louis encephalitis, Western equine encephalitis, Eastern equine encephalitis, Venezuelan equine encephalitis, La Crosse encephalitis, and Zika fever [1–7]. It is worth noting that MBDs represent the leading cause of morbidity and mortality among the poorest populations in developing countries [8,9]. MBDs are emerging or resurging as a result of insecticide resistance, genetic modifications in pathogens, climate changes, demographic and societal changes, and/or alterations in public health policy [10–17].

Insecticides play a central role in controlling populations of mosquitoes. In the early 2000s, implementation of the European Directive 98/8/EC [18] obliged manufacturers to perform and submit to the European Commission a complete toxicological, ecotoxicological, and biological evaluation of all substances and biocidal products marketed in Europe, especially those intended for public health. These substances include insecticides, acaricides, and other compounds used to fight against arthropods, as well as the insecticides used for mosquito control.

The cost of such an approach and its requirements in terms of human and environmental protection, which reflect societal demands in terms of hazard and risk, are so important that, although they are effective, biocides that do not fit these demands are not supported and, hence, are withdrawn from the market. However, if the withdrawal of existing products is not followed by the introduction of biocides showing more favorable toxicological and ecotoxicological profiles, this will undoubtedly lead to the reduction of the range of molecules available and suitable for vector control.

This situation is particularly annoying because the alternate use of insecticides is necessary to have a more effective vector control action. In addition, this is the best strategy to reduce the phenomena of resistance to insecticides [19,20].

Currently, in France, vector control is made almost exclusively with *Bacillus thuringiensis* subspecies *israelensis* (*Bti*) and deltamethrin to fight larvae and adults of mosquitoes, respectively. Unfortunately, cases of resistance to pyrethroids and also organophosphorus insecticides used as adulticides are commonly observed [10,11,20–24]. These resistances decrease the toxicity of the substances against the targeted organisms and, as a result, reduce the efficacy of the treatment campaigns.

The search for new insecticides can be made through the use of quantitative structure–activity relationship (QSAR) models and related *in silico* approaches [25; see also Chapters 8 to 11].

This can also be made by using molecules already marketed for other uses, such as in agriculture or veterinary medicine. This supposes that suitable formulations for vector control exist.

On the frontline of fighting the diseases, there is also a lack of effective, safe, and affordable drugs to control mosquito-borne diseases. Unfortunately, the situation is the same for all neglected diseases. Trouiller et al. [26] found that, between 1975 and 1999, among the 1,393 new pharmaceutical entities marketed, only 16 were for tropical diseases and tuberculosis. Pedrique et al. [27] showed that among the 850 new therapeutic products registered between 2000 and 2011, only 37 (4%) were indicated for neglected diseases, including 25 products with a new indication or formulation and eight vaccines or biological products. Only four new chemical entities were approved for neglected diseases (three for malaria, one for diarrheal disease), accounting for 1% of the 336 new chemical entities approved during the study period. In the same way, among the 148,445 clinical trials registered until the end of 2011, only 2,016 (1%) were for neglected diseases [27]. The pharmaceutical industry argues that research and development of new drugs is expensive, lengthy, complex, and risky (with most new compounds only reaching patients after a 10- to 12-year development process) to invest in low-return pharmaceuticals for neglected diseases. Different complementary strategies are used to overcome this problem. Use of two-dimensional (2D) and three-dimensional (3D) QSARs and other *in silico* techniques allows us to propose candidate drugs rapidly and at reduced cost [28–40]. The open source drug discovery process also provides a valuable strategy for discovering and developing new drugs. It offers the opportunity to speed up the discovery process while keeping expenditures to a minimum by encouraging collaborations between volunteer scientists. To be considered as open source, a research project must comply with the following set of attributes [41]:

1. The project's data must be open access, meaning that anyone can view the protocol, and free of charge.
2. The project must provide a forum for open collaboration across organizations.
3. The project must be governed by a defined set of rules that mandates the project's "openness."

There are numerous advantages to sharing knowledge and resources without any kind of boundary. Thus, for example, in the drug discovery process, numerous candidate molecules are discarded due to excessive toxicity, because they have poor metabolic profiles, and so on. Access to these structures is pivotal for driving computational drug discovery processes. Many industrial companies actively discourage publication of discontinued drug discovery projects to reduce the likelihood of providing a competitor any kind of advantage that such information could provide [42]. On the other hand, open research initiatives guarantee access to all tested molecules.

Thus, for example, for about a decade, more than six million compounds were screened against asexual-stage *Plasmodium falciparum*, at GlaxoSmithKline, Novartis, and two academic centers (St. Jude, Memphis, Tennessee, and Eskitis, Australia), resulting in the identification of more than 20,000 compounds active in the low- to submicromolar range. The structures of the 20,000 antimalaria hits were made available in ChEMBL (www.ebi.ac.uk/chembl). Cluster analysis and commercial availability reduced this to a set of 400 representative compounds assembled in the 'Malaria Box,' which was made freely accessible to researchers [43,44]. Another interesting collaborative action is the OpenZika project (http://openzika.ufg.br/), based on distributed computing on millions of computers, Android-based tablets, or smartphones in over 80 countries, to dock tens of millions of drug-like compounds against crystal structures and homology models of Zika proteins [45].

The third strategy that can be used to find new drugs for mosquito-borne diseases relies on the use of drugs marketed for other purposes. Drug repurposing, also termed drug repositioning, is the identification of secondary uses for existing and developmental drugs, with or without the help of *in silico* tools. It offers significant potential benefits such as a reduction of costs, a lower risk of toxicity, and opportunities for patent protection of the new use [46,47].

In this context, the goal of this chapter is to focus on strategies used for insecticide and drug repurposing in order to control mosquitoes and the diseases they carry.

1.2 LARVICIDE AND ADULTICIDE REPURPOSING

1.2.1 Marketed Insecticides

In the frame of a collective expertise directed by ANSES (French Agency for Food, Environmental and Occupational Health and Safety), an attempt was made to evaluate whether insecticides already marketed for specific uses could be used in vector control [48]. The SIRIS (system of integration of risk with interaction of scores) method [49] was used to rationalize the selection of these potential insecticides. Thus, 129 substances potentially active on mosquitoes were described by their (eco)toxicity (i.e., oral LD_{50} in rat, carcinogenicity, endocrine disruption potential; LD_{50} in honey bee; EC_{50} in *Daphnia magna*; LC_{50} in fish [mainly *Danio rerio* and *Oncorhynchus mykiss*]) and environmental fate (1-octanol/water partition coefficient [log *P*], vapor pressure, level of use, primary and ultimate biodegradation potential).

Data were retrieved from literature, obtained from expert judgments, or computed from QSAR and quantitative structure–property relationship (QSPR) models. In the SIRIS method [49], the values of the selected criteria (variables), which can be qualitative or quantitative, are transformed into modalities coded as favorable (f) or unfavorable (d) or as favorable (f), moderately favorable (m), or unfavorable (d). The variables are then ranked according to their decreasing order of importance. Indeed, unlike other multicriteria methods, it is not necessary to introduce coefficients, which are very often difficult to justify, in the calculation procedure to modify the final weight of the selected variables. Finally, a min/max scale of scores is calculated according to specific incremental rules [49]. Both categories of parameters were considered separately to construct two scales of SIRIS scores, from which a SIRIS map can be drawn. In addition, two scenarios, one for the larvicides and another for the adulticides, were considered [48] due to the constraints imposed by the corresponding vector control strategies.

Thus, for example, the typology of the 129 insecticides characterized by their two SIRIS scores calculated from the two hierarchies selected for the larvicide scenario is displayed in Figure 1.1. The SIRIS scores range from 0 to 81 on the toxicity/ecotoxicity scale and from 10 to 52 on the exposure/environmental fate scale. The insecticides located in the left bottom part of Figure 1.1 are the most interesting because they present low score values on the toxicity/ecotoxicity and exposure/environmental fate scales. This is the case, for example, for insecticides numbers 58 (cyromazine) and 86 (dicyclanil). Conversely, the insecticides located in the top right part of Figure 1.1 are not suited for the selected scenario due to the high values of their SIRIS scores on both scales. Thus, for example, insecticide number 117 (dichlofenthion) shows bad scores on both toxicity/ecotoxicity and exposure/environmental fate scales. A substance can be correctly classified on one scale but with a high score on the other scale. Thus, formetanate (61) is well located on the exposure/environmental fate scale but shows a rather high score on the toxicity/ecotoxicity scale. This is also the case for deltamethrin (11), located in the right bottom part of Figure 1.1. *Bacillus sphaericus* str. 2362 (*Bs*) (4) and *Bti* (5) are perfectly ranked on the toxicity/ecotoxicity scale, having SIRIS scores equal to zero, but they are a slightly less well ranked on the other scale. This is even more so the case for insecticide numbers 88 (azadirachtin) and 127 (silafluofen). The insecticides located in

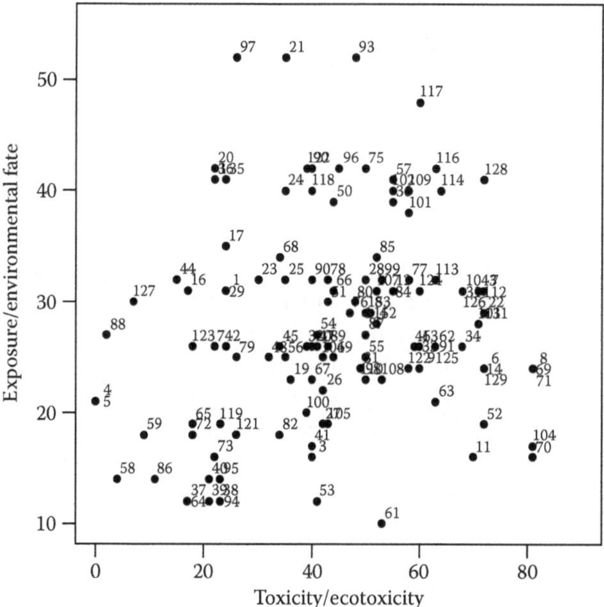

FIGURE 1.1 SIRIS map of the 129 insecticides used or potentially usable as larvicides. (J. Devillers, in *Juvenile Hormones and Juvenoids. Modeling Biological Effects and Environmental Fate*, J. Devillers, ed., CRC Press, Boca Raton, FL, 2013, pp. 341–381.)

the middle part of Figure 1.1, such as metaflumizone (68) or tau-fluvalinate (85), are less interesting because they show rather high SIRIS scores on both scales. However, this does not mean that they have to be excluded from a future selection. Indeed, a map of SIRIS scores has to be considered as a decision support tool. There are obvious selections for insecticides with low SIRIS score values on both scales, revealing favorable modalities for most of the criteria. Conversely, a rejection is also obvious for insecticides with high SIRIS score values on the toxicity/ecotoxicity and exposure/environmental fate scales. Between these two extremities, all the intermediate situations exist. An insecticide with a low SIRIS score on only one scale can be selected after a reasoned decision. In this case, the SIRIS map allows us to pinpoint the criteria showing the worst modalities and, hence, for which it will be necessary to pay special attention. In order to interpret more thoroughly the SIRIS map, specific information can be graphically projected on it.

The same strategy was used for the scenario for the adulticides (results not shown). However, the SIRIS method was applied to 114 insecticides because, in our initial list, some insecticides can be used only as larvicides (e.g., *Bti*). The criteria were the same but if their hierarchy for the exposure/environmental scale did not change, the one for the toxicity/ecotoxicity scale was different from the hierarchy used for the larvicides [48].

Use of this methodological strategy allowed us to propose a list of insecticides that could be used as adulticides and/or larvicides (e.g., formothion, acetamiprid, thiacloprid, cyromazine, thiamethoxam). Obviously, the different insecticides that have emerged from the SIRIS analysis are only potential candidates, and risk assessment procedures are required before their use can be envisaged. In addition, it is obvious that even if the work is of interest, it needs to be updated for at least the three following reasons:

1. In the case of missing data for one criterion, the modality was considered as unfavorable. Unfortunately, for some toxicity/ecotoxicity criteria, the information was scarce, leading to a penalization of the score of the chemical on the toxicity/ecotoxicity scale, the importance of which depended on the location of the corresponding criterion in the hierarchy. Filling these gaps by new experimental data or from the use of new available QSAR models will lead to a refinement of the scores and to a more accurate location of a number of chemicals on the SIRIS maps.
2. The list of criteria for classifying the larvicides and adulticides is perfectible. Thus, for example, it should be absolutely necessary to also consider the toxicity of chemicals against aquatic and terrestrial birds in the SIRIS selection of larvicides and adulticides, respectively.
3. It is obvious that it would be fruitful to extend the list of chemicals by including natural compounds, some main metabolites, and so on.

1.2.2 Marketed Drugs

G protein-coupled receptors (GPCRs) are highly targeted for the development of human drugs. They account for the main best-selling drugs and about 40% of all prescription pharmaceuticals on the market [50]. GPCRs have been identified in the genome of *Aedes aegypti* [51] and other mosquitoes [52]. These findings have provided a basis for the design of new insecticides. Biogenic amine-binding GPCRs are components of the nervous system of eukaryotes and include receptors that bind various neurotransmitters, among them dopamine, which plays key roles in arthropods. Andersen et al. [53] showed that dopamine levels increased in females of *Ae. aegypti* following a blood meal, suggesting that dopamine might be involved in ovarian and/or egg development. Dopamine receptors (DRs) are classified as either D1- or D2-like, based on their functional roles. Both have been identified in the *Ae. aegypti* genome and used as a starting point by Meyer et al. [54] for the development of new mode-of-action insecticides. Molecular analysis was used to functionally characterize the putative DRs in *Ae. aegypti*, named AaDOP1 and AaDOP2. Their deduced amino acid (AA) sequences were analyzed to identify conserved structural features typically associated with biogenic amine-binding GPCRs and unique regions that could be potentially used for the design of new insecticides. Only a modest level of similarity was observed between the *Ae. aegypti* and human D1-like DRs, which shared between 47% and 54% AA identity among the transmembrane (TM) domains that represent the most conserved regions of GPCRs. In addition, comparison of the predicted TM domains from numerous invertebrate and vertebrate D1-like DRs showed that only 34% of the AAs were shared among all species included in the alignment. Heterologous expression of AaDOP1 and AaDOP2 in HEK293 cells showed dose-dependent responses to dopamine. Only AaDOP1 exhibited sensitivity to epinephrine and norepinephrine. AaDOP1 and AaDOP2 were not stimulated by histamine, octopamine, serotonin, or tyramine. Meyer et al. [54] selected the AaDOP2 receptor for an antagonist screen of the Library of Pharmacologically Active Compounds (LOPAC 1,280) due to its low constitutive activity and strong dopamine response. Fifty-one potential antagonists of the AaDOP2 receptor were identified; among them, twenty are known antagonists of mammalian dopamine receptors. Ten compounds were selected and tested for their activity in cAMP accumulation assays. Three of them revealed no significant antagonistic effects against AaDOP2. The 10 selected hit compounds were also tested against the human D1 receptor (hD1) for species comparison purposes. Analysis of the results showed that amitriptyline and doxepin presented more than 30-fold selectivity for AaDOP2 compared with hD1. Both compounds were tested in *Ae. aegypti* larval bioassays. A single dose point test at 400 μM led to 93% mortality for amitriptyline and 72% mortality for doxepin, both after 24 hours of exposure. LC_{50} = 78 μM and LC_{90} = 185 μM were observed for amitriptyline. The lead strategy was successfully applied to *Culex quinquefasciatus* and *Anopheles gambiae* and new candidate pharmacological molecules have been proposed as larvicides for the three species of mosquitoes [55–57]. Even if the lead discovery strategy is interesting, it is obvious that the ecotoxicological impact of the candidate molecules must not be occulted. Thus, for example, if the antipsychotic drug chlorpromazine (Figure 1.2) presents LC_{50} values of 88 ± 9 μM and 64 ± 9 μM in 72 hours against larvae of *Ae. aegypti* and *Cx. quinquefasciatus* [56], the molecule is very toxic against fish. Thus, Li et al. [58] showed that the LC_{50} values of chlorpromazine on the goldfish (*Carassius auratus*) were equal to 1.11, 0.43, and 0.32 mg/L after 24, 48, and 96 hours of exposure. The activity of

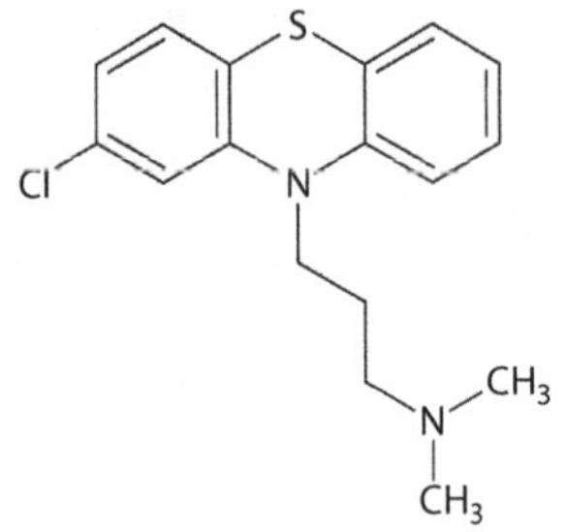

FIGURE 1.2 Structure of chlorpromazine.

superoxide dismutase and catalase in goldfish livers was significantly influenced by chlorpromazine. Munro [59] revealed that the behavior of *Aequidens pulcher*, including aggressiveness and activity, was changed by chlorpromazine. Adverse effects on aquatic invertebrates have also been observed. Thus, Oliveira et al. [60] showed that the catalase activity was significantly altered in *Daphnia magna* at a concentration of 0.001 mg/L of chlorpromazine.

1.3 DRUG REPURPOSING

1.3.1 Drugs Repurposed for Malaria

In 2015, the World Health Organization (WHO) estimated that 212 million cases of malaria occurred worldwide, leading to about 429,000 deaths [61]. Malaria, which is transmitted by *Anopheles* mosquitoes, is caused by parasites belonging to the genus *Plasmodium*. The majority of deaths (99%) are due to *P. falciparum*. *P. vivax* is estimated to have been responsible for 3,100 deaths in 2015 with 86% occurring outside Africa [61]. A number of drugs have shown their efficacy against malaria. They include the historical quinine and its derivatives, such as chloroquine (Figure 1.3), primaquine, and mefloquine, and also antifolates (e.g., sulfadoxine–pyrimethamine) and artemisinin derivatives [62]. The therapeutic potential of these drugs has decreased

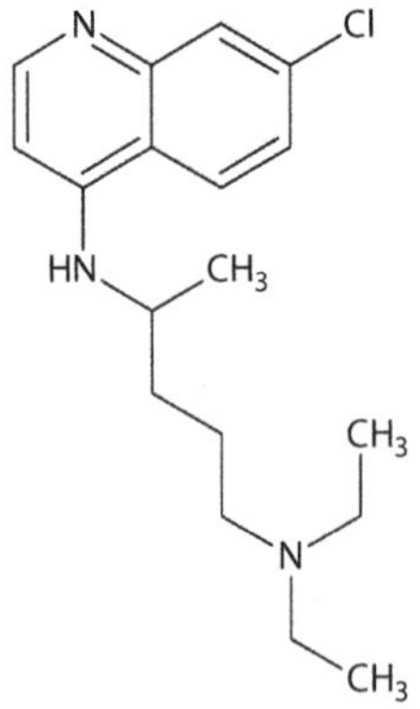

FIGURE 1.3 Structure of chloroquine.

due to the emergence of resistance in *Plasmodium*. Artemisinin-based combination therapy has been recommended by WHO [61] as the first line of treatment for multidrug-resistant *P. falciparum* infection, but, unfortunately, the frequent use of artemisinin and its derivatives has also resulted in phenomena of resistance, including for the partner drugs [63]. Different strategies have been used to find new active molecules for antimalarial chemotherapy. They have been thoroughly presented in recent valuable reviews [64–66]. Among them, the repurposing of drugs has been proposed. Thus, Yuan et al. [67] used quantitative high-throughput screening to identify antimalarial drugs. Sixty-one parasite lines were screened against the National Institutes of Health (NIH) Chemical Genomics Center pharmaceutical collection containing 2,816 chemicals registered or approved for human or animal use. The compounds were tested at eight fivefold serial dilutions using a growth inhibition assay against 61 parasite lines. From 171,776 assays, 32 compounds were identified that inhibited the growth of at least 45 parasite lines with IC_{50} values ≤ 1 μM (Table 1.1). Genome-wide association and linkage analysis were then used to identify compounds that had similar or different mechanisms of action, which might predict cross-resistance and hence guide combination therapy [67].

In order to investigate the likely best candidates for practical use as new potential antimalarial therapy, Oprea and Overington [68] checked whether:

- the 32 drugs were already approved for human use,
- for the dosage/absorption route, the oral dosage was preferred,
- special clinical monitoring should be avoided.

Among the 32 compounds listed in Table 1.1, only 8 presented the desired profile of being orally dosed and currently approved for human usage. Of these, six showed restrictive uses or appeared to be poorly tolerated. This left two potential candidates—namely, dextroamphetamine saccharate and orlistat. The former is a psychostimulant subject to highly controlled distribution to avoid abuse problems. Orlistat is used in the treatment of obesity. Its primary function is to prevent the absorption of fats from human food. Consequently, it is not suited to be used in underweight, malnourished populations that very often live in malaria-endemic regions. Oprea and Overington [68] concluded that even if some of the compounds identified by Yuan et al. [67] could serve as prototypes for further development, there was no candidate for immediate clinical studies.

Compound libraries of GlaxoSmithKline, Pfizer, and AstraZeneca—including drugs that have undergone clinical studies in other therapeutic areas, but not achieved approval—and a set of US FDA-approved drugs and other bioactives were tested *in vitro* against different strains of *P. falciparum* [69]. Chemicals with low micromolar activity were evaluated for toxicity, clinical safety, and human pharmacokinetics on the basis of available data. Compounds that were active *in vitro* and with an acceptable safety/pharmacokinetic profile were tested *in vivo*.

About 4,140 compounds were tested *in vitro* and, among them, 37 showed EC_{50} values at the micromolar level or below, but only two, lestaurtinib and UK-112,214, satisfied the conditions to be tested *in vivo*. The former is a protein kinase inhibitor with EC_{50} values of 0.51 and 0.30 μM on *P. falciparum* 3D7 strain (chloroquine

TABLE 1.1
Compounds Highly Active against Multiple *P. falciparum* Isolates

Compound	Lines Active	Mean IC_{50} (M)	Chiral Center
Ecteinascidin 743	46	7.15E-10	Yes
Gramicidin	51	1.52E-09	Yes
Decoquinate	47	2.40E-09	No
Epothilone B	61	3.22E-09	Yes
Actinomycin D	57	4.89E-09	Yes
Homoharringtonine	61	2.27E-08	Yes
Monensin sodium salt	61	3.29E-08	Yes
Docetaxel	58	5.18E-08	Yes
Buquinolate	59	1.03E-07	No
Narasin	61	1.12E-07	Yes
Bortezomib	61	1.50E-07	Yes
Mupirocin	58	1.88E-07	Yes
Fumagillin	45	2.03E-07	Yes
Alazanine triclofenate	61	2.36E-07	No
Aclarubicin hydrochloride	61	3.00E-07	Yes
Emetine	60	3.28E-07	Yes
Lestaurtinib	60	3.41E-07	Yes
Deserpidine	61	3.50E-07	Yes
Dextroamphetamine saccharate	60	3.52E-07	Yes
Plicamycin	51	3.52E-07	Yes
Demecarium bromide	59	3.90E-07	No
Paclitaxel	59	5.29E-07	Yes
Acriflavinium hydrochloride	60	6.72E-07	No
Zinc pyrithione	60	6.82E-07	No
Suberoylanilide[a]	58	7.30E-07	No
Orlistat	46	7.43E-07	Yes
Lasalocid sodium	60	8.18E-07	Yes
Clotrimazole	54	8.20E-07	No
Salinomycin	57	8.83E-07	Yes
Puromycin hydrochloride	60	8.92E-07	Yes
Lauryl isoquinolinium bromide	57	9.26E-07	No
Ciclesonide	50	9.36E-07	Yes

Source: J. Yuan et al., Science 333 (2011), pp. 724–729.

[a] In fact, suberoylanilide hydroxamic acid (T.I. Oprea and J.P. Overington, Assay Drug Develop. Technol. 13 (2015), pp. 299–306).

sensitive) and K1 strain (chloroquine resistant), respectively. UK-112,214 is a dual platelet activating factor receptor/histamine H1 receptor antagonist with EC_{50} values of 0.55 and 0.60 μM on *P. falciparum* 3D7 and K1 strains, respectively. The *in vivo* efficacy of these two compounds was evaluated with the *P. falciparum* huSCID mouse model. CEP-1347, a related compound of lestaurtinib and valspodar, a nonimmunosuppressive cyclosporine derivative developed primarily as a p-glycoprotein inhibitor, was also tested against *P. falciparum* Pf3D7$^{0087/N9}$ in the humanized mouse model. Last, two discontinued drugs, AZ-1 and AZ-3, were also evaluated *in vivo* with the *P. berghei* 4-day suppression test with the aim to find lead compounds for an optimization program in case of efficacy. Unfortunately, none of these compounds met the candidate selection criteria, hence warranting further developments [69].

Other attempts have been made to reposition existing drugs for use as antimalarial compounds [64,70]. It is interesting to note that the antihelminthic drug ivermectin (Figure 1.4), when administered to humans to control onchocerciasis and lymphatic filariasis, also showed an effect on malaria vectors. Thus, Kobylinski et al. [71] showed that ivermectin reduced the proportion of *P. falciparum* infectious *An. gambiae sensu stricto* in treated villages in southeastern Senegal.

1.3.2 Drugs Repurposed for Dengue Virus

Flaviviridae is a family of positive, single-stranded, enveloped RNA viruses including three genera: *Flavivirus*, *Hepacivirus*, and *Pestivirus*. The dengue virus belongs to the genus *Flavivirus* and has four distinct serotypes (DENV-1, DENV-2, DENV-3, and DENV-4). Dengue is an *Aedes*-borne disease causing a severe flu-like illness leading sometimes to a lethal complication called severe dengue. Females of *Ae. aegypti* and *Ae. albopictus* can transmit the virus [72]. Bhatt et al. [73] estimate that there are 390 million (95% credible interval 284–528) dengue infections per year, of which 96 million (67–136) manifest apparently (any level of disease severity). Brady et al. [74] estimate that 3.97 billion people, in 128 countries, are at risk of infection with dengue viruses. A large number of medicinal chemistry studies have been performed for discovering antidengue drugs. These studies have been thoroughly analyzed in excellent reviews [75–78]. Some of them focus on the repurposing of existing drugs. The antimalarial chloroquine (Figure 1.3) was tested by Farias et al. [79] against Vero and C6/36 cells infected with DENV-2. When compared to untreated cells, a statistically significant inhibition of virus production in infected Vero cells was obtained at 50 μg/mL of chloroquine. Moreover, the concentration was not toxic to the cells. In C6/36 cells, chloroquine does not induce a statistically significant difference in viral replication when compared to the control [79]. A randomized, double-blind clinical study was performed by administering chloroquine or placebo for 3 days to 129 patients with dengue-related symptoms [80]. Of these patients, 37 were confirmed as having dengue and completed the study. Thus, 19 dengue patients received chloroquine versus 18 that received a placebo. There was no significant difference in the duration of the disease or the degree and days of fever. However, 12 patients (63%) with confirmed dengue reported a substantial decrease in pain intensity and a significant improvement in their ability to perform daily activities [80]. In another study [81] performed in a hospital in Ho Chi Minh City (Vietnam), 153 dengue patients received the 3-day treatments while 154 patients received a placebo. No measurable impact of chloroquine on virological or immunological parameters of dengue virus infection in young adults was observed. No convincing evidence that chloroquine reduced the time to fever resolution in adults with dengue was noted [81]. Boonyasuppayakorn et al. [82] showed that the antimalarial drug amodiaquine inhibited DENV-2 infectivity measured by plaque assays, with

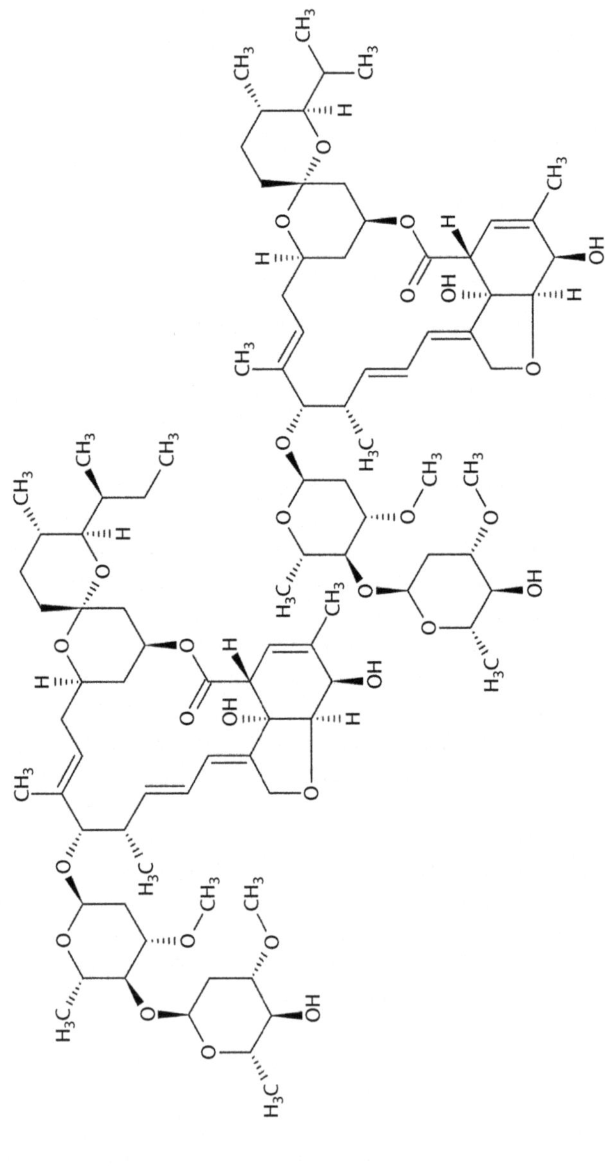

FIGURE 1.4 Structure of ivermectin (B1a: left; B1b: right).

EC_{50} and EC_{90} values of 1.08 ± 0.09 μM and 2.69 ± 0.47 μM, respectively, and DENV-2 RNA replication measured by *Renilla luciferase* reporter assay, with EC_{50} value of 7.41 ± 1.09 μM in the replicon expressing cells. Cytotoxic concentration (CC_{50}) in BHK-21 cells was equal to 52.09 ± 4.25 μM [82].

Dengue virus nonstructural (NS) 5 protein plays various functions in the cytoplasm of infected cells, enabling viral RNA replication and counteracting host antiviral responses [83]. Although dengue viral RNA replication takes place in the cytoplasm of the infected cell, NS5 is predominantly located in the nucleus [83–85]. There are functional signals for NS5 nuclear import and export NS5 under the dependence of the transporters importin (IMP) heterodimer IMPα/β1 and CRM1 (exportin 1) [84,85]. Tay et al. [85] showed that the antihelminthic drug ivermectin (Figure 1.4) inhibited recognition of both DENV-1 and DENV-2 NS5 by IMPα/β1 with similar half-maximal inhibitory concentration (IC_{50}) values of 2.3 and 1.5 μM, respectively. Antiviral action of ivermectin was confirmed in Huh-7 cells for DENV 1,2 using an established plaque assay [85]. Vero cells were treated with or without ivermectin for 3 hours before infection with DENV-2. The antihelminthic drug almost completely abolished virus production when used at 50 μM and significantly reduced virus production at 25 μM [86]. Mastrangelo et al. [87] showed that the concentration of ivermectin required to inhibit viral RNA synthesis by 50% (EC_{50}) in Vero cells infected with DENV-2 New Guinea C was equal to 0.7 μM. For comparison purposes, an EC_{50} = 0.0005 μM was obtained in Vero cells infected with yellow fever virus 17 D vaccine strain. Using a radiolabeled assay, the authors demonstrated that ivermectin inhibited NS3 helicase activity with an IC_{50} = 500 ± 70 nM versus 122 ± 10 nM for the yellow fever virus helicase [87].

Suramin (Figure 1.5), used for the treatment of human trypanosomiasis and onchocerciasis [88], also inhibits NS3 helicase activity of dengue virus with an inhibitory constant Ki of 0.75 ± 0.03 μM in a noncompetitive manner [89].

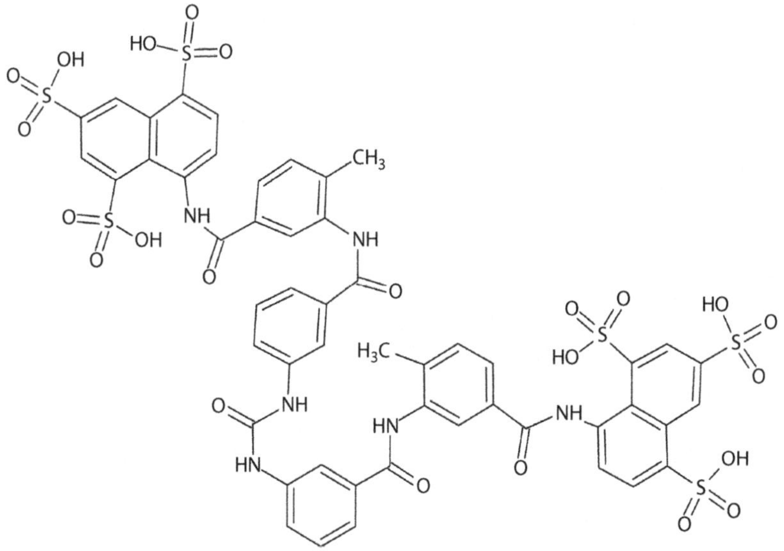

FIGURE 1.5 Structure of suramin.

Because pentoxifylline has been shown to blunt the proinflammatory actions of tumor necrosis factor-α (TNFα), it was tested by Salgado et al. [90] on children aged 7 months to 13 years with dengue hemorrhagic fever. Intravenous infusion of pentoxifylline at 12.5 mg/kg/day, during three consecutive days was performed in 28 children, while 27 other children received a placebo. A statistically significant decline in TNFα values was recorded 24 hours after pentoxifylline administration relative to the control group. This effect was more pronounced in the subgroup of patients with >grade III dengue hemorrhagic fever. No significant difference was found at 48 hours [90].

Prochlorperazine is clinically approved to treat headache, nausea, and vomiting, which are also symptoms commonly found in dengue patients. It is a dopamine D2 receptor antagonist and belongs to the phenothiazine class of antipsychotic agents. Simanjuntak et al. [91] showed that prochlorperazine (up to 30 μM) had no significant effect on cell viability, cell proliferation, or cytotoxicity in several cell lines, such as human kidney HEK293T, lung A549, microglia CHME3, and monocytic THP-1 cells and mouse neuroblastoma N18 cells. Treatment with prochlorperazine significantly reduced DENV-2 viral protein expression and viral progeny production in a dose-dependent manner in HEK293T cells, with an EC_{50} value of 88 nM [91]. The molecule can block DENV-2 infection by targeting viral binding and viral entry through D2R- and clathrin-associated mechanisms, respectively [91].

Zhang et al. [92] demonstrated that the aminoglycoside geneticin prevented the cytopathic effect resulting from DENV-2 infection of BHK cells in a dose-dependent manner with an EC_{50} = 3 μg/mL. Geneticin also inhibited DENV-2 viral yield with an EC_{50} = 2 μg/mL and an EC_{90} = 20 μg/mL. Furthermore, 25 μg/mL of geneticin nearly completely blocked plaque formation induced by DENV-2 [92]. Narasin is a polyether antibiotic active *in vitro* against Gram-positive bacteria, anaerobic bacteria, and fungi and is effective in protecting chickens from coccidial infections [93]. Low et al. [94] showed that narasin presented an IC_{50} < 1 μM against the four serotypes of dengue virus. They demonstrated that the chemical was highly effective in reducing infectious virus production by inhibiting dengue virus replication at the postentry early phase [94]. Minocycline (Figure 1.6) is

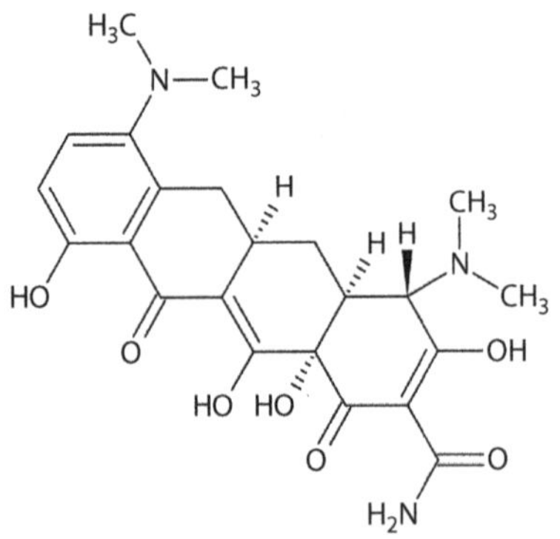

FIGURE 1.6 Structure of minocycline.

a second-generation semisynthetic derivative of tetracycline and has well-known antibacterial effects. The molecule possesses anti-inflammatory, anti-oxidant, anti-apoptotic, and immunomodulatory effects [95]. Minocycline was found to be effective in reducing dengue virus infection, and this antiviral effect was confirmed in all four serotypes. It reduces viral RNA synthesis, intracellular viral protein synthesis, and infectious virus production. The molecule reduces the phosphorylation of ERK1/2, which is associated with enhanced pathogenesis and organ injury in dengue virus infection [96].

Other drug repurposing examples for dengue virus are discussed in the next subsection.

1.3.3 Drugs Repurposed for West Nile Virus

West Nile virus, which also belongs to the genus *Flavivirus*, was first isolated in 1937 from the blood of a febrile female patient in the West Nile district of northern Uganda [97]. It is maintained in nature in a mosquito–bird–mosquito cycle primarily involving *Culex* sp. (especially *Cx. pipiens* and *Cx. quinquefasciatus*). Birds represent the natural reservoir of the virus. Bird infection is not species dependent even if West Nile virus seems to be particularly virulent against *Corvidae* that are consequently used as sentinels for detecting and tracking the virus [98]. A rather high number of mammalian species are susceptible to be naturally or experimentally infected with the West Nile virus, but naturally acquired diseases in mammals have been conclusively demonstrated in human beings and equines only [99,100]. Although most people acquire the West Nile virus following the bite of an infected female mosquito, other routes of transmission have been described, including via blood transfusion, solid organ transplant, and congenital infection, as well as laboratory accidents [101]. Most West Nile viral infections are symptomless. Otherwise, the incubation period is approximately 2–14 days. Uncomplicated West Nile fever typically begins with sudden onset of fever, headache, and myalgia, often accompanied by gastrointestinal symptoms [99]. Overall, less than 1% of individuals infected with epidemic West Nile virus will develop neuroinvasive disease, although the proportion increases with age [101]. To date there is neither effective human vaccine nor operative antiviral therapy available to treat West Nile virus infections. Efforts have been made for designing West Nile virus inhibitors [102,103]. Attempts have been also made to use drugs normally dedicated to other purposes.

Mastrangelo et al. [87] showed that the concentration of ivermectin (Figure 1.4) required to inhibit viral RNA synthesis by 50% (EC_{50}) in Vero cells infected with West Nile virus strain NY99 was equal to 4 μM. Using a radiolabeled assay, the authors demonstrated that ivermectin inhibited NS3 helicase activity with an IC_{50} = 350 ± 40 nM.

The effect of minocycline (Figure 1.6) on West Nile virus strain NY385-99 was investigated by Michaelis et al. [104] in different human brain-derived cell types including human brain neurons (HBNs), human retinal pigment epithelial (HRPE) cells, and T98G cells derived from a patient with stage IV glioma. The antibiotic inhibited cytopathic effect formation in T98G cells and significantly reduced infectious West Nile virus titers in all three cell types tested in a dose-dependent manner.

IC_{50} values obtained were equal to 5.56 ± 0.45, 5.01 ± 1.23, and 4.23 ± 0.99 µg/mL for HBN, HRPE, and T98G cells, respectively. Treatment of T98G cells with 40 µg/mL minocycline resulted in a low but significant decrease in cell viability. Lower concentrations were not effective on T98G cell viability. Conversely, minocycline did not influence cellular viability of HBN or HRPE cells at concentrations up to 40 µg/mL. Minocycline inhibited West Nile virus-induced apoptosis and suppressed virus-induced activation of c-Jun N-terminal kinase (JNK) and its target c-jun [104].

It has been envisaged that inhibitory compounds that act on related viruses, such as the hepatitis C virus (HCV), may also be effective on West Nile virus. The subject has been reviewed by Parkinson and Pryde [105] also including the case of dengue virus. Moreover, the recent review of de Clercq and Li [106] represents a gold mine of information on antiviral drugs.

1.3.4 Drugs Repurposed for Zika Virus

Zika virus is also a member of the *Flaviviridae* family, whose natural transmission cycle involves mainly vectors from the *Aedes* genus and monkeys. Zika virus was first isolated in April 1947 from the serum of a pyrexial rhesus monkey caged in the canopy of Zika Forest near Entebbe in Uganda. The second isolation was made from a lot of *Ae.* (*Stegomyia*) *africanus* taken in January 1948 in the same forest [107]. Serologic studies indicated a possible transmission to humans [108]. This was confirmed rapidly. Thus, during an epidemic of jaundice suspected of being yellow fever, in eastern Nigeria, infection with Zika virus was shown to have occurred in three patients, one by isolation of the virus and two by a rise in serum antibodies [109]. Inoculation of the eastern Nigerian strain of Zika virus to a volunteer led to febrile symptoms after an incubation period of 82 hours. Zika virus was isolated from the blood during the febrile period [110]. Transmission of the Zika virus by *Ae. aegypti* mosquitoes artificially fed to mice and a rhesus monkey in a laboratory was demonstrated by Boorman and Porterfield [111]. Thereafter, serologic evidence of human infection by Zika virus was reported in most of the African and Asian countries [112–114]. The first major outbreak of Zika virus disease occurred in 2007 on Yap Island (Micronesia), where 49 confirmed and 59 probable cases of Zika virus disease were identified. Rash, fever, arthralgia, and conjunctivitis were the common symptoms. No hospitalizations, hemorrhagic manifestations, or deaths due to the Zika virus were reported [114,115]. A subsequent outbreak associated with cases of Guillain–Barré syndrome occurred in 2013 and 2014 in French Polynesia [116,117]. From the start of 2015, patients with Zika symptoms began to be reported in the northeast of Brazil [118]. At the beginning of 2016, Zika virus infection expanded in the majority of Latin American and Caribbean countries, with estimated cases of 440,000—1,300,000 in Brazil alone [119]. Moreover, the virus was linked to an explosion of microcephaly in Brazil [120]. Faced with this situation, there is special urgency to repurpose drugs for Zika because of a lack of rapidly available new drugs against the virus [121].

Delvecchio et al. [122] showed that chloroquine (Figure 1.3) exhibited antiviral activity against Zika virus in Vero cells and human brain microvascular endothelial and neural stem cells. Chloroquine reduced the number of infected cells, virus production, and cell death promoted by Zika virus infection without cytotoxic effects. Chloroquine analogs with C-4 modification appear more potent than chloroquine against Zika virus replication [123].

Tang et al. [124] showed that Zika virus efficiently targets human neural progenitor cells (hNPCs) with an increase in caspase-3 activation in hNPCs 3 days after infection, leading to their death. This provides a potential mechanism for explaining microcephaly induced by Zika virus because hNPCs drive the development of the cortex in the human brain. Consequently, Xu et al. [125] performed a screening investigation using the caspase-3 activity test and SNB-19 cells, with the Library of Pharmacologically Active Compounds (LOPAC, 1,280 compounds), the National Center for Advancing Translational Sciences (NCATS) pharmaceutical collection of approved drugs (2,816 compounds), and a collection of clinical candidate compounds (2,000 compounds). Primary hits led to 116 compounds that suppressed Zika virus-induced caspase-3 activity in SNB-19 cells. An independent primary screen in hNPCs using the same libraries was also carried out [125]. This second screen resulted in 173 primary hits that included all 116 compounds from the initial caspase-3 screen in SNB-19 cells. Next, the activity of these primary hits from the caspase-3 activity assay was reevaluated in Zika virus-infected SNB-19 cells, hNPCs, and astrocytes. A cytotoxicity assay was also performed in parallel. Emricasan (Figure 1.7) was identified as the most potent anti-cell-death compound, with half-maximal inhibitory concentration (IC_{50}) values of 0.13–0.9 μM in both caspase activity and cell viability assays for SNB-19 cells against three different Zika virus strains: 1947 Ugandan, 2010 Cambodian, and 2015 Puerto Rican. However, emricasan does not suppress Zika virus replication. On the other hand, two compounds were identified that substantially inhibited Zika virus replication—namely, niclosamide (Figure 1.8)—a category B anthelmintic drug approved by the US FDA and PHA-690509 (Figure 1.9), a compound acting as a cyclin-dependent kinase inhibitor (CDKi). Other CDKis were

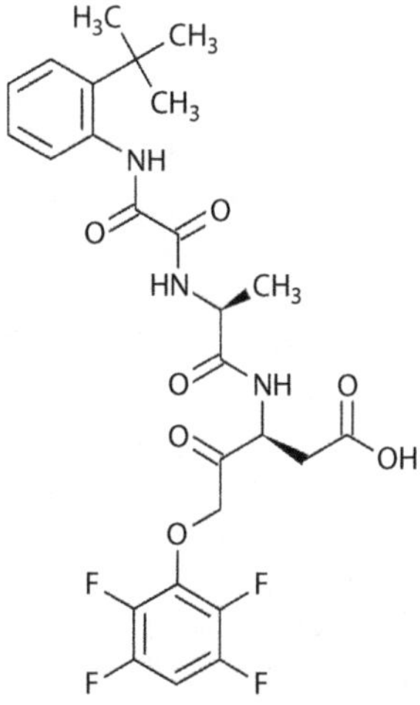

FIGURE 1.7 Structure of emricasan.

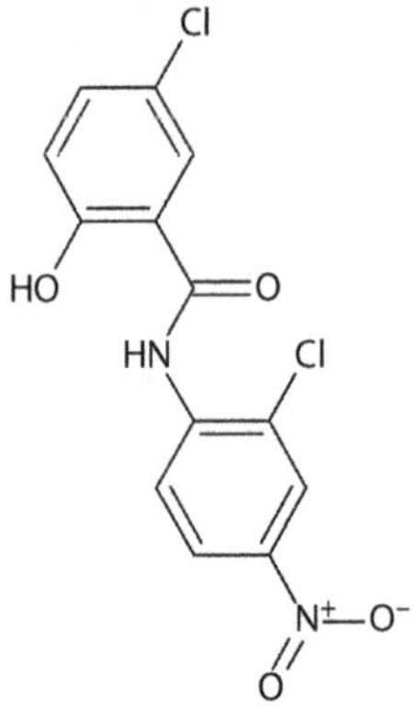

FIGURE 1.8 Structure of niclosamide.

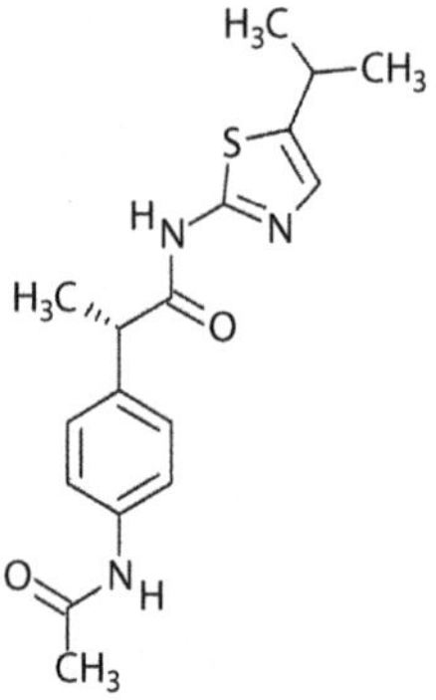

FIGURE 1.9 Structure of PHA-690509.

also identified. Last, combination of emricasan and PHA-690509 showed an additive effect in inhibiting caspase-3 activity in SNB-19 cells [125].

Barrows et al. [126] screened 774 FDA-approved drugs for their ability to block infection of HuH-7 cells by a newly isolated Zika virus strain (ZIKV MEX_1_7). Forty-five drugs significantly inhibited Zika virus infection. Among them, 30 were selected based on their activity and considerations of their clinical use. Six drugs, which reduced percentage of infection values by less than half or only eliminated cells, were excluded from further testing, leaving a list of 24 drugs (Table 1.2) with validated anti-Zika virus activity. Due to the evidence for sexual transmission [127] and for placental infection [128], ivermectin (Figure 1.4), daptomycin, mycophenolic acid (Figure 1.10), sertraline, pyrimethamine, cyclosporine A, azathioprine, and mefloquine were tested on infected HeLa cells derived from a cervical adenocarcinoma and on JEG3 cells, which were derived from a placental choriocarcinoma. Different responses were found. Thus, 1 μM mycophenolic acid completely inhibited infection in HeLa cells, as did 10 μM ivermectin. In JEG3 cells, four

TABLE 1.2
Selected Drugs with Anti-Zika Virus Activity, Their Known Mechanism of Action, and Pregnancy Category (PG)

Drug Name	Known Mechanism	PG[a]
Auranofin	Inhibits thioredoxin reductase	C
Azathioprine	Inhibits purine synthesis	D
Bortezomib	Proteasome inhibitor	D
Clofazimine	Unknown (antimicrobial)	C
Cyclosporine A	Cyclophilin inhibitor	C
Dactinomycin	Transcription inhibitor	D
Daptomycin	Unknown (antimicrobial)	B
Deferasirox	Chelator of intracellular iron	C
Digoxin	Na^+K^+ ATPase inhibitor; impacts Ca signaling	C
Fingolimod	Sphingosine-1-phosphate receptor modulator	C
Gemcitabine HCl	Nucleoside analog; blocks DNA replication	D
Ivermectin	Unknown (antiparasitic)	C
Mebendazole	Unknown (antihelminthic)	C
Mefloquine HCl	Disrupts autophagy (antiparasitic)	B
Mercaptopurine hydrate	Inhibits purine synthesis	D
Methoxsalen	DNA synthesis inhibitor	C
Micafungin	Unknown (antifungal)	C
Mycophenolate mofetil	Prodrug of mycophenolic acid	D
Mycophenolic acid	Inosine-5′-monophosphate dehydrogenase inhibitor	D
Palonosetron HCl	5-Hydroxytryptamine-3 receptor antagonist	B
Pyrimethamine	Dihydrofolate reductase antagonist	C
Sertraline HCl	Selective serotonin reuptake inhibitor	C
Sorafenib tosylate	Multitarget tyrosine kinase inhibitor	D
Thioguanine	Inhibits purine synthesis	n.a.

Source: N.J. Barrows et al., Cell Host Microbe 20 (2016), pp. 259–270.

[a] A = adequate and well-controlled studies have failed to demonstrate a risk to the fetus in the first trimester of pregnancy (and there is no evidence of risk in later trimesters). B = animal reproduction studies have failed to demonstrate a risk to the fetus and there are no adequate and well-controlled studies in pregnant women. C = animal reproduction studies have shown an adverse effect on the fetus and there are no adequate and well-controlled studies in humans, but potential benefits may warrant use of the drug in pregnant women despite potential risks. D = there is positive evidence of human fetal risk based on adverse reaction data from investigational or marketing experience or studies in humans, but potential benefits may warrant use of the drug in pregnant women despite potential risks (N.J. Barrows et al., Cell Host Microbe 20 (2016), pp. 259–270; https://www.drugs.com/pregnancy-categories.html).

out of eight drugs tested revealed antiviral effects, and mycophenolic acid was the most potent [126]. Barrows et al. [126] also investigated the effects of mycophenolic acid, ivermectin, cyclosporine A, and mefloquine on Zika virus MEX_I_7 infection of a human fetal brain-derived neural stem cell (hNSC) cell line (K048), which was obtained without genetic modifications. Sertraline and bortezomib were tested

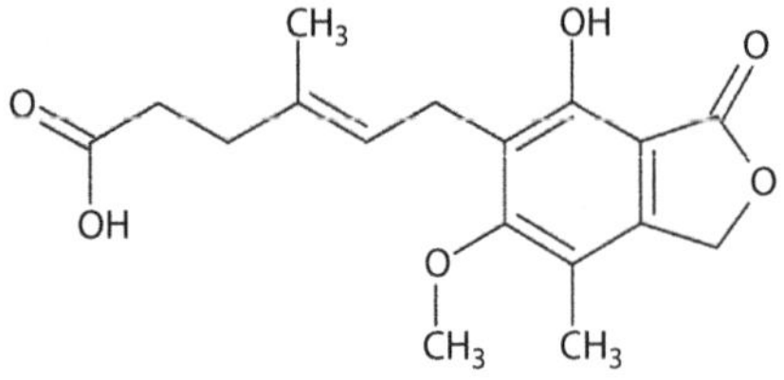

FIGURE 1.10 Structure of mycophenolic acid.

by using the Zika virus strain from the African lineage (ZIKV DAK_41525). At 1 μM concentrations, mycophenolic acid, ivermectin, and bortezomib all exerted antiviral effects, and only the latter showed a moderate toxicity [126]. Ivermectin, daptomycin, mycophenolic acid, sertraline, pyrimethamine, cyclosporine A, azathioprine, and mefloquine were also tested on primary human amnion epithelial cells (HAECs) from the lining of the amniotic sac. These cells were initially infected *in vitro* by Zika virus MEX_I_7. At 16 μM, none of these drugs caused significant reduction of cell numbers. A moderate inhibition of the virus infectivity was observed with daptomycin and strong inhibition was recorded for ivermectin, sertraline, and mefloquine at 16 μM. Mycophenolic acid inhibited Zika virus MEX_I_7 infection at 1.6 μM. Barrows et al. [126] showed that it was possible to identify FDA-approved drugs that can inhibit Zika virus infection in several human cells, including those of genitourinary and neural origin.

It is important to note that similar strategies to identify drugs inhibiting Zika virus have been applied by others [129–131].

1.3.5 Drugs Repurposed for Japanese Encephalitis Virus

Japanese encephalitis virus, which belongs to the family *Flaviviridae*, is an arbovirus transmitted in an enzootic cycle involving birds, particularly wading ardeids (e.g., herons egrets). Pigs can become infected and act as amplifying hosts, bringing the virus closer to human habitats. Many mosquito species are potential vectors, but *Culex* sp. such as *Cx. tritaeniorhynchus* and *Cx. vishnui*, which breed in rice paddies and other dirty water, are especially important. Humans become infected when they are bitten by infected mosquitoes [132]. Japanese encephalitis affects about 25 Asian countries but also northern Australia. The annual incidence of Japanese encephalitis is estimated to be in the range of 50,000–175,000 cases, depending on age group, geographical area, and immunization status. About 20%–30% of clinical Japanese encephalitis cases are fatal, and ~30%–50% of survivors experience serious neurologic, cognitive, or psychiatric complications even years later. Vaccination is the only strategy to develop long-term sustainable protection because there are no specific drugs available to treat Japanese encephalitis [133,134].

Dexamethasone is a corticosteroid drug that has numerous therapeutic applications as anti-inflammatory medication. Dexamethasone was evaluated in the treatment of people with acute Japanese encephalitis [135]. Twenty-five patients, randomly selected, received 0.6 mg/kg of dexamethasone intravenously as a loading dose followed by 0.2 mg/kg every 6 hours during 5 days. The 30 remaining patients received a saline solution as placebo. Dexamethasone was totally inefficient. Thus, for example, mortality through 25 days after admission in the dexamethasone group was 24% compared with 27% in the placebo group [135].

Ribavirin is a large spectrum antiviral agent used, for example, in the treatment of hepatitis C. A total of 153 children (6 months to 15 years) with Japanese encephalitis virus were enrolled during a 3-year period; 70 patients received ribavirin (10 mg/kg per day in four divided doses for 7 days), and 83 received a placebo [136]. There was no statistically significant difference between the two groups in the early mortality rate: 19 (27.1%) of 70 ribavirin recipients and 21 (25.3%) of 83 placebo recipients died (odds ratio, 1.10; 95% confidence interval, 0.5–2.4). No statistically significant differences in secondary outcome measures (e.g., time to resolution of fever, duration of hospitalization) were found [136].

A group of 44 patients, positive for anti-Japanese encephalitis IgM in cerebrospinal fluid and/or serum, were split into two equal groups, A and B, of 7.65 ± 2.40 and 6.31 ± 3.42 years old, respectively. Group A patients received minocycline (Figure 1.6) in the dose of 5–6 mg/kg in two divided doses for 10 days; group B patients received a placebo [137]. The mean duration of all symptoms was less in group A in comparison to group B. However, duration of fever and unconsciousness was statistically significant, while duration of vomiting and convulsions was not. The mean duration of stay in hospital for patients from group A (13.84 days) was significantly lower than that for group B (21.59 days). This study, which needs to be reproduced on a larger sample size, shows that minocycline has significant beneficial effects on patients, but that mortality and prevalence of neurological deficits and behavioral problems on follow-up of these patients remain unchanged [137].

1.3.6 Drugs Repurposed for Chikungunya Virus

Chikungunya virus is a member of the family *Togaviridae*, genus *Alphavirus*, which comprises enveloped, positive, single-stranded RNA viruses [138,139]. The first epidemic was described in 1952 on the Makonde Plateau in the southern province of Tanganyika (Tanzania). It was clinically indistinguishable from dengue. Owing to the distinctive severity of joint pains and the sudden onset, it was locally termed chikungunya, meaning "that which bends up" [140,141]. Since that date, chikungunya fever has reemerged extensively, resulting in 1.4 to 6.5 million infected individuals between 2004 and 2014 in Africa, as well as regions within the Indian Ocean, Southeast Asia, the Pacific Islands, and Europe [142]. At the end of 2013, an outbreak was first detected in the French West Indies and then the virus dispersed to other Caribbean islands and Latin America [143,144]. Transmission of chikungunya virus occurs through the bite of a mosquito belonging to the *Aedes* genus (*Ae. aegypti* and *Ae. albopictus*). Human beings are virus reservoirs during epidemic periods, but outside them, the main reservoirs are wild vertebrates such as monkeys, rodents, and birds [138,139].

Rashad et al. [145] have reviewed the key target proteins of the chikungunya virus and their molecular functions that can be used in drug design, as well as some examples of research of chemotherapeutics against chikungunya virus. Different attempts have been made to use existing drugs to fight this virus.

Albulescu et al. [146] studied the effect of suramin (Figure 1.5) on chikungunya virus RNA synthesis using an *in vitro* assay based on the RNA-synthesizing activity of replication and transcription complexes (RTCs) isolated from chikungunya virus-infected cells. This assay measures the incorporation of $[\alpha]^{32}$P-CTP into viral RNA [147], which was severely impaired by suramin in a dose-dependent manner, with an IC_{50} value of about 5 μM. To estimate the antiviral efficacy of suramin in cell culture, Vero E6 cells were infected with different chikungunya virus strains and treated with serial dilutions of the compound in a cytopathic effect protection assay. Viability assays on uninfected cells were performed in parallel to determine the half maximal cytotoxic concentration (CC_{50}). The EC_{50} values for infectious clone-derived CHIKV LS3, a natural isolate from Italy (ITA07-RA1) and a Caribbean CHIKV strain (STM35) were equal to 79 ± 11.6 μM, 76 ± 7 μM, and 79 ± 12.9 μM, respectively. The CC_{50} of suramin was >5 mM, but the compound presented a cytotoxic effect at high concentrations, as viability dropped to 65% at 5 mM, the highest tested concentration. Albulescu et al. [146] also showed that suramin inhibited an early step of the chikungunya virus replicative cycle, possibly attachment or entry.

Ho et al. [148] demonstrated that suramin inhibited chikungunya virus entry and transmission through binding onto E1/E2 glycoproteins. They also showed that the chemical inhibited infection in early stages. Among their different investigations, they tested on three different cell types, the antiviral effects of suramin on chikungunya virus S27 strain and three clinically isolated strains: 0611aTw, 0810bTw, and 0706aTw. The EC_{50} values of the chikungunya virus strains were as follows: 8.8 to 28.9 μM (BHK-21 cells), 17.9 to 59.6 μM (U2OS cells), and 18.1 to 62.1 μM (MRC-5cells). These authors also showed that suramin was nontoxic to zebrafish (*Danio rerio*) following treatment with 700 μM. After exposure, the embryos did not present significant variation in survival, body length, malformation, and hatching [148]. In another study [149], C57BL/6 mice were infected with chikungunya viruses to evaluate antivirus activities of suramin in terms of histopathology, viral burden, and disease score. Suramin treatment reduced viral loads in 0810bTw-, 0611aTw-, and 0706aTw-infected mice, which might be due to suramin treatment to block virus entry and egress. However, 0810bTw was the most sensitive strain to suramin treatment among the three isolate strains *in vivo*, which corresponded to their EC_{50} activities *in vitro* [148]. Suramin treatment also ameliorated foot swelling and reduced inflammatory infiltration, which corresponded to reduced viremia and viral antigen expression in infected tissues. This study highlights the potential ability of suramin to treat chikungunya virus infection in clinical settings [149].

A baby hamster kidney BHK-21 cell line containing a stable chikungunya virus replicon with a luciferase reporter was used in a high-throughput platform to screen about 3,000 compounds [150]. Following initial validation, 25 compounds were chosen as primary hits for secondary validation with wild type and reporter chikungunya virus infection, which identified abamectin (Figure 1.11), ivermectin (Figure 1.4), and berberine (Figure 1.12) as promising compounds. The EC_{50} (CC_{50}) of abamectin, berberine, and ivermectin against chikungunya virus in Huh-7.5 cells was equal to 1.4 ± 0.9 μM (15.2 ± 1 μM), 1.9 ± 0.9 μM (>100 μM), and 1.9 ± 0.8 μM

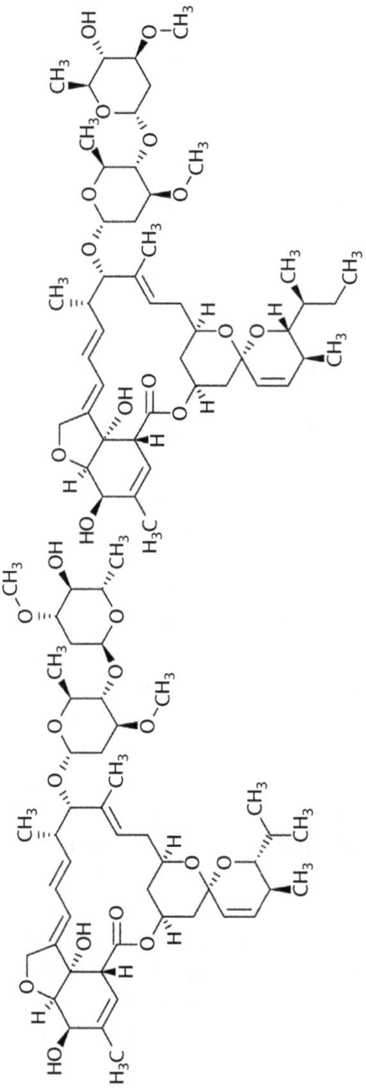

FIGURE 1.11 Structure of abamectin (B1b: left; B1a: right).

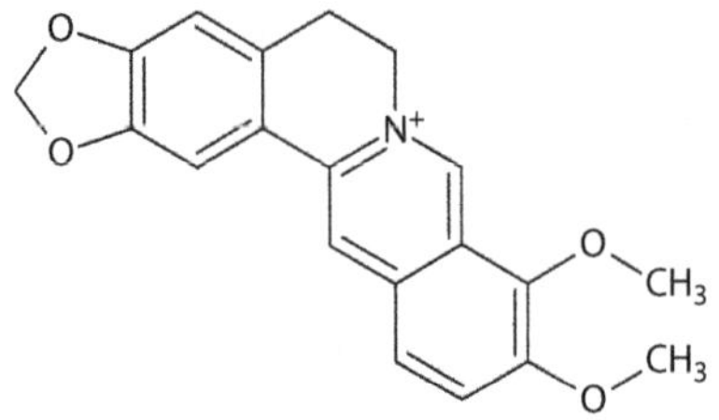

FIGURE 1.12 Structure of berberine.

(8 ± 0.2 μM), respectively. In the same way, the EC_{50} (CC_{50}) of the three compounds against chikungunya virus in BKK-21 cells was equal to 1.5 ± 0.6 μM (28.2 ± 1.1 μM), 1.8 ± 0.5 μM (>100 μM), and 0.6 ± 0.1 μM (37.9 ± 7.6 μM), respectively [150].

1.3.7 Drugs Repurposed for Rift Valley Fever Virus

Rift Valley fever virus belongs to the genus *Phlebovirus* in the family *Bunyaviridae.* The virus is transmitted to vertebrate hosts by the bite of infected mosquitoes belonging to *Aedes* spp. and *Culex* spp. Rift Valley fever mainly affects domestic animals (cattle, goats, sheep, and camels, among others) and generally causes abortions in pregnant females and high mortality in young animals. Humans acquire Rift Valley fever through bites from infected mosquitoes or through exposure to infected animal material. Infection usually causes a self-limiting, acute, and febrile illness; however, a small number of cases progress to neurological disorders, partial or complete blindness, hemorrhagic fever, or thrombosis [151,152].

Rift Valley fever is endemic to sub-Saharan African countries; it was first isolated in Kenya in 1930 [153] and has caused major outbreaks in several countries including Kenya, Tanzania, Somalia, South Africa, Madagascar, Egypt, Sudan, Mauritania, Senegal, Saudi Arabia, and Yemen [152,154].

Plaque assays were used by Ellenbecker et al. [155] to measure the ability of suramin (Figure 1.5) to inhibit the production of viable Rift Valley fever virus in function of time. Suramin (100 μM) decreased the number of plaque-forming units (PFUs) per milliliter by approximately 1 log at 2 and 3 days postinfection and was not cytotoxic to human 293 cells. The viral EC_{50} of suramin was equal to 58.5 μM, and the CC_{50} equaled 200 μM, yielding a therapeutic index of about 3.4. Suramin seems to be able to prevent Rift Valley fever virus replication at multiple stages during its replication. It inhibits an early step during viral replication, as well as one or more later steps that have yet to be determined. The authors showed that inhibition of RNA binding to Rift Valley fever virus nucleocapsid protein could represent an attractive antiviral therapeutic strategy [155]. In the same way, Narayanan et al. [156] clearly showed that curcumin (Figure 1.13), a polyphenol derived from the *Curcuma longa* plant, exerted an inhibitory effect on Rift Valley fever virus replication. The molecule was used as positive control in a recent FDA-approved drug screening to find therapeutics effective against Rift Valley fever virus [157]. In order to determine which of the 420 FDA-approved drugs were effective at inhibiting Rift Valley

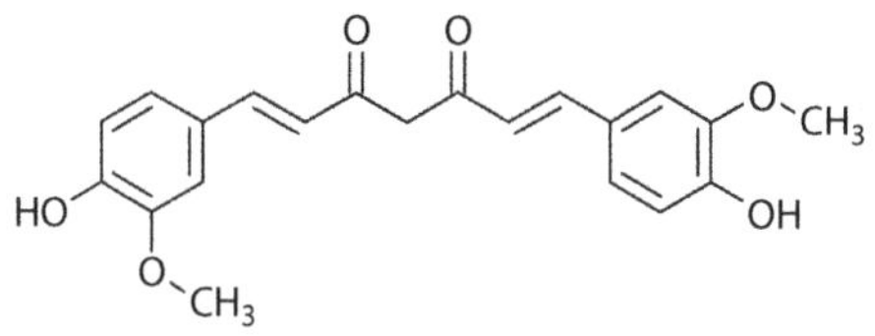

FIGURE 1.13 Structure of curcumin.

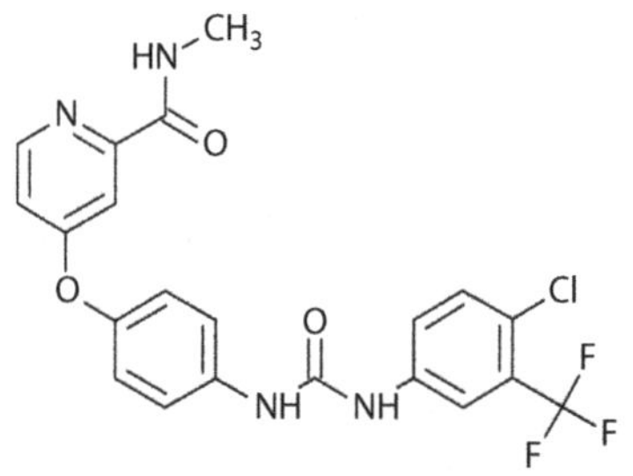

FIGURE 1.14 Structure of sorafenib.

fever virus infection, a high-throughput assay was performed. The reporter virus, Rift Valley fever virus M12-Luc, encoding the *Renilla luciferase*, was used to measure virus replication. A list of 21 compounds was retrieved. They reduced *Renilla luciferase* reporter activity while cell viability was maintained ≥80%. Ivermectin (Figure 1.4) belonged to the list but was definitively less interesting than sorafenib (Figure 1.14), a hepatocellular and renal cell carcinoma drug, which was able to reduce Rift Valley fever virus replication with the greatest efficacy, leading to 93% reduction in luciferase luminescence with no toxicity. Mechanism of action studies showed that this drug targeted at least two stages in the virus cycle, RNA synthesis and viral egress [157].

1.4 CONCLUDING REMARKS

Developing a new drug is costly and time consuming in research and development. A cost-effective reduced-risk strategy avoiding the high failure rates linked to the launch of a new drug is to identify new therapeutic uses for molecules that were already approved to treat diseases but might be relevant against other diseases. This constitutes the basis of the concept of drug repurposing that can be done from various strategies including or not including *in silico* tools. Such an approach relies on the fact that in addition to a target of reference for which a drug is commercialized and used in therapeutics, there is no logical reason to exclude the possibility that this drug could be active on one or several other targets. Thus, for example, thalidomide originally introduced to prevent nausea and insomnia in pregnant women, but withdrawn from the market due to important teratogenic effects, has made a successful comeback and now the molecule and its derivatives are used for the treatment of leprosy and multiple myeloma [158]. Another example is sildenafil, a vasoactive agent that has evolved from a potential antiangina drug to an oral treatment for erectile dysfunction (Viagra™) and, more recently, to a new orally active treatment for pulmonary hypertension (Revatio™) [159].

Drug repurposing appears particularly appealing for neglected diseases in response to the disinterest of the pharmaceutical companies that prefer to focus only on the discovery of drugs suitable for people living in rich countries and who have the possibility to pay prohibitive prices for new drugs potentially able to increase, even more, their life expectancy. In theory, drug repurposing is also able to respond much more rapidly to a situation of crisis than a classical drug discovery process. Indeed, faced with the emergence or reemergence of a disease that can become rapidly pandemic, it is necessary to be able to rapidly offer a therapeutic solution.

The situation is exactly the same for the insecticides used as larvicides and adulticides. Indeed, there is an urgent need to find new molecules targeting the mosquitoes in order to overcome the problems of resistance to insecticides. Unfortunately, the market of vector control agents occupies a too limited niche to expect efficient insecticide discovery programs coming from the industry. For the chemical industry, it is more profitable to work for agriculture than vector control.

In these conditions, does the repurposing of insecticides and drugs represent a panacea for the control of mosquitoes and their diseases or might the tree hide the forest?

Undoubtedly, the paradigm is interesting in both situations, but it does not exclude inherent problems linked to the launch of a new insecticide or a new drug.

Repurposing a xenobiotic in mosquito control requires having the assurance that the molecule is not too toxic against the nontarget organisms or that, at least, it will be possible to provide efficient instructions to limit the adverse effects. Use of an insecticide in vector control also requires a formulation adapted for the type of treatment and material [160] and it is a condition very often difficult to satisfy in practice. Thus, for example, it has recently been shown (C. Lagneau, EID Méditerranée, Fr., personal communication) that some insecticides selected as potential adulticides [48] could not be used in practice because it was impossible to find a formulation adapted to space spraying [160].

The same type of problem exists for repurposed drugs. Oprea and Overington [68] have proposed a classification scheme for drug repurposing based on scientific evidence according to five levels ranging from 0 (*in silico* predictions without confirmation) to 4 (well-documented clinical endpoints observed for the repurposed drug at doses within safety limits). Their examples in neglected disease research show some of the additional problems we are faced with for such drugs, which are very often used in difficult conditions. These problems deal with the dosage/absorption route, the conservation and distribution processes, the use of these drugs by malnourished, even famished, people suffering of multiple pathologies, and so on.

This review attempted to show a few examples of repurposing of insecticides and marketed drugs to find new mosquitocidal chemicals able to fight larvae and adults. Examples of repurposing of drugs as potential therapeutics for malaria, dengue, West Nile fever, Zika, Japanese encephalitis, chikungunya, and Rift Valley fever have also been shown. Our aim was primarily to account for the diversity of the studies while focusing more specifically on some of them. Thus, we do not claim to have been exhaustive and certainly interesting studies have been forgotten.

The repurposing of an existing insecticide no longer in use or used in agriculture, for controlling populations of mosquitoes, is certainly the easier way to obtain rapidly a new adulticide or a larvicide. The constraints to overcome mainly deal with problems of regulation, formulation, and production. The recent temporary use of malathion in French Guiana to fight the chikungunya outbreak illustrates this situation [161]. The repurposing of a xenobiotic, especially if it presents pharmacological effects, is much more difficult and can even be impossible. It is interesting to note that the search for a synergist is less constrainable.

There is also a scalable difficulty in the repurposing of drugs to find rapidly potential treatments against mosquito borne diseases. We cannot equally consider the repositioning of a compound already tested on the hepatitis C virus for its potential use against a *Flavivirus* or the multifaceted drug ivermectin (for which Satoshi Ōmura and William Campbell received the 2015 Nobel prize in physiology or medicine [162]) with molecules that have no chance of fulfilling the specifications of the drugs that can be used against mosquito-borne diseases.

Thus, on one side, we are faced with so many challenges for the control and elimination of mosquito-borne diseases and, on the other side, the availability of funding is so terrifically limited in the domain, so all research initiatives have to strongly be encouraged. Diversity in science, as elsewhere, can be only fruitful!

REFERENCES

[1] R.B. Marí and R.J. Peydró, *Re-emergence of malaria and dengue in Europe*, in *Current Topics in Tropical Medicine*, A.J. Rodriguez-Morales, ed., InTech, Rijeka, Croatia, 2012, pp. 483–512.

[2] D.J. Gubler, *Dengue and dengue hemorrhagic fever*, Clin. Microbiol. Rev. 11 (1998), pp. 480–496.

[3] A.J. Mathew, A. Ganapati, J. Kabeerdoss, A. Nair, N. Gupta, P. Chebbi, S.K. Mandal, and D. Danda, *Chikungunya infection: A global public health menace*, Curr. Allergy Asthma Rep. 17 (2017), pp. 13.

[4] E. Krow-Lucal, N.P. Lindsey, J. Lehman, M. Fischer, and J.E. Staples, *West Nile virus and other nationally notifiable arboviral diseases — United States*, 2015, CDC Morbid. Mortal. Weekly Rep. 66 (2017), pp. 51–55.

[5] N. Aréchiga-Ceballos and A. Aguilar-Setién, *Alphaviral equine encephalomyelitis (Eastern, Western and Venezuelan)*, Rev. Sci. Tech. Off. Int. Epiz. 34 (2015), pp. 491–501.

[6] M.C. Harris, E.J. Dotseth, B.T. Jackson, S.D. Zink, P.E. Marek, L.D. Kramer, S.L. Paulson, and D.M. Hawley, *La Crosse virus in* Aedes japonicus japonicus *mosquitoes in the Appalachian Region, United States*, Emerg. Infect. Dis. 21 (2015), pp. 646–649.

[7] N. Armstrong, W. Hou, and Q. Tang, *Biological and historical overview of Zika virus*, World J. Virol. 6 (2017), pp. 1–8.

[8] M.A. Tolle, *Mosquito-borne diseases*, Curr. Probl. Pediatr. Adolesc. Health Care 39 (2009), pp. 97–140.

[9] A.D. LaBeaud, *Why arboviruses can be neglected tropical diseases*, PLoS Neglect. Trop. Dis. 2 (2008), pp. e247.

[10] J. Hemingway and H. Ranson, *Insecticide resistance in insect vectors of human disease*, Ann. Rev. Entomol. 45 (2000), pp. 371–391.

[11] S. Marcombe, R. Poupardin, F. Darriet, S. Reynaud, J. Bonnet, C. Strode, C. Brengues, A. Yébakima, H. Ranson, V. Corbel, and J.P. David, *Exploring the molecular basis of insecticide resistance in the dengue vector* Aedes aegypti*: A case study in Martinique Island (French West Indies)*, BMC Genom. 10 (2009), pp. 494.

[12] D.J. Gubler, *Resurgent vector-borne diseases as a global health problem*, Emerg. Infect. Dis. 4 (1998), pp. 442–450.

[13] P. Reiter, *Climate change and mosquito-borne disease*, Environ. Health Perspectives 109 (2001), pp. 141–161.

[14] P.R. Epstein, H.F. Diaz, S. Elias, G. Grabherr, N.E. Graham, W.J.M. Martens, E. Mosley-Thompson, and J. Susskind, *Biological and physical signs of climate change: Focus on mosquito-borne diseases*, Bull. Am. Meteorol. Soc. 79 (1998), pp. 409–417.

[15] A. Rossati, *Global warming and its health impact*, Int. J. Occup. Environ. Med. 8 (2017), pp. 7–20.

[16] U. Bulugahapitiya, S. Siyambalapitiya, S.L. Seneviratne, and D.J.S. Fernando, *Dengue fever in travellers: A challenge for European physicians*, Eur. J. Int. Med. 18 (2007), pp. 185–192.

[17] T.M. Breslin, U.N. Lonmhain, C. Bergin, D. Gallagher, N. Collins, N. Kinsella, and G. McMahon, *Malarial cases presenting to a European urban Emergency Department*, Eur. J. Emerg. Med. 20 (2013), pp. 115–119.

[18] Anonymous, *Directive 98/8/EC of the European Parliament and of the Council of 16 February 1998 concerning the placing of biocidal products on the market*, Off. J. Europ. Comm. April 24 (1998), L123/1.

[19] IRAC, *Prevention and Management of Insecticide Resistance in Vectors of Public Health Importance*, Second Edition, Insecticide Resistance Action Committee, 2011.

[20] M.A. Osta, Z.J. Rizk, P. Labbé, M. Weill, and K. Knio, *Insecticide resistance to organophosphates in* Culex pipiens *complex from Lebanon*, Parasit. Vectors 5 (2012), pp. 132.

[21] I. Dusfour, V. Thalmensy, P. Gaborit, J. Issaly, R. Carinci, and R. Girod, *Multiple insecticide resistance in* Aedes aegypti *(Diptera: Culicidae) populations compromises the effectiveness of dengue vector control in French Guiana*, Mem. Inst. Oswaldo Cruz. 106 (2011), pp. 346–352.

[22] S. Marcombe, A. Carron, F. Darriet, M. Etienne, P. Agnew, M. Tolosa, M.M. Yp-Tcha, C. Lagneau, A. Yébakima, and V. Corbel, *Reduced efficacy of pyrethroid space sprays for dengue control in an area of Martinique with pyrethroid resistance*, Am. J. Trop. Med. Hyg. 80 (2009), pp. 745–751.

[23] G. Munhenga, H.T. Masendu, B.D. Brooke, R.H. Hunt, and L.K. Koekemoer, *Pyrethroid resistance in the major malaria vector* Anopheles arabiensis *from Gwave, a malaria endemic area in Zimbabwe*, Malar. J. 28 (2008), pp. 247.

[24] J.G. Scott, M.H. Yoshimizu, and S. Kasai, *Pyrethroid resistance in* Culex pipiens *mosquitoes*, Pestic. Biochem. Physiol. 120 (2015), pp. 68–76.

[25] J. Devillers, C. Lagneau, A. Lattes, J.C. Garrigues, M.M. Clemente, and A. Yébakima, *In silico models for predicting vector control chemicals targeting* Aedes aegypti, SAR QSAR Environ. Res. 25 (2014), pp. 805–835.

[26] P. Trouiller, P. Olliaro, E. Torreele, J. Orbinski, R. Laing, and N. Ford, *Drug development for neglected diseases: A deficient market and a public-health policy failure*, Lancet 359 (2002), pp. 2188–2194.

[27] B. Pedrique, N. Strub-Wourgaft, C. Some, P. Olliaro, P. Trouiller, N. Ford, B. Pécoul, and J.H. Bradol, *The drug and vaccine landscape for neglected diseases (2000–11): A systematic assessment*, Lancet Glob. Health 1 (2013), pp. e371–379.

[28] P. Shah and M.I. Siddiqi, *3D-QSAR studies on triclosan derivatives as* Plasmodium falciparum *enoyl acyl carrier reductase inhibitors*, SAR QSAR Environ. Res. 21 (2010), pp. 527–545.

[29] S.P. Kumar, L.B. George, Y.T. Jasrai, and H.A. Pandya, *Prioritization of active antimalarials using structural interaction profile of* Plasmodium falciparum *enoyl-acyl carrier protein reductase (PfENR)-triclosan derivatives*, SAR QSAR Environ. Res. 26 (2015), pp. 61–77.

[30] A. Hasan, H.H. Mazumder, A.S. Chowdhury, A. Datta, and A. Khan, *Molecular-docking study of malaria drug target enzyme transketolase in* Plasmodium falciparum *3D7 portends the novel approach to its treatment*, Source Code Biol. Med. 10 (2015), pp. 14.

[31] S. Gupta, A. Jadaun, H. Kumar, U. Raj, P.K. Varadwaj, and A.R. Rao, *Exploration of new drug like inhibitors for serine/threonine protein phosphatase 5 of* Plasmodium falciparum*: A docking and simulation study*, J. Biomol. Struct. Dyn. 13 (2015), pp. 1–68.

[32] M. Kumari, S. Chandra, N. Tiwari, and N. Subbarao, *3D QSAR, pharmacophore and molecular docking studies of known inhibitors and designing of novel inhibitors for M18 aspartyl aminopeptidase of* Plasmodium falciparum, BMC Struct. Biol. 16 (2016), pp. 12.

[33] X. Hou, X. Chen, M. Zhang, and A. Yan, *QSAR study on the antimalarial activity of* Plasmodium falciparum *dihydroorotate dehydrogenase (PfDHODH) inhibitors*, SAR QSAR Environ. Res. 27 (2016), pp. 101–124.

[34] M. Thillainayagam, A. Anbarasu, and S. Ramaiah, *Comparative molecular field analysis and molecular docking studies on novel aryl chalcone derivatives against an important drug target cysteine protease in* Plasmodium falciparum, J. Theor. Biol. 403 (2016), pp. 110–128.

[35] H.M. Faidallah, S.S. Panda, J.C. Serrano, A.S. Girgis, K.A. Khan, K.A. Alamry, T. Therathanakorn, M.J. Meyers, F.M. Sverdrup, C.S. Eickhoff, S.G. Getchell, and A.R. Katritzky, *Synthesis, antimalarial properties and 2D-QSAR studies of novel triazole-quinine conjugates*, Bioorg. Med. Chem. 24 (2016), pp. 3527–3539.

[36] J.E. Knox, N.L. Ma, Z. Yin, S.J. Patel, W.L. Wang, W.L. Chan, K.R.R. Rao, G. Wang, X. Ngew, V. Patel, D. Beer, S.P. Lim, S.G. Vasudevan, and T.H. Keller, *Peptide inhibitors of West Nile NS3 protease: SAR study of tetrapeptide aldehyde inhibitors*, J. Med. Chem. 49 (2006), pp. 6585–6590.

[37] S. Fernando, T. Fernando, M. Stefanik, L. Eyer, and D. Ruzek, *An approach for Zika virus inhibition using homology structure of the envelope protein*, Mol. Biotechnol. 58 (2016), pp. 801–806.

[38] P. Wang, L.F. Li, Q.Y. Wang, L.Q. Shang, P.Y. Shi, and Z. Yin, *Anti-dengue-virus activity and structure-activity relationship studies of lycorine derivatives*, ChemMedChem. 9 (2014), pp. 1522–1533.

[39] V. Frecer and S. Miertus, *Design, structure-based focusing and in silico screening of combinatorial library of peptidomimetic inhibitors of dengue virus NS2B-NS3 protease*, J. Comput. Aided Mol. Des. 24 (2010), pp. 195–212.

[40] T. Knehans, A. Schüller, D.N. Doan, K. Nacro, J. Hill, P. Güntert, M.S. Madhusudhan, T. Weil, and S.G. Vasudevan, *Structure-guided fragment-based in silico drug design of dengue protease inhibitors*, J. Comput. Aided Mol. Des. 25 (2011), pp. 263–274.

[41] C. Årdal and J.A. Røttingen, *Open source drug discovery in practice: A case study*, PLoS Negl. Trop. Dis. 6 (2012), pp. e1827.

[42] M.P. Pollastri, *Finding new collaboration models for enabling neglected tropical disease drug discovery*, PLoS Negl. Trop. Dis. 8 (2014), pp. e2866.

[43] T. Spangenberg, J.N. Burrows, P. Kowalczyk, S. McDonald, T.N.C. Wells, and P. Willis, *The Open Access Malaria Box: A drug discovery catalyst for neglected diseases*, PLoS ONE 8 (2013), pp. e62906.

[44] W.C. Van Voorhis, J.H. Adams, R. Adelfio, V. Ahyong, M.H. Akabas, P. Alano, A. Alday, Y. Alemán Resto, A. Alsibaee, A. Alzualde, K.T. Andrews, S.V. Avery, V.M. Avery, L. Ayong, M. Baker, S. Baker, C. Ben Mamoun, S. Bhatia, Q. Bickle, L. Bounaadja,

T. Bowling, J. Bosch, L.E. Boucher, F.F. Boyom, J. Brea, M. Brennan, A. Burton, C.R. Caffrey, G. Camarda, M. Carrasquilla, D. Carter, M.B. Cassera, K.C.C. Cheng, W. Chindaudomsate, A. Chubb, B.L. Colon, D.D. Colón-López, Y. Corbett, G.J. Crowther, N. Cowan, S. D'Alessandro, N. Le Dang, M. Delves, J.L. DeRisi, A.Y. Du, S. Duffy, S.A. El-Salam El-Sayed, M.T. Ferdig, J.A.F. Robledo, D.A. Fidock, I. Florent, P.V.T. Fokou, A. Galstian, F.J. Gamo, S. Gokool, B. Gold, T. Golub, G.M. Goldgof, R. Guha, W.A. Guiguemde, N. Gural, R.K. Guy, M.A.E. Hansen, K.K. Hanson, A. Hemphill, R.H. van Huijsduijnen, T. Horii, P. Horrocks, T.B. Hughes, C. Huston, I. Igarashi, K. Ingram-Sieber, M.A. Itoe, A. Jadhav, A.N. Jensen, L.T. Jensen, R.H.Y. Jiang, A. Kaiser, J. Keiser, T. Ketas, S. Kicka, S. Kim, K. Kirk, V.P. Kumar, D.E. Kyle, M.J. Lafuente, S. Landfear, N. Lee, S. Lee, A.M. Lehane, F. Li, D. Little, L. Liu, M. Llinás, M.I. Loza, A. Lubar, L. Lucantoni, I. Lucet, L. Maes, D. Mancama, N.R. Mansour, S. March, S. McGowan, I.M. Vera, S. Meister, L. Mercer, J. Mestres, A.N. Mfopa, R.N. Misra, S. Moon, J.P. Moore, F.M.R. da Costa, J. Müller, A. Muriana, S.N. Hewitt, B. Nare, C. Nathan, N. Narraidoo, S. Nawaratna, K.K. Ojo, D. Ortiz, G. Panic, G. Papadatos, S. Parapini, K. Patra, N. Pham, S. Prats, D.M. Plouffe, S.A. Poulsen, A. Pradhan, C. Quevedo, R.J. Quinn, C.A. Rice, M.A. Rizk, A. Ruecker, R. St. Onge, R.S. Ferreira, J. Samra, N.G. Robinett, U. Schlecht, M. Schmitt, F.S. Villela, F. Silvestrini, R. Sinden, D.A. Smith, T. Soldati, A. Spitzmüller, S.M. Stamm, D.J. Sullivan, W. Sullivan, S. Suresh, B.M. Suzuki, Y. Suzuki, S.J. Swamidass, D. Taramelli, L.R.Y. Tchokouaha, A. Theron, D. Thomas, K.F. Tonissen, S. Townson, A.K. Tripathi, V. Trofimov, K.O. Udenze, I. Ullah, C. Vallieres, E. Vigil, J.M. Vinetz, P.V. Vinh, H. Vu, N.A. Watanabe, K. Weatherby, P.M. White, A.F. Wilks, E.A. Winzeler, E. Wojcik, M. Wree, W. Wu, N. Yokoyama, P.H.A. Zollo, N. Abla, B. Blasco, J. Burrows, B. Laleu, D. Leroy, T. Spangenberg, T. Wells, and P.A. Willis, *Open source drug discovery with the malaria box compound collection for neglected diseases and beyond*, PLoS Pathog. 12 (2016), pp. e1005763.

[45] S. Ekins, A.L. Perryman, and C. Horta Andrade, *OpenZika: An IBM World Community Grid Project to accelerate Zika virus drug discovery*, PLoS Negl. Trop. Dis. 10 (2016), pp. e0005023.

[46] D. Cavalla, *Predictive methods in drug repurposing: Gold mine or just a bigger haystack?* Drug Discov. Today 18 (2013), pp. 523–532.

[47] D. Savoia, *New antimicrobial approaches: Reuses of old drugs*, Curr. Drug Targ. 17 (2016), pp. 731–738.

[48] J. Devillers, L. Lagadic, O. Yamada, F. Darriet, R. Delorme, X. Deparis, J.P. Jaeg, C. Lagneau, B. Lapied, F. Quiniou, and A. Yébakima, *Use of multicriteria analysis for selecting candidate insecticides for vector control*, in *Juvenile Hormones and Juvenoids. Modeling Biological Effects and Environmental Fate*, J. Devillers, ed., CRC Press, Boca Raton, FL, 2013, pp. 341–381.

[49] M. Vaillant, J.M. Jouany, and J. Devillers, *A multicriteria estimation of the environmental risk of chemicals with the SIRIS method*, Toxicol. Model. 1 (1995), pp. 57–72.

[50] D. Filmore, *It's a GPCR world*, Mod. Drug Discov. 7 (2004), pp. 24–28.

[51] V. Nene, J.R. Wortman, D. Lawson, B. Haas, C. Kodira, Z.J. Tu, B. Loftus, Z. Xi, K. Megy, M. Grabherr, Q. Ren, E.M. Zdobnov, N.F. Lobo, K.S. Campbell, S.E. Brown, M.F. Bonaldo, J. Zhu, S.P. Sinkins, D.G. Hogenkamp, P. Amedeo, P. Arensburger, P.W. Atkinson, S. Bidwell, J. Biedler, E. Birney, R.V. Bruggner, J. Costas, M.R. Coy, J. Crabtree, M. Crawford, B. Debruyn, D. Decaprio, K. Eiglmeier, E. Eisenstadt, H. El-Dorry, W.M. Gelbart, S.L. Gomes, M. Hammond, L.I. Hannick, J.R. Hogan, M.H. Holmes, D. Jaffe, J.S. Johnston, R.C. Kennedy, H. Koo, S. Kravitz, E.V. Kriventseva, D. Kulp, K. Labutti, E. Lee, S. Li, D.D. Lovin, C. Mao, E. Mauceli, C.F. Menck, J.R. Miller, P. Montgomery, A. Mori, A.L. Nascimento, H.F. Naveira, C. Nusbaum, S. O'leary, J. Orvis, M. Pertea, H. Quesneville, K.R. Reidenbach, Y.H. Rogers, C.W. Roth,

J.R. Schneider, M. Schatz, M. Shumway, M. Stanke, E.O. Stinson, J.M. Tubio, J.P. Vanzee, S. Verjovski-Almeida, D. Werner, O. White, S. Wyder, Q. Zeng, Q. Zhao, Y. Zhao, C.A. Hill, A.S. Raikhel, M.B. Soares, D.L. Knudson, N.H. Lee, J. Galagan, S.L. Salzberg, I.T. Paulsen, G. Dimopoulos, F.H. Collins, B. Birren, C.M. Fraser-Liggett, and D.W. Severson, *Genome sequence of* Aedes aegypti, *a major arbovirus vector*, Science 316 (2007), pp. 1718–1723.

[52] C.A. Hill, A.N. Fox, R.J. Pitts, L.B. Kent, P.L. Tan, M.A. Chrystal, A. Cravchik, F.H. Collins, H.M. Robertson, and L.J. Zwiebel, *G protein-coupled receptors in* Anopheles gambiae, Science 298 (2002), pp. 176–178.

[53] J.P. Andersen, A. Schwartz, J.B. Gramsbergen, and V. Loeschcke, *Dopamine levels in the mosquito* Aedes aegypti *during adult development, following blood feeding and in response to heat stress*, J. Insect Physiol. 52 (2006), pp. 1163–1170.

[54] J.M. Meyer, K.F.K. Ejendal, L.V. Avramova, E.E. Garland-Kuntz, G.I. Giraldo-Calderón, T.F. Brust, V.J. Watts, and C.A. Hill, *A "genome-to-lead" approach for insecticide discovery: Pharmacological characterization and screening of* Aedes aegypti *D1-like dopamine receptors*, PLoS Negl. Trop. Dis. 6 (2012), pp. e1478.

[55] C.A. Hill, J.M. Meyer, K.F.K. Ejendal, D.F. Echeverry, E.G. Lang, L.V. Avramova, J.M. Conley, and V.J. Watts, *Re-invigorating the insecticide discovery pipeline for vector control: GPCRs as targets for the identification of next gen insecticides*, Pest. Biochem. Physiol. 106 (2013), pp. 141–146.

[56] A.B. Nuss, K.F.K. Ejendal, T.B. Doyle, J.M. Meyer, E.G. Lang, V.J. Watts, and C.A. Hill, *Dopamine receptor antagonists as new mode-of-action insecticide leads for control of* Aedes *and* Culex *mosquito vectors*, PLoS Negl. Trop. Dis. 9 (2015), pp. e0003515.

[57] C.A. Hill, T. Doyle, A.B. Nuss, K.F.K. Ejendal, J.M. Meyer, and V.J. Watts, *Comparative pharmacological characterization of D1-like dopamine receptors from* Anopheles gambiae, Aedes aegypti *and* Culex quinquefasciatus *suggests pleiotropic signaling in mosquito vector lineages*, Parasites Vect. 9 (2016), pp. 192.

[58] T. Li, O. Zhou, N. Zhang, and Y. Luo, *Toxic effects of chlorpromazine on* Carassius auratus *and its oxidative stress*, J. Environ. Sci. Health B 43 (2008), pp. 638–643.

[59] A.D. Munro, *The effects of apomorphine, d-amphetamine and chlorpromazine on the aggressiveness of isolated* Aequidens pulcher *(Teleostei, Cichlidae)*, Psychopharmacology 88 (1986), pp. 124–128.

[60] L.L. Oliveira, S.C. Antunes, F. Gonçalves, O. Rocha, and B. Nunes, *Evaluation of ecotoxicological effects of drugs on* Daphnia magna *using different enzymatic biomarkers*, Ecotoxicol. Environ. Saf. 119 (2015), pp. 123–131.

[61] WHO, *World Malaria Report 2016*: World Health Organization, Geneva, Switzerland, 2016.

[62] M. Mishra, V.K. Mishra, V. Kashaw, A.K. Iyer, and S.K. Kashaw, *Comprehensive review on various strategies for antimalarial drug discovery*, Eur. J. Med. Chem. 125 (2017), pp. 1300–1320.

[63] H. Matthews, M. Usman-Idris, F. Khan, M. Read, and N. Nirmalan, *Drug repositioning as a route to anti-malarial drug discovery: Preliminary investigation of the in vitro anti-malarial efficacy of emetine dihydrochloride hydrate*, Malaria J. 12 (2013), pp. 359.

[64] C. Teixeira, N. Vale, B. Pérez, A. Gomes, J.R.B. Gomes, and P. Gomes, *"Recycling" classical drugs for malaria*, Chem. Rev. 114 (2014), pp. 11164–11220.

[65] B. Aneja, B. Kumar, M.A. Jairajpuri, and M. Abid, *A structure guided drug-discovery approach towards identification of* Plasmodium *inhibitors*, RSC Adv. 6 (2016), pp. 18364–18406.

[66] C. Hobbs and P. Duffy, *Drugs for malaria: Something old, something new, something borrowed*, F1000 Biol. Rep. 3 (2011), pp. 24.

[67] J. Yuan, K.C.C. Cheng, R.L. Johnson, R. Huang, S. Pattaradilokrat, A. Liu, R. Guha, D. Fidock, J. Inglese, T.E. Wellems, C.P. Austin, and X.Z. Su, *Chemical genomic profiling for antimalarial therapies, response signatures and molecular targets*, Science 333 (2011), pp. 724–729.
[68] T.I. Oprea and J.P. Overington, *Computational and practical aspects of drug repositioning*, Assay Drug Develop. Technol. 13 (2015), pp. 299–306.
[69] J. Lotharius, F.J. Gamo-Benito, I. Angulo-Barturen, J. Clark, M. Connelly, S. Ferrer-Bazaga, T. Parkinson, P. Viswanath, B. Bandodkar, N. Rautela, S. Bharath, S. Duffy, V.M. Avery, J.J. Möhrle, R.K. Guy, and T. Wells, *Repositioning: The fast track to new anti-malarial medicines?* Malaria J. 13 (2014), pp. 143.
[70] M. Kaiser, P. Mäser, L.P. Tadoori, J.R. Ioset, and R. Brun, *Antiprotozoal activity profiling of approved drugs: A starting point toward drug repositioning*, PLoS ONE 10 (2015), pp. e0135556.
[71] K.C. Kobylinski, M. Sylla, P.L. Chapman, M.D. Sarr, and B.D. Foy, *Ivermectin mass drug administration to humans disrupts malaria parasite transmission in Senegalese villages*, Am. J. Trop. Med. Hyg. 85 (2011), pp. 3–5.
[72] N. Gyawali, R.S. Bradbury, and A.W. Taylor-Robinson, *The epidemiology of dengue infection: Harnessing past experience and current knowledge to support implementation of future control strategies*, J. Vector Borne Dis. 53 (2016), pp. 293–304.
[73] S. Bhatt, P.W. Gething, O.J. Brady, J.P. Messina, A.W. Farlow, C.L. Moyes, J.M. Drake, J.S. Brownstein, A.G. Hoen, O. Sankoh, M.F. Myers, D.B. George, T. Jaenisch, G.R. Wint, C.P. Simmons, T.W. Scott, J.J. Farrar, and S.I. Hay, *The global distribution and burden of dengue*, Nature 496 (2013), pp. 504–507.
[74] O.J. Brady, P.W. Gething, S. Bhatt, J.P. Messina, J.S. Brownstein, A.G. Hoen, C.L. Moyes, A.W. Farlow, T.W. Scott, and S.I. Hay, *Refining the global spatial limits of dengue virus transmission by evidence-based consensus*, PLoS Negl. Trop. Dis. 6 (2012), pp. e1760.
[75] A.J. Stevens, M.E. Gahan, S. Mahalingam, and P.A. Keller, *The medicinal chemistry of dengue fever*, J. Med. Chem. 52 (2009), pp. 7911–7926.
[76] J. Green, U. Bandarage, K. Luisi, and R. Rijnbrand, *Recent advances in the discovery of dengue virus inhibitors*, in *Annual Reports In Medicinal Chemistry*, M.C. Desai, ed. Vol. 47, Academic Press, Burlington, 2012, pp. 297–317.
[77] S.P. Lim, Q.Y. Wang, C.G. Noble, Y.L. Chen, H. Dong, B. Zou, F. Yokokawa, S. Nilar, P. Smith, D. Beer, J. Lescar, and P.Y. Shi, *Ten years of dengue drug discovery: Progress and prospects*, Antiviral Res. 100 (2013), pp. 500–519.
[78] D.N. Fusco and R.T. Chung, *Review of current dengue treatment and therapeutics in development*, J. Bioanal. Biomed. S8 (2014), pp. 1–10.
[79] K.J. Farias, P.R. Machado, and B.A. da Fonseca, *Chloroquine inhibits dengue virus type 2 replication in Vero cells but not in C6/36 cells*, Sci. World J. ID 282734 (2013), pp. 1–5.
[80] M.C. Borges, L.A. Castro, and B.A.L. da Fonseca, *Chloroquine use improves dengue-related symptoms*, Mem. Inst. Oswaldo Cruz, Rio de Janeiro 108 (2013), pp. 596–599.
[81] V. Tricou, N.N. Minh, T.P. Van, S.J. Lee, J. Farrar, B. Wills, H.T. Tran, and C.P. Simmons, *A randomized controlled trial of chloroquine for the treatment of dengue in Vietnamese adults*, PLoS Negl. Trop. Dis. 4 (2010), pp. e785.
[82] S. Boonyasuppayakorn, E.D. Reichert, M. Manzano, K. Nagarajan, and R. Padmanabhan, *Amodiaquine, an antimalarial drug, inhibits dengue virus type 2 replication and infectivity*, Antiviral Res. 106 (2014), pp. 125–134.
[83] F.A. De Maio, G. Risso, N.G. Iglesias, P. Shah, B. Pozzi, L.G. Gebhard, P. Mammi, E. Mancini, M.J. Yanovsky, R. Andino, N. Krogan, A. Srebrow, and A.V. Gamarnik, *The dengue virus NS5 protein intrudes in the cellular spliceosome and modulates splicing*, PLoS Pathog. 12 (2016), pp. e1005841.

[84] M.J. Pryor, S.M. Rawlinson, R.E. Butcher, C.L. Barton, T.A. Waterhouse, S.G. Vasudevan, P.G. Bardin, P.J. Wright, D.A. Jans, and A.D. Davidson, *Nuclear localization of dengue virus nonstructural protein 5 through its importin α/β-recognized nuclear localization sequences is integral to viral infection*, Traffic 8 (2007), pp. 795–807.
[85] M.Y. Tay, J.E. Fraser, W.K. Chan, N.J. Moreland, A.P. Rathore, C. Wang, S.G. Vasudevan, and D.A. Jans, *Nuclear localization of dengue virus (DENV) 1-4 non-structural protein 5; protection against all 4 DENV serotypes by the inhibitor Ivermectin*, Antiviral Res. 99 (2013), pp. 301–306.
[86] K.M. Wagstaff, H. Sivakumaran, S.M. Heaton, D. Harrich, and D.A. Jans, *Ivermectin is a specific inhibitor of importin α/β-mediated nuclear import able to inhibit replication of HIV-1 and dengue virus*, Biochem. J. 443 (2012), pp. 851–856.
[87] E. Mastrangelo, M. Pezzullo, T. De Burghgraeve, S. Kaptein, B. Pastorino, K. Dallmeier, X. de Lamballerie, J. Neyts, A.M. Hanson, D.N. Frick, M. Bolognesi, and M. Milani, *Ivermectin is a potent inhibitor of flavivirus replication specifically targeting NS3 helicase activity: New prospects for an old drug*, J. Antimicrob. Chemother. 67 (2012), pp. 1884–1894.
[88] F. Hawking, *Suramin: With special reference to onchocerciasis*, Adv. Pharmacol. 15 (1978), pp. 289–322.
[89] C. Basavannacharya and S.G. Vasudevan, *Suramin inhibits helicase activity of NS3 protein of dengue virus in a fluorescence-based high throughput assay format*, Biochem. Biophys. Res. Commun. 453 (2014), pp. 539–544.
[90] D. Salgado, T.E. Zabaleta, S. Hatch, M.R. Vega, and J. Rodriguez, *Use of pentoxifylline in treatment of children with dengue hemorrhagic fever*, Pediatr. Infect. Dis. J. 31 (2012), pp. 771–773.
[91] Y. Simanjuntak, J.J. Liang, Y.L. Lee, and Y.L. Lin, *Repurposing of prochlorperazine for use against dengue virus infection*, J. Infect. Dis. 211 (2015), pp. 394–404.
[92] X.G. Zhang, P.W. Mason, E.J. Dubovi, X. Xu, N. Bourne, R.W. Renshaw, T.M. Block, and A.V. Birk, *Antiviral activity of geneticin against dengue virus*, Antiviral Res. 83 (2009), pp. 21–27.
[93] D.H. Berg and R.L. Hamill, *The isolation and characterization of narasin, a new polyether antibiotic*, J. Antibiot. 31 (1978), pp. 1–6.
[94] J.S. Low, K.X. Wu, K.C. Chen, M.M. Ng, and J.J. Chu, *Narasin, a novel antiviral compound that blocks dengue virus protein expression*, Antivir. Ther. 16 (2011), pp. 1203–1218.
[95] S. Nagarakanti and E. Bishburg, *Is Minocycline an antiviral agent? A review of current literature*, Basic Clin. Pharmacol. Toxicol. 2016, 118, pp. 4–8.
[96] S.L. Leela, C. Srisawat, G.P. Sreekanth, S. Noisakran, P.T. Yenchitsomanus, and T. Limjindaporn, *Drug repurposing of minocycline against dengue virus infection*, Biochem. Biophys. Res. Comm. 478 (2016), pp. 410–416.
[97] K.C. Smithburn, T.P. Hughes, A.W. Burke, and J.H. Paul, *A neurotropic virus isolated from the blood of a native of Uganda*, Am. J. Trop. Med. 20 (1940), pp. 471–492.
[98] M. Eidson, N. Komar, F. Sorhage, R. Nelson, T. Talbot, F. Mostashari, and R. McLean, *Crow deaths as a sentinel surveillance system for West Nile virus in the northeastern United States, 1999*, Emerg. Infect. Dis. 7 (2001), pp. 615–620.
[99] G.L. Campbell, A.A. Marfin, R.S. Lanciotti, and D.J. Gubler, *West Nile virus*, Lancet Infect. Dis. 2 (2002), pp. 519–529.
[100] I.G. Bouzalas, N. Diakakis, S.C. Chaintoutis, G.D. Brellou, M. Papanastassopoulou, K. Danis, I. Vlemmas, T. Seuberlich, and C.I. Dovas, *Emergence of equine West Nile encephalitis in central Macedonia, Greece, 2010*, Transbound. Emerg. Dis. 63 (2016), pp. e219–e227.

[101] T.J. Gray and C.E. Webb, *A review of the epidemiological and clinical aspects of West Nile virus*, Int. J. Gen. Med. 7 (2014), pp. 193–203.

[102] S.P. Lim and P.Y. Shi, *West Nile virus drug discovery*, Viruses 5 (2013), pp. 2977–3006.

[103] S.A. Elseginy, A. Massarotti, G.A.M. Nawwar, K.M. Amin, and A. Brancale, *Small molecule inhibitors of West Nile virus*, Antivir. Chem. Chemother. 23 (2014), pp. 179–187.

[104] M. Michaelis, M.C. Kleinschmidt, H.W. Doerr and J. Cinatl, *Minocycline inhibits West Nile virus replication and apoptosis in human neuronal cells*, J. Antimicrob. Chemoth. 60 (2007), pp. 981–986.

[105] T. Parkinson and D.C. Pryde, *Small molecule drug discovery for dengue and West Nile viruses: Applying experience from hepatitis C virus*, Future Med. Chem. 2 (2010), pp. 1181–1203.

[106] E. De Clercq and G. Li, *Approved antiviral drugs over the past 50 years*, Clin. Microbiol. Rev. 29 (2016), pp. 695–747.

[107] G.W. Dick, S.F. Kitchen, and A.J. Haddow, *Zika virus. I. Isolations and serological specificity*, Trans. R. Soc. Trop. Med. Hyg. 46 (1952), pp. 509–520.

[108] G.W. Dick, *Zika virus. II. Pathogenicity and physical properties*, Trans. R. Soc. Trop. Med. Hyg. 46 (1952), pp. 521–534.

[109] F.N. Macnamara, *Zika virus: A report on three cases of human infection during an epidemic of jaundice in Nigeria*, Trans. R. Soc. Trop. Med. Hyg. 48 (1954), pp. 139–145.

[110] W.G. Bearcroft, *Zika virus infection experimentally induced in a human volunteer*, Trans. R. Soc. Trop. Med. Hyg. 50 (1956), pp. 442–448.

[111] J.P. Boorman and J.S. Porterfield, *A simple technique for infection of mosquitoes with viruses transmission of Zika virus*, Trans. R. Soc. Trop. Med. Hyg. 50 (1956), pp. 238–242.

[112] P. Brès, *Données récentes apportées par les enquêtes sérologiques sur la prévalence des arbovirus en Afrique, avec référence spéciale à la fièvre jaune*, Bull. Org. Mond. Santé 43 (1970), pp. 223–267.

[113] N.J. Marchette, R. Garcia, and A. Rudnick, *Isolation of Zika virus from* Aedes aegypti *mosquitoes in Malaysia*, Am. J. Trop. Med. Hyg. 18 (1969), pp. 411–415.

[114] H.J. Posen, J.S. Keystone, J.B. Gubbay, and S.K. Morris, *Epidemiology of Zika virus, 1947–2007.* BMJ Global Health 1 (2016), pp. e000087.

[115] M.R. Duffy, T.H. Chen, W.T. Hancock, A.M. Powers, J.L. Kool, R.S. Lanciotti, M. Pretrick, M. Marfel, S. Holzbauer, C. Dubray, L. Guillaumot, A. Griggs, M. Bel, A.J. Lambert, J. Laven, O. Kosoy, A. Panella, B.J. Biggerstaff, M. Fischer, and E.B. Hayes, *Zika virus outbreak on Yap Island, Federated States of Micronesia*, N. Engl. J. Med. 360 (2009), pp. 2536–2543.

[116] V.M. Cao-Lormeau, C. Roche, A. Teissier, E. Robin, A.L. Berry, H.P. Mallet, A.A. Sall, and D. Musso, *Zika virus, French Polynesia, South Pacific, 2013*, Emerg. Infect. Dis. 20 (2014), pp. 1085–1086.

[117] E. Oehler, L. Watrin, P. Larre, I. Leparc-Goffart, S. Lastère, F. Valour, L. Baudouin, H.P. Mallet, D. Musso, and F. Ghawche, *Zika virus infection complicated by Guillain-Barré syndrome — Case report, French Polynesia, December 2013*, Euro Surveill. 2014, 19(9), pii=20720. Available online: http://www.eurosurveillance.org/ViewArticle.aspx?ArticleId=20720

[118] C. Zanluca, V.C.A. de Melo, A.L.P. Mosimann, G.I.V. dos Santos, C.N.D. dos Santos, and K. Luz, *First report of autochthonous transmission of Zika virus in Brazil*, Mem. Inst. Oswaldo Cruz, Rio de Janeiro 110 (2015), pp. 569–572.

[119] N. Gyawali, R.S. Bradbury, and A.W. Taylor-Robinson, *The global spread of Zika virus: Is public and media concern justified in regions currently unaffected*? Infect. Dis. Poverty 5 (2016), pp. 37.

[120] A.S. Fauci and D.M. Morens, *Zika virus in the Americas — Yet another arbovirus threat*, N. Engl. J. Med. 374 (2016), pp. 601–604.

[121] E. Kincaid, *A second look: Efforts to repurpose old drugs against Zika cast a wide net*, Nature Med. 22 (2016), pp. 824–825.

[122] R. Delvecchio, L.M. Higa, P. Pezzuto, A.L. Valadão, P.P. Garcez, F.L. Monteiro, E.C. Loiola, A.A. Dias, F.J.M. Silva, M.T. Aliota, E.A. Caine, J.E. Osorio, M. Bellio, D.H. O'Connor, S. Rehen, R.S. de Aquiar, A. Savarino, L. Campanati, and A. Tanuri, *Chloroquine inhibits Zika virus infection in different cellular models*, Viruses 8 (2016), pp. E322.

[123] G. Barbosa-Lima, L.S. da Silveira Pinto, C.R. Kaiser, J.L. Wardell, C.S. De Freitas, Y.R. Vieira, A. Marttorelli, J.C. Neto, P.T. Bozza, S.M.S.V. Wardell, M.V.N. de Souza, and T.M.L. Souza, *N-(2-(arylmethylimino)ethyl)-7-chloroquinolin-4-amine derivatives, synthesized by thermal and ultrasonic means, are endowed with anti-Zika virus activity*, Eur. J. Med. Chem. 127 (2017), pp. 434–441.

[124] H. Tang, C. Hammack, S.C. Ogden, Z. Wen, X. Qian, Y. Li, B. Yao, J. Shin, F. Zhang, E.M. Lee, K.M. Christian, R.A. Didier, P. Jin, H. Song, and G.L. Ming, *Zika virus infects human cortical neural progenitors and attenuates their growth*, Cell Stem Cell 18 (2016), pp. 587–590.

[125] M. Xu, E.M. Lee, Z. Wen, Y. Cheng, W.K. Huang, X. Qian, J. Tcw, J. Kouznetsova, S.C. Ogden, C. Hammack, F. Jacob, H.N. Nguyen, M. Itkin, C. Hanna, P. Shinn, C. Allen, S.G. Michael, A. Simeonov, W. Huang, K.M. Christian, A. Goate, K.J. Brennand, R. Huang, M. Xia, G.L. Ming, W. Zheng, H. Song, and H. Tang, *Identification of small-molecule inhibitors of Zika virus infection and induced neural cell death via a drug repurposing screen*, Nature Med. 22 (2016), pp. 1101–1107.

[126] N.J. Barrows, R.K. Campos, S.T. Powell, K.R. Prasanth, G. Schott-Lerner, R. Soto-Acosta, G. Galarza-Muñoz, E.L. McGrath, R. Urrabaz-Garza, J. Gao, P. Wu, R. Menon, G. Saade, I. Fernandez-Salas, S.L. Rossi, N. Vasilakis, A. Routh, S.S. Bradrick, and M.A. Garcia-Blanco, *A screen of FDA-approved drugs for inhibitors of Zika virus infection*, Cell Host Microbe 20 (2016), pp. 259–270.

[127] D. Musso, C. Roche, E. Robin, T. Nhan, A. Teissier, and V.M. Cao-Lormeau, *Potential sexual transmission of Zika virus*, Emerg. Infect. Dis. 21 (2015), pp. 359–361.

[128] L. de Noronha, C. Zanluca, M.L.V. Azevedo, K.G. Luz, and C.N.D. dos Santos, *Zika virus damages the human placental barrier and presents marked fetal neurotropism*, Mem. Inst. Oswaldo Cruz, Rio de Janeiro 111 (2016), pp. 287–293.

[129] R. Narayanan, *Zika virus therapeutics: Drug targets and repurposing medicine from the human genome*, MOJ Proteom. Bioinf. 3 (2016), pp. 00084.

[130] B.S. Pascoalino, G. Courtemanche, M.T. Cordeiro, L.H.V.G. Gil, and L. Freitas-Junior, *Zika antiviral chemotherapy: Identification of drugs and promising starting points for drug discovery from an FDA-approved library*, [version 1; referees: 2 approved], F1000Res. 5 (2016), pp. 2523.

[131] K. Rausch, B. Hackett, N. Weinbren, S. Reeder, Y. Sadovsky, C. Hunter, D.C. Schultz, C. Coyne, and S. Cherry, *Screening bioactives reveals nanchangmycin as a broad spectrum antiviral active against Zika virus*, Cell Rep. 18 (2017), pp. 804–815.

[132] T. Solomon, *Control of Japanese encephalitis — Within our grasp?*, N. Engl. J. Med. 355 (2006), pp. 869–871.

[133] D. Ghosh and A. Basu, *Japanese encephalitis. A pathological and clinical perspective*, PLoS Neglect. Trop. Dis. 3 (2009), pp. e437.

[134] S.I. Yun and Y.M. Lee, *Japanese encephalitis. The virus and vaccines*, Hum. Vacc. Immunother. 10 (2014), pp. 263–279.

[135] C.H. Hoke, D.W. Vaughn, A. Nisalak, P. Intralawan, S. Poolsuppasit, V. Jongsawas, U. Titsyakorn, and R.T. Johnson, *Effect of high-dose dexamethasone on the outcome of acute encephalitis due to Japanese encephalitis virus*, J. Infect. Dis. 165 (1992), pp. 631–637.

[136] R. Kumar, P. Tripathi, M. Baranwal, S. Singh, S. Tripathi, and G. Banerjee, *Randomized, controlled trial of oral ribavirin for Japanese encephalitis in children in Uttar Pradesh, India*, Clin. Infect. Dis. 48 (2009), pp. 400–406.

[137] A. Singh, A. Mehta, K.P. Kushwaha, A.K. Pandey, M. Mittal, B. Sharma, and J. Pandey, *Minocycline trial in Japanese encephalitis: A double blind, randomized placebo study*, Pediat. Rev. Intern. J. Pediat. Res. 3 (2016), pp. 371–377.

[138] G. Pialoux, B.Al. Gaüzère, S. Jauréguiberry, and M. Strobel, *Chikungunya, an epidemic arbovirosis*, Lancet Infect. Dis. 7 (2007), pp. 319–327.

[139] O. Schwartz and M.L. Albert, *Biology and pathogenesis of chikungunya virus*, Nat. Rev. Microbiol. 8 (2010), pp. 491–500.

[140] M.C. Robinson, *An epidemic of virus disease in Southern Province, Tanganyika Territory, in 1952–53. I. Clinical features*, Trans. R. Soc. Trop. Med. Hyg. 49 (1955), pp. 28–32.

[141] M.C. Robinson, *An epidemic of virus disease in Southern Province, Tanganyika Territory, in 1952–53. II. General description and epidemiology*, Trans. R. Soc. Trop. Med. Hyg. 49 (1955), pp. 33–57.

[142] E. Javelle, A. Ribera, I. Degasne, B.A. Gaüzère, C. Marimoutou, and F. Simon, *Specific management of post-chikungunya rheumatic disorders: A retrospective study of 159 cases in Reunion Island from 2006–2012*, PLoS Negl. Trop. Dis. 9 (2015), pp. e0003603.

[143] H. Zeller, W. van Bortel, and B. Sudre, *Chikungunya: Its History in Africa and Asia and its spread to new regions in 2013–2014*, J. Infect. Dis. 214 (2016), pp. S436–S440.

[144] P. Gallian, I. Leparc-Goffart, P. Richard, F. Maire, O. Flusin, R. Djoudi, J. Chiaroni, R. Charrel, P. Tiberghien, and X. de Lamballerie, *Epidemiology of chikungunya virus outbreaks in Guadeloupe and Martinique, 2014: An observational study in volunteer blood donors*, PLoS Negl. Trop. Dis. 11 (2017), pp. e0005254.

[145] A.A. Rashad, S. Mahalingam, and P.A. Keller, *Chikungunya virus: Emerging targets and new opportunities for medicinal chemistry*, J. Med. Chem. 57 (2014), pp. 1147–1166.

[146] I.C. Albulescu, M. van Hoolwerff, L.A. Wolters, E. Bottaro, C. Nastruzzi, S.C. Yang, S.C. Tsay, J.R. Hwu, E.J. Snijder, and M.J. van Hemert, *Suramin inhibits chikungunya virus replication through multiple mechanisms*, Antiviral. Res. 121 (2015), pp. 39–46.

[147] I.C. Albulescu, A. Tas, F.E. Scholte, E.J. Snijder, and M.J. van Hemert, *An in vitro assay to study chikungunya virus RNA synthesis and the mode of action of inhibitors*, J. Gen. Virol. 95 (2014), pp. 2683–2692.

[148] Y.J. Ho, Y.M. Wang, J.W. Lu, T.Y. Wu, L.I. Lin, S.C. Kuo, and C.C. Lin, *Suramin inhibits chikungunya virus entry and transmission*, PLoS ONE 10 (2015), pp. e0133511.

[149] S.C. Kuo, Y.M. Wang, Y.J. Ho, T.Y. Chang, Z.Z. Lai, P.Y. Tsui, T.Y. Wu, C.C. Lin, *Suramin treatment reduces chikungunya pathogenesis in mice*, Antiviral Res. 134 (2016), pp. 89–96.

[150] F.S. Varghese, P. Kaukinen, S. Gläsker, M. Bespalov, L. Hanski, K. Wennerberg, B.M. Kümmerer, and T. Ahola, *Discovery of berberine, abamectin and ivermectin as antivirals against chikungunya and other alphaviruses*, Antiviral Res. 126 (2016), pp. 117–124.

[151] M. Pepin, M. Bouloy, B.H. Bird, A. Kemp, and J. Paweska, *Rift Valley fever virus (Bunyaviridae: Phlebovirus): An update on pathogenesis, molecular epidemiology, vectors, diagnostics and prevention*, Vet. Res. 41 (2010), pp. 61.

[152] T. Ikegami and S. Makino, *The pathogenesis of Rift Valley fever*, Viruses 3 (2011), pp. 493–519.

[153] R. Daubney, J.R. Hudson, and P.C. Garnham, *Enzootic hepatitis or Rift Valley fever. An undescribed virus disease of sheep cattle and man from east Africa*, J. Pathol. Bacteriol. 34 (1931), pp. 545–579.

[154] A.M. Samy, A.T. Peterson, and M. Hall, *Phylogeography of Rift Valley fever virus in Africa and the Arabian peninsula*, PLoS Negl. Trop. Dis. 11 (2017), pp. e0005226.

[155] M. Ellenbecker, J.M. Lanchy, and J.S. Lodmell, *Inhibition of Rift Valley fever virus replication and perturbation of nucleocapsid-RNA interactions by suramin*, Antimicrob. Agents Chemother. 58 (2014), pp. 7405–7415.

[156] A. Narayanan, K. Kehn-Hall, S. Senina, L. Lundberg, R. Van Duyne, I. Guendel, and F. Kashanchi, *Curcumin inhibits Rift Valley fever virus replication in human cells*, J. Biol. Chem. 287 (2012), pp. 33198–33214.

[157] A. Benedict, N. Bansal, S. Senina, I. Hooper, L. Lundberg, C. de la Fuente, and K. Kehn-Hall, *Repurposing FDA-approved drugs as therapeutics to treat Rift Valley fever virus infection*, Front. Microbiol. 6 (2015), pp. 676.

[158] G. Greenstone, *The revival of thalidomide: From tragedy to therapy*, BC Med. J. 53 (2011), pp. 230–233.

[159] H.A. Ghofrani, I.H. Osterloh, and F. Grimminger, *Sildenafil: From angina to erectile dysfunction to pulmonary hypertension and beyond*, Nat. Rev. Drug Discov. 5 (2006), pp. 689–702.

[160] Anonymous, *Space Spray Application of Insecticides for Vector and Public Health Pest Control. A Practitioner's Guide*, World Health Organization, Geneva, Communicable Disease Control, Prevention and Eradication WHO Pesticide Evaluation Scheme (WHOPES), WHO/CDS/WHOPES/GCDPP/2003.5, 2003.

[161] ANSES, *Avis de l'Agence nationale de sécurité sanitaire de l'alimentation, de l'environnement et du travail relatif aux substances actives biocides pouvant être utilisées dans le cadre de la prévention d'une épidémie de chikungunya en Guyane*, Saisine n° 2014-SA-0060, ANSES, Maisons–Alfort, France, 2014.

[162] H. Długońska, *The Nobel Prize 2015 in physiology or medicine for highly effective antiparasitic drugs*, Ann. Parasitol. 61 (2015), pp. 299–301.

2 Insect Olfactory System as Target for Computer-Aided Design of Mosquito Repellents

Polina V. Oliferenko, Alexander A. Oliferenko, Gennady Poda, Girinath G. Pillai, Ulrich R. Bernier, Natasha M. Agramonte, and Kenneth J. Linthicum

CONTENTS

ABSTRACT

Blood-feeding insects are common vectors of a number of diseases such as dengue fever, malaria, West Nile encephalitis, meningitis, and poliomyelitis, which affect public life throughout the world. Prevention of mosquito-borne diseases primarily depends on vector control, which requires knowledge of insect sensory biology for the development of efficient methods to manipulate their behavior. The need for better repellents is increasingly urgent in view of the heightening risk of mosquito expansion caused by global climate changes and a growing resistance to existing repellents. Insect olfaction is a sophisticated, multistep process, which involves interactions between odorants and specific proteins, generation of electric signals in neurons, and propagation of the signals to the brain for processing. Each step represents a separate line of research, which needs to be taken into account in a rational approach to repellent design. A full cycle of rational repellent design should include the following key steps: molecular modeling and design, synthesis, and behavioral assays.

KEYWORDS

- Mosquito repellents
- Insect olfactory system
- QSAR
- Computer-aided design
- Molecular docking

2.1 INTRODUCTION

Botanicals and plant extracts have been used as insecticides since antiquity. Despite a huge body of knowledge of traditional mosquito-repelling plants obtained through ethnobotanical studies and increasing interest in insecticides with low risk of health and environmental impacts, commercial repellent products containing plant-based ingredients are quite limited [1]. The main impediments to the development of plant-derived repellents on a commercial scale include limited natural resources and high expenses on standardization and refinement of the plant preparation, as well as difficulties with regulatory approvals [2]. Synthetic repellents generally do not face such shortages, but may have limitations associated with expanding insecticide resistance and potential health and environmental issues. Significant research into synthetic repellents was conducted by the military to find long-lasting compounds in the 1940s, since repellents containing plant essential oils had a limited duration [3]. In 1953, DEET (*N,N*-diethyl-3-methylbenzamide), the most effective broad-spectrum repellent, was discovered. It remains in use and is considered the gold standard for insect repellents. DEET has some shortcomings, though, such as

- Limited efficacy against *Anopheles albimanus* (the principal malaria vector in Central America and the Caribbean) [4], tolerant varieties of *Aedes aegypti* [5], and some other vectors [6]
- Skin irritation
- Possible neurotoxicity [7]
- A plasticizing action on some plastics and synthetic fabrics
- Relatively high cost

A few other repellent active ingredients, such as piperidine derivatives KBR 3023 (picaridin) and AI3-37220, are considered almost as effective as DEET and were reported to remain effective longer and have more desirable cosmetic properties. The repellent *N,N*-diethyl phenylacetamide (DEPA) is comparable to DEET in its broad-spectrum repellency and effectiveness and can be produced at about half the cost of DEET. Ethyl ester of 3-[*N*-butyl-*N*-acetyl]-aminopropionic acid (IR3535), although less effective than DEET, has been favored by some consumers because of a low incidence of side effects. The naturally and synthetically available compound 2-undecanone (2-U) was recently approved as a protection against mosquitoes and ticks [8–10]. Another recently registered PMD (*p*-menthane-3,8-diol), an active ingredient originally obtained from essential oil of the Australian *Eucalyptus citriodora* tree, was found to provide efficacy similar to DEET and Picaridin against mosquitoes [11].

The search for new mosquito repellents never ends. What is now needed is access to more potent and less toxic repellents with a subtle environmental footprint. Ideally, a repellent should be effective, target specific, and long lasting, very safe to humans, animals, and the environment. Similarly to computer-aided drug design (CADD), computer-aided approaches to the design of new molecular entities that can serve as effective mosquito repellents would be helpful and thought provoking. Such approaches can help save time and resources required for classical trial-and-error methods involving screening of large compound libraries.

2.2 COMPUTATIONAL STUDIES OF MOSQUITO REPELLENCY

A number of attempts have previously been made to apply the fundamental principles of the quantitative structure–activity relationships (QSARs) to link the chemical structure of repellents to their activity and to develop models for prediction of new arthropod repellents. Since the discovery of DEET, efforts have been devoted to finding a superior repellent by making use of DEET analogs or other structurally similar benzamides. Recent examples include stereoelectronic analysis and *in silico* pharmacophore modeling performed for repellent activities of 10 DEET-like compounds and 9 PMD analogs using the CATALYST methodology [12,13]. The methodology was developed by incorporation of a semiautomated quasi-multiway partial least squares approach and applied for a set of 40 *N*-alkyl substituted benzamides and benzylamides [14]. Mosquito bite protection times of a large set of carboxamides and *N*-acylpiperidines were also analyzed using artificial neural networks and multiple linear regression methods [15,16]. Another group of intensively studied compounds are natural terpenoids, long known for their insect-deterring properties. A prediction model built for 12 sesquiterpenes containing hydroxyl and carbonyl moieties against mosquitoes *Ae. aegypti* employed vapor pressure and molecular polarizability as the main parameters describing spatial and contact mosquito repellency [17]. Mosquito repellency was modeled for a set of 20 pinene derivatives containing hydroxyl, ether, and ester groups using a large number of descriptors of the optimized repellents and repellent-lactic acid complexes [18,19]. The same data set was studied by the so-called topological-mathematical approach employing only topological and topochemical descriptors [20]. Another recent study focused on the effect of hydrocarbon chain length on the biting-deterrent activity of saturated and unsaturated fatty acids against *Ae. aegypti* [21].

One can notice that the vast majority of computational studies on repellent activity were based on utilizing sets of structurally close compounds, although modeling of diverse chemical classes would be more beneficial. The latter is challenging for a reason: ligand-based approaches, such as QSAR, do not take into account the interactions between odorant molecules and the insect chemosensory system. Despite an increase in research efforts over the last several decades, the mechanism of repellency is not yet fully understood. The repellency effect can be caused by inhibition of acetylcholinesterase (AChE), modulation of sodium channels and nicotinic acetylcholine receptors, or, most likely, due to the interference with the insect olfactory system that governs behavioral patterns such as host-seeking, oviposition, and fleeing from chemical irritants. For example, DEET was evidenced to modulate olfaction in insects [22], inhibit acetylcholinesterase activity [7,23], and affect gustatory receptors [24]. Recent breakthroughs in understanding the mechanisms of insect olfaction brought new opportunities in identification of putative molecular targets associated with odor reception, which could be used in structure-based approaches for more efficient design of target-specific repellents.

2.3 OLFACTORY SYSTEM AS TARGET OF RATIONAL REPELLENT DESIGN

The insect olfactory system is believed to be the prime target for many natural and synthetic repellents. A recent increase in the number of research projects on functional genomics, sensory physiology, structural biology, and chemical ecology of blood-feeding insects has added extensively to knowledge of the molecular basis of insect olfaction [25]. The most important features of the early olfactory processing of blood-feeding insects are outlined in the following subsections. State-of-the-art information on insect odor perception can be found in the literature [25–28].

2.3.1 Physiology of Olfaction in Insects

Olfaction is a complex, multistep process that includes absorption of odorants at the periphery, their transportation to the sites of action, and interaction with the olfactory transmembrane proteins. Interaction with the olfactory receptors results in activation of neurons and transmission of electric signals to the antennal lobes followed by integration and processing of the olfactory signals in the higher brain centers and, ultimately, translation of the olfactory signals into behavioral responses [25,29]. The periphery chemosensory system of hematophagous insects responsible for olfactory and gustatory perception consists of numerous highly diverse and specialized sensory hairs—sensilla—located mainly in antennae, parts of the mouth, and legs. The olfactory sensilla usually house two to three olfactory sensory neurons (OSNs) with dendrites expressing the olfactory receptors. The axons of OSNs from various sensilla join the antennal nerve and terminate in the antennal lobe, the first processing center of the olfactory signals in the brain [27,30]. The antennal lobe is composed of spheroid functional units—glomeruli—each of them receiving input from a single class of OSNs expressing the same receptors. Hence, the activity of an odorant receptor (OR) population responding to certain odorants is located in one glomerulus [31]. Upon integration and coding of olfactory information in glomeruli, a vast receptor-neuronal input is transferred to a relatively few projection (output) neurons of the antennal lobe and transmitted further to higher brain areas for odor perception and behavioral readout. Each odor stimulus evokes a unique response pattern in an odorant receptor neuron (ORN) ensemble with excitatory and inhibitory responses of various intensities from different odorant receptors, which all together encode the quality and quantity pattern of the odor stimulus. The target receptors for primary insect repellents, including DEET, as well as how excitatory and inhibitory responses of olfactory neurons contribute to repellency, are unknown [32].

2.3.2 Supramolecular Chemistry of Odor Reception

The major molecular components identified so far as mediating reception of general odors in insects are odorant-binding proteins (OBPs) and three classes of olfactory receptors: odorant receptors, gustatory receptors, and ionotropic receptors. Here we consider the odorant-binding proteins and odorant receptors as currently most extensively studied functional blocks of the olfaction system in insects.

2.3.2.1 Odorant Binding Proteins

Odorants reaching through the pores in the sensillum cuticle to the lymph, which bathes dendrites of the olfactory receptor neurons, are solubilized by carrier proteins (OBPs) present in the lymph and activate odorant receptors residing in the ORN membrane [4,29,33,34]. There are 66 putative odorant-binding proteins in *Ae. aegypti* [35], and more than 80 OBP encoding genes in the *Anopheles gambiae* genome [36]. High affinities of some mosquitoes' OBPs for one or more ligands have been demonstrated by resolving the X-ray structures of the crystallized protein–ligand complexes. Highly abundant in mosquitoes and overly expressed in the female antennae, OBP1 is one of the most well studied odorant binding proteins [37,38]. Electroantennogram responses and *in vitro* binding assays revealed that OBP1 had high binding affinities toward certain human skin effluents such as indole and 1-octen-3-ol, as well as toward repellents, including geranylacetone, 2-undecanone, and octanal and some other elongated hydrophobic molecules. X-ray structures were solved for OBP1 in *An. gambiae* [39,40], *Cx. quinquefasciatus* [41], and *Ae. aegypti* [42]. Complexation of diverse small molecule odorants was reported for *Agam*OBP07 [43], *Agam*OBP22a [44], *Agam*OBP4 [45], and *Agam*OBP47 [46]. Affinities of some benzoates and phthalates toward *Aaeg*OBP22 were reported as measured experimentally with the use of immunofluorescence and fluorescent probe methods [47].

OBPs represent the first "selection gate" in a multistage odor perception process. Recently, successful expression, purification, and crystallization of OBPs made tertiary structures of some OBPs available for computer modeling. For example, affinities to *Agam*OBP1 were studied for a set of 29 carboxamides and piperidines using AutoDock [40] and several components of the clove oil with potential repellent activities using MVD [48].

2.3.2.2 Odorant Receptors

Odorant receptor proteins are a family of specific highly divergent transmembrane proteins that can interact and recognize certain odor molecules. Usually, ORs are complexed with a nonconventional conservative co-receptor (Orco) subunit, which is believed to form nonselective ligand-gated cation channels [25]. Stimulation of an

OR by an odorant initiates a sequence of biochemical events amplifying an action potential that is forwarded along the axons to the brain [49]. The odor code reflects the odorant chemical structure and concentration [50,51] and the presence of other volatiles [52].

The odorant-induced activity of an individual odorant receptor neuron can be measured by recording a single cell electrophysiological current with an electrode inserted into an insect's olfactory sensillum [53]. In order to study the specific functional contribution of a specific receptor, an *ex vivo* method is used: The odorant receptor is expressed in the outer cell membrane of the *Xenopus laevis* frog oocytes, which are then exposed to a flow of the odorant water/DMSO solutions, and recordings of current responses are taken with two microelectrodes impaled into the oocyte. Studying of OR functions using *in vivo* and *ex vivo* approaches made it possible to study OR responses to a broad spectrum of odorants and identify the ORs tuned to certain odorants such as components of human skin effluents, pheromones, and repellents [54]. A systematic analysis of a large collection of odor responses demonstrated that a given odorant can elicit responses of different intensities from different ORs, whereas ORs can be broadly or narrowly tuned for a wide or restricted panel of odors [55,56], thus attesting to a combinatorial model of odor coding [50,51]. For example, a three-dimensional map of 50 OR responses to 110 odorants was constructed for *An. gambiae* [54], and a similar analysis was carried out for *Drosophila* [50,56]. ORN currents were also studied for several repellents (DEET, 2-U, S220, callicarpenal, pyrethroid) in *Ae. aegypti* [8,57].

While tertiary structures of OBPs have been successfully deduced using X-ray crystallography, structure elucidation of odorant receptors and odorant–OR complexes is a challenging problem, since no three-dimensional (3D) structure has been experimentally determined so far [58–61]. Currently, molecular modeling of odorant–OR complexes is done by two distinct approaches: (1) *ab initio* protocols for construction of 3D structures on the basis of a known amino acid sequence, and (2) homology modeling for prediction of OR structure from a set of experimentally known homologous protein structures. So far, docking algorithms for prediction of the ligand–OR interactions have been developed only for G protein-coupled receptors in mammals since their 3D structures have been studied in sufficient details [62] (Figure 2.1).

2.3.3 Computational Modeling of Olfactory System

Modeling of various functional aspects of insect olfactory systems using computational, mathematical, and statistical approaches has been continuously attempted and reported in the literature [63]. Recent examples include but are not limited to (1) multistep biophysical models of single ORNs for insects [64–66], including a statistical distribution of collective firing rates [67]; (2) use of artificial neural networks and molecular parameters for the prediction of responses in *Drosophilla melanogaster* [68]; or (3) mimicking neurocomputational principles of the whole olfaction process [69,70].

Olfaction perception is an integral process that combines the effects from (a) a sequence of intermolecular interactions between the odorant and specific components (OBPs, ORs) of the external olfactory system transduced into (b) a cascade of electric signals in an insect sensory network. Because of the complexity of the

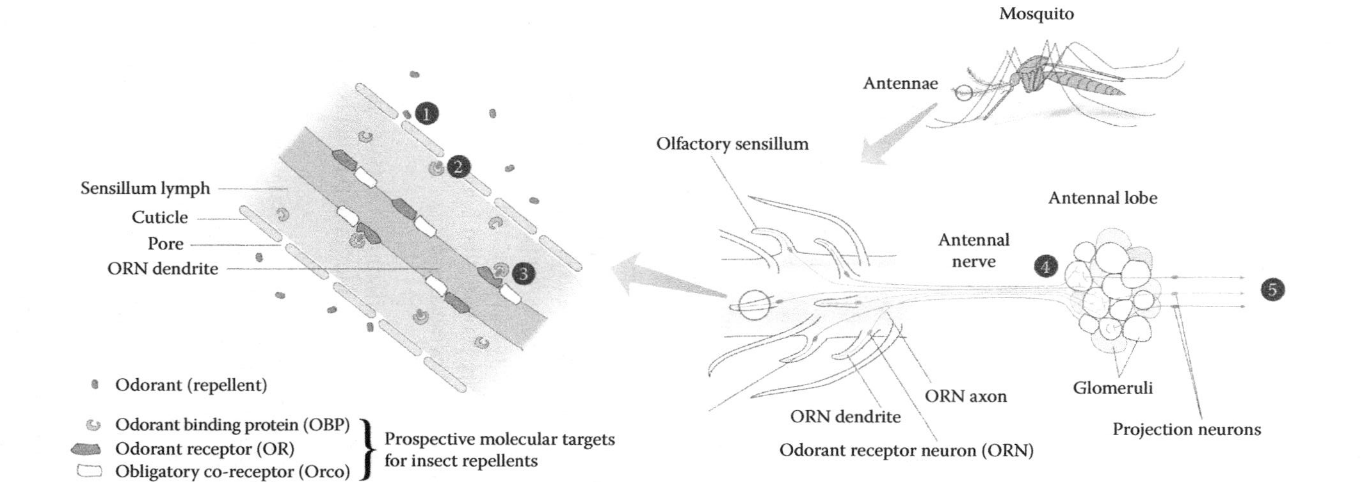

1. Odorants in the air penetrate through the pores in the cuticle of a mosquito antennae.
2. Hydrophobic odorants are recognized and bound by specific odorant binding proteins (OBPs).
3. Odorant-OBP complex approaches odorant receptor embedded in a dendrite of odorant receptor neuron (ORN). Interaction between the odor molecule and receptor results in generation of odor-specific electrical signal in the neuron which can be experimentally measured by single cell recording technique.
4. Signals generated by ORNs are sent to antennal lobe of the brain where they are partially processed in glomeruli.
5. Signals from ORNs holding the same ORs converge in their own glomerulus to be transmitted further by projection neurons to higher-order olfactory centers.

FIGURE 2.1 **(See color insert.)** Schematic illustration of odor reception in mosquitoes. (Modified from C.I. Bargmann, Nature 444 (2006), pp. 295–301; E. Jacquin-Joly and C. Merlin, J. Chem. Ecol. 30 (2004), pp. 2359–2397; A.F. Carey and J.R. Carlson, PNAS 108 (2011), pp. 12987–12995.)

olfaction system, individual studies are normally focused on specific aspects of the olfactory process. Thus, modeling interaction of odorant molecules to the odorant binding proteins and odorant receptors, approximation of neurons' action potential curves, construction of multidimensional odorant/OR space representations, modeling of a response of a population of ORNs to an odorant, building models of post-receptor transduction events, and some others represent separate fields of research. Each of these employs intricate and elaborate techniques that enable one to partially reproduce natural conditions, which is somewhat incomplete, although critical for a successful prediction of odorant effects. For example, an odorant molecule can form a tight binding with OBP, but this does not guarantee that it will translate into a strong response through binding with OR. Similarly, identification of a chemical that elicits highest firing rates or silencing of one or more ORs is insufficient for concluding whether this chemical possesses deterring properties or not. In turn, if odorant–OR interaction studies are performed in an artificial environment, an identified chemical substance possessing strong activation or inhibition of ORNs may lack *in vivo* activity due, for example, to low volatility. Ideally, computational models that predict behavioral response to a sensory input would need to simulate all stages of the odorant transduction cascade: from odorant solubilization by sensillum lymph to insect body movements elicited by motor neurons [63]. In reality, not all stages of the olfaction process are deciphered yet. Additionally, assembling all pieces of the puzzle would require agreed-upon efforts of multiple research groups in different research fields. Outcome of such collaboration could be a multivariable model predicting repellent efficiency that accounts for physical chemical properties of a repellent, tightness of binding to OBPs, an activation/inhibition profile of elicited signals in a body of ORs, and some other relevant characteristics.

Recently we have reported a study of repellent activities of assorted compounds against *Ae. aegypti* using integrated computational and experimental approaches—namely, QSAR, molecular docking, organic synthesis, and bioassay [71]. The goal of the study was to identify links among molecular structure and electronic properties, binding affinities to a specific odorant-binding protein, and biological activity. The rational design approach incorporated the following steps:

- Building computational topological models utilizing electronic, structural, and physical chemical characteristics of known mosquito repellents belonging to different chemotypes
- Prioritization of the training set compounds based on their affinity to OBP1 calculated with a molecular docking software
- Computational hit expansion based on the structures of the most potent repellents identified, followed up by a virtual screening against OBP1

The computational tools used in the mentioned study are briefly discussed next.

2.3.4 Molecular Field Topology Analysis

Molecular field topology analysis (MFTA) is a computational approach and software for analysis of structure–activity relationships of structurally related compounds.

It may be considered as a topological analog of comparative molecular field analysis (CoMFA) [72]. In MFTA, the spatial alignment of structures is replaced by a superimposition of molecular graphs (topological entities in which only atom types and molecular connectivity are taken into account, but not molecular geometry). The superimposed molecular graphs form the so-called "molecular supergraph," abbreviated as MSG [73,74]. The MSG's vertices and edges corresponding to atoms and bonds are loaded with numerical values of local various atomic descriptors or "molecular fields." Distribution of these fields is subjected to a special type of statistical analysis, which results in an equation that ties experimental activity to chemical structure. Typical molecular fields generated by MFTA are atomic charges, van der Waals radii, electronegativities, hydrogen bonding, and lipophilicity indices. So far, MFTA has been successfully applied to several medicinal chemistry problems [75–77].

MFTA analysis was done over 43 previously published carboxamides and 27 assorted compounds retrieved from the USDA archive. The repellent activity in this study was defined as the minimum effective dosage (MED, μmol/cm^2). In essence, this is an estimate of the effective dose for 99% repellency (ED_{99}), as the failure threshold is selected to be 1% (five mosquito bites out of approximately 500 mosquitoes in 30 min). For the purposes of MFTA, logarithmic values of the MED were used to directly relate the MED values to changes of the free energy of interaction. The training set structures and respective MED values are given in Tables 2.1 and 2.2, respectively.

Here we briefly describe the main results to avoid overloading the text with technical details; an in-depth description can be found in the original paper [69]. The molecular supergraph image is shown in Figure 2.2.

2.3.5 Molecular Docking

Compounds represented in Tables 2.1 and 2.2, as well as results of hit expansion, were docked in the OPB1 structure (PDB code 3K1E) using the Glide docking program developed by Schrödinger, Inc. Up to 1,000,000 docking poses were generated in one docking run, and then 100 best poses were selected for postdocking minimization, with the top five poses saved for each ligand. The docking poses were carefully inspected for the presence of good hydrogen bonding and van der Waals interactions as well as steric clashes. Compounds that fit the best in the OBP1 structure were selected for prioritization.

To date, only a handful of X-ray crystal structures of insect OBP1 have been resolved: 3N7H, 3K1E, 3R1O, 3R1P, 3R1V, and 2L2C [78]. In 3N7H, DEET taken as a reference binds at the interface formed by two OBP1 monomeric units. Interestingly, it does not form any direct hydrogen bonds with the protein. The only directional polar interaction at this site is the hydrogen bond between the carbonyl oxygen of DEET and the structural water molecule, which also forms two additional hydrogen bonds with the Trp114 indole NH and the backbone carbonyl of Cys95. On the other hand, based on our expanded binding site docking results against the 3K1E OBP1 structure, the major part of the hit expansion compounds consistently dock in the middle of the OBP1 channel (occupied by PEG in the 3K1E structure)

TABLE 2.1
Structure and Repellent Activity of a Set of Carboxamides (*Ae. aegypti*)

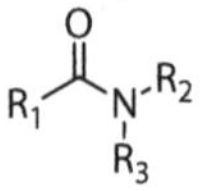

ID	Compound Name	MED (μmol/cm²)	R1	R2	R3
5d	Hexahydro-1-(1-oxohexyl)-1*H*-azepine	0.033			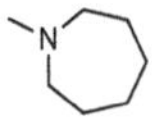
5t	(E)-1-(1-azepanyl)-2-methyl-2-penten-1-one	0.098	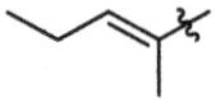		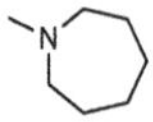
5h	1-(1-Azepanyl)-2-methyl-1-pentanone	0.102	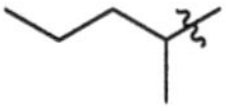		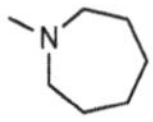
5g	*N*-butyl-*N*-ethyl-2-methylpentanamide	0.104	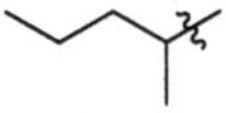	C_2H_5	
5a	*N*-butyl-*N*-methyl-hexanamide	0.117		CH_3	
5r	(E)-*N*-butyl-*N*-ethyl-2-methyl-2-pentenamide	0.117	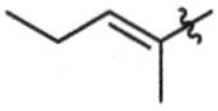	C_2H_5	
5i	*N*-butyl-*N*,2-diethylbutanamide	0.125	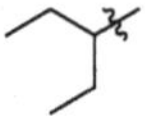	C_2H_5	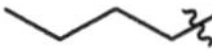
5j′	(E)-*N*,2-dimethyl-*N*-octylpent-2-enamide	0.125	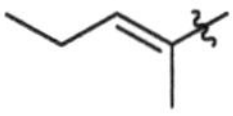	CH_3	
5k	*N*-butyl-*N*-ethyl-3-methylbutanamide	0.125	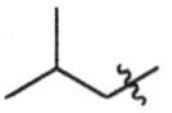	C_2H_5	
5l′	(E)-*N*-cyclohexyl-*N*-ethyl-2-methylpent-2-enamide	0.14	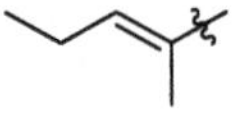	C_2H_5	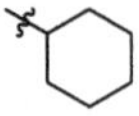
5a′	Hexahydro-1-(3-methylcrotonoyl)-1*H*-azepine	0.14	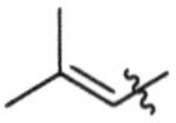		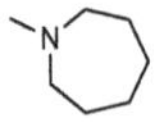
5v	*N*-ethyl-2-methyl-*N*-(2-methyl-2-propenyl) benzamide	0.145	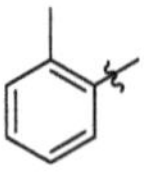	C_2H_5	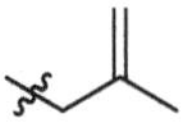
5b	*N*-butyl-*N*-ethylhexanamide	0.156		C_2H_5	
5q	*N*-butyl-*N*-ethyl-2-methylbenzamide	0.156	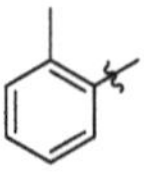	C_2H_5	
5m	*N*-cyclohexyl-*N*-ethyl-3-methylbutanamide	0.172	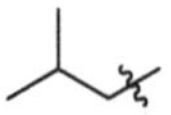	C_2H_5	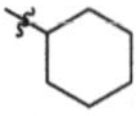

(Continued)

TABLE 2.1 (CONTINUED)
Structure and Repellent Activity of a Set of Carboxamides (*Ae. aegypti*)

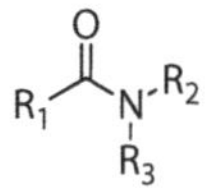

ID	Compound Name	MED (μmol/cm²)	R1	R2	R3
5k′	*N*-cyclohexyl-*N*-methylheptanamide	0.172		CH_3	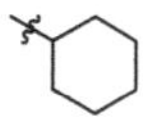
5s	(E)-*N*-ethyl-2-methyl-*N*-(2-methyl-2-propenyl)-2-pentenamide	0.182	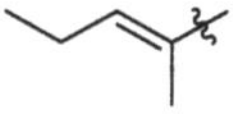	C_2H_5	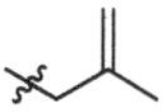
5h′	*N*-butyl-*N*-methyl-5-hexynamide	0.182	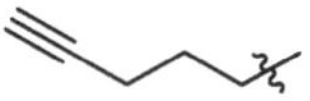	CH_3	
5x	*N*-butyl-*N*-ethyl-3-methyl-2-butenamide	0.192	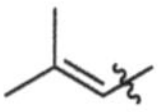	CH_3	
5c	*N*,*N*-diallylhexanamide	0.195		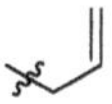	
5z	*N*,*N*-diisobutyl-3-methylcrotonamide	0.219	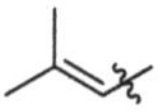	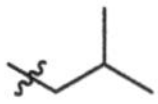	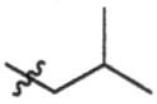
5e	*N*-cyclohexyl-*N*-ethylhexanamide	0.266	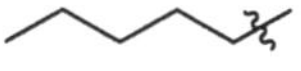	C_2H_5	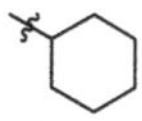
5e′	(E)-*N*-n-butyl-*N*-ethyl-2-hexenamide	0.274	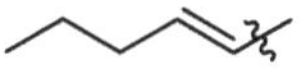	C_2H_5	
5n	*N*-butyl-*N*-ethyl-2,2-dimethylpropanamide	0.286	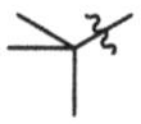	C_2H_5	
5p	1-(1-Azepanyl)-2,2-dimethyl-1-propanone	0.313			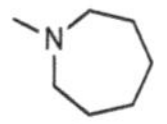 N
5y	*N*-ethyl-3-methyl-*N*-(2-methyl-2-propenyl)-2-butenamide	0.313	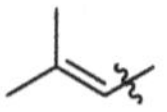	C_2H_5	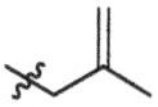
5j	*N*,2-diethyl-*N*-(2-methyl-2-propenyl) butanamide	0.375	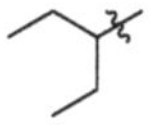	C_2H_5	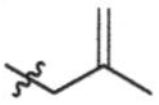
5l	*N*,*N*-diisobutyl-3-methylbutanamide	0.406	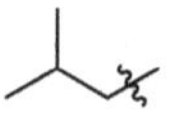	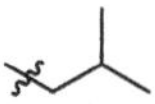	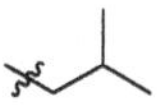
5u	(E)-2-methyl-*N*,*N*-di-2-propenyl-2-pentenamide	0.417	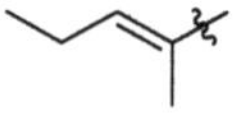	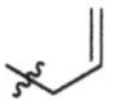	
5o	*N*-ethyl-2,2-dimethyl-*N*-(2-methyl-2-propenyl)propanamide	0.469		C_2H_5	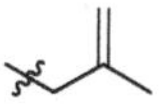

(Continued)

TABLE 2.1 (CONTINUED)
Structure and Repellent Activity of a Set of Carboxamides (*Ae. aegypti*)

ID	Compound Name	MED (μmol/cm²)	R1	R2	R3
5f′	(E)-*N,N*-di-(2-methylpropyl)-2-hexenamide	0.625			
5f	*N*-ethyl-*N*-phenylhexanamide	0.625		C_2H_5	
5g′	(E)-*N*-cyclohexyl-*N*-ethyl-2-hexenamide	0.651		C_2H_5	
5w	*N*-ethyl-2-methyl-*N*-phenylbenzamide	5.160		C_2H_5	
5b′	*N*-butyl-*N*-ethyl-cinnamamide	10.75		C_2H_5	
5c′	*N,N*-bis(2-methylpropyl)-3-phenyl-2-propenamide	20.125			
5d′	*N*-ethyl-*N*,3-diphenyl-2-propenamide	20.250		C_2H_5	
5i′	*N*,3-dicyclohexyl-*N*-ethylpropanamide	20.500		C_2H_5	
C39	3-Cyclohexyl-*N*-methyl-*N*-octylpropanamide	25.000		CH_3	
C40	4-Methyl-*N*-phenylbenzamide	25.000		H	
C41	2-Methyl-*N*-phenylbenzamide	25.000		H	
C42	*N*-cyclohexyl-*N*-isopropyl-4-methyloctanamide	25.000			
C43	*N,N*-dicyclohexyl-4-methyloctanamide	25.000			

TABLE 2.2
Experimental Repellency Data for 27 Assorted Compounds

ID	Compound Structure	Compound Name	MED (μmol/cm²)	ID	Compound Structure	Compound Name	MED (μmol/cm²)
YF2	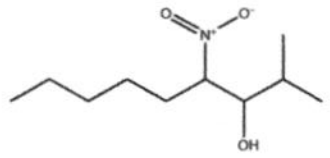	2-Methyl-4-nitro-3-nonanol	0.047	YF12	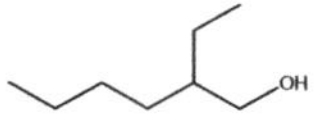	2-Ethyl-1-hexanol	1.875
YF4	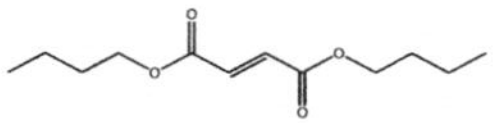	Dibutyl fumarate	0.047	YF15	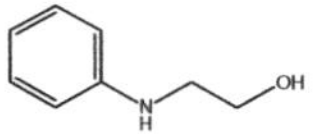	2-Anilinoethanol	1.875
YF24	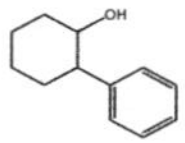	2-Phenyl-cyclohexanol	0.047	YF6	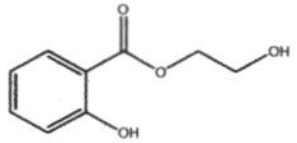	2-Hydroxyethyl 2-hydroxybenzoate	2.500
YF8	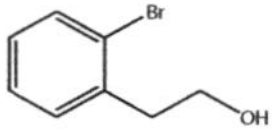	2-Bromophenethyl alcohol	0.049	YF18	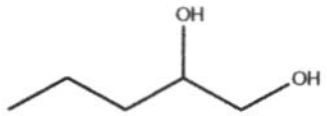	1,2-Pentanediol	2.500
YF25	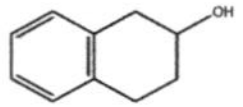	1,2,3,4-Tetrahydro-2-naphthol	0.062	NR5	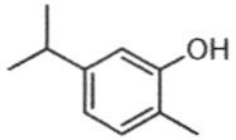	Carvacrol	0.013
YF39	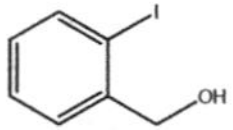	(2-Iodophenyl)methanol	0.070	NR6	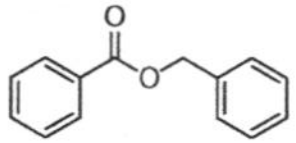	Benzyl benzoate	0.023
YF19	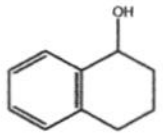	1,2,3,4-Tetrahydro-1-naphthol	0.078	NR7	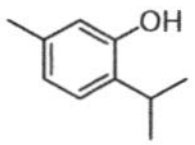	Thymol	0.031
YF7	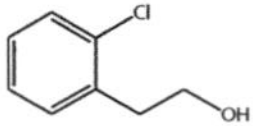	2-Chlorophenethyl alcohol	0.101	NR8	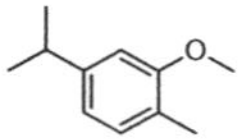	Carvacrol methyl ether	0.063

(*Continued*)

TABLE 2.2 (CONTINUED)
Experimental Repellency Data for 27 Assorted Compounds

ID	Compound Structure	Compound Name	MED (μmol/cm²)	ID	Compound Structure	Compound Name	MED (μmol/cm²)
YF21	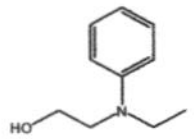	2-(*N*-ethylanilino)-ethanol	0.156	NR9	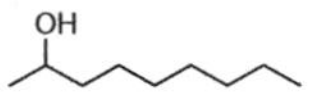	2-Nonanol	0.066
YF27	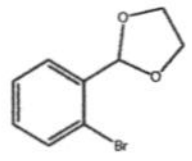	2-(2-Bromophenyl)-1,3-dioxolane	0.156	NR2	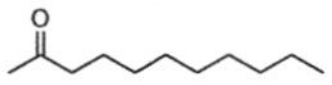	2-Undecanone	0.109
YF22	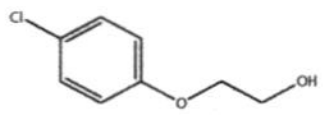	2-(*p*-Chlorophenoxy)-ethanol	0.219	NR3	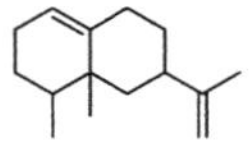	Valencene	0.138
YF16	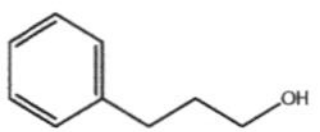	3-Phenyl-1-propanol	0.406	NR4	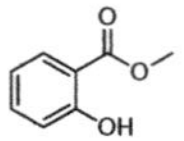	Methyl salicylate	0.312
YF23	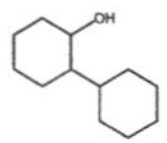	2-Cyclohexyl-cyclohexanol	0.437	NR1	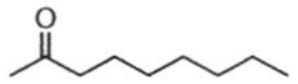	2-Nonanone	0.437
YF20	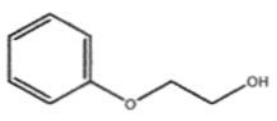	2-Phenoxyethanol	0.563	ref	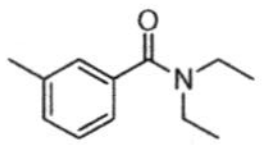	DEET	0.052

Notes: Bioassay was done with *Ae. aegypti*. DEET is given as a positive control.

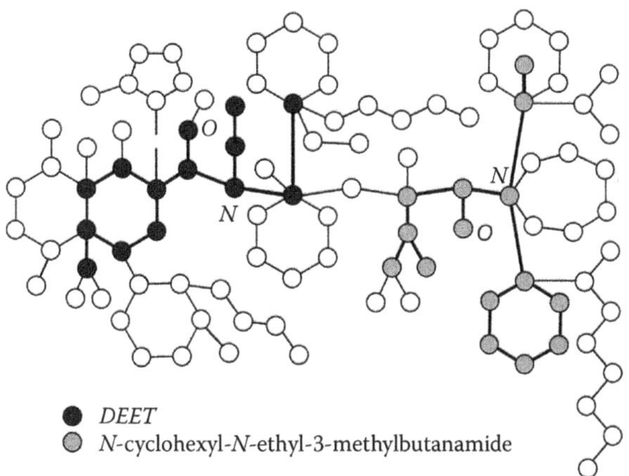

FIGURE 2.2 MFTA model: molecular supergraph. The molecular supergraph is shown with two subgraphs of superimposed structures: DEET and *N*-cyclohexyl-*N*-ethyl-3-methylbutanamide (5m). The algorithm of the structure's superimposition on the molecular supergraph is based on searching for a maximum complete subgraph followed by vertex-by-vertex expansion; the process is optimized by taking into account similarities in the distribution of local properties and special limitations imposed by a researcher.

in the proximity of Phe123, and they form hydrogen bonds to the backbone of either C=O or NH groups (or both) of Phe123. It was generally observed that the docked molecules were ranked higher if located deep inside the protein cavity (similarly to the PEG molecule in 3N7H). This can be explained by a larger van der Waals contact surface between the protein and the ligand, which leads to more favorable interactions and, accordingly, to higher docking scores.

Due to the mostly hydrophobic nature of protein–ligand interactions and a possibility of a conformational change upon binding, it is rather difficult to explain subtle differences in activity. Generally, the binding affinity is defined by the shape complementarity between the protein and ligand. Very small compounds, such as **YF15**, may be binding at the binding site in several locations with partial occupancies. The high activity of **YF19** and **YF24** can be explained by the similarity of their shapes to DEET. Molecular volumes and shapes of these molecules are very close to each other, and they are docked in the same region close to Phe123.

A plausible pharmacophore scheme would include at least two features: (1) a hydrogen bond to backbone NH or carbonyl of Phe123 and (2) a deep-inside aromatic or hydrophobic moiety (e.g., corresponding to tolyl in DEET) bound in the hydrophobic pocket formed by a set of the aromatic and hydrophobic residues Phe59, Leu76, Trp114, Tyr122, and Phe123.

One of the best of the docked structures in terms of the Glide score, **YF24**, is shown in Figures 2.3 and 2.4. The 2D protein–ligand interaction diagram was generated using the ligand interaction script in Maestro software (Schrödinger, Inc.). It outlines a highly hydrophobic cavity consisting of a number of proximate hydrophobic residues (shown in light gray circles; Figure 2.3) where **YF24** binds. **YF24** is

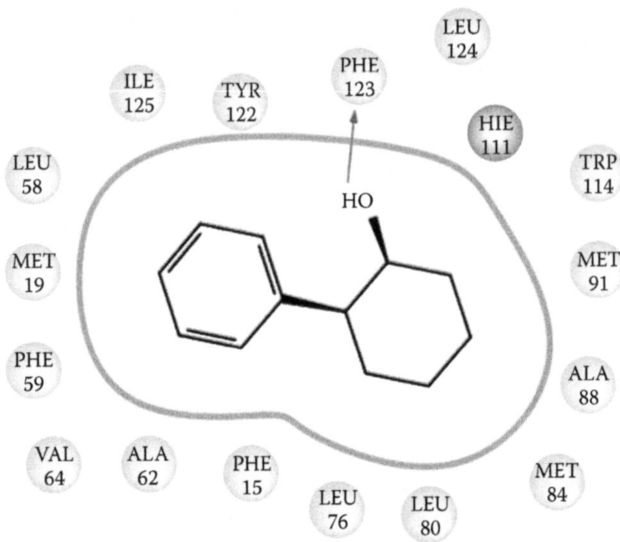

FIGURE 2.3 Predicted binding mode for **YF24** ((1S,2S)-2-phenylcyclohexanol).

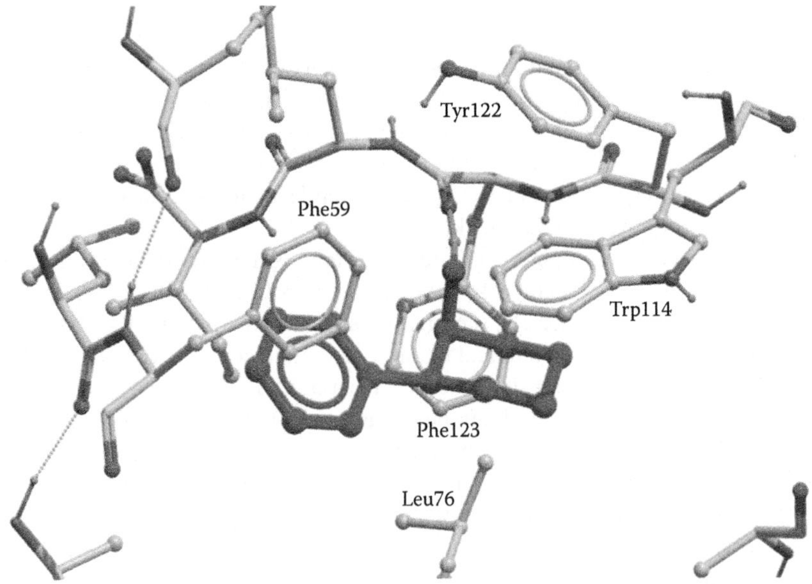

FIGURE 2.4 **(See color insert.)** Atomic details on how **YF24** binds to OBP1, as depicted by ICM Browser (MolSoft, www.molsoft.com). An anchoring interaction that defines the position and orientation of the ligand is the hydrogen bond between the hydroxyl group of **YF24** and the backbone carbonyl group of Phe123. The rest of the interaction is driven by a set of aromatic and hydrophobic residues, Phe59, Leu76, Trp114, Tyr122, and Phe123, that accommodates the cyclohexylbenzene core. Only proximate residues making contact with **YF24** are shown.

represented as a 2D chemical sketch. A hydrogen bond between the ligand and Phe123 is shown by an arrow.

The (1S,2S)-2-phenylcyclohexanol enantiomer is expected to have the highest binding affinity to OBP1 and also has the highest Glide SP score of –8.52. Other enantiomers have minor steric clashes, more strained geometry, and accordingly lower Glide scores (–8.39, –8.37, –8.16). One sees how well the molecule sits inside the OBP1 cavity and also makes a strong hydrogen bond by its hydroxyl group with the backbone carbonyl of Phe123. Although purely computational, the docking results can serve as a rough but still good guidance for rational design. If a molecule fits poorly to the binding site in an *in silico* model, it is highly probable that its physiological effect will be negligible (if any).

Comparison of experimental activity with the docking score and binding interaction energies has not been straightforward in the field of computer-aided molecular design. One should be aware of the limitations and applicability ranges of the theoretical models implemented in the docking software. Looking at the data in Table 2.3, one can notice a clear correlation between MED_{obs} and Glide SP scores for a subset of the first seven most active compounds (boldface in the table). These structures are all compact enough, contain a hydroxyl or a carboxylic oxygen, and can be matched to either a biphenyl or naphthalene scaffold. A correlation coefficient as high as 0.922 suggests that a possible chemotype can exist that binds strongly to *Aaeg*OBP1 and also acts as a strong repellent. For the first time, this would give us a direct link between predicted ligand–OBP affinity and insect behavioral response.

TABLE 2.3
Comparison of Best SP Glide Docking Results with Experimental MED

Compound	Glide SP Score	MED_{obs}
YF24	–8.491	0.039
YF19	–8.363	0.065
YF25	–8.21	0.078
5h	–7.508	0.102
5l′	–7.394	0.140
5m	–7.203	0.172
YF21	–7.226	0.208
YF23	–8.52	0.625
5f	–6.944	0.625
5w	–8.273	5.160
5b′	–7.019	10.75
5d′	–7.222	20.25
5i′	–8.334	20.50
C40	–8.367	25.00
C41	–7.64	25.00

Note: Boldface indicates the most active compounds with MED_{obs} values highly correlated with Glide SP scores (correlation coefficient: 0.92).

This relationship between OBP binding and repellency is not general, of course, because in the olfactory cascade, OBP is just a gateway and it is believed that the odorant receptors play the major role. Table 2.3 gives also examples of compounds with high affinity to OBP, which are nevertheless inactive as repellents.

The fact that a group of the least active compounds (**5i′**, **C40**, and **5w**) are predicted to have similar docking scores is an interesting result by itself. One plausible explanation is that although these compounds have a high potential to be tightly bound in OBP1, they may occur completely inactively on the second step of the odorant transduction—that is, the interaction with and stimulation of the odorant receptors. Another potential reason could be formation of a very tight complex with OBP that may lead to long residence times, which interferes with the compound's ability to effectively interact with the ORs.

2.3.6 Hit Expansion: An Integrated Approach

The hit expansion was done based on the structures of the four most potent repellents: DEET, Picaridin (KBR 3023), **5d**, and **YF23**. Two-dimensional structures of molecules have been sketched in ChemDraw and saved in MOL format for further processing. We defined 29 mol files to conduct substructure and similarity searches against the eMolecules Plus database of approximately five million unique commercially available compounds [79] using Pipeline Pilot 8.0 from Accelrys. Similarity searches were conducted with the 0.65 Tanimoto similarity cutoff based on FCFP4 Pipeline Pilot fingerprints. After applying the OICR HTS filters and molecular weight cutoff of 250 Da, the search resulted in 47 analogs of DEET, 30 analogs of Picaridin, 59 analogs of **YF23**, and 208 analogs of **5d**. The hit expansion procedure accomplished through the eMolecules Plus database resulted in 344 commercially available analogs of the four starting scaffolds. To evaluate this large number of compounds and do a more rational selection of candidates for experimental testing, we docked all the 344 virtual hits (including the original four lead compounds) in the A monomer of the *Ae. aegypti* OBP1 structure (PDB 3K1E, 1.85 Å) [42]. Based on the Glide SP docking of these analogs against the 3K1E OBP1 structure and thorough visual inspection, 36 compounds were identified that looked promising as potential OBP1 binders.

Behavioral bioassay carried out to determine the MED values for selected compounds from the hit expansion data set is given in Table 2.4 along with the MFTA predictions and the Glide scores.

Two of the identified compounds, **X4** and **X23**, exhibited high activity. **X4** is structurally similar to 2-phenylcyclohexanol (**YF24**), which has been known as a repellent since 1945, but only methyl-substituted analogs of it were studied [80]. **X4** is therefore a result of the successful hit expansion of the existing scaffold, and its repellent activity is reported here for the first time. **X23** was first identified as a repellent in 1975 by McGovern et al. [81], but no effective dosage was reported at that time. In fact, it was determined in our previous study [73] to be equal to 0.039 μg/cm^2, as high as that of DEET.

TABLE 2.4
Comparison of Glide SP Scoring, MFTA Predicted MED Values, and Bioassay Results for Selected Compounds

Compound	Structure	Glide Score	MED_{pred}	MED_{obs}[a]
X23	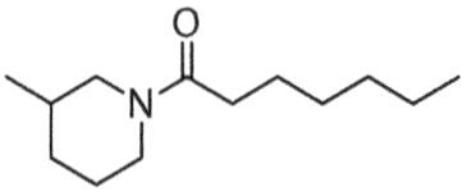	–6.832	0.061	0.039
X4	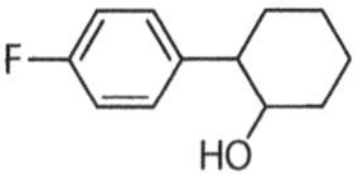 F; HO	–8.715	0.030	0.078
X5	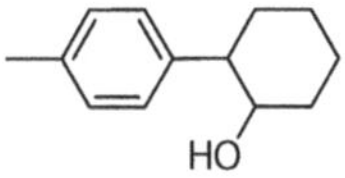 HO	–8.643	0.067	0.156
X16	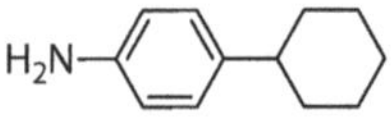 H_2N	–8.405	0.303	0.261
X7	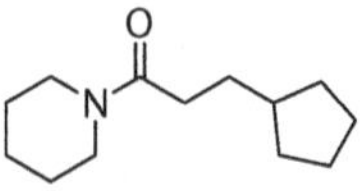 O; N	–7.797	0.112	0.417
X10	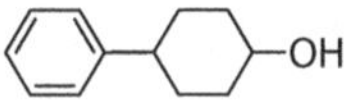 OH	–	0.330	1.25
X17	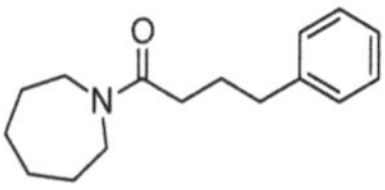 O; N	–	0.082	1.667
X25	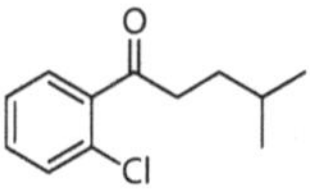 O; Cl	–	0.198	1.87
X2	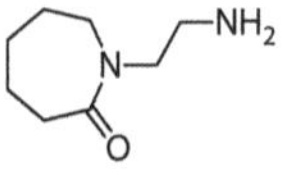 N; NH_2; O	–8.166	–	>2.5
X9	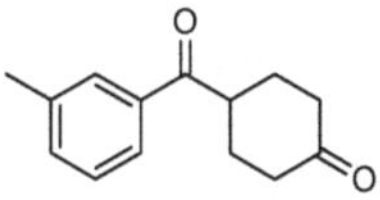 O; O	–8.106	0.182	>2.5
X12	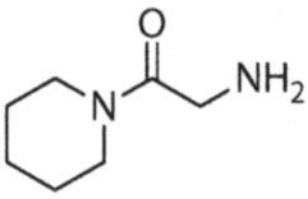 O; N; NH_2	–7.757	0.096	>2.5
X21	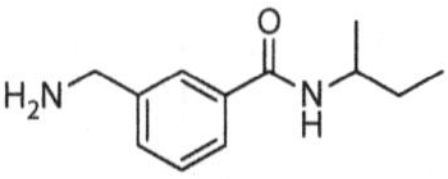 H_2N; O; N; H	–7.529	2.786	>2.5

(Continued)

TABLE 2.4 (CONTINUED)
Comparison of Glide SP Scoring, MFTA Predicted MED Values, and Bioassay Results for Selected Compounds

Compound	Structure	Glide Score	MED_{pred}	MED_{obs}[a]
X24	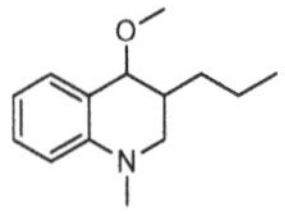	–	0.036	>2.5
X26	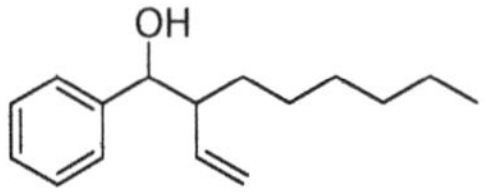	–	0.010	>2.5
X1	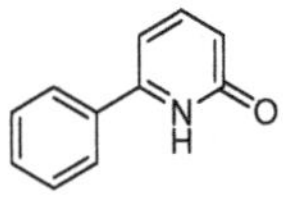	–9.109	0.307	>2.5

[a] Repellent activity was not observed at the highest dosage of 2.5 μg/cm^2.

The minimum effective dosages of **X4** and **X23** were slightly higher than that of **YF24**, and this observation led to the conclusion that 2-phenylcyclohexanol was a viable scaffold for developing more diverse active compounds. At least one structural feature was elucidated: Substituents in the *para*-position of the 2-phenylcyclohexanol ring are not favored. A fluorine substitution in this position results in a slight decrease in activity, while a methyl substituent decreases the activity by a factor of two compared to the fluorine.

A high correlation between measured repellency and Glide SP scores was again observed for the first four most active compounds (**X4**, **X5**, **X16**, **X7**), with the correlation coefficient equal to 0.97, except for **X23**, which was predicted to bind much more weakly than the others.

Although the quantitative correspondence between the MFTA predicted and observed MED values was not perfect, the qualitative trend was quite satisfactory. As can be seen in Table 2.4, MFTA predicted very low values for **X4** and **X23**, whereas for the other compounds, the MFTA predictions were one order of magnitude larger. The docking score was also in qualitative agreement with the experimental data despite the very favorable docking scores assigned to inactive compounds **X1** and **X2**. This apparent mismatch implies that not every compound having a high affinity to OBP1 is a good repellent. The mosquito olfactory mechanism is complex, and OBP1 is just the first step in developing a response. OBP1 either brings the odorant molecule into direct contact with the olfactory receptors or exerts an allosteric action upon the OR. In both cases, the initial binding state of the *Aaeg*OBP1–odorant complex can change dramatically as the recognition event proceeds. It is quite reasonable to assume that the compounds sharing the 2-phenylcyclohexanol scaffold are bound in such a favorable mode that they are able to activate the OR machinery.

2.4 CONCLUDING REMARKS AND FUTURE PERSPECTIVES

The goal of this chapter is twofold: (1) to familiarize the community with recent advances in rational design of mosquito repellents using computational tools and (2) to demonstrate the potential of computer-aided methods in the design of novel or modified known repellents. The computational case studies discussed here represent the very first attempt to integrate the key aspects of repellent perception in mosquitoes into a model with a satisfactory predictive ability. It turn, a viable predictive model will be a giant step toward the *de novo* design. Reaching this goal is not easy, since insect olfactory systems have not been studied well—definitely not as well as many systems of the human body. Combined efforts from, at a first glance, distant fields such as insect physiology, molecular biology, genetics, synthetic chemistry, and computer-aided design methods seem to afford a way to a really rational, far from "trial and error" method of design of insect repellents with desirable properties. The proposed integrated approach to the rational design of mosquito repellents is somewhat similar to that widely used in drug discovery, and it should borrow ideas, methods, and tools from this flourishing area. Thus, it seems feasible that translation of the computer-aided drug design rich experience into the rational design of mosquito repellents will help develop vector control systems of the future.

ACKNOWLEDGMENTS

G. Poda would like to acknowledge the support of the Ontario Institute for Cancer Research and funding from the government of Ontario.

REFERENCES

[1] O. Koul and G.S. Dhaliwal, *Phytochemical Biopesticides*, Harwood Academic Publishers, Amsterdam, 2005.

[2] C. Regnault-Roger, B.J.R. Philigene, and C. Vincent, *Biopesticides of Plant Origin*, Intercept Ltd, Hampshire, 2005.

[3] M. Debboun, S.P. Frances, and D. Strickman, *Insect Repellents. Principles, Methods, and Uses*, CRC Press, Boca Raton, FL, 2007.

[4] W. van der Goes van Naters and J.R. Carlson, *Insects as chemosensors of humans and crops*, Nature 444 (2006), pp. 302–307.

[5] N.M. Stanczyk, J.F.Y. Brookfield, R. Ignell, J.G. Logan, and L.M. Field, *Behavioral insensitivity to DEET in* Aedes aegypti *is a genetically determined trait residing in changes in sensillum function*, PNAS 107 (2010), pp. 8575–8580.

[6] O.A. Belova, L.A. Burenkova, and G.G. Karganova, *Different tick-borne encephalitis virus (TBEV) prevalences in unfed versus partially engorged ixodid ticks—Evidence of virus replication and changes in tick behavior*, Ticks Tick-borne Dis. 3 (2012), pp. 240–246.

[7] V. Corbel, M. Stankiewicz, C. Pennetier, D. Fournier, J. Stojan, E. Girard, M. Dimitrov, J. Molgó, J. Hougard, and B. Lapied, *Evidence for inhibition of cholinesterases in insect and mammalian nervous systems by the insect repellent DEET*, BMC Biology 7 (2009), pp. art47.

[8] J.D. Bohbot and J. Dickens, *Insect repellents: Modulators of mosquito odorant receptor activity*, PLoS ONE 5 (2010), pp. e12138.

[9] G. Paluch, L. Bartholomay, and J. Coats, *Mosquito repellents: A review of chemical structure diversity and olfaction*, Pest. Manag. Sci. 66 (2010), pp. 925–935.

[10] Z. Syed, J. Pelletier, E. Flounders, R.F. Chitolina, and W.S. Leal, *Generic insect repellent detector from the fruit fly* Drosophila melanogaster, PLoS ONE 6 (2011), pp. e17705.

[11] S.P. Carroll and J. Loye, *PMD, a registered botanical mosquito repellent with DEET-like efficacy*, J. Am. Mosq. Control Assoc. 22 (2006), pp. 507–514.

[12] A.K. Bhattacharjee, W. Dheranetra, D.A. Nichols, and R.K. Gupta, *3D pharmacophore model for insect repellent activity and discovery of new repellent candidates*, QSAR Comb. Sci. 24 (2005), pp. 593–602.

[13] A. Bhattacharjee, *In silico stereo-electronic analysis of PMD (p-menthane-3-8-diol) and its derivatives for pharmacophore development may aid discovery of novel insect repellents*, Curr. Comput. Aided Drug Des. 9 (2013), pp. 308–316.

[14] J.B. Bhonsle, A.K. Bhattacharjee, and R.K. Gupta, *Novel semi-automated methodology for developing highly predictive QSAR models: Application for development of QSAR models for insect repellent amides*, J. Mol. Model. 13 (2007), pp. 179–208.

[15] A.R. Katritzky, Z. Wang, S. Slavov, M. Tsikolia, D. Dobchev, N.G. Akhmedov, C.D. Hall, U.R. Bernier, G.G. Clark, and K.J. Linthicum, *Synthesis and bioassay of improved mosquito repellents predicted from chemical structure*, PNAS 105 (2008), pp. 7359–7364.

[16] A.R. Katritzky, Z. Wang, S. Slavov, D. Dobchev, C.D. Hall, M. Tsikolia, U.R. Bernier, N.M. Elejalde, G.G. Clark, and K.J. Linthicum, *Novel carboxamides as potential mosquito repellents*, J. Med. Ent. 47 (2010), pp. 924–938.

[17] G. Paluch, J. Grodnitzky, L. Bartholomay, and J. Coats, *Quantitative structure-activity relationship of botanical sesquiterpenes: Spatial and contact repellency to the yellow fever mosquito* Ae. aegypti, J. Agric. Food Chem. 57 (2009), pp. 7618–7625.

[18] Z. Wang, J. Song, J. Chen, Z. Song, S. Shang, Z. Jiang, and Z. Han, *QSAR study of mosquito repellents from terpenoid with a six-member ring*, Bioorg. Med. Chem. Let. 18 (2008), pp. 2854–2859.

[19] J. Song, Z. Wang, A. Findlater, Z. Han, Z. Jiang, J. Chen, W. Zheng, and S. Hyde, *Terpenoid mosquito repellents: A combined DFT and QSAR study*, Bioorg. Med. Chem. Let. 23 (2013), pp. 1245–1248.

[20] R. García-Domenech, J. Aguliera, A. El Moncef, S. Pocovi, and J. Gálvez, *Application of molecular topology to the prediction of mosquito repellents of a group of terpenoid compounds*, Mol. Divers. 14 (2010), pp. 321–329.

[21] A. Ali, C.L. Cantrell, U.R. Bernier, S.O. Duke, J.C. Schneider, N.M. Agramonte, and I. Khan, Aedes aegypti *(Diptera: Culicidae) biting deterrence: Structure–activity relationship of saturated and unsaturated fatty acids*, J. Med. Entomol. 49 (2012), pp. 1370–1378.

[22] M. Ditzen, M. Pellegrino, and L.B. Vosshall, *Insect odorant receptors are molecular targets of the insect repellent DEET*, Science 319 (2008), pp. 1838–1842.

[23] Y. Pang, F. Ekström, G.A. Polsinelli, Y. Gao, S. Rana, D.H. Hua, B. Andersson, P.O. Andersson, L. Peng, S.K. Singh, R.K. Mishra, K.Y. Zhu, A.M. Fallon, D.W. Ragsdale, and S. Brimijoin, *Selective and irreversible inhibitors of mosquito acetylcholinesterases for controlling malaria and other mosquito-borne diseases*, PLoS ONE 4 (2009), pp. e6851.

[24] Y. Lee, S.H. Kim, and C. Montell, *Avoiding DEET through insect gustatory receptors*, Neuron 67 (2010), pp. 555–561.

[25] W.S. Leal, *Odorant receptors in insects: Roles of receptors, binding proteins, and degrading enzymes*, Ann. Rev. Entomol. 58 (2013), pp. 373–391.

[26] E.A. Hallem, A. Dahanukar, and J.R. Carlson, *Insect odor and taste receptors*, Annu. Rev. Entomol. 51 (2006), pp. 113–135.

[27] F. Guidobaldi, I.J. May-Concha, and P.G. Guerenstein, *Morphology and physiology of the olfactory system of blood-feeding insects*, J. Physiol. 108 (2014), pp. 96–111.

[28] C.G. Galizia, *Olfactory coding in the insect brain: Data and conjectures*, Eur. J. Neurosci. 39 (2014), pp. 1784–1795.

[29] K. Sato and K. Touhara, *Insect olfaction: Receptors, signal transduction, and behavior,* in *Chemosensory Systems in Mammals, Fishes, and Insects*, W. Meyerhof and S. Korsching, eds., Springer, Berlin, 2009, pp. 121–138.

[30] H. Mustaparta, *Introduction IV: Coding mechanisms in insect olfaction,* in *Olfaction in Mosquito-host Interactions*, G.R. Bock and G, Cardew, eds., Ciba Foundation, Wiley, 1996, pp. 149–157.

[31] S. Sachse and J. Krieger, *Olfaction in insects. The primary processes of odor recognition and coding*, e-Neuroforum 2 (2011), pp. 49–60.

[32] P. Kain, S.M. Boyle, S.K. Tharadra, T. Guda, C. Pham, A. Dahanukar, and A. Ray, *Odour receptors and neurons for DEET and new insect repellents*, Nature 502 (2013), pp. 507–514.

[33] R.G. Vogt, *Biochemical diversity of odor detection: OBPs, ODEs and SNMPs*, in *Insect Pheromone Biochemistry and Molecular Biology*, G.J. Blomquist and R.G. Vogt, eds., Elsevier, London, 2003, pp. 391–446.

[34] T.S. Ha and D.P. Smith, *Odorant and pheromone receptors in insects*, Front. Cell. Neurosci. 3 (2009), pp. art10.

[35] G. Yang, G. Winberg, H. Ren, and S. Zhang, *Expression purification and functional analysis of an odorant binding protein AaegOBP22 from* Ae. aegypti, Protein Expression Purif. 75 (2011), pp. 165–171.

[36] A. Schultze, D. Schymura, M. Forstner, and J. Krieger, *Expression pattern of a 'Plus-C' class odorant binding protein in the antenna of the malaria vector* Anopheles gambiae, Insect Mol. Biol. 21 (2012), pp. 187–195.

[37] H. Biessmann, E. Andronopoulou, M.R. Biessman, V. Douris, S.D. Dimitratos, E. Eliopoulus, P.M. Guerin, K. Iatrou, R.W. Justice, T. Kröber, O. Marinotti, P. Tsitoura, D.F. Woods, and M.F. Walter, *The* Anopheles gambiae *odorant binding protein 1 (AgamOBP1) mediates indole recognition in the antennae of female mosquitoes*, PLoS ONE 5 (2010), pp. e9471.

[38] W. Xu, J. Cornel, and W.S. Leal, *Odorant-binding proteins of the malaria mosquito* Anopheles funestus *sensu stricto*, PLoS ONE 5 (2010), pp. e15403.

[39] M. Wogulis, T. Morgan, Y. Ishida, W.S. Leal, and D.K. Wilson *The crystal structure of an odorant binding protein from* Anopheles gambiae: *Evidence for a common ligand release mechanism*, Biochem. Biophys. Res. Commun. 339 (2006), pp. 157–164.

[40] K.E. Tsitsanou, T. Thireou, C.E. Drakou, K. Koussis, M.V. Keramioti, D.D. Leonidas, E. Eliopoulos, K. Iatrou, and S.E. Zographos, Anopheles gambiae *odorant binding protein crystal complex with the synthetic repellent DEET: Implication for structure-based design of novel mosquito repellents*, Cell. Mol. Life Sci. 69 (2012), pp. 283–297.

[41] Y. Mao, X. Xu, W. Xu, Y. Ishida, W.S. Leal, J.B. Ames, and J. Clardy, *Crystal and solution structures of an odorant-binding protein from the southern house mosquito complexed with an oviposition pheromone*, PNAS 107 (2010), pp. 19102–19107.

[42] N.R. Leite, R. Krogh, W. Xu, Y. Ishida, J. Iulek, W.S. Leal, and G. Oliva, *Structure of an odorant-binding protein from the mosquito* Ae. aegypti *suggests a binding pocket covered by a pH-sensitive 'lid,'* PLoS ONE 4 (2009), pp. e8006.

[43] A. Lagarde, S. Spinelli, M. Tegoni, X. He, L. Field, J. Zhou, and C. Cambillau, *The crystal structure of odorant binding protein 7 from* Anopheles gambiae *exhibits an outstanding adaptability of its binding site*, J. Mol. Biol. 414 (2011), pp. 401–412.

[44] H. Ren and S. Zhang, Protein data bank refcodes 3QME 3L4A 3L4L. http://www.pdb.org/pdb/home/home.do

[45] F. Darvazou, E. Dong, E.J. Murphy, H.T. Johnson, and D.N.M. Jones, *New insights into the mechanism of odorant detection by the malaria-transmitting mosquito* Anopheles gambiae, J. Biol. Chem. 286 (2011), pp. 34175–34183.
[46] A. Lagarde, S. Spinelli, H. Qiao, M. Tegoni, P. Pelosi, and C. Cambillau, *Crystal structure of a novel type of odorant-binding protein from* Anopheles gambiae *belonging to the C-plus class*, Biochem. J. 437 (2007), pp. 423–430.
[47] S. Li, J. Picimbon, S. Ji, Y. Kan, Q. Chuanling, J. Zhou, and P. Pelosi, *Multiple functions of an odorant-binding protein in the mosquito* Aedes aegypti, *Biochem. Biophys. Res. Commun.* 372 (2008), pp. 464–468.
[48] R.S. Affonso, A.P. Guimarães, A.A. Oliveira, G.B.C. Slana, and T.C.C. França. *Applications of molecular modeling in the design of new insect repellents targeting the odorant binding protein of* Anopheles gambiae, J. Braz. Chem. Soc. 24 (2013), pp. 473–482.
[49] E. Jacquin-Joly and C. Merlin, *Insect olfactory receptors: Contributions of molecular biology to chemical ecology*, J. Chem. Ecol. 30 (2004), pp. 2359–2397.
[50] E.A. Hallem and J.R. Carlson, *Coding of odors by a receptor repertoire*, Cell 125 (2006), pp. 143–160.
[51] B. Malnic, J. Hirono, T. Sato, and L.B. Buck, *Combinatorial receptor modes for odors*, Cell 96 (1999), pp. 713–723.
[52] J.P. Martin, A. Beyerlein, A.M. Dacks, C.E. Reisenman, J.A. Riffell, H. Lei, and J.D. Hildebrand, *The neurobiology of insect olfaction: Sensory processing in a comparative context*, Prog. Neurobiol. 95 (2011), pp. 427–447.
[53] J.C. Dickens, J.D. Bohbot, and A.J. Grant, *Analysis of odorant receptor protein function in the yellow fever mosquito*, Aedes aegypti, in *Protein Purification and Analysis III—Methods and Applications*. ISBN: 978-1-922227-64-5. iConcept Press, 2013.
[54] A.F. Carey, G. Wang, C. Su, L.J. Zwiebel, and J.R. Carlson, *Odorant reception in the malaria mosquito* Anopheles gambiae, Nature 464 (2010), pp. 66–71.
[55] E.A. Hallem, M.G. Ho, and J.R. Carlson, *The molecular basis of odor coding in the* Drosophila *antenna*, Cell 117 (2004), pp. 965–979.
[56] G. Wang, A.F. Carey, J.R. Carlson, and L.J. Zwiebel, *Molecular basis of odor coding in the malaria vector mosquito* Anopheles gambiae, PNAS 107 (2010), pp. 4418–4423.
[57] J.D. Bohbot, L. Fu, T.C. Le, K.R. Chauhan, C.L. Cantrell, and J.C. Dickens, *Multiple activities of insect repellents on odorant receptors in mosquitoes*, Med. Vet. Entomol. 25 (2011), pp. 436–444.
[58] L. Charlier, J. Topin, C.A. de March, P.C. Lai, C.J. Crasto, and J. Golebowski, *Molecular modelling of odorant/olfactory receptor complexes*, in *Olfactory Receptors, Methods and Protocols*, C.J. Crasto, ed., Humana Press, New York, 2013, pp. 53–66.
[59] C.J. Crasto, *Computational biology of olfactory receptors*, Curr. Bioinf. 4 (2009), pp. 8–15.
[60] M. Persuy, G. Sanz, A. Tromelin, T. Thomas-Danguin, J. Gibrat, E. Pajot-Augy, *Mammalian olfactory receptors: Molecular mechanisms of odorant detection, 3D-modeling, and structure-activity relationship*, in *Molecular Basis of Olfaction*, R. Glatz, ed., Prog. Mol. Biol. Transl. Sci. Elsevier, 130 (2015), pp. 1–36.
[61] C.I. Bargmann, *Comparative chemosensation from receptors to ecology*, Nature, 444 (2006), pp. 295–301.
[62] A.F. Carey and J.R. Carlson, *Insect olfaction from model systems to disease control*, PNAS 108 (2011), pp. 12987–12995.
[63] G.C. Rains, D. Kulasiri, Z. Zhou, S. Samarasinghe, J.K. Tomberlin, and D.M. Olson, *Synthesizing neurophysiology, genetics, behavior and learning to produce whole-insect programmable sensors to detect volatile chemicals*, Biotechnol. Genet. Eng. 26 (2009), pp. 191–216.

[64] Y. Gu, P. Lucas, and J. Rospars, *Computational model of the insect pheromone transduction cascade*, PLoS Comp. Biol. 5 (2009), pp. e1000321.

[65] J. Rospars, P. Lucas, and M. Coppey, *Modeling the early steps of transduction in insect olfactory receptor neurons*, BioSystems 89 (2007), pp. 101–109.

[66] K. Kaissling, *Olfactory perireceptor and receptor events in moths: A kinetic model revised*, J. Comp. Phys A 195 (2009), pp. 895–922.

[67] M. Sandström, A. Lansner, J. Hellgren-Kotaleski, and J. Rospars, *Modeling the response of a population of olfactory receptor neurons to an odorant*, J. Comput. Neurosci. 27 (2009), pp. 337–355.

[68] M. Schmuker, M. Bruyne, M. Hähnel, and G. Schneider, *Predicting olfactory receptor neuron responses from odorant structure*, Chem. Cen. J. 1 (2007), pp. 11.

[69] M. Schmuker and G. Schnider, *Processing and classification of chemical data inspired by insect olfaction*, PNAS 104 (2007), pp. 20285–20289.

[70] S. Namiki, S.S. Haupt, T. Kazawa, A. Takashima, H. Ikeno, and R. Kanzaki, *Reconstruction of virtual neural circuits in an insect brain*, Front. Neurosci. 3 (2009), pp. 206–213.

[71] P.V. Oliferenko, A.A. Oliferenko, G.I. Poda, D.I. Osolodkin, G.G. Pillai, U.R. Bernier, M. Tsikolia, N.M. Agramonte, G.G. Clark, K.L. Linthicum, and A.R. Katritzky, *Promising* Aedes aegypti *repellent chemotypes identified through integrated QSAR, virtual screening, synthesis, and bioassay*, PLoS ONE 8 (2013), pp. e64547.

[72] R.D. Cramer, D.E. Patterson, and J.D. Bunce, *Comparative molecular field analysis (CoMFA). 1. Effect of shape on binding of steroids to carrier proteins*, J. Am. Chem. Soc. 110 (1988), pp. 5959–5967.

[73] A.A. Mel'nikov, V.A. Palyulin, and N.S. Zefirov *Generation of molecular graphs for QSAR studies*, Dokl. Chem. 402 (2005), pp. 81–85.

[74] V.A. Palyulin, E.V. Radchenko, and N.S. Zefirov, *Molecular field topology analysis method in QSAR studies of organic compounds*, J. Chem. Inf. Comput. Sci. 40 (2000), pp. 659–667.

[75] W. Zhan, Z. Liang, A. Zhu, S. Kurtkaya, H. Shim, J.P. Snyder, and D.C. Liotta, *Discovery of small molecule CXCR4 antagonists*, J. Med. Chem. 50 (2007), pp. 5655–5664.

[76] E.V. Radchenko, G.F. Makhaeva, V.V. Malygin, V.B. Sokolov, V.A. Palyulin, and N.S. Zefirov, *Modeling of the relationships between the structure of o-phosphorylated oximes and their anticholinesterase activity and selectivity using molecular field topology analysis (MFTA)*, Dokl. BioChem. Biophys. 418 (2008), pp. 47–51.

[77] V.I. Chupakhin, S.V. Bobrov, E.V. Radchenko, V.A. Palyulin, and N.S. Zefirov, *Computer-aided design of selective ligands of the benzodiazepine-binding site of the GABAA receptor*, Dokl. Chem. 422 (2008), pp. 227–230.

[78] RCSB Protein Data Bank, http://www.rcsb.org/pdb/home/home.do

[79] www.emolecules.com

[80] H.A. Jones and B.V. Travis, Insect repellent compositions, US Patent 2,396,013, 1946.

[81] T.P. McGovern, C.E. Schreck, J. Jackson, and M. Beroza, *N-acylamides and N-alkylsulfonamides from heterocyclic amines as repellents or yellow fever mosquitoes*, Mosquito News 35 (1975), pp. 204–210.

3 OBP Structure-Aided Repellent Discovery

An Emerging Tool for Prevention of Mosquito-Borne Diseases

Spyros E. Zographos, Elias Eliopoulos, Trias Thireou, and Katerina E. Tsitsanou

CONTENTS

ABSTRACT

The rational discovery and development of novel insect repellents has challenged researchers for more than six decades. While a number of ligand-based computational methods have been utilized so far, a structure-based approach was until recently unfeasible due to lack of a three-dimensional (3D) structure of an olfactory macromolecule in complex with a repellent. To this end, the report of *Anopheles gambiae* odorant binding protein 1 (OBP1) crystal structure in complex with the most widely used repellent DEET opened the way for OBP structure-aided repellent discovery. Herein, we provide an overview of the ligand-based quantitative structure–activity relationship (QSAR) studies on repellent amides conducted so far, and we further focus on the impact of AgamOBP1-DEET crystal structure on the design of novel mosquito repellents. Based on this structure, pharmacophore models were constructed to incorporate the essential features of DEET interaction with the protein. The pharmacophores were then used for the screening of a divergent set of compounds of known bioactivity followed by molecular docking on the crystal structure of AgamOBP1. Our analysis suggests that a link can be established between the binding mode of a repellent and the observed repellency.

KEYWORDS

- Odorant binding protein
- OBP1, crystal structure
- DEET
- Structure-based design
- Repellents
- 3D pharmacophore screening
- Molecular docking

3.1 INTRODUCTION

Mosquitoes are the main vectors of several deadly diseases such as malaria, dengue fever, West Nile fever, and encephalitis, which affect the health and livelihood of people and suppress the economy and industrial growth of nations. Malaria alone is responsible for millions of human deaths each year, mostly of children under 5 years old [1]. Today, due to the lack of safe and effective vaccines for many vector-borne diseases, including malaria [2], alternative approaches are urgently needed to prevent disease transmission. Current strategies to control malaria rely on prompt diagnosis, treatment with antimalarial drugs, and control of mosquito vectors using genetically modified mosquitoes, indoor residual spraying, long-lasting insecticidal nets, and repellents. However, most of the measures taken to control vector-borne diseases are becoming less effective as insecticide and drug resistance increases [3–5].

To date, the use of repellents has been considered one of the most successful means to control transmission of mosquito-borne diseases, especially during outdoor activities, as it reduces the frequency of contact between mosquito vectors and their human targets. During the past few years, there has been an increasing interest in the development of improved new chemical repellents/attractants, due to reemergence and resurgence of vector-borne diseases in areas where they had been eliminated before, including the European Union and the United States, and to the increased public awareness that insect bites are not only annoying, but can also transmit serious diseases.

Many research efforts have been focused on finding novel repellents superior to the existing ones. During the last 70 years, the discovery of repellents was based exclusively on both random screening and rational ligand-based approaches. In the 1950s, the USDA performed extensive random screenings and constructed a large library of bioactive compounds [6,7]. The qualitative analysis of structure–activity relationships of active entities belonging to this library clearly indicated the diethylamides as potential leads for further development. The exploration of this class of compounds led to *N*,*N*-diethyl-benzamide, which was patented as an effective mosquito repellent in 1946 [8], and to *N,N*-diethyl-3-methylbenzamide (DEET) [9], which became the gold standard of insect repellents and remains the most widely used repellent worldwide.

It took about 40 years to discover another synthetic repellent as effective as DEET. Icaridin (1-piperidinecarboxylic acid 2-(2-hydroxyethyl)-1-methylpropylester) was developed in the 1980s by Bayer AG during a 12-year research project and was first patented in 1988 [10]. According to the product brochure, icaridin's development was achieved by applying molecular modeling techniques on various derivatives of already known repellents in order to optimize their interactions with an insect receptor model. Little was known about the exact methodology that led to this discovery until the publication by Boeckh and colleagues, in 1996, on the molecular modeling and biological studies of acylated 1,3-aminopropanols [11].

3.2 LIGAND-BASED QSAR STUDIES ON NOVEL REPELLENT AMIDES

The structural and electronic features of DEET, DEPA (*N,N*-diethyl phenylacetamide), and *N*-acylpiperidine analogs, all bearing an amide moiety (Figure 3.1), have been studied extensively toward gaining an understanding of their mechanism of action in order to facilitate the development of more efficient repellents.

In 1991, Suryanarayana et al. synthesized a series of DEET and DEPA analogs and subsequently evaluated them in behavioral assays in order to define the relationship between their structural and physical properties, and repellency [12]. The study demonstrated acceptable correlation (correlation coefficient $r = 0.551$) between a set of three physical properties (vapor pressure, *VP*; lipophilicity, *P*; and molecular

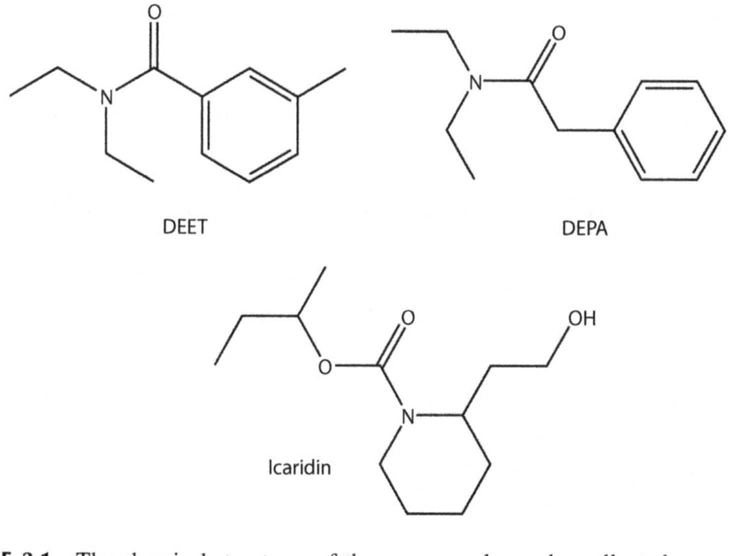

DEET

DEPA

Icaridin

FIGURE 3.1 The chemical structures of three commonly used repellents bear an amide group.

length, *ML*) and the observed protection times according to the following QSAR equation:

$$\text{Prot. time} = \text{a} \log P + \text{b} \log VP + \text{c} \log ML + \text{d}$$

where a, b, c, and d are constants.

However, none of these parameters alone appeared to correlate with protection time.

Furthermore, Ma and colleagues [13] used quantum chemical calculations to examine the influence of the stereoelectronic properties of these amide analogs of DEET to the duration of protection against mosquito bites. This analysis showed that each of the electronic properties, van der Waals (vdW) surface, dipole moments, electrostatic potential of the amide nitrogen and oxygen, and charge of the amide nitrogen must fall within a specific numerical range to achieve high efficiency. It was proposed that a compound could be highly effective in terms of long protection time (at least 4 h) only if all its properties are simultaneously within the optimal ranges. Notably, the amide group seemed to play a key role in the length of protection; therefore, its electronic properties may be used as discriminators of repellent efficacy in the synthesis of new compounds. Its calculated properties with their optimal values are the overall maximum positive (16.2 to 21.1 kcal/mol) and negative (–75.0 to –73.1 kcal/mol) potentials, the negative potential of the amide nitrogen (–25.1 to –22.8 kcal/mol), the dipole moment (3.25 to 3.82 Debye), the atomic charges at the carbonyl oxygen (–0.55 to –0.50 e), the carbonyl carbon (0.50 to 0.68 e), and the amide nitrogen (–0.51 to –0.30 e).

In subsequent studies, Bhattacharjee et al. [14,15] used quantum chemical calculations to show that DEET, a number of its analogs [12], and a synthetic insect juvenile hormone mimic (JH-mimic, undecen-2-yl carbamate), which all possess an amide group, share common stereoelectronic characteristics. Based on the concept that both DEET and JH-mimic may target the juvenile receptor using a common mechanism, they suggested that the molecular recognition is facilitated by three major factors common in the examined molecules:

1. A negative electrostatic potential which is localized on the amide surface. The oppositely charged nitrogen and carbonyl oxygen atoms of amide could be considered as the source of the strong electronic interactions with the residues of the receptor's binding site.
2. A weakly charged, vdW surface region that comprises a long hydrocarbon chain complementary to the binding site of the receptor that mediates short-range hydrophobic interactions.
3. A negative electrostatic potential that extends beyond the vdW surface in the vicinity of the carbonyl group, which therefore participates in long-distance interactions with the receptor.

By utilizing their earlier suggestions, the same research group generated three-dimensional pharmacophore models for repellent activity using the CATALYST software suite [16] and a training set of 11 known repellents, including DEET [17]. The members of this set were characterized by structure divergence and they covered a wide range of potency (ED50 ~1 μg to 50 mg/cm^2). The desirable stereoelectronic features of a repellent compound determined in their previous work [14] were also incorporated into the new model. The selected pharmacophore model that showed the best predictive performance in terms of protection time (correlation coefficient $r = 0.91$) was constituted of one aromatic and two aliphatic hydrophobic sites separated by a hydrogen-bond acceptor site. This pharmacophore was further evaluated as a screening tool, in a search for repellents within a database containing 290,000 compounds. The screening resulted in the identification of 138 hit compounds, 4 of which, after ADME/toxicity and shape similarity filtering steps, were shortlisted and tested. Protection times were determined to be between 1.6 and 9.3 h, suggesting the validity of this model.

A newer 3D-QSAR model was also built based on 40 amide derivatives synthesized earlier by Suryanarayana et al. [12], which were then split into a training set and a test set, consisting of 30 and 10 compounds, respectively. The model was generated by conformational analysis, sequential two-way partial least squares (PLS) modeling, and the use of 30 selected descriptors [18] that were classified in the following nine categories:

1. ADME
2. Electrotopological state [19]
3. Kier's shape indices [20]
4. Jurs' charged partial surface area [21]
5. Shadow shape [22]
6. Quantum mechanical and electronic
7. Conformational energy
8. 3D spatial: density [23,24] and principal moments of inertia magnitude (PMI-mag) [25]
9. Others such as Balaban's relative electronegativities (JX) [26], water desolvation energy (Fh2o), and atomic contribution to log *P* (Atype_C_5 and Atype_H_47) [23,24]

The descriptors with the highest positive impact (descriptor significance percentage; DSP, given in parentheses) to bioactivity were found to be the following:

- Jurs' relative hydrophobic surface area (Jurs-RASA), which is defined as the ratio between the total hydrophobic surface area and the total solvent-accessible surface area (25%)
- Jurs' fractional positive surface area-3 (Jurs-FPSA-3), which is obtained by dividing the sum of the products of solvent-accessible surface area and partial charge of all positively charged atoms by the total molecular solvent-accessible surface area (11%)
- Jurs' fractional positive surface area-1 (Jurs-FPSA-1), which results from the division of the sum of the solvent-accessible area of all partial positively charged atoms by the total molecular solvent-accessible surface area (2.3%),

- Balaban index JX, which is inversely proportional to the electronegativities and covalent radii of the atoms and increases with the length of the carbon chain to which multiple methyl groups are attached (8.7%)
- Kier's shape index, Kappa-3-AM, which represents the branching degree at the molecular center and has higher values for larger and denser molecules (7.6%)
- Atype_H_47 atomic contribution to log *P*, which is related to hydrophobicity/lipophilicity (3.5%)
- DIPOLE_MOPAC of dipole moment (3.5%)
- ADME_Absorption_T2_2D (3.3%)

On the other hand, the highest negative contributions (given in parentheses) came from the following descriptors:

- ADME_solubility_level (–8.5%)
- Shadow indices (shadow-Xlength: length of molecule in the X dimension [–7.7%]; shadow-nu: the ratio of largest to smallest dimension [–2.3%]; shadow-XY: area of the molecular shadow in the XY plane [–1.1%]; shadow-Zlength: length of molecule in the Z dimension [–0.8%)])
- Energy, which gives the energy of the conformer (–4.9%)
- Jurs' differential partial positive solvent-accessible surface area-3 (Jurs-DPSA-3) (–1.6%)
- FH2O, which represents the desolvation free energy from water molecules derived from a hydration shell (–1.1%)
- ADME_BBB_level_2D, which correlates with blood–brain barrier permeability (–2%)
- PMI-mag, which is the magnitude of the principal moments of inertia about the principal axes of a molecule and evaluates the symmetry of the top of the molecule (–0.99%)

All the preceding quantitative data could be interpreted in qualitative terms as follows:

1. Molecules with large hydrophobic and small solvent-accessible surface areas would be more potent as judged by the combination of Jurs-RASA (+), ADME_solubility_level (–), Atype_H_47 atomic contribution to hydrophobicity (+), and FH2O water desolvation energy (–).
2. Large partial positive charges diffused on a large surface area or existence of soft positively charged moieties would contribute positively to the activity of the molecules as supported by Jurs-FPSA-3 (+), Jurs-FPSA-1(+), and Balaban index JX (+) descriptors.
3. The DIPOLE_MOPAC (+) and Jurs-DPSA-3 (–) values indicated that the optimal dipole moments created by appropriate charge separation are important for enhanced activity.
4. Molecules with an elongated parallelepiped-like shape would be more potent, as concluded by combining the shadow indices (–) with the Kappa-3-AM positive contribution.

These conclusions are in good agreement with the observations reported previously by Suryanarayana et al. [12], Bhattacharjee et al. [13], and Ma et al. [17] (*vide supra*). In addition, the predictions obtained from the best pharmacophore model upon its validation on the test set containing 8 out of 10 compounds (two were excluded due to their low vapor pressure) showed good correlation (predictive r^2 value of 0.845) and were proved to be statistically significant with an F-value of 32.8, at the 99.5% confidence level.

Furthermore, Katritzky et al. [27] modeled 31 analogs from Suryanarayana's set [12] using the Codessa Pro software and devised a QSAR formula that demonstrated high correlation values ($r^2 = 0.78$, $F = 23.9$). This formula, constituted by only four descriptors, is reflecting, as defined in Karelson [28], the:

- Size and shape of the molecule (principal moment of inertia A and structure information content, SIC)
- Shape and branching (Kier and Hall index, order 2)
- Charge distribution (total hybrid component of the molecular dipole moment)

To be effective, an insect repellent should be volatile enough (high vapor pressure) to be detected, but if it is too volatile, the duration of protection will dramatically decrease. Therefore, an additional model for the prediction of compounds' vapor pressure was created based on the vapor pressures (V_p) of the 31 analogs, which were taken from the SciFinder catalog. Continuing, the authors constructed a QSAR equation where the $(\log V_p)^2$ descriptor of vapor pressure was included as an independent variable alongside the principal moment of inertia A—the HA-dependent HDSA-2 (Zefirov) descriptor that accounts for the area-weighted surface charge of hydrogen bonding donor atoms—and, finally, the minimum atomic orbital electronic population. This model satisfactorily predicted the repellent protection time ($r^2 = 0.70$) by employing only two descriptors (principal moment of inertia A and $\log V_p^2$), highlighting the significance of molecules' volatility on the duration and efficacy of repellents, as has been previously extensively reported [29–33]. However, this work received some criticism from Natarajan et al. [34] because the generation of the model for vapor pressure prediction was based on theoretical rather than experimental data. Moreover, Natarajan et al. developed a model using hierarchical QSAR (HiQSAR), where the descriptor classes were used in increasing order of complexity, to show that not only mainly the topochemical (TC), but also the quantum chemical (QC) descriptors were sufficient to account for most of the variance in the data, while the inclusion of the calculated lipophilicity (Clog$P$ and log P) and volatility (V_p) descriptors did not result in any significant improvement. Hence, they suggested that the information derived from these calculated secondary descriptors encoding physicochemical properties is already provided by the primary descriptors of molecular structure and, therefore, the use of secondary descriptors could lead to over-fitting.

Artificial neural network (ANN) modeling has also been employed toward prediction of repellents' efficacy [35,36]. The ANN approach got considerable attention after the introduction of the back-propagation learning algorithm in the 1980s [37] and is now considered a sophisticated alternative to conventional multivariate

QSAR analysis. The major advantage of the method includes its ability to detect nonlinear connections between the predictor (independent) and output (dependent) variables, as well as to capture the interactions between the predictor variables. On the other hand, ANN has some drawbacks compared to standard logistic regression techniques, including the difficulty to interpret the magnitude of a predictor's contribution to the output (black box) and the need for greater computational resources [38]. The method was first applied in a data set of 200 *N*-acylpiperidines previously evaluated by the USDA [39–41], which were divided into training (150 molecules) and test (50 molecules) subsets [36]. The predictors (input) used were the total number of bonds, molecular weight, Kier and Hall index (order 3), molecular surface area, total dipole of the molecule, total molecular electrostatic interaction, and surface area for atoms C and N. The resultant ANN model showed high prediction accuracy of the repellency class—for example, 71% of the predicted compounds belonged to classes 4 and 5 (10 to >21 days of protection using cloth patch assays), and it was further used in the selection of 23 novel candidate repellents from a set of ~2,000 untested analogs. These compounds together with 11 of the previously examined analogs were synthesized and tested in two different concentrations (25 and 2.5 μmol/cm^2). The repellency assays showed that a number of the selected compounds were found to display considerably higher protection time (PT) compared to DEET, thus demonstrating the usefulness of the ANN model. However, no significant correlation was found between the experimental and the predicted repellency classes ($r^2 = 0.007$ and 0.06 for 25 and 2.5 μmol/cm^2, respectively). Nevertheless, when the same data set was remodeled by using the best multilinear regression (BMLR) algorithm implemented in Codessa Pro software, a significant correlation of predicted with experimental protection times expressed in days was observed ($r^2 = 0.729$ and 0.689 for 25 and 2.5 μmol/cm^2, respectively). Furthermore, this linear regression QSAR model highlighted the significance of the presence of a carbonyl group and of suitable steric interactions in the emergence of enhanced potency.

Similarly, 167 carboxamide derivatives from the USDA archives, divided into 120 training and 47 test molecules were used to construct a predictive ANN model [35]. Based on this model, nine compounds that had been predicted and experimentally proved to belong to repellency class 4 or 5 were selected and used as templates for the design of structurally similar novel derivatives. Subsequently, 144 novel analogs were designed and evaluated by the ANN model, and the best 34 of them were selected for synthesis and further assessment. As in the case of *N*-acylpiperidines, a number of novel compounds showed good performance, including the (*E*)-*N*-cyclohexyl-*N*-ethyl-2-hexenamide that provided three and two times longer protection compared to DEET at 25 and 2.5 μmol/cm^2 doses, respectively. Despite this fact, when the minimum effective dosage (MED) was used for comparison, DEET turned out to be effective at a 20 times lower dosage. MED is the measurement of the minimum surface concentration of a compound required to produce a repellent effect. Thus, it seems clear from these two works that beyond the usefulness of the ANN approach for prediction and design of novel repellents, there is a need for using a common test protocol and selecting a suitable expression for the potency of repellency in the comparative studies.

All of the QSAR studies presented in this section have undoubtedly made a great contribution to better understanding the crucial molecular properties that result in gain of activity and have assisted significantly in the discovery and design of novel repellent amides. However, little can be said accurately about the mode of action of mosquito repellents with their macromolecular targets. Thus, a structure-based approach similar to that being actively exploited in drug discovery was, until recently, unfeasible due to the absence of structural data on repellents in complex with a macromolecular target. Knowledge of the 3D structures of target proteins in complex with ligands can be used very efficiently to complement QSAR techniques and to help both computational and synthetic chemists to optimize compounds. In general, the first step of a structure-based approach is the identification of an appropriate physiological pathway and the selection of key macromolecules that participate in this pathway as targets for the development of activity modulators. Seeking new repellents, the insect's olfactory pathway is clearly the target system of choice and the macromolecules involved in repellents' recognition and processing constitute, without exception, putative molecular targets for structure-based approaches.

3.3 MOLECULAR TARGETS IN THE OLFACTORY SYSTEMS OF INSECTS

Insects rely on their olfactory sensory system to detect and localize olfactory cues present in the environment that are critical for a number of behaviors such as orientation, food and habitat selection, social interactions, and mating. The main path of odor perception begins at the olfactory sensilla of insect antennae that house the dendrites of olfactory receptor neurons (ORNs) [42], each of which generally expresses a single functional type of odorant receptor [43]. Several protein classes have been identified to be involved in olfactory processes, including odorant receptors (ORs) and 83b subtype receptors [44,45], ionotropic receptors (IRs) [46,47], odorant binding proteins (OBPs) [48], sensory neuron membrane proteins (SNMPs) [49,50], arrestins [51], and odorant degrading enzymes [52–54].

In order to trigger the olfactory signal transduction cascade, airborne hydrophobic odors penetrate into the aqueous sensilla lymph through numerous cuticular pores. Once inside, they are transported to the olfactory neurons through the aqueous environment, with the assistance of OBPs secreted in high concentrations (up to 10 mM) around olfactory dendrites [55]. OBPs capture the odorants in order to solubilize, carry, and deliver them to their cognate ORs, while protecting them against degradation by the odorant degrading enzymes [56]. Upon reaching the dendritic membrane of ORNs, OBPs release the odorants, leading to the activation of ORs and thereby triggering a cascade of intracellular events that give rise to nervous activity. Although the available information on odor release remains limited, several mechanisms have been proposed. OBPs may either undergo low pH-induced conformational changes in the vicinity of the ORs—thus releasing the odorant molecules [57–59]—or may stimulate the receptors by functioning as an OBP-odor complex as in the case of *Drosophila* OBP LUSH [60]. However, this second model is questioned by a recent study showing that sex pheromone 11-*cis*-vaccenyl acetate can directly activate the OR67d receptor without requiring the presence of LUSH protein [61].

Today, there is increasing evidence that odorant discrimination is achieved by the combinatorial action of OBPs and ORs which probably act as a two-step odorant filter, sharing one or very few common ligands and thus contributing significantly to both specificity and sensitivity of the insect olfactory system [62].

Therefore, it has become clear that ORs and OBPs constitute potential targets for the design of vector control agents.

So far, ORs have be shown to be responsible for the reception of a wide range of chemically diverse odorants, including volatiles from human emanations and oviposition sites [63–67] as well as insect repellents [68,69]. While they are considered the prime targets in the search for new insect repellents, they are extremely difficult to crystallize, and the structural insights into their ligand-binding sites are considerably limited [70]. Furthermore, attempts for construction of ORs' homology models are hindered by the lack of sequence and topology similarity with GPCRs of known structure [71].

On the other hand, OBPs are so far the best characterized olfactory macromolecules. It has been suggested that each OBP may specifically recognize certain classes of structurally related odorants and also distinguish semiochemicals of different chemical classes [72–76], thus selecting which odors to transport. The determination of their 3D structures has provided valuable information on their structure–function relationships that, in conjunction with binding studies [77,78], behavioral and electrophysiological data [79–81], and knockdown experiments [81–83], strongly supports that OBPs play an active role in olfaction that makes them promising targets for the structure-based design of novel insect repellents or attractants.

At this point, the report of the OBP1-DEET crystal structure [84], 58 years after the discovery of DEET, revealed the binding mode of the most famous repellent to an olfactory macromolecule and paved the way for structure-based approaches that were not possible until then. On the one hand, since OBPs are not the endpoint macromolecular targets in the odorant signal transduction cascade, the question that arises is whether an OBP structure-based approach can lead to OR-active ligands. On the other hand, the synergistic selection of one or very few common ligands by the matched OBP–OR pairs could only be accomplished if OBPs and ORs possess binding sites of similar 3D shapes and physicochemical properties. Therefore, OBP structure-aided approaches constitute an appealing tool for the identification and/or design of OBP-specific ligands and, consequently, of possible OR binders, potential repellents or attractants.

3.4 MOSQUITO ODORANT BINDING PROTEINS

The completion of the genome sequences of several insect species prompted the identification of more than 300 putative genes encoding OBPs [48,85], increasing the number of candidate proteins that can be used in structure-based approaches that target the insect olfactory system. In the case of the malaria vector *Anopheles gambiae*, the yellow fever mosquito *Aedes aegypti*, and the southern house mosquito *Culex quinquefasciatus*, the total number of OBP genes discovered so far is 69, 111, and 109, respectively [86].

OBPs are small (15–20 kDa), globular proteins that exhibit high sequence variability between different species as well as within the same insect family. The major common feature among all OBP sequences is the presence of six highly conserved cysteines with invariant spacing, which form disulfide bonds to facilitate proper folding and stabilization of protein structure. OBPs have been classified into three main groups according to their sequence similarity: the "classic" OBPs with a typical six-cysteine signature, the "atypical" OBPs, and the "plus-C" OBPs, which have at least two extra conserved cysteines and a characteristic proline [48,87,88].

In the last decade, several 3D structures of mosquito OBPs have been determined, such as AgamOBP1 [84,89,90], AgamOBP4 [91], AgamOBP7 [92], AgamOBP20 [93], AgamOBP47 [94], and AgamOBP48 [95] from *An. gambiae*; AaegOBP1 [59] from *Ae. aegypti*; and CquiOBP1 [96] from *Cx. quinquefasciatus*, either alone or in the form of complexes with selected compounds.

Structurally, the majority of OBPs share a common core of at least six α-helices connected by loops and interlinked by disulfide bonds that enclose a hydrophobic cavity (Figure 3.2).

Despite their similar 3D fold, they bear a wide variety of cavities that differ in size, shape, and position; solvent accessibility; and nature of amino acids involved in their wall formation, which enables them to selectively recognize and bind a diverse range of organic molecules and naturally occurring odorants [97] (Figure 3.3).

Almost all the determined OBP crystal structures deposited in the Protein Data Bank (http://www.rcsb.org/pdb/home/home.do) are dimers. The "plus-C" AgamOBP48, especially, represents the first example of a 3D domain swapped dimer in dipteran species [95]. The dimerization of OBPs has been proposed to create new binding sites at the subunit interface leading to the formation of dimers with binding features different

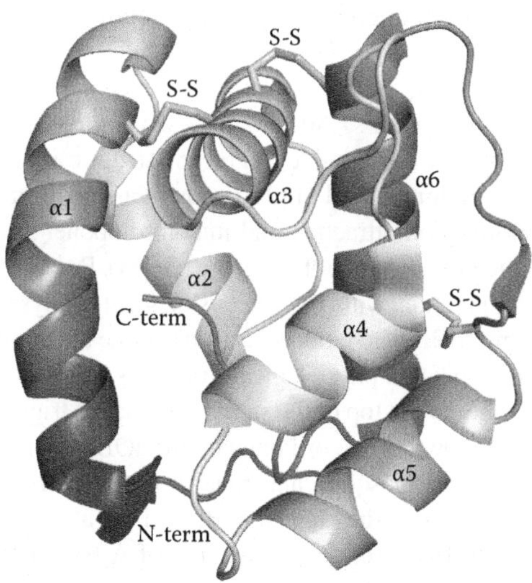

FIGURE 3.2 The overall structure of the "classic" AgamOBP1 monomer (PDB id 3N7H). The three disulfide bonds are indicated by S-S.

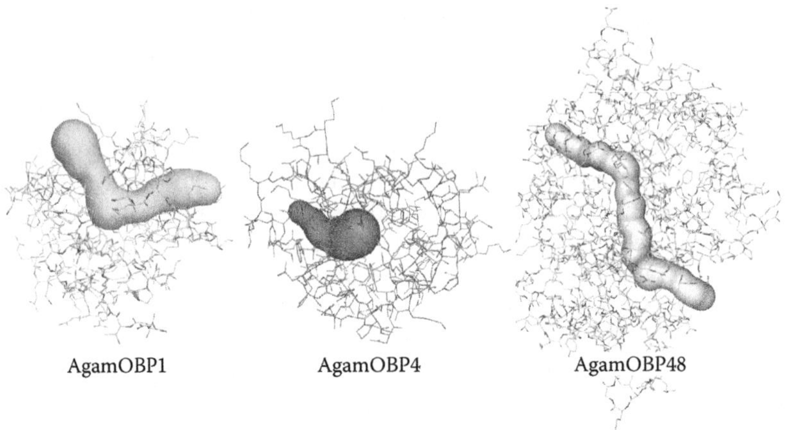

FIGURE 3.3 Comparison of the binding cavities of AgamOBP1, AgamOBP4, and AgamOBP48. These cavities differ in their shape and environment suggesting different odorant specificities.

from those of the monomeric proteins and therefore must be taken into account in the discovery and design of disruptors of mosquitoes' host-seeking behavior [75,84,98].

Among the OBPs encoded in the genome of *An. gambiae*, 10 were found to be expressed at a high level in female antennae. Specifically, for one of them, AgamOBP1, experiments have shown that its mRNA is significantly more abundant in the female relative to the male antennae and also that its abundance is reduced after a blood meal, suggesting that this protein is an important determinant of the host detection behavior of female mosquitoes [99].

AgamOBP1 shares a high degree of sequence identity with its orthologous proteins AaegOBP1 (82.4%) and CquiOBP1 (90.4%), while the amino acids that line up the ligand/pheromone binding cavity are, with three exceptions (for AaegOBP1), well conserved among the three OBPs. Additionally, crystallographic studies revealed that these proteins exhibit a remarkable degree of structural similarity, suggesting that these OBPs might recognize and bind ligands in a similar manner.

Indole, a known oviposition attractant and major component of human sweat, was shown to be a specific ligand for AgamOBP1 ($K_d = 2.3$ μM). RNAi-mediated gene silencing experiments in conjunction with electrophysiological analyses demonstrated that when the levels of AgamOBP1 were reduced drastically, the antennal responses to indole and, consequently, the associated physiological responses were severely impaired, indicating that AgamOBP1 lies at the top of the signaling cascade triggered by this olfactory cue [81]. Likewise, knockdown of the orthologous CquiOBP1 gene caused a significant decrease in the sensitivity of mosquito antennae to *Cx. quinquefasciatus* oviposition pheromone (5R,6S)-6-acetoxy-5-hexadecanolide and the attractant molecules indole and 3-methyl-indole [82]. However, gene silencing of AgamOBP1 or CquiOBP1 did not affect the antennae responsiveness to terpene and geranylacetone, or to nonanal, respectively, signifying that these proteins participate in the perception of specific odorants.

In 2012, Tsitsanou et al. reported the high-resolution crystal structure of AgamOBP1 in complex with the most widely used repellent DEET, which represents the first example of a repellent recognized by an OBP [84]. Analyses revealed that DEET binds in each subunit, with high shape complementarity, at a site located at the edge of the long hydrophobic tunnel near the dimer interface (Figure 3.4). It exploits numerous nonpolar interactions with the protein residues and one hydrogen bond,

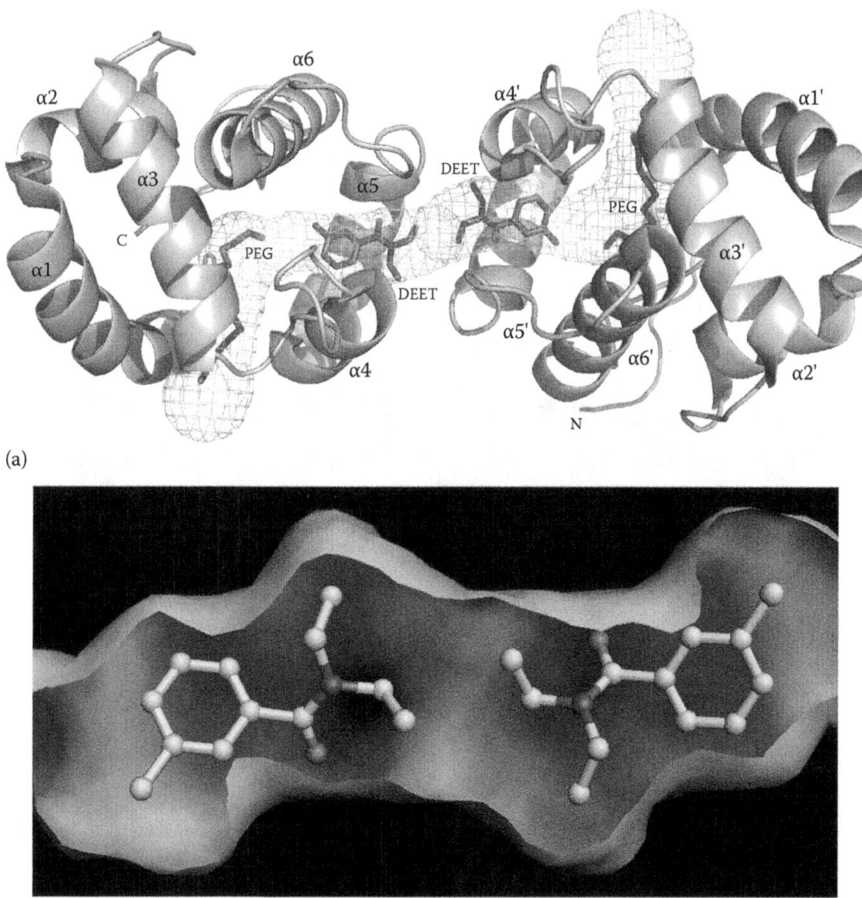

FIGURE 3.4 (a) Cartoon representation of AgamOBP1 dimer with molecules of DEET and PEG bound. DEET's binding site is located at the center of a long hydrophobic tunnel (represented as a mesh) running through the dimer interface. The tunnel is bordered by residues Leu15, Ala18, Leu19, Leu22 (from α1 helix), Ala62 (from α3/10 helix), Lys63 (residue in isolated beta-bridge), Leu73, Leu76, His77, Ser79, Leu80 (from α4 helix), Met84, Ala88, Met89 (from α5 helix), Leu96 (loop), His111, Trp114 (from α6 helix), and Phe123 from the C-terminal loop (K.E. Tsitsanou et al., Cell. Mol. Life Sci. 69 (2012), pp. 283–297). (With kind permission from Springer Science and Business Media.) (b) The Connolly surface of AgamOBP1 at the region of DEET-binding site showing the high complementarity of the protein-binding pocket to DEET molecule. Both subunits are shown.

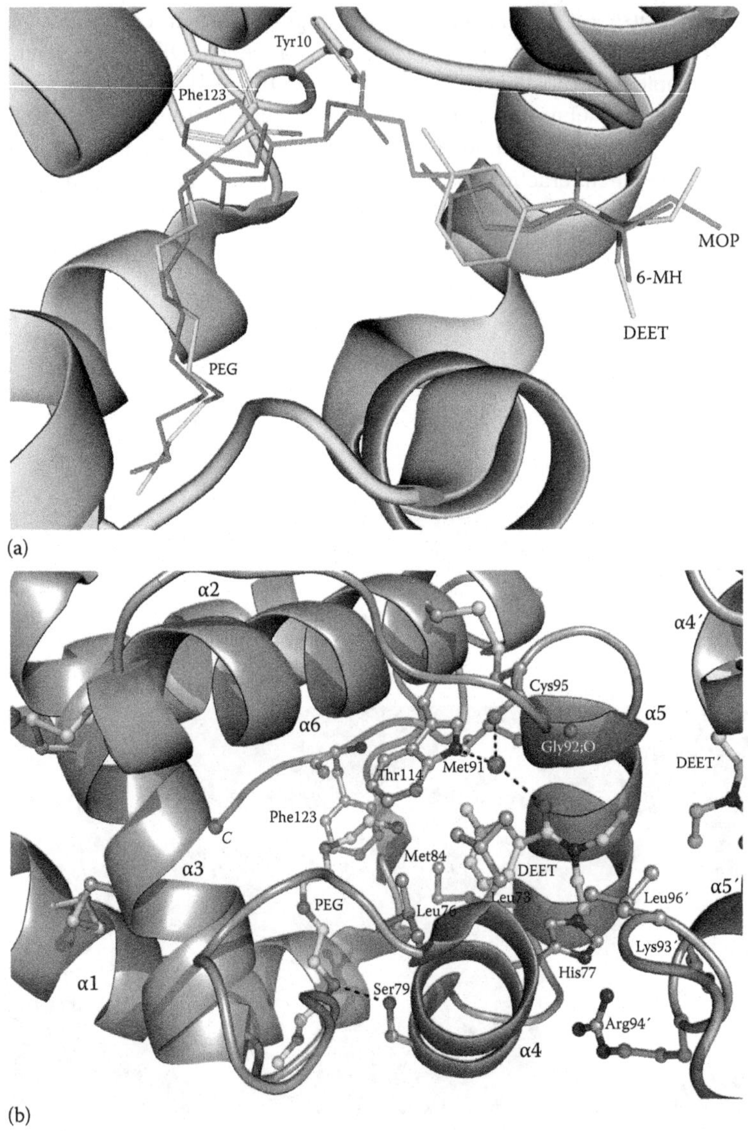

FIGURE 3.5 (See color insert.) (a) Schematic presentation illustrating the binding mode of DEET and PEG (yellow, PDB id 3N7H), 6-MH, and PEG (green, PDB id 4FQT), and MOP (cyan, PDB id 3OGN) to the binding site of AgamOBP1 and CquiOBP1, respectively. (b) DEET and PEG (with carbon atoms colored in yellow) form one hydrogen bond each, with a water molecule and the side chain of Ser79, respectively. Hydrogen bonds are shown as dashed lines. The contacts between ligands and AgamOBP1 are dominated by nonpolar vdW interactions. In particular, DEET makes vdW interactions with residues of α4, α5 helices and residues from the other monomer as well as with the neighboring DEET's molecule. The three disulfide bonds are represented in light orange sticks (K.E. Tsitsanou et al., Cell. Mol. Life Sci. 69 (2012), pp. 283–297). (With kind permission from Springer Science and Business Media.) *(Continued)*

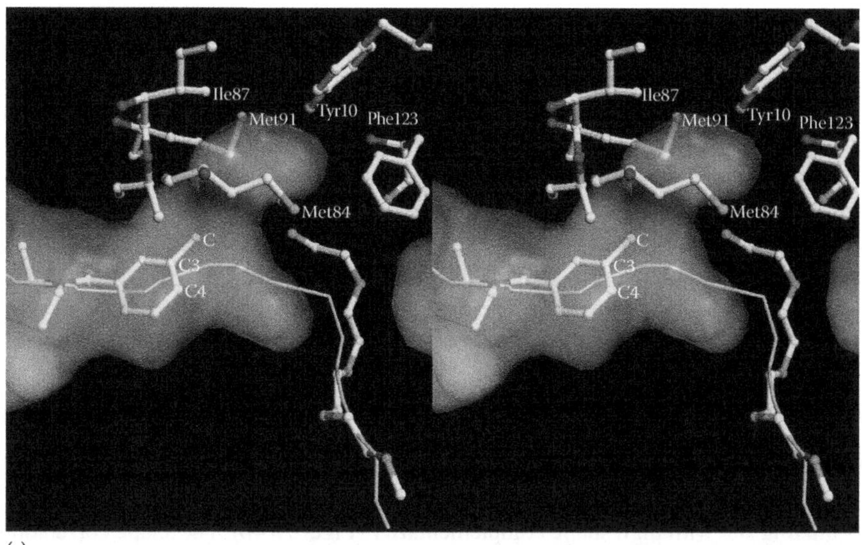

(c)

FIGURE 3.5 (CONTINUED) **(See color insert.)** (c) A stereodiagram of the Connolly molecular surface (yellow) in the vicinity of DEET's binding site. The corresponding path of the binding tunnel is represented as a white line. The existence of an empty pocket that diverges from the tunnel direction and emerges above the 3-methyl group of DEET is apparent. The shape of this cavity suggests that substitution at 3 (*meta-*) and 4 (*para-*) positions of the benzene ring could be exploited for the design of DEET analogs with improved affinity for AgamOBP1 (K.E. Tsitsanou et al., Cell. Mol. Life Sci. 69 (2012), pp. 283–297). (With kind permission from Springer Science and Business Media.)

which is perceived to be critical for DEET's recognition. The remaining area of each cavity is filled with a PEG molecule that originates from the crystallization solution. It is noteworthy that the natural human-derived mosquito repellent 6-methyl-5-heptene-2-one (6-MH) [100] was also found to bind to the DEET-binding site of AgamOBP1 [89], although it belongs to a different chemical class. Superimposition of the AgamOBP1–DEET–PEG and AgamOBP1–6-MH–PEG complexes with the structure of CquiOBP1–MOP [96] revealed that the mosquito oviposition pheromone (MOP) is also bound to the same region of the hydrophobic tunnel, where it overlaps with DEET or 6-MH, and a part of the PEG molecule (Figure 3.5a).

Fluorescence binding studies of AgamOBP1 with various natural and synthetic volatile compounds showed that the protein displays a preference for medium-sized terpenoids [77]. Interestingly, the plant-based mosquito repellent citronellal [101] was found to bind to AgamOBP1 with a relative dissociation constant of 5.3 μM. Furthermore, when 34 ligands, including oviposition and human-derived attractants as well as repellents, were assayed against CquiOBP1, it was found that, except for the MOP, the known repellent icaridin binds to the protein (K_d = 6.4 μM) [78].

Recently, using docking and molecular dynamics stimulation studies, Affonso et al. [102] carried out prediction of the binding mode of known attractants such as lactic acid, 1-octen-3-ol, indole, and major components of the Indian clove oil with repellent activity to AgamOBP1. All ligands, attractants, and repellents were predicted to bind to the same binding site as DEET by adopting similar conformations and to interact with the same protein residues. Analysis suggested that AgamOBP1 might have higher affinity for repellent than attractant molecules. Thus, it was proposed that the ability of a repellent to block the binding of an attractant to AgamOBP1 could lead to repellency.

The fact that two compounds with repellent activity and a natural pheromone are able to bind to mosquito OBP1—which may also recognize and bind other attractant and repellent molecules according to gene silencing experiments, *in vitro* binding, and *in silico* modeling studies—defines this protein as a valuable molecular target for the structure-based design of novel compounds. These agents will be able to interfere with the olfactory response in mosquitoes, reducing the risk of mosquito-borne disease transmission.

3.5 OBP1–DEET CRYSTAL COMPLEX: IMPLICATIONS FOR OBP STRUCTURE-AIDED APPROACHES

Analysis of the AgamOBP1-DEET crystal structure [84] revealed that DEET binds to its binding site with high shape complementarity (Figure 3.4b). The shape correlation statistic Sc for the protein–ligand interface was calculated to be 0.834, a relatively high value compared to 1, which represents a perfect interface [103]. Shape complementarity is directly correlated with binding specificity [104]. Comparison of the available OBP1 crystal structures from different mosquito species complexed with various ligands showed no conformational changes in the region of the DEET-binding site, suggesting that this site is in all likelihood preorganized. Therefore, OBP1 ligand selectivity may be governed by the "key-and-lock" principle. This has a significant implication in the OBP1-structure-based screening or in the design of DEET-like analogs, as it essentially suggests that the selection or design of ligands preshaped to the geometry of the DEET-binding site is more likely to result in their recognition by the cognate OR binding site. Size and shape descriptors of candidate compounds were found to be important in all the QSAR studies described before. Significantly, a detailed delimitation of the DEET-binding site could provide additional size, shape, and conformational constraints toward enhanced specificity and selectivity.

Another important feature of the DEET-binding site is the existence of a structural water molecule (Figure 3.5b). Structural waters could be considered "prosthetic groups." They shape the architecture of the binding sites and play an important role in protein function and stability [105,106]. Upon DEET binding, the only hydrogen bond that is observed is formed between the nucleophilic carbonyl oxygen of the amide group and this particular water molecule. The necessity of an amide moiety with optimal charge separation that could participate in an electronic interaction with the receptor has been previously proposed by Bhattacharjee et al. [14] and Bhonsle et al. [18]. It is clear that inclusion of the water molecule into 3D-QSAR analysis can significantly improve the scoring of protein ligand interactions [107,108] and the accuracy of the predictive model. In addition, new ligands bearing a suitably positioned nucleophilic group can be designed to maximize the favorable interactions with the water molecule [109].

The binding of DEET is accompanied by a drastic decrease in its solvent accessible surface area (90%). The 92% of the vdW surface that becomes buried consists of nonpolar atoms. As a result, the binding of DEET is dominated by numerous hydrophobic interactions between the m-toluene and diethyl moieties of DEET and several residues of the binding cavity. It is therefore obvious that hydrophobicity provides the main binding force of DEET to AgamOBP1. The contribution of hydrophobicity and solvent accessibility to a compound's bioactivity has been quantified in the work of Bhonsle et al. [18]. Moreover, the orientation of DEET's benzene ring relative to the Trp114 ring (Figure 3.5b) suggests that larger aromatic systems (e.g., indole or naphthalene ring) may be accommodated and develop more favorable π–π stacking interactions with the Trp114 residue, thus exhibiting enhanced binding affinity.

Analysis of the binding cavity revealed the presence of an empty space extending from the toluene ring toward the direction of the PEG-binding site. An extra empty pocket formed by the residues Met84, Met91, Ile87, Phe123, and Tyr10 was also observed (Figure 3.5c).

The expansion of the DEET-binding site facilitates the specific recognition of more elongated compounds such as the MOP [96]. The long lipid chain of MOP occupies the binding region of DEET, while its lactone ring is located on the edge of the PEG-binding site (first ethylene unit) (Figure 3.5a). Interestingly, MOP's acetyl ester branch extends into the extra pocket and is also in hydrogen bond distance from the hydroxyl group of Tyr10.

Based on the preceding analysis of the available structural data, Tsitsanou et al. [84] proposed that new molecules can be searched or/and designed to

- Utilize shape complementarity with parts or the full length of the available space, which extends from the entrance of the DEET-binding site to the start of the PEG-binding site including or not the bottom area of the extra pocket
- Make a hydrogen bond either to a water molecule in the DEET-binding site or with polar atoms located in the extra pocket, or both
- Contain an aromatic system (benzyl or a larger ring) appropriately oriented to make aromatic interactions with Trp114 in the DEET-binding site or with Phe123 at the PEG-binding site
- Bear a hydrophobic moiety localized at the entrance of the DEET-binding site that corresponds either to the diethyl moiety of DEET or to the lipophilic chains of MOP and 6-MH compounds

In order to assess whether the preceding predictions could serve as the basis for AgamOBP1-based modeling of novel compounds, a test set of 29 carboxamide and piperidine derivatives was studied by docking simulations. Twenty-four of these analogs had been previously reported to show repellent activity against the *Periplaneta americana, Blattella germanica*, and *Blatta orientalis* cockroach species [110] and the *Ae. aegypti* mosquito [35,36]; five ligands, including piperine and capsaicin, were naturally occurring molecules known to be active against insects [110]. The free energy of binding (FEB), K_i, LE, and FQ calculated values for DEET and these compounds' conformations were calculated with AutoDock [111]. The binding of ligands to AgamOBP1 was predicted to occur at the expected region of the binding pocket, which expands between the entrance of the DEET-binding site and the top edge of the PEG-binding site. Some characteristic scaffolds with better FEB than DEET (FEB = –5.86 kcal/mol) are shown in Figure 3.6.

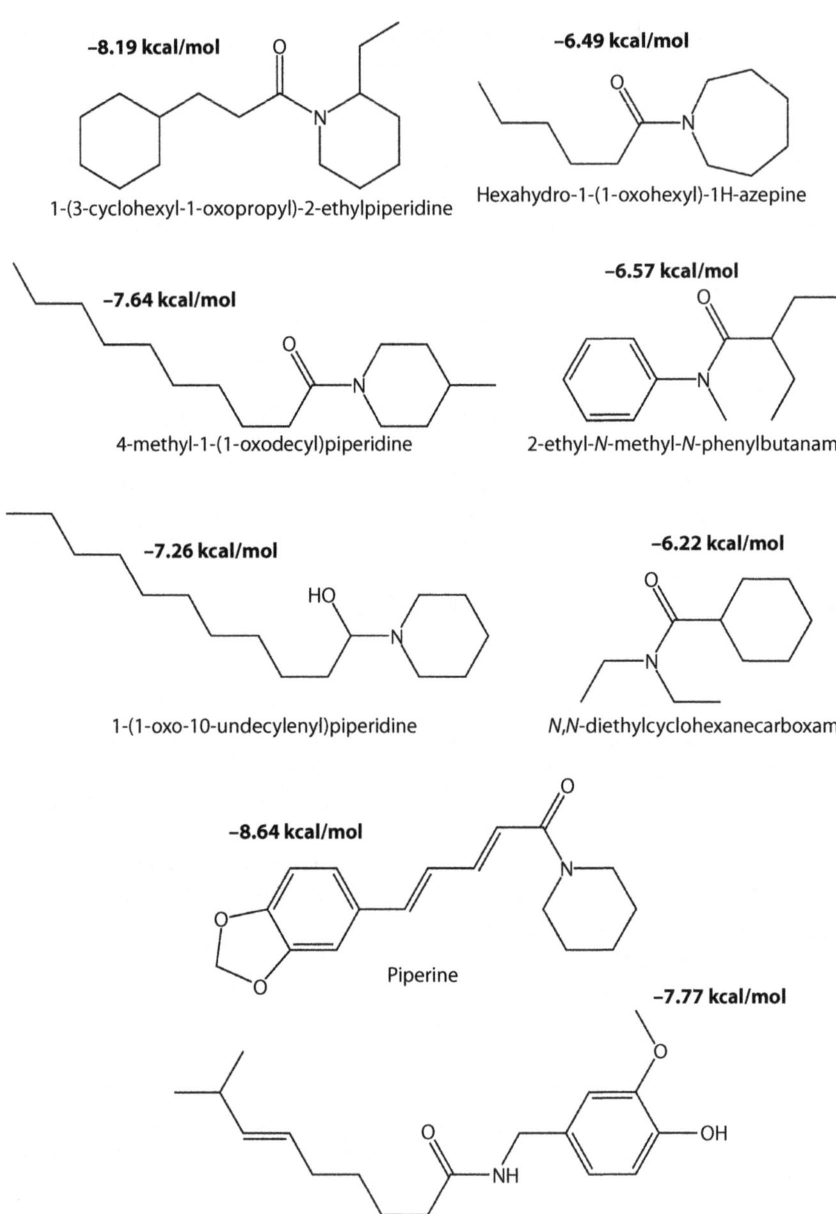

FIGURE 3.6 Examples of the chemical structures of compounds with docking scores better than DEET. Their calculated FEB to AgamOBP1 is also shown.

To validate the modeling predictions, four of the test compounds—1-(3-cyclohexyl-1-oxopropyl)-2-ethylpiperidine, 4-methyl-1-(1-oxodecyl)piperidine (referred to as 4o′ and 4j, respectively, by Katritzky et al. [36]), piperine, and capsaicin were acquired, and their K_ds were experimentally determined by a fluorescence competition assay (Table 3.1).

Although a perfect linear correlation between the predicted (K_i) and the experimental (K_d) binding affinities was not observed, it is clear that docking calculations follow the trend of the experimental values and therefore can guide the selection of OBP1 binders. However, the experimental values should be preferred over calculations in order to conduct more accurate and reliable QSAR studies. Significantly, the compounds 1-(3-cyclohexyl-1-oxopropyl)-2-ethylpiperidine and 4-methyl-1-(1-oxodecyl)piperidine were found to display 13 and 24 times higher affinity for AgamOBP1 than DEET and have both been reported to provide about five times longer protection compared to DEET against *Ae. aegypti* mosquitoes (Table 3.1). In contrast, piperine is ineffective even though it was found to be a potent binder of AgamOBP1.

This finding stresses the fact that a good OBP1 binder is not necessarily a good repellent because it must also be recognized by the cognate OR. To this end, the combination of the stereoelectronic attributes and physicochemical properties of repellents with their binding mode can serve as an additional filter in the selection of compounds, increasing the probability to obtain an OR-active ligand. The predicted modes of binding of the tested molecules are illustrated in Figure 3.7.

TABLE 3.1
Docking Parameters of Selected Compounds and Comparison of Their Predicted and Experimental Binding Affinities with Protection Time and Repellency Class

Compound	HA	LE	FQ	K_i, pred (μM)	K_d, obs (μM)	PT	Converted Class
DEET	14	0.31	0.50	50.51	31.30	2.5	2
Capsaicin	22	0.26	0.60	2.00	13.0	N/A	N/A
1-(3-Cyclohexyl-1-oxopropyl)-2-ethylpiperidine (4o′)	18	0.33	0.66	0.99	2.4	10.5	4
Piperine	21	0.30	0.67	0.46	2.1	–	1[a]
4-Methyl-1-(1-oxodecyl)piperidine (4j)	18	0.31	0.60	2.99	1.3	11.5	4

Notes: HA: number of heavy atoms; K_i: predicted inhibition constant; LE: ligand efficiency; FQ: fit quality; PT: days of protection at 2.5 μmol/cm^2 on cloth against *Ae. aegypti* (data from A.R. Katritzky et al., Proc. Natl. Acad. Sci. USA 105 (2008), pp. 7359–7364); N/A: data not available.

[a] Data from USDA, *Chemicals evaluated as insecticides and repellents at Orlando, Fla,* compiled by W.V. King (US Dept Agriculture, Washington, DC), Agriculture Handbook No. 69, 1954.

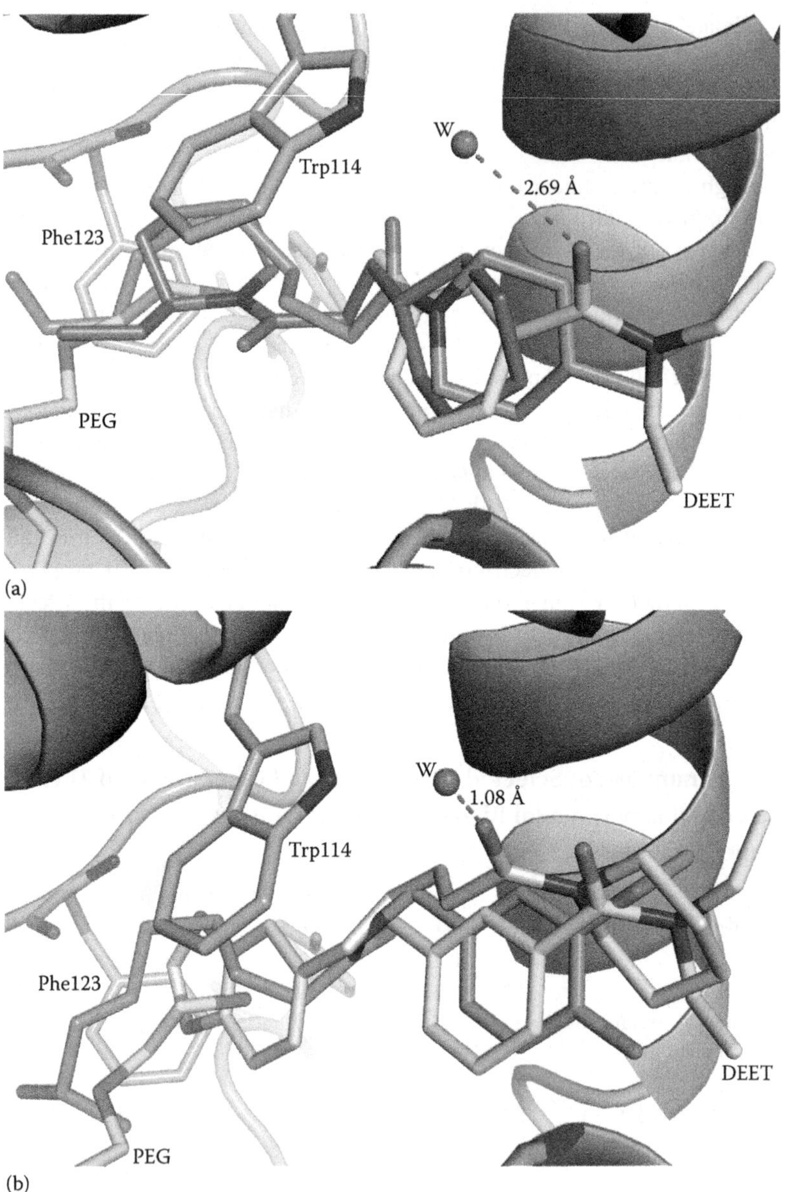

FIGURE 3.7 (See color insert.) Representation of the AgamOBP1 binding site with DEET and PEG (yellow) superimposed with the tested compounds, showing the predicted binding modes of (a) the active compounds 4-methyl-1-(1-oxodecyl)piperidine (orange) and 1-(3-cyclohexyl-1-oxopropyl)-2-ethylpiperidine (blue) as well as of (b) piperine (gray) and capsaicin (raspberry).

Analysis of the predicted models of the highly active compounds 4j (orange) and 4o′ (blue) revealed the following common characteristics (Figure 3.7a):

- The cyclohexyl ring of 4o′ and the piperidine ring of 4j overlap with the benzene ring of DEET.
- The piperidine ring of 4o′ and the aliphatic chain of 4j are placed between the aromatic rings of Phe123 and Trp114, where they are overlaid with the first ethylene unit of PEG at the PEG-binding site.
- 4o′ and 4j neither make a hydrogen bond to the structural water (in contrast to DEET but similarly to MOP and 6-MH) nor enter the extra pocket (unlike MOP).

The binding mode of capsaicin has all the preceding characteristics (Figure 3.7b). Consequently, it can be considered a promising scaffold for the design of novel repellents. Nevertheless, capsaicin is the heaviest of all tested compounds and is predicted to be nonvolatile with a V_p of 1.32×10^{-8} mmHg, threefold higher to that of 4o′ or 4j. Interestingly, piperine is the only one of the four compounds with an aromatic ring that does not overlap with that of DEET (Figure 3.7b). Furthermore, its carbonyl group is in proximity to that of DEET but its binding requires the displacement of the structural water ($O^{W}–O^{PIP.}$= 1.08 Å). In contrast to MOP's lactone ring and 4o′s piperidine ring, which are arranged in parallel and sandwiched between Trp114 and Phe123 residues, piperine's benzodioxol aromatic ring is placed almost perpendicular. All of these characteristics could account for a poor fitting of piperine to its cognate OR receptor and, consequently, lead to loss of efficient repellent activity. The high affinity of piperine, which has been determined experimentally, indicates that the ligand adopts an alternative conformation into the DEET-binding site, binds to an adjacent site, or displaces the structural water to minimize the observed steric clash. These deviations from optimal shape and stereochemical complementarity may be tolerated by OBP1 but not by its OR partner. It is likely, for example, that the carbonyl group of piperine may enter into a steric conflict with its electrophilic counterpart of the OR-binding site.

Recently, Oliferenko and colleagues conducted a molecular field topology analysis (MFTA) of 70 compounds in combination with a study of their molecular docking into the crystal structure of AaegOBP1 [112]. The compounds belonging to divergent chemical classes included 43 carboxamides previously studied by the authors [35], and 27 compounds selected from the USDA internal records containing hydroxyl, ether, ester, amine, nitro, and halogen functionalities. Moreover, DEET, icaridin, hexahydro-1-(1-oxohexyl)-1H-azepine with a MED (minimum effective dosage) of 0.033 μmol/cm² [35], and the class 4 repellent 2-cyclohexylcyclohexanol [39] were used as lead compounds for similarity screening against a database containing five million commercially available compounds. Thereby, 344 additional virtual hits were selected for docking to AaegOBP1 toward selection of novel active scaffolds.

In their study the full-length channel running through the OBP1 was investigated (Figure 3.4a) without applying any positional or hydrogen-bonding constraint. The molecules showed a tendency to dock to the middle of the channel, the most hydrophobic portion of the protein, and to form hydrogen bonds with the backbone amide nitrogen or the carbonyl oxygen of Phe123 rather than with the structural water molecule, as DEET does.

Within the set of the 70 compounds, a high correlation ($r = 0.922$) was observed between the MED and docking scores for the seven most active ones. The two top ones with MED values of 0.039 and 0.065 μmol/cm^2 share high shape similarity with DEET and were found to have the second and fourth highest docking scores, respectively (Figure 3.8a).

Furthermore, docking of the obtained virtual hits to AaegOBP1 in combination with repellency bioassays resulted in the identification of two additional potent compounds (Figure 3.8b). The 2-(4-fluorophenyl)cyclohexanol with a MED of 0.078 μmol/cm^2 is a derivative of 2-phenyl-cyclohexanol (Figure 3.8a, right) and possess the third highest docking score in the hits set. It is noteworthy that the piperidine derivative 1-(3-methylpiperidin-1-yl)heptan-1-one (Figure 3.8b, left) was found to be the most active in this series (MED = 039 μmol/cm^2) and performed better than DEET (MED = 0.052 μmol/cm^2), closely resembling the structure of the previously mentioned highly active 4j analog (Figure 3.6 and Table 3.1).

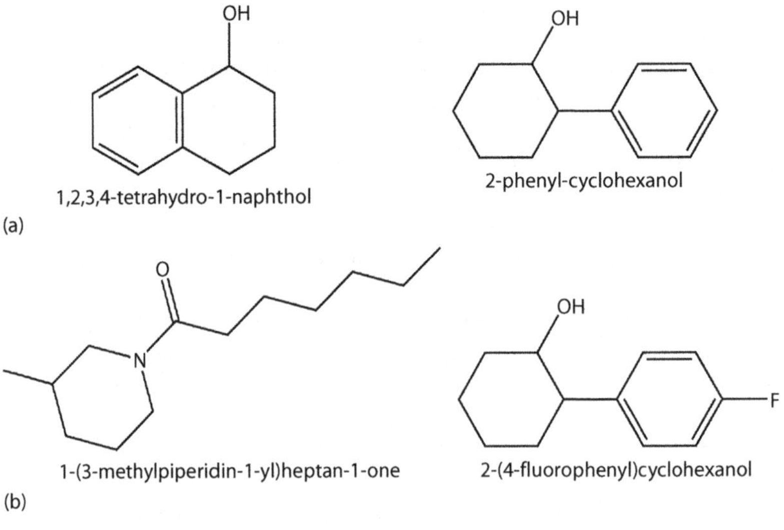

FIGURE 3.8 The chemical structures of the most active compounds from P.V. Oliferenko et al., (PLoS ONE 8 (2013), pp. e64547) study that were found to exhibit high docking scores for AaegOBP1 crystal structure (a) belonging to the test set (b) identified by lead hopping virtual screening.

Based on the structural features of the active compounds and on their predicted binding modes, the authors suggested a plausible pharmacophore model that includes

- A hydroxyl or carbonyl group that makes a hydrogen bond to the backbone amide nitrogen and/or the carbonyl oxygen of Phe123
- Either an aromatic biphenyl or a naphthalene group (corresponding to toluene ring of DEET) or a hydrophobic moiety located in the hydrophobic cavity at the start of the PEG-binding site (two first ethylene units of PEG)

The study of Oliferenko et al. [112] pointed out instances of molecules that displayed very high docking scores but were inactive, in agreement with the observation for piperine in the work of Tsitsanou et al. [84].

The opposite could be also true. Although OBP1 has been proved to bind DEET and 6-MH, it is possible that repellents of other chemical classes could be transferred by different OBPs. Therefore, more experimental data on the structure of different OBPs in complex with repellents and the determination of their binding specificities are an urgent need for OBP-structure-aided discovery approaches. Importantly, determination of ORs' binding specificities and identification of their OBP partners could significantly narrow the selection of appropriate chemical classes for each OBP target. Nevertheless, we anticipate that combination of QSAR, molecular docking, and 3D-QSAR studies on divergent chemical libraries in association with experimental structural, binding, and behavioral data could significantly accelerate the discovery of novel effective repellents.

3.6 OBP1-BASED PHARMACOPHORE MODEL

Recently, we performed a lead hopping screening based on shape similarity with natural repellents against a divergent chemical database of commercially available compounds. The resultant hits were docked into the AgamOBP1 crystal structure (PDB id 3N7H), using molecular mechanics techniques, and 20 of these compounds were acquired. Subsequently, their binding affinities for AgamOBP1 as well as their activities against the *An. gambiae* mosquito were determined (unpublished data). Interestingly, we have identified compounds that display significant repellency and combine several desired structural features with direct hydrogen bonds to the structural water and the backbone N and O of Phe123, and aromatic interactions with Trp114.

We decided to construct and validate pharmacophore models based on two distinct data sets. Using these models we undertook a new lead hopping screening against the USDA archival database [39,40]:

- Data set 1 consisted of 81 molecules including the previously acquired 20 shape-similarity screening hits as well as DEET and natural and synthetic compounds with known binding affinities for AgamOBP1 that were also tested for repellency against *An. gambiae* (unpublished data).
- Data set 2 was used by Oliferenko et al. (2013) and consisted of 70 compounds (43 carboxamides and 27 other evaluated analogs selected from USDA archives). In this data set, repellent activity was characterized by the MED [112].

The external test set consisted of 1,656 selected compounds classified according to their repellency against yellow fever (Yf) and/or malaria (M) mosquitoes (Table 3.2).

For each entity of these three data sets, an .smi file containing the SMILES string was constructed. Geometries of the ligands were converted to 3D coordinate files (.mol) with LigPrep [113] using the OPLS2005 force field and, finally, energy minimized with default LigPrep settings. Defined chiral centers were preserved, while all undefined chiral centers were generated as separate entries. In the case of the external test set, the total generated 2,748 3D-structures were processed through Confgen (Schrödinger Inc.) for global minimum energy conformation. A detailed conformational search of compounds was performed by multiple rotations of single bonds in the compounds, thereby generating several low-energy conformers with varying population densities. Representative conformers with energies <10 kcal/mol of the global energy minimum were selected and the most abundant and the lowest energy conformers were identified.

All ligands were studied for their binding to AgamOBP1 using the molecular docking programs AutoDock Vina [114] and Glide [115] running on an HP DL360 G8 server. Using AutoDock Vina, the search space was defined by a grid box centered on DEET with dimensions of $25 \times 25 \times 25$ Å^3. The maximum number of binding modes to output and exhaustiveness were set to 20 and 10, respectively. No positional or hydrogen-bonding constraints were applied. The docking poses were carefully inspected for the presence of strong hydrogen bonding and vdW interactions, as well as steric clashes. Compounds that fit into the AgamOBP1 binding pocket were ordered according to the energy of binding.

Structure-based pharmacophore models were generated using LigandScout [116] and Phase [117] based on the AgamOBP1-DEET crystal structure (PDB id 3N7H). The program pharmacophore algorithms treat ligand molecular structures as templates composed of chemical functions localized in 3D space, interacting with complementary functions located in the binding site of the selected receptor protein. Structural information for both active and inactive compounds of data sets is employed to generate a pharmacophore representative of the binding capacity of lead compounds. Within LigandScout, the crystal structure of the protein–ligand complex is used to identify and locate the best 3D arrangement of assigned chemical functions, such as hydrophobic regions, hydrogen bond donor or acceptor, and positively or negatively ionizable sites, distributed over the 3D space of the protein's binding site, to accommodate the variation among the lead compound set. Hydrogen-bonding features are represented as vector functions, whereas all other features are point-centered radial functions. Each feature (e.g., hydrogen-bond acceptor [HBA], hydrophobic regions [H], aromatic ring [AR], etc.) contributes equally to estimate the score. Similarly, each chemical feature in the pharmacophore requires a match to a corresponding ligand atom to be within the same distance of tolerance. All pharmacophore features were marked as optional and feature tolerance was increased by 0.15 Å.

Features were combined in two separate pharmacophores (Figure 3.9) that were screened against the AutoDock Vina generated ligand conformations (Tables 3.3 and 3.4). The screening for each molecule was terminated when the first matched conformation was found. The pharmacophores were then used to estimate the fitness parameter (pharmacophore score) of the test set. These activities are

TABLE 3.2
Distribution of External Test Set Compounds According to USDA Repellency Tests

Screening Property	No. of Compounds/3D Structures	Class 1 Repellency	Class 2 Repellency	Class 3 Repellency	Class 4 Repellency	Class 4A Repellency	Total No. of Conformations
Yfs	620/649	140	194	110	140	36	98,506
Yfc	973/1066	230	198	159	86	300	124,327
Ms	597/623	154	259	110	68	6	90,818
Mc	352/410	238	78	16	11	9	37,283

Notes: Yfs (skin repellency test on *Ae. aegypti*), Yfc (cloth repellency test on *Ae. aegypti*), Ms (skin repellency test on *An. quadrimaculatus*), Mc (cloth repellency test on *An. quadrimaculatus*). Classes of repellency are according to the USDA Agriculture Handbook nos. 69 (1954) and 340 (1967). Last column indicates the number of initial 3D conformations tested.

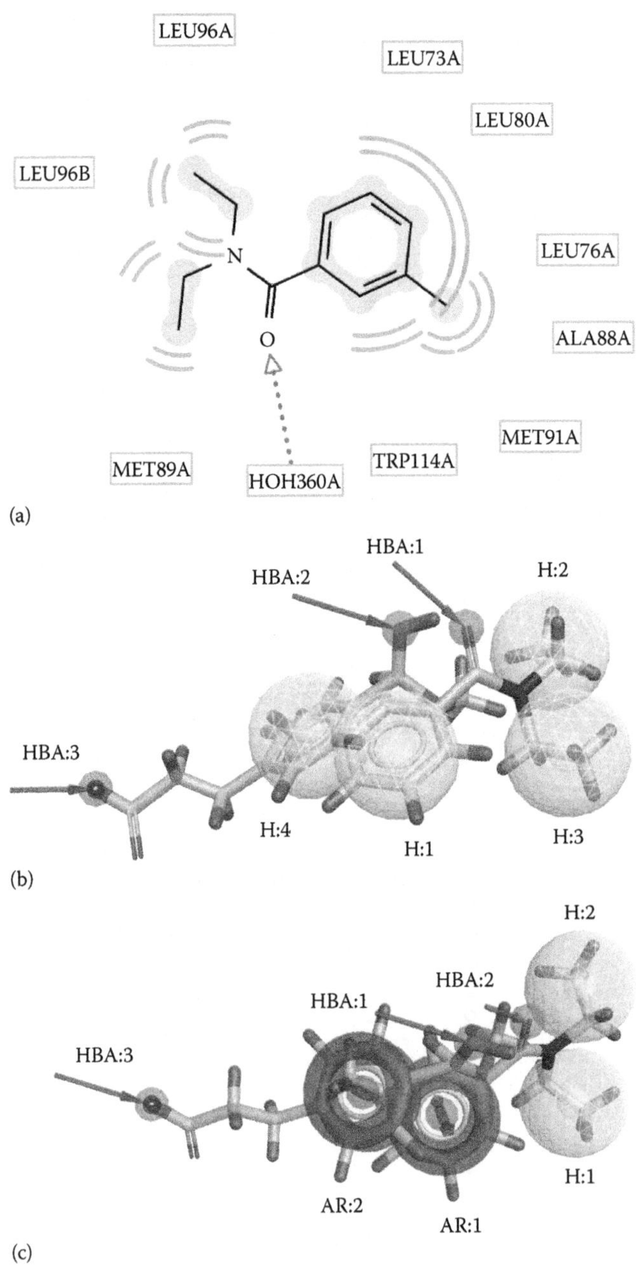

FIGURE 3.9 (a) The DEET interaction characteristics obtained from the AgamOBP1-DEET crystal structure. The two pharmacophores applied on the two data sets: (b) The "hydrophobic" pharmacophore, using hydrophobic features in the central area of the ligand and (c) the "aromatic" pharmacophore, containing aromatic features in the central area of the ligand. Spheres indicate hydrophobic features (H1, H2, H3, H4), dark gray circles indicate aromatic rings (AR1, AR2), and arrows point in the direction of hydrogen bond acceptors (HBA1, HBA2, HBA3).

TABLE 3.3
Top 25 USDA Compounds Fitted to AgamOBP1 Structure Using the "Hydrophobic" Pharmacophore

<table>
<tr><th>Ligand</th><th>Yfs[a]</th><th>Yfc[b]</th><th>Affinity (kcal/mol)[c]</th><th>Pharmacophore-Fit Score[d]</th><th>Matching Features[e]</th></tr>
<tr><td>m-Toluamide, N,N-diethyl (DEET)</td><td>–</td><td>4A</td><td>–7.4</td><td>66.76</td><td>H:1|H:2|H:3|H:4|HBA:1|HBA:2|</td></tr>
<tr><td>p-Toluamide, N,N-diethyl</td><td>–</td><td>4A</td><td>–6.9</td><td>65.69</td><td>H:1|H:2|H:3|H:4|HBA:1|HBA:2|</td></tr>
<tr><td>Benzamide, N,N-diethyl</td><td>4A</td><td>4A</td><td>–6.7</td><td>65.66</td><td>H:1|H:2|H:3|H:4|HBA:1|HBA:2|</td></tr>
<tr><td>2,4-Hexanediol, 3-ethyl</td><td>4</td><td>3</td><td>–5</td><td>57.21</td><td>H:1|H:2|H:3||HBA:1|HBA:2|</td></tr>
<tr><td>Phthalimide, N-sec-butyl</td><td>4</td><td>4</td><td>–7.5</td><td>56.87</td><td>H:1|H:2||H:4|HBA:1|HBA:2|</td></tr>
<tr><td>o-Toluic acid benzyl ester</td><td>1</td><td>–</td><td>–8.9</td><td>56.81</td><td>|H:2|H:3|H:4|HBA:1|HBA:2|</td></tr>
<tr><td>Sorbamide, N,N-diethyl</td><td>–</td><td>3</td><td>–6.3</td><td>56.62</td><td>|H:2|H:3|H:4|HBA:1|HBA:2|</td></tr>
<tr><td>2,4-Heptanediol, 3-methyl</td><td>3</td><td>–</td><td>–5.5</td><td>56.47</td><td>H:1|H:2||H:4|HBA:1|HBA:2|</td></tr>
<tr><td>Tiglic acid benzyl ester</td><td>3</td><td>–</td><td>–7.6</td><td>56.31</td><td>H:1||H:3|H:4|HBA:1|HBA:2|</td></tr>
<tr><td>Isobutyranilide, N-ethyl</td><td>4</td><td>4</td><td>–6.4</td><td>56.28</td><td>H:1|H:2|H:3||HBA:1|HBA:2|</td></tr>
<tr><td>Benzoic acid o-tolyl ester</td><td>3</td><td>2</td><td>–8.3</td><td>56.28</td><td>H:1||H:3|H:4|HBA:1|HBA:2|</td></tr>
<tr><td>2,4-Hexanediol, 3,4-dimethyl</td><td>4</td><td>4</td><td>–5.8</td><td>56.19</td><td>H:1||H:3|H:4|HBA:1|HBA:2|</td></tr>
<tr><td>Piperonylamide, N,N-diethyl</td><td>–</td><td>2</td><td>–7.1</td><td>56.19</td><td>H:1|H:2|H:3||HBA:1|HBA:2|</td></tr>
<tr><td>o-Toluic acid ethyl ester</td><td>1</td><td>–</td><td>–6.3</td><td>56.03</td><td>H:1|H:2|H:3||HBA:1|HBA:2|</td></tr>
<tr><td>2,4-Octanediol, 3-ethyl</td><td>4</td><td>4A</td><td>–5.8</td><td>56.01</td><td>H:1|H:2|H:3||HBA:1|HBA:2|</td></tr>
<tr><td>4-Cyclohexene-1,2-dicarboximide, N-allyl-3-methyl</td><td>4</td><td>4A</td><td>–7</td><td>56.01</td><td>H:1|H:2||H:4|HBA:1|HBA:2|</td></tr>
</table>

(Continued)

TABLE 3.3 (CONTINUED)
Top 25 USDA Compounds Fitted to AgamOBP1 Structure Using the "Hydrophobic" Pharmacophore

<table>
<tr><th>Ligand</th><th>Yfs[a]</th><th>Yfc[b]</th><th>Affinity (kcal/mol)[c]</th><th>Pharmacophore-Fit Score[d]</th><th>Matching Features[e]</th></tr>
<tr><td>Malonic acid, benzyl-, diethyl ester</td><td>1</td><td>–</td><td>–6.8</td><td>55.99</td><td>|H:2|H:3|H:4|HBA:1|HBA:2|</td></tr>
<tr><td>Thymol chloro</td><td>–</td><td>4A</td><td>–6.5</td><td>55.97</td><td>H:1|H:2|H:3||HBA:1|HBA:2|</td></tr>
<tr><td>Phthalic acid dipropyl ester</td><td>1</td><td>–</td><td>–6.5</td><td>55.96</td><td>|H:2|H:3|H:4|HBA:1|HBA:2|</td></tr>
<tr><td>4-Cyclohexene-1,2-dicarboximide,*N*-isopropyl-3-methyl</td><td>2</td><td>4A</td><td>–6.9</td><td>55.92</td><td>H:1|H:2||H:4|HBA:1|HBA:2|</td></tr>
<tr><td>Oxazolidine, 4,4-dimethyl-2-phenyl</td><td>–</td><td>4A</td><td>–7.5</td><td>55.77</td><td>H:1|H:2||H:4|HBA:1|HBA:2|</td></tr>
<tr><td>Quinaldine chloro</td><td>2</td><td>4A</td><td>–7.2</td><td>55.7</td><td>H:1|H:2|H:3||HBA:1|HBA:2|</td></tr>
<tr><td>Benzamide, *o*-chloro-*N*,*N*-diethyl</td><td>–</td><td>4A</td><td>–6.9</td><td>55.68</td><td>H:1|H:2|H:3||HBA:1|HBA:2|</td></tr>
<tr><td>Phthalic acid diisopropyl ester</td><td>–</td><td>4A</td><td>–6.8</td><td>55.58</td><td>H:1|H:2|H:3||HBA:1|HBA:2|</td></tr>
</table>

Note: Text in boldface is the common reference repellent. It is used as control in comparative repellency studies.

[a] Yfs is the column for repellency efficiency on yellow fever mosquito when ligand is applied to skin (0–4 scale) (USDA Agriculture Handbooks nos. 69 (1954) and 340 (1967)).

[b] Yfc is the column for repellency efficiency on yellow fever mosquito when ligand is applied to cloth (0–4A scale).

[c] Affinity (in kcal/mol) is the *in silico* calculated binding affinity on OBP1.

[d] Pharmacophore fit score is similarity of ligand with the applied pharmacophore (see Figure 3.9b).

[e] Pharmacophore matching features are described in the caption of Figure 3.9.

TABLE 3.4
Top 25 USDA Compounds Fitted to AgamOBP1 Structure Using the "Aromatic" Pharmacophore

<table>
<tr><th>Ligand</th><th>Yfs[a]</th><th>Yfc[b]</th><th>Affinity (kcal/mol)[c]</th><th>Pharmacophore-Fit Score[d]</th><th>Matching Features[e]</th></tr>
<tr><td>Benzamide, N,N-diethyl</td><td>4A</td><td>4A</td><td>–6.7</td><td>65.74</td><td>AR:1|AR:2|H:1|H:2|HBA:1||HBA:2</td></tr>
<tr><td>Oxazolidine, 4,4-dimethyl-2-phenyl</td><td>–</td><td>4A</td><td>–7.5</td><td>65.48</td><td>AR:1|AR:2|H:1|H:2|HBA:1||HBA:2</td></tr>
<tr><td>m-Toluamide, N,N-diethyl (DEET)</td><td>–</td><td>4A</td><td>–7.4</td><td>56.84</td><td>AR:1||H:1|H:2|HBA:1||HBA:2</td></tr>
<tr><td>o-Toluic acid benzyl ester</td><td>1</td><td>–</td><td>–8.9</td><td>56.1</td><td>|AR:2|H:1|H:2|HBA:1||HBA:2</td></tr>
<tr><td>p-Toluamide, N,N-diethyl</td><td>–</td><td>4A</td><td>–6.9</td><td>55.95</td><td>|AR:2|H:1|H:2|HBA:1||HBA:2</td></tr>
<tr><td>Butyric acid alpha-benzoyl-ethyl ester</td><td>2</td><td>4A</td><td>–6.9</td><td>55.04</td><td>AR:1|AR:2|H:1|H:2|HBA:1||</td></tr>
<tr><td>2,4-Hexanediol, 3-ethyl</td><td>4</td><td>3</td><td>–5</td><td>47.13</td><td>||H:1|H:2|HBA:1||HBA:2</td></tr>
<tr><td>1,3-Hexanediol, 2-ethyl (Rutgers 612)</td><td>4A</td><td>4A</td><td>–5.3</td><td>47.09</td><td>||H:1|H:2|HBA:1||HBA:2</td></tr>
<tr><td>Benzaldehyde, 3,4-diethoxy</td><td>2</td><td>4A</td><td>–6.2</td><td>46.91</td><td>AR:1||H:1|H:2|HBA:1||</td></tr>
<tr><td>Propyl phosphate, di</td><td>–</td><td>1</td><td>–4.6</td><td>46.89</td><td>||H:1|H:2|HBA:1||HBA:2</td></tr>
<tr><td>Acetoacetamide, N,N-diethyl</td><td>2</td><td>4</td><td>–5.1</td><td>46.84</td><td>||H:1|H:2|HBA:1||HBA:2</td></tr>
<tr><td>Benzene, 1-iodo-2-nitro</td><td>–</td><td>3</td><td>–5.8</td><td>46.84</td><td>||H:1|H:2|HBA:1||HBA:2</td></tr>
<tr><td>Butyl sulfone</td><td>–</td><td>4A</td><td>–5.1</td><td>46.75</td><td>||H:1|H:2|HBA:1||HBA:2</td></tr>
<tr><td>3,4-Hexanediol, 3,4-diethyl</td><td>3</td><td>–</td><td>–5.3</td><td>46.71</td><td>||H:1|H:2|HBA:1||HBA:2</td></tr>
<tr><td>o-Toluic acid</td><td>–</td><td>4</td><td>–6.1</td><td>46.69</td><td>||H:1|H:2|HBA:1||HBA:2</td></tr>
<tr><td>Ethanol 2-toloxy</td><td>3</td><td>4A</td><td>–5.3</td><td>46.66</td><td>||H:1|H:2|HBA:1||HBA:2</td></tr>
</table>

(Continued)

TABLE 3.4 (CONTINUED)
Top 25 USDA Compounds Fitted to AgamOBP1 Structure Using the "Aromatic" Pharmacophore

<table>
<tr><th>Ligand</th><th>Yfs[a]</th><th>Yfc[b]</th><th>Affinity (kcal/mol)[c]</th><th>Pharmacophore-Fit Score[d]</th><th>Matching Features[e]</th></tr>
<tr><td>Acrylamide, N,N-diallyl</td><td>–</td><td>3</td><td>–5.4</td><td>46.59</td><td>||H:1|H:2|HBA:1||HBA:2</td></tr>
<tr><td>1,3-Dioxolane-4-methanol, 2-(p-chlorophenyl)-2-methyl</td><td>3</td><td>–</td><td>–6.2</td><td>46.53</td><td>||H:1|H:2|HBA:1||HBA:2</td></tr>
<tr><td>Phenethyl alcohol, o-methyl</td><td>4</td><td>4</td><td>–5.8</td><td>46.53</td><td>||H:1|H:2|HBA:1||HBA:2</td></tr>
<tr><td>1,2-Propanediol, 3-chloro-2-methyl</td><td>2</td><td>3</td><td>–4.3</td><td>46.5</td><td>||H:1|H:2|HBA:1||HBA:2</td></tr>
<tr><td>Butyraldehyde alpha-(2-cyanoethyl)-alpha-ethyl</td><td>4A</td><td>4A</td><td>–5.1</td><td>46.49</td><td>||H:1|H:2|HBA:1||HBA:2</td></tr>
<tr><td>Citronellal oxime</td><td>2</td><td>4A</td><td>–6.2</td><td>46.49</td><td>||H:1|H:2|HBA:1||HBA:2</td></tr>
<tr><td>Butyric acid, 2-octynyl ester</td><td>2</td><td>4A</td><td>–6.4</td><td>46.48</td><td>||H:1|H:2|HBA:1||HBA:2</td></tr>
<tr><td>Crotonic acid, beta-diethylamino-, methyl ester</td><td>4</td><td>2</td><td>–4.9</td><td>46.45</td><td>||H:1|H:2|HBA:1||HBA:2</td></tr>
</table>

Note: Text in boldface is the common reference repellent. It is used as control in comparative repellency studies.

[a] Yfs is the column for repellency efficiency on yellow fever mosquito when ligand is applied to skin (0–4 scale) (USDA Agriculture Handbooks nos. 69 (1954) and 340 (1967)).

[b] Yfc is the column for repellency efficiency on yellow fever mosquito when ligand is applied to cloth (0–4A scale).

[c] Affinity (in kcal/mol) is the *in silico* calculated binding affinity on OBP1.

[d] Pharmacophore fit score is similarity of ligand with the applied pharmacophore (see Figure 3.9c).

[e] Pharmacophore matching features are described in the caption of Figure 3.9.

derived from the best conformation generation model of the conformers displaying the smallest root mean square (RMS) deviations when projected onto the pharmacophore.

The ordering of compounds according to their pharmacophore scores clearly indicates the capability of the two pharmacophore models to select preferably ligands belonging to the 4A and 4 classes of repellency. Overall, 21% of the compounds of the initial screening data set were class 4 and 4A repellents (Yfs or Yfc) (Table 3.2). The corresponding values for the 25 top-ranked predicted compounds were found to be 60% and 68% for the hydrophobic and aromatic models, respectively. Given the divergent chemical classes and the distribution of activities within the test set, the pharmacophore-based selection performance could be considered equal or even superior compared to the ANN model constructed by Katritzky et al. [36]. Importantly, beyond the DEET analogs, which occupy the top positions of both hit lists as was expected, new active scaffolds not belonging to carboxamides or *N*-acylpiperidines were also highly scored (Figure 3.10).

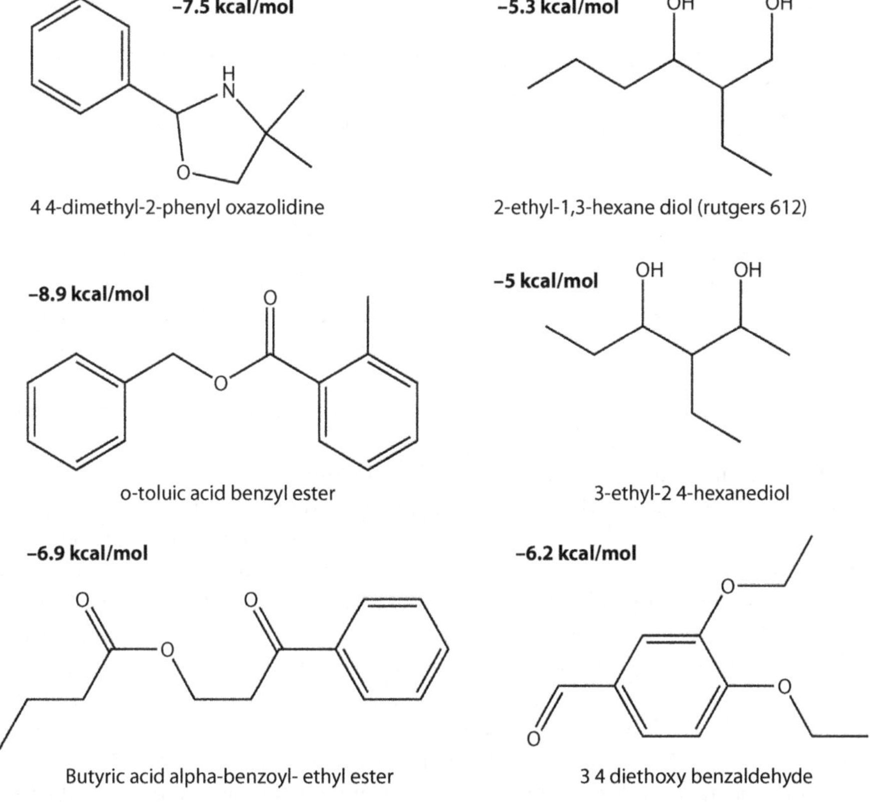

FIGURE 3.10 Representative examples of top-ranked structures identified through AgamOBP1 structure-based pharmacophore screening against USDA archives. Their calculated FEB to AgamOBP1 is also shown.

Docking analysis revealed that these molecules bind in a DEET-like manner to AgamOBP1. For example, oxazolidine,4,4-dimethyl-2-phenyl, which displays the second highest "aromatic" pharmacophore fit and docking score (Table 3.4), overlaps quite well with the DEET molecule and makes a hydrogen bond (2.7 Å) with the structural water molecule (Figure 3.11a). Another characteristic case is the repellent Rutgers 612 (2-ethyl-1,3-hexane diol) that was discovered at the Rutgers School of Environmental and Biological Sciences, the State University of New Jersey; in 1942, it was one of the three chemicals approved for use by the armed forces for the prevention of mosquito-borne diseases. Rutgers 612 is predicted to be able to form two hydrogen bonds from its hydroxyl groups (2.3 and 2.7 Å) to the water molecule. Although the experimental K_d of Rutgers 612 is not known, the relatively low docking score predicted for this compound (–5 kcal/mol) could be explained by the limited number of nonpolar vdW interactions with the surrounding protein residues. In contrast, the *o*-toluic acid benzyl ester, a methyl derivative of the insect repellent benzyl benzoate, showed a very high docking score (–8.9 kcal/mol), but it is inactive according to the USDA skin repellency tests on *Ae. aegypti*. This compound is predicted to bind with its benzyl ring oriented almost parallel to Trp114, thus making favorable aromatic interactions (Figure 3.11b). However, the molecule is placed close to the α5 helix, where its C2 carbon atom is in close proximity (2.36 Å) to the α-carbon of Gly92 and the carbonyl oxygen is located in close contact to the structural water (2.0 Å). Actually, the molecular shape of this biphenyl compound does not conform to the shape complementarity feature of the binding pocket, but rather extends beyond its boundaries (Figure 3.4b), giving rise to considerations for possible occurrence of steric clashes with the OBP1's partner receptor. Despite these findings, benzyl benzoate is a known repellent for clothing impregnation that has been used widely by the US military [118]. Most recently, Oliferenko et al. have determined the MED of benzyl benzoate to be 0.023 μmol/cm^2 in a bioassay against *Ae. aegypti* [112]. Remarkably, in a fluorescence binding assay of various ligands to AgamOBP1, we found that benzyl benzoate is a very weak binder (unpublished data). In all likelihood, as previously mentioned in this chapter, the activity of benzyl benzoate, as that of other repellents, might be mediated by another OBP–OR pair.

Our molecular docking results showed that binding affinity is correlated to shape and chemical feature complementarity between the protein and the ligand, in agreement with Oliferenko et al. [112]. However, in the molecular docking studies of Oliferenko et al., where the whole length of the protein channel was exploited in the analysis, the compounds consistently docked into the middle of the long tunnel where they interact with Phe123. Binding to this highly hydrophobic region of the channel resulted in higher docking scores due to the increase of favorable interactions between the protein and the compound that possibly derive from a larger vdW contact surface area. In our analysis, the pharmacophoric approach is centered on the experimental DEET location in the binding site of AgamOBP1 and the features selected expand from this position to the PEG binding site, permitting small compounds to emulate DEET binding and longer compounds, which have a more linear chemical structure, to bind and expand into the tunnel toward the start of the binding site of PEG.

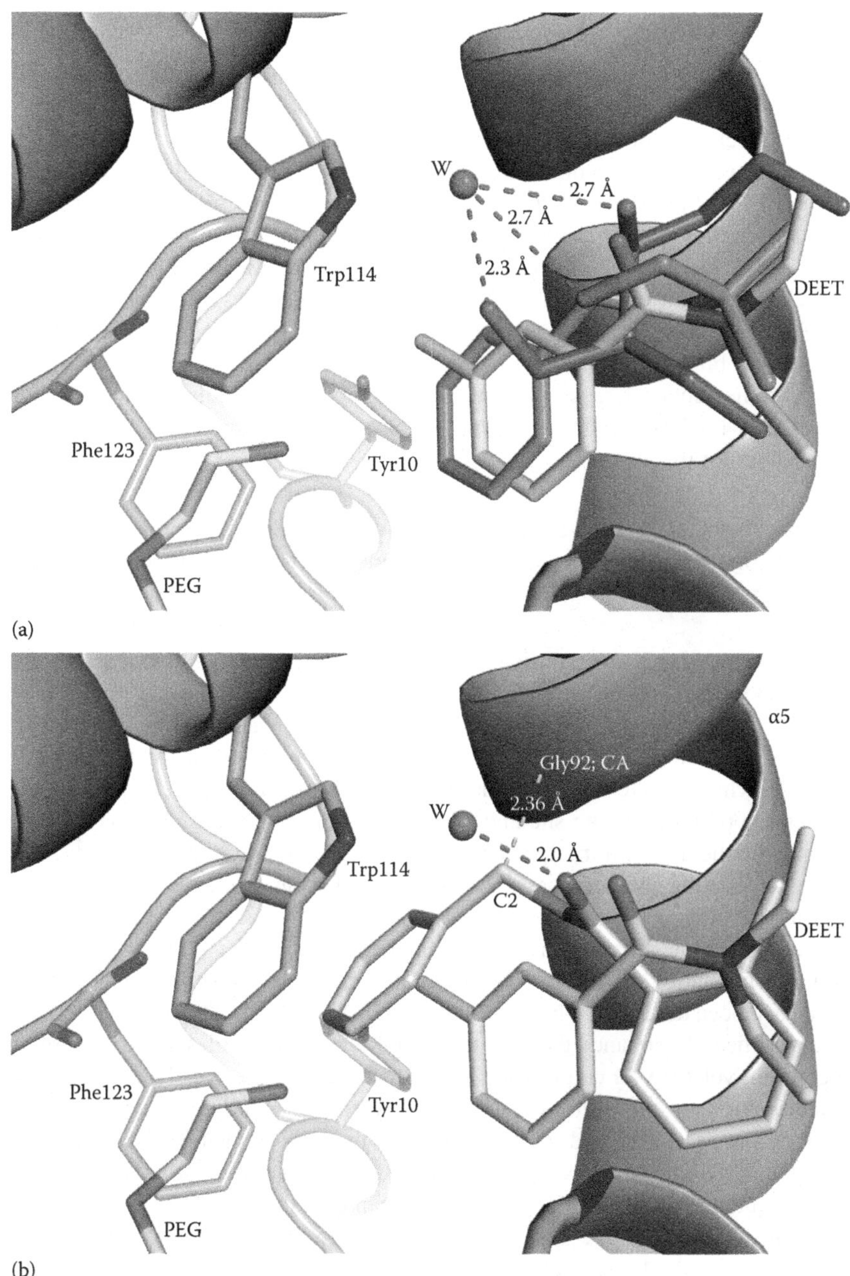

FIGURE 3.11 **(See color insert.)** Representation of the AgamOBP1 binding site with DEET and PEG (yellow) superimposed with tested compounds. (a) The predicted binding modes of the active compounds oxazolidine,4,4-dimethyl-2-phenyl (orange) and Rutgers 612 (2-ethyl-1,3-hexane diol) (blue); (b) the predicted binding mode of *o*-toluic acid benzyl ester, a methyl derivative of benzyl benzoate (gray).

3.7 CONCLUSIONS AND PROSPECTIVE

Over the past 25 years several ligand-based QSAR studies have successfully provided useful guidelines for the required stereoelectronic and physicochemical properties of effective mosquito repellents, thus contributing significantly to the rational design of novel agents against mosquito vectors. The determination of the crystal structure of AgamOBP1 in complex with DEET in 2012 led the way to structure-based approaches similar to those being actively exploited in drug discovery. Soon after, in 2013, the first combined ligand-based QSAR and AaegOBP1-based molecular docking study was performed. Through these two studies, Tsitsanou et al. [84] and Oliferenko et al. [112] provided evidence for a direct link between the predicted OBP1-ligand binding affinity and mosquito behavioral response. However, in both studies, some reported cases prevent the generalization of this relation—that is, strong OBP1 binders that do not promote a physiological response. Several reasons may account for this discrepancy:

- The ligand cannot be recognized by the cognate odorant receptor.
- The ligand is not volatile enough to be detected by the mosquito.
- The predicted affinity of the ligand for the OBP1 is overestimated.

To overcome the OR recognition problem, we have made the rational assumption that the interplay between ORs and OBPs could only be accomplished if OBPs and their cognate ORs possess binding sites of similar 3D shapes and physicochemical properties. Therefore, we proposed that the binding mode of candidate ligands to the OBP1, which is determined by the shape and chemical feature complementarity between the protein binding site and the ligand, should be taken into consideration as it may be the determining factor for ligand recognition by the OR. The fact that different modeling methodologies can lead to different predicted binding modes indicates the urgent need for additional experimental structural data on OBPs in complex with repellents of various chemical classes. To this point, we anticipate that the combination of ligands' binding mode with descriptors and functionalities which have been determined by SAR and QSAR studies to be critical for repellency, could significantly enhance the selection of OR-active molecules. Furthermore, we suggest that volatility of compounds is a very critical factor and should be used as an additional restriction filter during ligand selection. Finally, predicted affinities obviously have to be treated with a reasonable measure of caution, like every predicted result. Determination of experimental K_ds that requires synthesis or purchase of small quantities of the candidate ligands can reveal actual OBP targets before embarking on an OBP-specific structure-based design strategy and the synthesis of new compounds. In general, acquisition of more experimental binding data, in crystal and solution, and their correlation with behavioral data could strikingly optimize the next generations of *in silico* screening protocols. This rational approach can drastically shorten the number of ligands to be synthesized in large scale and subsequently tested in time- and money-consuming behavioral or field studies. With the identification of a considerable number of genes encoding mosquito OBPs so far, and an increasing accumulation of data on their 3D-structures, binding specificities, and interactions with their cognate ORs, the OBP structure-aided approach is expected to give a significant impetus to discovery of novel repellents in the very near future.

ACKNOWLEDGMENTS

We would like to acknowledge Dr. Kostas Iatrou (Institute of Biosciences & Applications NCSR "Demokritos," Greece) for helpful discussions, Prof. Patrick M. Guerin and Dr. Thomas Kröber (Institute of Biology, University of Neuchâtel, Switzerland) for kindly providing behavioral data on lead compounds and screening hits, and Ms. Christina Drakou, MSc (Institute of Biology, Medicinal Chemistry and Biotechnology, NHRF, Greece) for providing the binding data of benzyl benzoate to AgamOBP1.

This work was supported by funding provided under the NSRF-Bilateral Greece-Turkey R&D cooperation—2013 project "PREVENT" (GSRT 14TUR), cofinanced by the European Union and the Greek State, Ministry of Education and Religious Affairs/General Secretariat for Research and Technology (O. P. Competitiveness & Entrepreneurship (EPAN II), ROP Macedonia—Thrace, ROP Crete and Aegean Islands, ROP Thessaly—Mainland Greece—Epirus, ROP Attica). It was also supported by the European Commission for the FP7-HEALTH-2007-2.3.2.9 project "ENAROMaTIC" (GA-222927), the FP7-REGPOT-2008-1 project "EUROSTRUCT" (GA-230146), the FP7-REGPOT-2009-1 project "ARCADE" (GA-245866). Work at the Synchrotron Radiation Sources, MAX-lab, Lund, Sweden; EMBL Hamburg Outstation, Germany; and ALBA, Barcelona, Spain, was supported by funding provided by the European Community's Seventh Framework Program (FP7/2007-2013) under BioStruct-X (grant agreement 283570).

REFERENCES

[1] WHO, *World Health Organization: Malaria fact sheet N. 94.* (2015), WHO website [online] http://www.who.int/mediacentre/factsheets/fs094/en/

[2] K. Karunamoorthi, *Malaria vaccine: A future hope to curtail the global malaria burden*, Int. J. Prev. Med. 5 (2014), pp. 529–538.

[3] T. Chareonviriyaphap, M.J. Bangs, W. Suwonkerd, M. Kongmee, V. Corbel, and R. Ngoen-Klan, *Review of insecticide resistance and behavioral avoidance of vectors of human diseases in Thailand*, Parasit. Vectors 6 (2013), pp. 280.

[4] WHO, *World Health Organization: A global brief on vector-borne diseases* (2014), WHO website [online] http://www.who.int/campaigns/world-health-day/2014/global-brief/en/

[5] A.M. Dondorp, S. Yeung, L. White, C. Nguon, N.P. Day, D. Socheat, and L. von Seidlein, *Artemisinin resistance: Current status and scenarios for containment*, Nat. Rev. Microbiol. 8 (2010), pp. 272–280.

[6] B.V. Travis, F.A. Morton, H.A. Jones, and J.H. Robinson, *The more effective mosquito repellents tested at the Orlando, Fla, Laboratory, 1942–47*, J. Econ. Entomol. 42 (1949), pp. 686–694.

[7] C.N. Smith and D. Burnett, *Effectiveness of repellents applied to clothing for protection against salt-marsh mosquitoes*, J. Econ. Entomol. 42 (1949), pp. 439–444.

[8] S.I. Gertler, *N,N-diethylbenzamide as an insect repellent*, US Patent 2408389 A, Google Patents, 1946.

[9] E.T. Mccabe, W.F. Barthel, S.I. Gertler, and S.A. Hall, *Insect repellents. 3. N,N-diethylamides*, J. Org. Chem. 19 (1954), pp. 493–498.

[10] B.W.D. Krueger, K.D. Sasse, F.P.D. Hoever, G.D. Nentwig, and W.D. Behrenz, *Mittel zur Insekten und Milbenabwehr (Agents for repelling insects and mites)*, DE Patent 3801082 A1, Google Patents, 1988.

[11] J. Boeckh, H. Breer, M. Geier, F.P. Hoever, B.W. Kruger, G. Nentwig, and H. Sass, *Acylated 1,3-aminopropanols as repellents against bloodsucking arthropods*, Pest. Sci. 48 (1996), pp. 359–373.

[12] M.V.S. Suryanarayana, K.S. Pandey, S. Prakash, C.D. Raghuveeran, R.S. Dangi, R.V. Swamy, and K.M. Rao, *Structure activity relationship studies with mosquito repellent amides*, J. Pharm. Sci.-US 80 (1991), pp. 1055–1057.

[13] D. Ma, A.K. Bhattacharjee, R.K. Gupta, and J.M. Karle, *Predicting mosquito repellent potency of N,N-diethyl-m-toluamide (DEET) analogs from molecular electronic properties*, Am. J. Trop. Med. Hyg. 60 (1999), pp. 1–6.

[14] A.K. Bhattacharjee, R.K. Gupta, D. Ma, and J.M. Karle, *Molecular similarity analysis between insect juvenile hormone and N,N-diethyl-m-toluamide (DEET) analogs may aid design of novel insect repellents*, J. Mol. Recognit. 13 (2000), pp. 213–220.

[15] A.K. Bhattacharjee and R.K. Gupta, *Analysis of molecular stereoelectronic similarity between N,N-diethyl-m-toluamide (DEET) analogs and insect juvenile hormone to develop a model pharmacophore for insect repellent activity*, J. Am. Mosquito Contr. Assoc. 21 (2005), pp. 23–29.

[16] S.D. Accelrys Inc., CA, *CATALYST Version 4.5 software.*

[17] A.K. Bhattacharjee, W. Dheranetra, D.A. Nichols, and R.K. Gupta, *3D pharmacophore model for insect repellent activity and discovery of new repellent candidates*, QSAR Comb. Sci. 24 (2005), pp. 593–602.

[18] J.B. Bhonsle, A.K. Bhattacharjee, and R.K. Gupta, *Novel semi-automated methodology for developing highly predictive QSAR models: Application for development of QSAR models for insect repellent amides*, J. Mol. Model. 13 (2007), pp. 179–208.

[19] L.H. Hall, B. Mohney, and L.B. Kier, *The electrotopological state—An atom index for QSAR*, Quant. Struc.–Act. Relat. 10 (1991), pp. 43–51.

[20] L.B. Kier, *A shape index from molecular graphs*, Quant. Struc.–Act. Relat. 4 (1985), pp. 109–116.

[21] D.T. Stanton and P.C. Jurs, *Development and use of charged partial surface-area structural descriptors in computer-assisted quantitative structure property relationship studies*, Anal. Chem. 62 (1990), pp. 2323–2329.

[22] R.H. Rohrbaugh and P.C. Jurs, *Descriptions of molecular shape applied in studies of structure activity and structure/property relationships*, Anal. Chim. Acta 199 (1987), pp. 99–109.

[23] A.K. Ghose and G.M. Crippen, *Atomic physicochemical parameters for 3-dimensional structure-directed quantitative structure–activity relationships. 1. Partition coefficients as a measure of hydrophobicity*, J. Comput. Chem. 7 (1986), pp. 565–577.

[24] A.K. Ghose and G.M. Crippen, *Atomic physicochemical parameters for 3-dimensional-structure-directed quantitative structure–activity relationships. 2. Modeling dispersive and hydrophobic interactions*, J. Chem. Inf. Comput. Sci. 27 (1987), pp. 21–35.

[25] T.L. Hill, *An Introduction to Statistical Thermodynamics*, Addison–Wesley Publishing Co, Reading, MA, 1960.

[26] A.T. Balaban, *Highly discriminating distance-based topological index*, Chem. Phys. Lett. 89 (1982), pp. 399–404.

[27] A.R. Katritzky, D.A. Dobchev, I. Tulp, M. Karelson, and D.A. Carlson, *QSAR study of mosquito repellents using CODESSA Pro*, Bioorg. Med. Chem. Lett. 16 (2006), pp. 2306–2311.

[28] M. Karelson, *Molecular Descriptors in QSAR/QSPR*, Wiley-Interscience, New York, 2000.

[29] H.L. Johnson, W.A. Skinner, D. Skidmore, and H.I. Maibach, *Topical mosquito repellents. 2. Repellent potency and duration in ring-substituted N,N-dialkyl- and aminoalkylbenzamides*, J. Med. Chem. 11 (1968), pp. 1265–1268.

[30] L.R. Garson and M.E. Winnike, *Relationships between insect repellency and chemical and physical parameters—A review*, J. Med. Entomol. 5 (1968), pp. 339–352.

[31] V.G. Dethier, *Repellents*, Annu. Rev. Entomol. 1 (1956), pp. 181–202.

[32] S.B. Mciver, *A model for the mechanism of action of the repellent DEET on* Aedes aegypti *(Diptera, Culicidae)*, J. Med. Entomol. 18 (1981), pp. 357–361.

[33] S.S. Rao and K.M. Rao, *Insect repellent N,N-diethylphenylacetamide—An update*, J. Med. Entomol. 28 (1991), pp. 303–306.

[34] R. Natarajan, S.C. Basak, D. Mills, J.J. Kraker, and D.M. Hawkins, *Quantitative structure–activity relationship modeling of mosquito repellents using calculated descriptors*, Croat. Chem. Acta 81 (2008), pp. 333–340.

[35] A.R. Katritzky, Z.Q. Wang, S. Slavon, D.A. Dobchev, C.D. Hall, M. Tsikolia, U.R. Bernier, N.M. Elejalde, G.G. Clark, and K.J. Linthicum, *Novel carboxamides as potential mosquito repellents*, J. Med. Entomol. 47 (2010), pp. 924–938.

[36] A.R. Katritzky, Z.Q. Wang, S. Slavov, M. Tsikolia, D. Dobchev, N.G. Akhmedov, C.D. Hall, U.R. Bernier, G.G. Clark, and K.J. Linthicum, *Synthesis and bioassay of improved mosquito repellents predicted from chemical structure*, Proc. Natl. Acad. Sci. USA 105 (2008), pp. 7359–7364.

[37] D.E. Rumelhart, G.E. Hinton, and R.J. Williams, *Learning representations by back-propagating errors*, Nature 323 (1986), pp. 533–536.

[38] J.V. Tu, *Advantages and disadvantages of using artificial neural networks versus logistic regression for predicting medical outcomes*, J. Clin. Epidemiol. 49 (1996), pp. 1225–1231.

[39] USDA, *Chemicals evaluated as insecticides and repellents at Orlando, Fla,* compiled by King WV (US Dept Agriculture, Washington, DC), *Agriculture Handbook No. 69*, 1954.

[40] USDA, *Materials evaluated as insecticides, repellents, and chemosterilants at Orlando and Gainesville, Fla, 1952–1964 (US Dept Agriculture, Washington, DC), Agriculture Handbook No. 340*, 1967.

[41] USDA, *Repellent activity of compounds submitted by Walter Reed Army Institute of Research Part I. Protection time and minimum effective dosage against* Aedes aegypti *mosquitoes* (US Dept Agriculture, Washington, DC), *Technical Bulletin No. 1549*, 1977.

[42] S.B. McIver, *Sensilla mosquitoes (Diptera: Culicidae)*, J. Med. Entomol. 19 (1982), pp. 489–535.

[43] A. Couto, M. Alenius, and B.J. Dickson, *Molecular, anatomical, and functional organization of the* Drosophila *olfactory system*, Curr. Biol. 15 (2005), pp. 1535–1547.

[44] K. Sato, M. Pellegrino, T. Nakagawa, L.B. Vosshall, and K. Touhara, *Insect olfactory receptors are heteromeric ligand-gated ion channels*, Nature 452 (2008), pp. 1002–1006.

[45] M.C. Larsson, A.I. Domingos, W.D. Jones, M.E. Chiappe, H. Amrein, and L.B. Vosshall, *Or83b encodes a broadly expressed odorant receptor essential for* Drosophila *olfaction*, Neuron 43 (2004), pp. 703–714.

[46] V. Croset, R. Rytz, S.F. Cummins, A. Budd, D. Brawand, H. Kaessmann, T.J. Gibson, and R. Benton, *Ancient protostome origin of chemosensory ionotropic glutamate receptors and the evolution of insect taste and olfaction*, PLoS Genet. 6 (2010), pp. e1001064.

[47] C. Liu, R.J. Pitts, J.D. Bohbot, P.L. Jones, G. Wang, and L.J. Zwiebel, *Distinct olfactory signaling mechanisms in the malaria vector mosquito* Anopheles gambiae, PLoS Biol. 8 (2010), pp. e1000467.

[48] J.J. Zhou, *Odorant-binding proteins in insects*, Vitam. Horm. 83 (2010), pp. 241–272.
[49] M.E. Rogers, J. Krieger, and R.G. Vogt, *Antennal SNMPs (sensory neuron membrane proteins) of Lepidoptera define a unique family of invertebrate CD36-like proteins*, J. Neurobiol. 49 (2001), pp. 47–61.
[50] R.G. Vogt, N.E. Miller, R. Litvack, R.A. Fandino, J. Sparks, J. Staples, R. Friedman, and J.C. Dickens, *The insect SNMP gene family*, Insect Biochem. Mol. Biol. 39 (2009), pp. 448–456.
[51] C.E. Merrill, J. Riesgo-Escovar, R.J. Pitts, F.C. Kafatos, J.R. Carlson, and L.J. Zwiebel, *Visual arrestins in olfactory pathways of* Drosophila *and the malaria vector mosquito* Anopheles gambiae, Proc. Natl. Acad. Sci. USA 99 (2002), pp. 1633–1638.
[52] R. Rybczynski, J. Reagan, and M.R. Lerner, *A pheromone-degrading aldehyde oxidase in the antennae of the moth* Manduca sexta, J. Neurosci. 9 (1989), pp. 1341–1353.
[53] R.G. Vogt, *Molecular basis of pheromone detection in insects*, in *Comprehensive Insect Physiology, Biochemistry, Pharmacology and Molecular Biology*, L.I. Gilbert, K. Latro, and S. Gill, eds., Elsevier, London, 2005, pp. 753–804.
[54] R.G. Vogt, L.M. Riddiford, and G.D. Prestwich, *Kinetic properties of a sex pheromone-degrading enzyme: The sensillar esterase of* Antheraea polyphemus, Proc. Natl. Acad. Sci. USA 82 (1985), pp. 8827–8831.
[55] U. Klein, *Sensillum-lymph proteins from antennal olfactory hairs of the moth* Antheraea polyphemus *(Saturniidae)*, Insect Biochem. 17 (1987), pp. 1193–1204.
[56] P. Pelosi, *Perireceptor events in olfaction*, J. Neurobiol. 30 (1996), pp. 3–19.
[57] C. Lautenschlager, W.S. Leal, and J. Clardy, *Coil-to-helix transition and ligand release of* Bombyx mori *pheromone-binding protein*, Biochem. Biophys. Res. Commun. 335 (2005), pp. 1044–1050.
[58] S. Zubkov, A.M. Gronenborn, I.J.L. Byeon, and S. Mohanty, *Structural consequences of the pH-induced conformational switch in* A. polyphemus *pheromone-binding protein: Mechanisms of ligand release*, J. Mol. Biol. 354 (2005), pp. 1081–1090.
[59] N.R. Leite, R. Krogh, W. Xu, Y. Ishida, J. Iulek, W.S. Leal, and G. Oliva, *Structure of an odorant-binding protein from the mosquito* Aedes aegypti *suggests a binding pocket covered by a pH-sensitive "Lid,"* PLoS ONE 4 (2009), pp. e8006.
[60] J.D. Laughlin, T.S. Ha, D.N. Jones, and D.P. Smith, *Activation of pheromone-sensitive neurons is mediated by conformational activation of pheromone-binding protein*, Cell 133 (2008), pp. 1255–1265.
[61] C. Gomez-Diaz, J.H. Reina, C. Cambillau, and R. Benton, *Ligands for pheromone-sensing neurons are not conformationally activated odorant binding proteins*, PLoS Biol. 11 (2013), pp. e1001546.
[62] W.S. Leal, *Pheromone reception*, in *Chemistry of Pheromones and Other Semiochemicals II*, S. Schulz, ed., Springer, Berlin, 240 (2005), pp. 1–36.
[63] G. Wang, A.F. Carey, J.R. Carlson, and L.J. Zwiebel, *Molecular basis of odor coding in the malaria vector mosquito* Anopheles gambiae, Proc. Natl. Acad. Sci. USA 107 (2010), pp. 4418–4423.
[64] A.F. Carey, G. Wang, C.Y. Su, L.J. Zwiebel, and J.R. Carlson, *Odorant reception in the malaria mosquito* Anopheles gambiae, Nature 464 (2010), pp. 66–71.
[65] J. Pelletier, D.T. Hughes, C.W. Luetje, and W.S. Leal, *An odorant receptor from the southern house mosquito* Culex pipiens quinquefasciatus *sensitive to oviposition attractants*, PLoS ONE 5 (2010), pp. e10090.
[66] D.T. Hughes, J. Pelletier, C.W. Luetje, and W.S. Leal, *Odorant receptor from the southern house mosquito narrowly tuned to the oviposition attractant skatole*, J. Chem. Ecol. 36 (2010), pp. 797–800.
[67] C.S. McBride, F. Baier, A.B. Omondi, S.A. Spitzer, J. Lutomiah, R. Sang, R. Ignell, and L.B. Vosshall, *Evolution of mosquito preference for humans linked to an odorant receptor*, Nature 515 (2014), pp. 222–227.

[68] P. Xu, Y.M. Choo, A. De La Rosa, and W.S. Leal, *Mosquito odorant receptor for DEET and methyl jasmonate*, Proc. Natl. Acad. Sci. USA 111 (2014), pp. 16592–16597.

[69] P. Tsitoura, K. Koussis, and K. Iatrou, *Inhibition of* Anopheles gambiae *odorant receptor function by mosquito repellents*, J. Biol. Chem. 290 (2015), pp. 7961–7972.

[70] P.X. Xu and W.S. Leal, *Probing insect odorant receptors with their cognate ligands: Insights into structural features*, Biochem. Biophys. Res. Commun. 435 (2013), pp. 477–482.

[71] P. Miszta, S.C. Basak, R. Natarajan, and W. Nowak, *How computational studies of mosquito repellents contribute to the control of vector-borne diseases*, Curr. Comput.-Aid. Drug 9 (2013), pp. 300–307.

[72] G.H. Du and G.D. Prestwich, *Protein-structure encodes the ligand-binding specificity in pheromone binding-proteins*, Biochemistry 34 (1995), pp. 8726–8732.

[73] Q.Y. Jiang, W.X. Wang, Z. Zhang, and L. Zhang, *Binding specificity of locust odorant binding protein and its key binding site for initial recognition of alcohols*, Insect Biochem. Mol. Biol. 39 (2009), pp. 440–447.

[74] R. Maida, J. Krieger, T. Gebauer, U. Lange, and G. Ziegelberger, *Three pheromone-binding proteins in olfactory sensilla of the two silkmoth species* Antheraea polyphemus *and* Antheraea pernyi, Eur. J. Biochem. 267 (2000), pp. 2899–2908.

[75] E. Plettner, J. Lazar, E.G. Prestwich, and G.D. Prestwich, *Discrimination of pheromone enantiomers by two pheromone binding proteins from the gypsy moth* Lymantria dispar, Biochemistry 39 (2000), pp. 8953–8962.

[76] J.J. Zhou, G. Robertson, X. He, S. Dufour, A.M. Hooper, J.A. Pickett, N.H. Keep, and L.M. Field, *Characterisation of* Bombyx mori *Odorant-binding proteins reveals that a general odorant-binding protein discriminates between sex pheromone components*, J. Mol. Biol. 389 (2009), pp. 529–545.

[77] H. Qiao, X. He, D. Schymura, L. Ban, L. Field, F.R. Dani, E. Michelucci, B. Caputo, A. della Torre, K. Iatrou, J.J. Zhou, J. Krieger, and P. Pelosi, *Cooperative interactions between odorant-binding proteins of* Anopheles gambiae, Cell. Mol. Life Sci. 68 (2011), pp. 1799–1813.

[78] J. Yin, Y.M. Choo, H. Duan, and W.S. Leal, *Selectivity of odorant-binding proteins from the southern house mosquito tested against physiologically relevant ligands*, Front Physiol. 6 (2015), pp. 56.

[79] L. Guha, T. Seenivasagan, S.T. Iqbal, O.P. Agrawal, and B.D. Parashar, *Behavioral and electrophysiological responses of* Aedes albopictus *to certain acids and alcohols present in human skin emanations*, Parasitol. Res. 113 (2014), pp. 3781–3787.

[80] Y.T. Qiu, R.C. Smallegange, S. Hoppe, J.J. van Loon, E.J. Bakker, and W. Takken, *Behavioural and electrophysiological responses of the malaria mosquito* Anopheles gambiae *Giles sensu stricto (Diptera: Culicidae) to human skin emanations*, Med. Vet. Entomol. 18 (2004), pp. 429–438.

[81] H. Biessmann, E. Andronopoulou, M.R. Biessmann, V. Douris, S.D. Dimitratos, E. Eliopoulos, P.M. Guerin, K. Iatrou, R.W. Justice, T. Krober, O. Marinotti, P. Tsitoura, D.F. Woods, and M.F. Walter, *The* Anopheles gambiae *odorant binding protein 1 (AgamOBP1) mediates indole recognition in the antennae of female mosquitoes*, PLoS ONE 5 (2010), pp. e9471.

[82] J. Pelletier, A. Guidolin, Z. Syed, A.J. Cornel, and W.S. Leal, *Knockdown of a mosquito odorant-binding protein involved in the sensitive detection of oviposition attractants*, J. Chem. Ecol. 36 (2010), pp. 245–248.

[83] M.S. Sengul and Z. Tu, *Expression analysis and knockdown of two antennal odorant-binding protein genes in* Aedes aegypti, J. Insect. Sci. 10 (2010), pp. 171.

[84] K.E. Tsitsanou, T. Thireou, C.E. Drakou, K. Koussis, M.V. Keramioti, D.D. Leonidas, E. Eliopoulos, K. Iatrou, and S.E. Zographos, Anopheles gambiae *odorant binding protein crystal complex with the synthetic repellent DEET: Implications for structure-based design of novel mosquito repellents*, Cell. Mol. Life Sci. 69 (2012), pp. 283–297.

[85] J. Fan, F. Francis, Y. Liu, J.L. Chen, and D.F. Cheng, *An overview of odorant-binding protein functions in insect peripheral olfactory reception*, Genet. Mol. Res. 10 (2011), pp. 3056–3069.
[86] M. Manoharan, M. Ng Fuk Chong, A. Vaitinadapoule, E. Frumence, R. Sowdhamini, and B. Offmann, *Comparative genomics of odorant binding proteins in* Anopheles gambiae, Aedes aegypti, *and* Culex quinquefasciatus, Genome Biol. Evol. 5 (2013), pp. 163–180.
[87] P.X. Xu, L.J. Zwiebel, and D.P. Smith, *Identification of a distinct family of genes encoding atypical odorant-binding proteins in the malaria vector mosquito*, Anopheles gambiae, Insect Mol. Biol. 12 (2003), pp. 549–560.
[88] J.J. Zhou, W. Huang, G.A. Zhang, J.A. Pickett, and L.M. Field, *"Plus-C" odorant-binding protein genes in two* Drosophila *species and the malaria mosquito* Anopheles gambiae, Gene 327 (2004), pp. 117–129.
[89] E.J. Murphy, J.C. Booth, F. Davrazou, A.M. Port, and D.N. Jones, *Interactions of* Anopheles gambiae *odorant-binding proteins with a human-derived repellent: Implications for the mode of action of N,N-diethyl-3-methylbenzamide (DEET)*, J. Biol. Chem. 288 (2013), pp. 4475–4485.
[90] M. Wogulis, T. Morgan, Y. Ishida, W.S. Leal, and D.K. Wilson, *The crystal structure of an odorant binding protein from* Anopheles gambiae: *Evidence for a common ligand release mechanism*, Biochem. Biophys. Res. Commun. 339 (2006), pp. 157–164.
[91] F. Davrazou, E. Dong, E.J. Murphy, H.T. Johnson, and D.N. Jones, *New insights into the mechanism of odorant detection by the malaria-transmitting mosquito* Anopheles gambiae, J. Biol. Chem. 286 (2011), pp. 34175–34183.
[92] A. Lagarde, S. Spinelli, M. Tegoni, X. He, L. Field, J.J. Zhou, and C. Cambillau, *The crystal structure of odorant binding protein 7 from* Anopheles gambiae *exhibits an outstanding adaptability of its binding site*, J. Mol. Biol. 414 (2011), pp. 401–412.
[93] B.P. Ziemba, E.J. Murphy, H.T. Edlin, and D.N. Jones, *A novel mechanism of ligand binding and release in the odorant binding protein 20 from the malaria mosquito* Anopheles gambiae, Protein Sci. 22 (2012), pp. 11–21.
[94] A. Lagarde, S. Spinelli, H. Qiao, M. Tegoni, P. Pelosi, and C. Cambillau, *Crystal structure of a novel type of odorant-binding protein from* Anopheles gambiae, *belonging to the C-plus class*, Biochem. J. 437 (2011), pp. 423–430.
[95] K.E. Tsitsanou, C.E. Drakou, T. Thireou, A. Vitlin Gruber, G. Kythreoti, A. Azem, D. Fessas, E. Eliopoulos, K. Iatrou, and S.E. Zographos, *Crystal and solution studies of the "Plus-C" odorant-binding protein 48 from* Anopheles gambiae*: Control of binding specificity through three-dimensional domain swapping*, J. Biol. Chem. 288 (2013), pp. 33427–33438.
[96] Y. Mao, X. Xu, W. Xu, Y. Ishida, W.S. Leal, J.B. Ames, and J. Clardy, *Crystal and solution structures of an odorant-binding protein from the southern house mosquito complexed with an oviposition pheromone*, Proc. Natl. Acad. Sci. USA 107 (2010), pp. 19102–19107.
[97] M. Tegoni, V. Campanacci, and C. Cambillau, *Structural aspects of sexual attraction and chemical communication in insects*, Trends Biochem. Sci. 29 (2004), pp. 257–264.
[98] E. Andronopoulou, V. Labropoulou, V. Douris, D.F. Woods, H. Biessmann, and K. Iatrou, *Specific interactions among odorant-binding proteins of the African malaria vector* Anopheles gambiae, Insect Mol. Biol. 15 (2006), pp. 797–811.
[99] H. Biessmann, Q.K. Nguyen, D. Le, and M.F. Walter, *Microarray-based survey of a subset of putative olfactory genes in the mosquito* Anopheles gambiae, Insect Mol. Biol. 14 (2005), pp. 575–589.
[100] J.G. Logan, N.M. Stanczyk, A. Hassanali, J. Kemei, A.E. Santana, K.A. Ribeiro, J.A. Pickett, and A.J. Mordue Luntz, *Arm-in-cage testing of natural human-derived mosquito repellents*, Malar. J. 9 (2010), pp. 239.

[101] L.S. Nerio, J. Olivero-Verbel, and E. Stashenko, *Repellent activity of essential oils: A review*, Bioresour. Technol. 101 (2010), pp. 372–378.
[102] R.D. Affonso, A.P. Guimaraes, A.A. Oliveira, G.B.C. Slana, and T.C.C. Franca, *Applications of molecular modeling in the design of new insect repellents targeting the odorant binding protein of* Anopheles gambiae, J. Brazil. Chem. Soc. 24 (2013), pp. 473–482.
[103] M.C. Lawrence and P.M. Colman, *Shape complementarity at protein–protein interfaces*, J. Mol. Biol. 234 (1993), pp. 946–950.
[104] E. Freire, *Thermodynamic rules to achieve high affinity and selectivity*, Abstracts of Papers of the American Chemical Society 244 (2012).
[105] V.A. Likic, N. Juranic, S. Macura, and F.G. Prendergast, *A "structural" water molecule in the family of fatty acid binding proteins*, Protein Sci. 9 (2000), pp. 497–504.
[106] T.E. Angel, S. Gupta, B. Jastrzebska, K. Palczewski, and M.R. Chance, *Structural waters define a functional channel mediating activation of the GPCR, rhodopsin*, Proc. Natl. Acad. Sci. USA 106 (2009), pp. 14367–14372.
[107] M. Pastor, G. Cruciani, and K.A. Watson, *A strategy for the incorporation of water molecules present in a ligand binding site into a three-dimensional quantitative structure–activity relationship analysis*, J. Med. Chem. 40 (1997), pp. 4089–4102.
[108] D.J. Huggins and B. Tidor, *Systematic placement of structural water molecules for improved scoring of protein–ligand interactions*, Protein Eng. Des. Sel. 24 (2011), pp. 777–789.
[109] R.A. Powers, F. Morandi, and B.K. Shoichet, *Structure-based discovery of a novel, noncovalent inhibitor of AmpC beta-lactamase*, Structure 10 (2002), pp. 1013–1023.
[110] J.M. Gaudin, T. Lander, and O. Nikolaenko, *Carboxamides combining favorable olfactory properties with insect repellency*, Chem. Biodivers. 5 (2008), pp. 617–635.
[111] G.M. Morris, D.S. Goodsell, R.S. Halliday, R. Huey, W.E. Hart, R.K. Belew, and A.J. Olson, *Automated docking using a Lamarckian genetic algorithm and an empirical binding free energy function*, J. Comput. Chem. 19 (1998), pp. 1639–1662.
[112] P.V. Oliferenko, A.A. Oliferenko, G.I. Poda, D.I. Osolodkin, G.G. Pillai, U.R. Bernier, M. Tsikolia, N.M. Agramonte, G.G. Clark, K.J. Linthicum, and A.R. Katritzky, *Promising* Aedes aegypti *repellent chemotypes identified through integrated QSAR, virtual screening, synthesis, and bioassay*, PLoS ONE 8 (2013), pp. e64547.
[113] *Schrödinger Release 2015-2: LigPrep, version 3.4,* Schrödinger, LLC, New York, NY, 2015.
[114] O. Trott and A.J. Olson, *Software news and update AutoDock Vina: Improving the speed and accuracy of docking with a new scoring function, efficient optimization, and multithreading*, J. Comput. Chem. 31 (2010), pp. 455–461.
[115] *Small-Molecule Drug Discovery Suite 2015-2: Glide, version 6.7*, Schrödinger, LLC, New York, NY, 2015.
[116] G. Wolber and T. Langer, LigandScout: *3-D pharmacophores derived from protein-bound ligands and their use as virtual screening filters*, J. Chem. Inf. Model. 45 (2005), pp. 160–169.
[117] *Small-Molecule Drug Discovery Suite 2015-2: Phase, version 4.3*, Schrödinger, LLC, New York, NY, 2015.
[118] W.C. McCain and G.J. Leach, *Repellents used in fabric: The experience of the U.S. military*, in *Insect repellents: Principles, Methods, and Uses*, S.P. Frances, M. Debboun, and D. Strickman, eds., CRC Press, Boca Raton, FL, 2007, pp. 261–273.

4 Molecular Topology as a Powerful Tool for Searching for New Repellents and Novel Drugs against Diseases Transmitted by Mosquitoes

Ramón García-Domenech, Riccardo Zanni, María Galvez-Llompart, and Jorge Galvez

CONTENTS

ABSTRACT

This chapter deals with thc application of molecular topology (MT) for the prediction, selection, and design of new repellents and/or compounds active against diseases transmitted by mosquitoes. After a brief introduction regarding the current methods used in drug design protocol, MT basic concepts are defined, including various examples of topological indices calculation. In addition, a bibliographic review about the latest implementation of MT in the fight against these diseases is conducted, along with a practical example of prediction for the repellent activity against *Anopheles albimanus* and *Aedes aegypti* through a multitarget (mt) quantitative structure–activity relationship (QSAR) study.

KEYWORDS

- Mosquito
- Repellent
- QSAR
- Molecular topology
- Malaria

4.1 INTRODUCTION

From the earliest days of civilization, man has invested time and resources to prevent or control arthropod pests, even without knowing the existing relationship between them and transmitting diseases. Nowadays, in spite of the knowledge acquired, they are still of great interest to public health, mainly due to the fact that chemicals are used to vanquish or control.

The danger of the mosquito in terms of morbidity and mortality is impressive. It represents the main carrier of pathogens such as viruses, bacteria, protozoa, and nematodes. It is also responsible for serious diseases such as malaria, dengue, yellow fever, chikungunya, filariasis, and rare forms of encephalitis. It may be argued that the mosquito is the most life-threatening of all insects [1].

In tropical and subtropical areas, the mosquito is a threat to more than three billion people. Its influence in history is such that it has not only affected our economy, society, and politics but also intervened in the fall of empires like the Greeks and Romans with devastating epidemics and pandemics. Malaria, "malus aria," in the flooded areas of Rome was one of the biggest health problems in the last days of the Roman Empire [2].

Mosquito control measures using insecticides, mosquito nets, antimosquito wall paint, etc. keep facing obstacles such as increased resistance or even appearance of new species. Recently, the interbreeding of two malaria mosquito species (*Anopheles gambiae* and *An. coluzzii*) in the West African country of Mali resulted in a "super mosquito" hybrid resistant to insecticide-treated bed nets [3].

Taking into account the lack of interest that exists in combating the mosquito's endemic diseases along with its resistance to actual drugs, there is a quite obvious urge to develop effective computer techniques able to introduce new compounds active against the vector and/or the parasite.

Among the *in silico* techniques, QSAR (quantitative structure–activity relationship) analysis is a powerful methodology that relates the chemical structure of a given compound with physicochemical, biological, or pharmacological activity using mathematical models. The QSAR-based modeling techniques enable one to determine in a quantitative way which molecular structure plays a major role on the activity of a molecule [4].

The development of a QSAR model can be summarized in the following stages (steps) (Figure 4.1):

- First, information about the compounds and their biological activities or physicochemical properties is collected.
- Subsequently, the molecular descriptors are calculated. These descriptors are related to the activity of the compounds using various mathematical-statistical regression methods in order to obtain the QSAR equations.
- Finally, the models obtained are subjected to processes of internal and external validation. The main purpose of these validations is to ensure, as far as possible, that the models are applicable in the search for new active molecules.

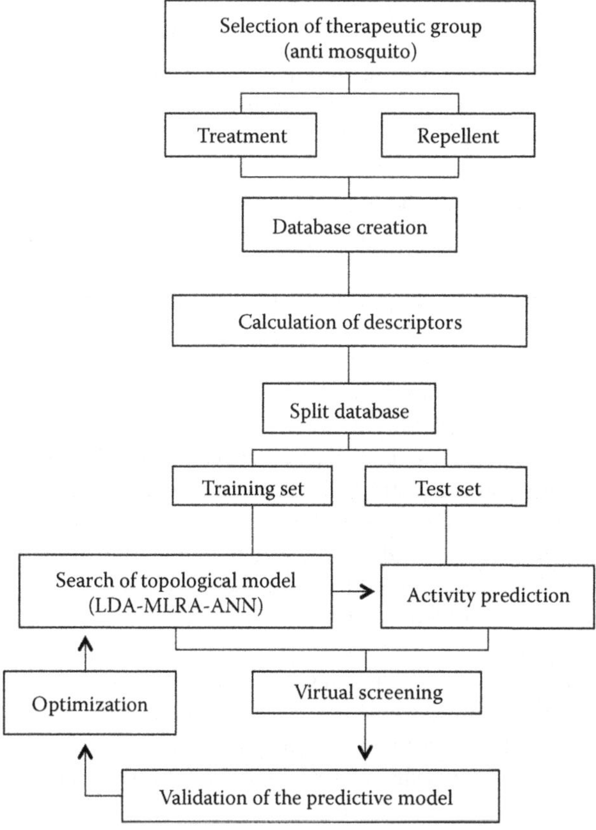

FIGURE 4.1 Steps to follow in the search of QSAR prediction models by molecular topology.

The main differences between the most common QSAR algorithms lie in the ways to generate descriptors. For example, comparative molecular field analysis (CoMFA) is based on the hypothesis that ligand–receptor interactions are of noncovalent type and that the variation of the biological activity correlates better with steric variations and electrostatic parameters of the molecule. Furthermore, other procedures, such as comparative analysis of molecular similarity indices (CoMSIA), are based on the calculation of similarity descriptors for molecules previously aligned in relation to certain physicochemical properties such as hydrophobic interactions or hydrogen bonds.

It is established that the success of a QSAR model lies in the selection of the adequate molecular descriptors. These descriptors, according to Basak et al. [5], can be classified as

- Topostructural indices, which encode information about vertices' connectivity, beyond their chemical nature
- Topochemical indices, which consider also their chemical characteristics (types of bonds)
- Three-dimensional (3D) geometrical parameters
- Quantum chemical descriptors, which take into account electronic features as well

It is surprising that the topostructural and topochemical information enables an accuracy of about 90% of the predicted properties, while the introduction of 3D properties results in a moderate increase of about 5% [6]. This can be explained by the fact that, contrary to the rest of the descriptors, these are purely mathematical. For that reason, molecular structure can be described in simple algebraic terms, enabling a straightforward and accurate prediction of many biological and physicochemical properties.

Molecular topology has emerged in recent years as a powerful approach for the *in silico* generation of new drugs. Within the last decade, its applications became increasingly popular among the leading research groups in the field of QSAR and drug design. This fact has contributed to the rapid development of new techniques and applications of MT in QSAR studies.

MT is basically related to the assimilation between molecules and graphs. Defined as a part of mathematical chemistry, it can depict molecular structures through graph theoretical indices [7]. Additionally, it deals with connectivity of atoms in molecules and rather than with geometrical features such as angles, distances, or tridimensional structure, which is common in standard/conventional approaches. So, MT's advantages can be summarized as follows:

- Molecular structure is depicted in simple mathematical terms—that is, as a set of elements (atoms) interacting by means of edges (bonds).
- It makes possible the fast and accurate analysis of large numbers of compounds as well as the design of novel ones by the reverse process (property → structure).
- It is easy to translate into computer language.

Thus, graph theory and surrounding disciplines are basic tools for MT development.

The use of molecular topology together with QSAR (MT-QSAR) has grown exponentially since the 1980s [8–16].

In this chapter molecular topology, along with statistical approaches, is used for the development of QSAR predictive models for mosquito repellent and/or antiprotozoal activity. Furthermore, a bibliographic review of the last decade about the latest implementation of MT in the fight against these diseases is conducted as well as a practical example of prediction for the repellent activity against *Anopheles albimanus* and *Aedes aegypti*, through a multitarget QSAR study.

4.2 METHODOLOGY AND APPLICATIONS

4.2.1 Molecular Structure Representation

MT's mathematical formalism consists in assigning a graph to each molecule, where the graph vertices correspond to atoms and its edges to bonds. Starting from the interconnections between the vertices, an adjacency topological matrix, A, can be built up, whose *ij* elements take the values of either one or zero, depending on whether the vertex *i* is connected or unconnected to the vertex *j*, respectively.

The valence or degree of each vertex $\delta\, i$ is the number of edges converging on it, which is equal to the sum of the terms that are in the row (or column) corresponding to that vertex.

Similarly, the distance matrix D can be constructed with each element *ij* corresponding to the minimum number of edges connecting the vertices *i* to *j* (Table 4.1). The manipulation of these matrices gives origin to a set of topological indices, TIs, also referred to as topological descriptors, which characterize each graph/molecule and allow the development of QSAR studies.

4.2.1.1 Topological Indices Calculation

A comprehensive review of topological descriptors is almost impossible due to the large number of such indices that are published in the literature and the number of them that are introduced every year, which continues to grow.

Table 4.1 illustrates the calculation of some representative TIs for 2,3-dimethylbutane.

One of the first descriptors was the Wiener index (W = path number), which is calculated from the distance matrix as

$$W = \frac{1}{2}\sum_{i=1}^{i=j} d_{ij} \qquad (4.1)$$

Despite its simplicity, this index showed excellent correlation with the boiling point of alkanes [17].

The first connectivity index was the branching index or Randić index [18], χ, introduced by professor Milan Randić at Drake University in 1975. This descriptor is

TABLE 4.1

Graph of 2,3-Dimethylbutane and Adjacency and Distance Matrices

Adjacency Matrix

$$A = \left|\begin{matrix} 0 & 1 & 0 & 0 & 0 & 0 \\ 1 & 0 & 1 & 1 & 0 & 0 \\ 0 & 1 & 0 & 0 & 0 & 0 \\ 0 & 1 & 0 & 0 & 1 & 1 \\ 0 & 0 & 0 & 1 & 0 & 0 \\ 0 & 0 & 0 & 1 & 0 & 0 \end{matrix}\right| \quad \delta_i = \begin{pmatrix} 1 \\ 3 \\ 1 \\ 3 \\ 1 \\ 1 \end{pmatrix}$$

Molecule

H_3C CH_3
CH—CH
H_3C CH_3

Graph

3 5
2—4
1 6

Distance Matrix

$$D = \left|\begin{matrix} 0 & 1 & 2 & 2 & 3 & 3 \\ 1 & 0 & 1 & 1 & 2 & 2 \\ 2 & 1 & 0 & 2 & 3 & 3 \\ 2 & 1 & 2 & 0 & 1 & 1 \\ 3 & 2 & 3 & 1 & 0 & 2 \\ 3 & 2 & 3 & 1 & 2 & 0 \end{matrix}\right| \quad d_i = \begin{pmatrix} 11 \\ 7 \\ 11 \\ 7 \\ 11 \\ 11 \end{pmatrix}$$

defined as the sum of the reciprocals of the square roots of products of the valences of the two vertices adjacent to each edge, extended to all edges of the graph.

$$\chi = \sum_i \sum_j (\delta_i \delta_j)^{-1/2} \tag{4.2}$$

The Randic index is a good measure of the branching of a graph and consequently of a given molecule. The intuitive idea of branching takes here a quantitative meaning and is therefore measurable.

In 1976 Kier and Hall extended the Randić index, introducing the connectivity indices of order *m* and type *t*. They are the first example of "family indices" and the entire set can be calculated from the adjacency matrix. They are normally written as, ${}^m\chi_t$, where m varies between 0 and *m*. Here the order is the number of connected edges (*m*) that appear in a given substructure or subgraph. Hence, the connectivity indices are defined as [19]

$${}^m\chi_t = \sum_{j=1}^{n_m} {}^m S_j \tag{4.3}$$

where

$${}^m S_j = \left[\prod_{i=1}^{m+1} \delta_i \right]^{-1/2} \tag{4.4}$$

where

δ_i is the number of simple bonds (σ bonds only) of the atom *i* to nonhydrogen atoms
S_j represents the *j*th substructure of order *m* and type *t*
${}^m n_t$ is the total number of subgraphs of order *m* and type *t* that can be identified in the molecular structure

Types used are path (*p*), cluster (*c*), and path–cluster (*pc*).

As for the presence of heteroatoms, this is recorded by assigning a weight on the main diagonal position (entry) corresponding to the heteroatom. This weight is no more than a numerical value assigned to the heteroatom to identify it. Although there are different algorithms for its calculation, the first was introduced by Kier and Hall, in which the valence of an heteroatom is defined as the difference between the number of valence electrons, *Zv*, and the number of hydrogens bound to the heteroatom, *h*, ($\delta^v = Z^v - h$).

In 1980, Alexander Balaban introduced the J index from the distance matrix D [20] as

$$J = \frac{B}{C+1} \sum_i \sum_j (d_i d_j)^{-1/2} \tag{4.5}$$

where

B = bounds number
C = cyclomatic number
d_i and d_j valence grade of vertices i, j

The values of these topological indices for the dimethylbutane are

$$W = 29;\ \chi = 2.643;\ {}^2\chi = 2.488;\ {}^3\chi_c = 1.1547;\ J = 2.993$$

Among other widely used topological descriptors are the kappa indices of molecular shape and flexibility index [21,22], electrotopological states indices [23], spectral moments, μ [24], topological charge indices [25], etc. The number of topological descriptors is actually huge and potentially much larger. Fortunately, a number of software programs are commercially available to calculate them. Among these programs are MOLCONNZ [26], DRAGON [27], POLLY [28], CODESSA [29], PADEL [30], etc.

One of MT's great advantages is that any of these programs can calculate a large number of TIs for a vast group of molecules in a very short time (minutes or even seconds). The type of format in which the information must be entered for each molecule could be MDL molfile (.mol), MDL multiple SD file (.sdf), smile (.smi), etc.

4.2.2 Statistical Techniques

Statistical calculations are essential to get a good outcome in the QSAR equations. Although there are many choices available, two types of analysis are commonly used: to predict quantitative properties (multilinear regression analysis, MLRA) or to recognize the category to which the compound belongs (linear discriminant analysis, LDA). These two kinds of statistical analysis can be refined with artificial neural networks (ANNs). ANNs take into account the nonlinear relationships existing between the analyzed property and the TIs.

4.2.2.1 Multilinear Regression Analysis

In MLRA, the TIs are correlated to the biological experimental values to get an equation:

$$P_i = A_o + \sum A_i X_i \tag{4.6}$$

where P_i is the experimental property, X_i is the TIs, and A_o and A_i are the regression coefficients of the equation obtained.

The predictability, quality, and robustness of the model can be verified by means of different types of criteria. Usually three strategies are adopted [31]:

1. Internal validation or cross-validation with leave-one-out, LOO
2. External validation
3. Data randomization or Y-scrambling

4.2.2.2 Linear Discriminant Analysis

The goal of linear discriminant analysis, LDA, is to find a linear combination of variables allowing the discrimination between two or more categories or objects. In our case, the "objects" are molecules. The final equation has the following form:

$$\mathrm{DF} = A_o + \sum A_i X_i \tag{4.7}$$

where DF is the value of the discriminant function related to a particular activity, X_i is the topological indices, and A_o and A_i are the regression coefficients relating one and others.

The most simple approach is the disjunctive, in which two sets of compounds are considered: one with proven biological activity that constitutes the "active" set and another one comprising inactive compounds. The quality of the discriminant function is evaluated by Wilks's λ [32].

The discriminant ability of the selected function is evaluated by

- The classification matrix, in which each case is classified into a group according to the classification function
- The jack-knifed classification matrix, in which each case is classified into a group according to the classification functions computed from all the data except the case being classified
- An external set test.

4.2.2.3 Artificial Neural Network

In order to account for nonlinear effects, an artificial neural network (ANN) was used as well. ANNs are information-processing algorithms that mimic the way that biological nervous systems work [33].

Most ANNs consist of at least three layers—input, hidden, and output—that are referred to as neurons. These are known as multilayer perceptrons (MLPs). The layers of input neurons receive the data from input files. The output layer sends information directly to the outside world. Between the input and output layers there may be many other hidden ones. These internal layers contain the neurons in several interconnected structures. In a supervised ANN, a training set of known output is used to train the network by assigning weights to each neuron. Based on the algorithm carrying out the training process, there are different types of neural networks, such as feed forward, back propagation; counter propagation; probabilistic neural network; or self-organizing map, just to mention some [34].

4.2.2.4 Other Statistical Techniques

In addition to the statistical techniques previously described, there are a variety of available statistical software that researchers can use for developing effective QSAR models, depending on the size of the data, the number of TIs, etc. Some of them are briefly discussed next.

The heuristic method (HM) is an advanced algorithm based on MLR, which is popular for building linear QSAR/quantitative structure–property relationship (QSPR) equations and also effective for descriptor selection before a linear or nonlinear model because of its convenience and high calculation speed. The advantage of HM is totally based on its unique strategy of selecting variables and the absence of software restrictions on the size of the data set [35].

A new method called GA-MLR is the result of a genetic algorithm (GA) combined with MLR. This method is becoming popular among current QSAR and QSPR studies. GA is performed to search the feature space and select the major descriptors relevant to the activities or properties of the compounds. GA, a well-estimated method for parameter selection, is embedded in the GA-MLR method so as to overcome the shortage of MLRs in variable selection. Like the MLR method, the regression tool in GA-MLR is a simple and classical regression method, which can provide explicit equations [35].

Partial least squares (PLS) is famous for its application to CoMFA and CoMSIA. Recently, PLS has evolved by combination with other mathematical methods to give better performance in QSAR/QSPR analyses. Three kinds of PLS are the most common:

- Genetic partial least squares (G/PLS)
- Factor analysis partial least squares (FA-PLS)
- Orthogonal signal correction partial least squares (OSC-PLS)

The interesting aspect of G/PLS is that it is derived from two QSAR calculation methods: genetic function approximation (GFA) and PLS. The G/PLS algorithm uses GFA to select appropriate basis functions to be used in a model of the data and PLS regression is used as the fitting technique to weigh the basis functions' relative contributions in the final model [35].

Support vector machine (SVM) is a novel type of machine learning method. SVMs map the input vector into a feature space of higher dimensionality in order to identify the hyperplane that separates the data points into two classes. The marginal distance between the decision hyperplane and the instances that are closest to the boundary is maximized. The resulting classifier achieves considerable generalizability and can therefore be used for the reliable classification of new samples [36].

Gene expression programming (GEP) was developed from genetic algorithms and genetic programming (GP). GEP is simpler than cellular gene progression. It mainly includes two sides: the chromosomes and the expression trees (ETs). The process of information of gene code and translation is very simple, such as a one-to-one relationship between the symbols of the chromosome and the functions or terminals they represent [35].

Project pursuit regression (PPR) and local lazy regression (LLR) are two kinds of novel regression methods. The first is a powerful tool for seeking interesting projections from high-dimensional data into lower dimensional space by means of linear projections. LLR is able to extract a prediction by locally interpolating the neighboring examples of the query, which are considered relevant according to a distance measure, rather than considering the whole data set [35].

CP-MLR is a method that enables extraction of the maximum and diverse structure–property relationship information from the parameter set considered in MLR. A combinatorial strategy with appropriately placed "filters" is adopted to recurrently select the nonrepetitive k independent variables, at a time, from a total of p variables for the model development [37].

4.2.3 Applications

In the field of the QSAR methods, researchers and scientists have striven for decades with the aim to find new and better solutions for the prevention, treatment, or vanquishing of parasitic diseases. Nowadays, the mosquito is considered the world's deadliest animal to humans, so it is not hard to understand the obligation on the part of scientific research to stem its life-threatening spread.

In Tables 4.2 and 4.3, a recompilation of the efforts made by researchers worldwide in the last decade is shown. The central theme concerns the application of molecular topology for developing QSAR models capable of predicting repellent and/or antiparasitic activity against the mosquito.

A quick analysis of the data in Table 4.2 shows that most of the studies are focused on antimalarial activity. Regarding vectors of other diseases, such as dengue, yellow fever and chikungunya, the data are missing.

Regarding the prediction of repellent or larvicidal activity, we are witnessing a diametrically opposite event. Virtually all studies are centered on the vector *Ae. aegypti* and its consequent diseases (dengue, yellow fever, and chikungunya), but only a few on malaria. An interesting hypothesis may center on the current inability to identify molecules that are active as drugs against *Aedes*—or, simply, a lack of information about molecules active against this vector. In other words, despite the numerous models for the prediction of repellent activity against *Aedes*, it is hard to find any models for the prediction of drugs to be used against it. It appears to be clear how to identify new repellent products to contain the plague, but not how to discover new drugs to treat its diseases.

4.2.3.1 Prediction Repellent and/or Antiprotozoal Activity and Search for Novel Drugs against Diseases Transmitted by Mosquitoes

In order to simplifying the reading and interpretation of the tables, the main results will be discussed (where possible) in chronological order.

In 2005, Gupta et al. [38] used topological descriptors in modeling the antimalarial activity of 4-(3′,5′-disubstituted anilino) quinolines. Authors analyzed two series of antimalarial agents using combinatorial protocol in multiple linear regression (CP-MLR) in order to identify key structural domains that may represent the core of action against *Plasmodium* transmitted diseases. The most interesting aspect of this study is the importance of topological descriptors as fundamental elements for the characterization of activity, such as MATS (Moran autocorrelation of topological structure), GATS (Geary autocorrelation of topological structure), and Galvez's class descriptor JG1, a mean topological charge index of path of length one.

The same year, our research group, led by Prof. J. Galvez, applied MT for the prediction of the potency of various insecticides active against the *Culex* mosquito [59].

TABLE 4.2
Papers Published during the Last Decade Applying Molecular Topology for Prediction of Activity and Design of New Active Compounds against Mosquito Diseases

Protozoa	Statistical Technique	Ref.
Plasmodium falciparum	CP-MLR, PLS	38
Plasmodium falciparum	MLR	39
Plasmodium falciparum	MLR, LDA	40
Plasmodium yoelii	MLR, LDA	41
Plasmodium falciparum	MLR, LDA, ANN	42,43,44
Plasmodium falciparum vivax ovale malariae	MLR	45
Plasmodium falciparum	CP-MLR, PLS	46
Plasmodium falciparum	LDA, ANN	47,48
Plasmodium falciparum	RR, PLS, PCR	49
Plasmodium falciparum vivax ovale malariae	MLR	50
Plasmodium falciparum	MLR, PLS	51
Plasmodium falciparum	LDA, MLR	52
Plasmodium falciparum	PLS	53
Plasmodium falciparum	CP-MLR	54
Plasmodium falciparum	MLR	55
Plasmodium falciparum	MLR	56
Plasmodium falciparum vivax ovale malariae	PLS	57
Plasmodium falciparum vivax ovale malariae	MLR	58

TABLE 4.3
Papers Published in the Last Decade Using Molecular Topology to Search QSAR Models for Predicting Repellent Activity and Design of New Active Compounds against Mosquitoes

Vector	Disease	Treatment	Statistical Technique	Ref.
Culex pipiens quinquefasciatus	West Nile fever, lymphatic filariasis	Larvicidal	MLR	59
Aedes aegypti	Dengue fever, chikungunya, yellow fever	Repellents	MLR	60
Aedes aegypti	Dengue fever, chikungunya, yellow fever	Repellents	RR, PLS, PCR	5,61
Aedes aegypti, albopictus	Dengue fever, chikungunya, yellow fever	Repellents	MLR	62
Aedes aegypti	Yellow fever	Repellents	MLR	63
Aedes aegypti, albopictus	Yellow fever	Repellents	MLR	64,65,66
Culex quinquefasciatus	West Nile fever	Larvicidal	MLR	67
Aedes aegypti	Dengue fever, chikungunya, yellow fever	Repellents	MFTA	68

The group of selected compounds was composed of a representative sample of the various classes of existing insecticides—for instance, the insect growth regulators (IGRs), among them methoprene, hydroprene, several butyl phenols, ureas (diflubenzuron), etc.

Insecticidal activity was expressed as LC_{50}, which is the lethal dose in parts per million causing 50% inhibition of adult emergence for larvae of *Culex pipiens quinquefasciatus*. The regression equation selected was

$$\log LC_{50} = -2.63 + 14.99\, J_3^v - 0.24\, V_4 \tag{4.8}$$

$n = 19$, $r^2 = 0.843$, $SEE = 0.467$, $F = 43.1$.

J_3^v is a topological charge index that takes into account the average charge transferred per atom at a topological distance three in the molecule. The fact that the index is weighted by the valence unveils the influence of heteroatoms such as N, O, and Cl in the insecticidal activity. The second index, V_4, is the simple sum of vertices (atoms different from hydrogen) with degree four. That includes quaternary carbons, carbonyl groups, and carbons substituted on aromatic rings. These results are all the most consistent since the activity seems to depend on the charge transfers between donor and acceptor groups as well as structural features such as steric hindrance or enhancement, encoded by the V4 index.

The following year, Katritzky et al. [60,39] published two QSAR studies in which they used CODESSA Pro software to predict respectively the repellent activity against *Ae. aegypti* and antimalarial activity against *Plasmodium*. The first study shows two statistically significant quantitative models with r^2 values above 80% for the prediction of the repellency time for a group of *N*,*N*-diethylphenylacetamide (DEPA), analogs. The second describes the antimalarial activity for several organic compounds against two strains of *Plasmodium falciparum*. Once again, the importance of the topological descriptors was confirmed, since five of the seven most significant indices of the predictive equations had a topological nature.

In 2006 our research group, continuing its research in the field of antimalarial agents, developed a QSAR study using a database of 395 compounds previously tested against chloroquine-susceptible strains of blood stage of *P. falciparum* to predict new *in vitro* antimalarial drugs [40]. Various topological indices were used as structural descriptors and were related to antimalarial activity by using linear discriminant analysis and multilinear regression. Two discriminant equations were obtained, which enabled a successful classification up above 85%. The values of IC_{50} were considered in order to find a multilinear regression equation model suitable to predict their *in vitro* activities.

By the application of the model, 27 drugs against a chloroquine-susceptible clone (3D7) of *P. falciparum* have been selected and evaluated *in vitro*. Among those drugs are monensin, nigericin, vincristine, vindesine, ehylhydrocupreine, and salinomycin with *in vitro* IC_{50} of 0.3, 0.4, 2, 6, 26, and 188 nM, respectively. Other compounds, such as hycantone, amsacrine, aphidicolin, bepridil, amiodarone, ranolazine, and triclocarban, showed IC_{50} *in vitro* values below 5 μM.

Along the same line, in collaboration with Mazier, we developed a novel QSAR study [41], using a database of 127 compounds previously tested against the liver stage of *P. yoelii* in order to build up a model capable of predicting the *in vitro* antimalarial activities of new compounds. A topological model consisting of two discriminant functions was then created. The first function discriminates between active and inactive compounds while the second is capable of identifying the most active compound.

The model was finally applied to a database of compounds with unknown activity against liver stages of *Plasmodium*. Seventeen drugs predicted as active or inactive were selected for testing *in vitro* against the hepatic stage of *P. yoelii*. Antiretroviral (delavirdine), antifungal (miconazole), and cardiotonic (dobutamine) drugs were found to be highly active (nanomolar 50% inhibitory concentration values), and two ionophores (monensin and nigericin) completely inhibited parasite development (see Figure 4.2). For both ionophores, the same *in vitro* assays as the ones for *P. yoelii*, confirmed their *in vitro* activities against *P. falciparum*.

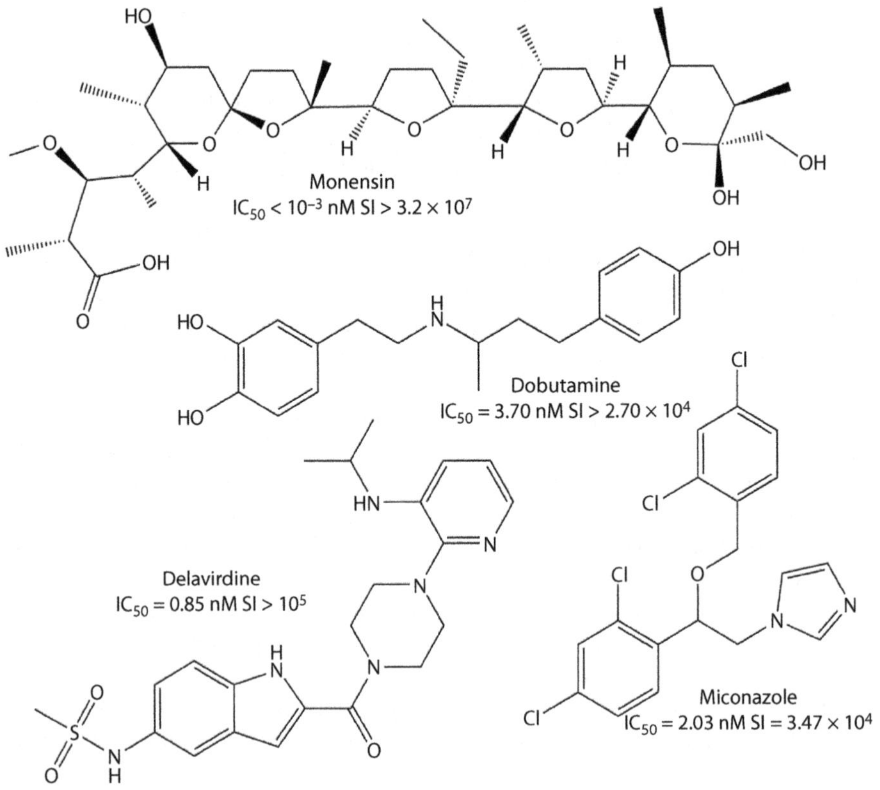

Monensin
$IC_{50} < 10^{-3}$ nM SI > 3.2×10^7

Dobutamine
$IC_{50} = 3.70$ nM SI > 2.70×10^4

Delavirdine
$IC_{50} = 0.85$ nM SI > 10^5

Miconazole
$IC_{50} = 2.03$ nM SI = 3.47×10^4

FIGURE 4.2 Drugs selected by molecular topology with high activity against the hepatic stage of *P. yoelii in vitro*, $IC_{50} < 5$nM, and excellent selectivity index, SI > 10^4.

Among the selected compounds, monensin, with its high activity on parasite development liver stage ($IC_{50} < 10^{-3}$ nM) and low toxicity (SI > 10^7), could be considered an excellent alternative as prophylactic for people who travel to malaria-endemic countries. As a matter of fact, in 2009 its corresponding patent application was filed [69].

It is evident that these results demonstrate the usefulness of MT for the selection and design of new lead drugs active against *P. falciparum*.

In 2007, Natarajan et al. [61] introduced a novel class of topological indices for the numerical characterization of molecular chirality. It is known that chiral compounds offer vast opportunities for the identification of pharmacologically active compounds. In this study, the authors identified new topological descriptors able to describe chirality as a continuous measure. These TIs were applied to develop various structure–activity relationship models, considering repellency data for the diastereomers of picaridin and AI3-37220.

Continuing the research in the field of mosquito repellents, in 2008, Natarajan and his team [61] identified a four-parameter model with a 20-fold cross-validated r^2 of 0.734 for the detection of active repellents against *Ae. aegypti* and *An. quadrimaculatus*. Topochemical and pure topological descriptors were the most significant ones.

In 2008, our team concluded a complex study for a group of 81 new uracil-based acyclic and deoxyuridine compound inhibitors of (dUTPase) with antimalarial activity against the chloroquine-resistant K1 strain of *P. falciparum* [42]. By applying linear discriminant and multilinear regression analysis, a predictive model with two functions was capable of adequately predicting the IC_{50} for each compound of the training and test series.

The discriminant function selected was

$$\mathrm{DF} = 29.76 + 9.60\, G_5 - 45.48\, J_1^v - 1.04 \frac{{}^4\chi_c}{{}^4\chi_c^v} \tag{4.9}$$

$n = 54$, $F = 121$, λ(Wilks's lambda) = 0.121.

The selected prediction function was

$$\log \mathrm{IC}_{50} = 3.15 - 0.34\ {}^1\chi^v + 0.38\ {}^4\chi_{pc} \tag{4.10}$$

$n = 54$, $r^2 = 0.842$, $Q^2 = 0.825$, $SEE = 0.308$, $F = 136.4$, $p < 0.0001$.

Based on the results from LDA and MLR, a topological model for the search for novel antimalarial agents can be set up, considering these requirements: if DF > 10, DF < 25, and log $IC_{50} < 1$, then the compound will be labeled as a potential antimalarial. Otherwise, the compound is classified as inactive.

After carrying out a virtual screening based upon such a model, new, potentially active structures against *P. falciparum* were proposed. Some of the compounds—CAS 104375-88-4, CAS 40615-39-2, CAS 6554-10-5, CAS 172469-19-1, and CAS 166829-96-5—previously showed some antiviral activity. Others, such as CAS

23669-79-6, showed anti-*Leishmania* activity. The discovery of novel applications for these compounds (i.e., antimalarial activity) could be very interesting since it may spread their therapeutical landscape.

Returning to the topic of mosquito repellents, 2008 and 2009 were fructiferous years. A relevant paper by Wang et al. [62] presented a statistical model by CODESSA software able to identify a new class of low-toxicity mosquito repellents against *Ae. aegypti* and *Ae. albopictus*. The new class was characterized as being terpenoids with a six member ring (pinene) (see Figure 4.3).

To endorse this study, our team, in a paper by García-Domenech et al. [64], employed MLRA in order to introduce a topological-mathematical model for searching new terpenoids as active repellents against mosquitoes. The structural description was obtained with topological indices and the model validated by cross-validation, internal validation, and randomization test. The results confirmed the model's capability to predict the analyzed property.

To summarize, a group of 20 compounds synthesized from α- and β-pinene were used to study the repellent activity against *Ae. albopictus* mosquito. Repellent ratio, RR, was determined by applying test compounds at a dose of 0.16 mg/cm^2 to the depilated abdomen of a white mouse. The best model was

$$\log \mathrm{CRR} = 1.387 - 0.535 J_1^v + 0.666 \frac{{}^2\chi}{{}^2\chi^v} + 0.112 \frac{{}^4\chi_c}{{}^4\chi_c^v} - 0.076 PR2 \tag{4.11}$$

$n = 20$, $r^2 = 0.967$, $SEE = 0.020$, $Q^2 = 0.943$, $F(4,15) = 110$, $p < 0.00000$, where CRR is the corrected repellent ratio (CRR = ((RR-CERR)/(100-CERR))*100) being CERR, the repellent ratio of the control experiment.

Once validated, the model was applied to the search of new terpenoids with potential repellent activity. Therefore, a virtual screening of the Scifinder Scholar database was carried out to find new molecular candidates. Figure 4.4 depicts the predicted values for the repellency activity in the selected compounds.

Among them are four candidates with repellent activity above 80%—namely, CAS 18368-91-7 with (classification repellent ratio) CRR_{pred} = 91.3%; CAS 111957-78-9 with CRR_{pred} = 90.7%; CAS 101666-14-2 with CRR_{pred} = 87.4%, and CAS 111957-81-4 with CRR_{pred} = 81.7%. Furthermore, none of them had been described

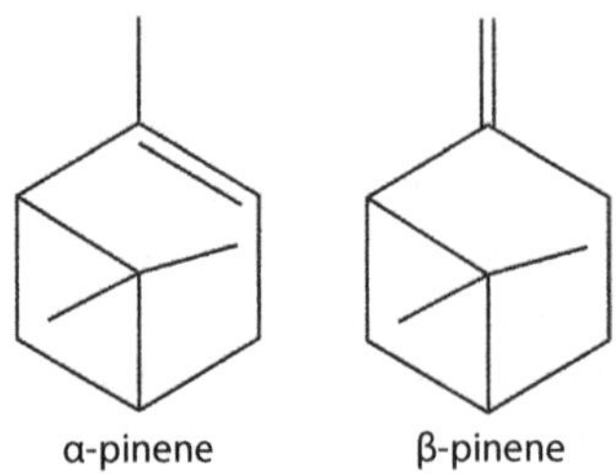

FIGURE 4.3 Structures of α-pinene and β-pinene.

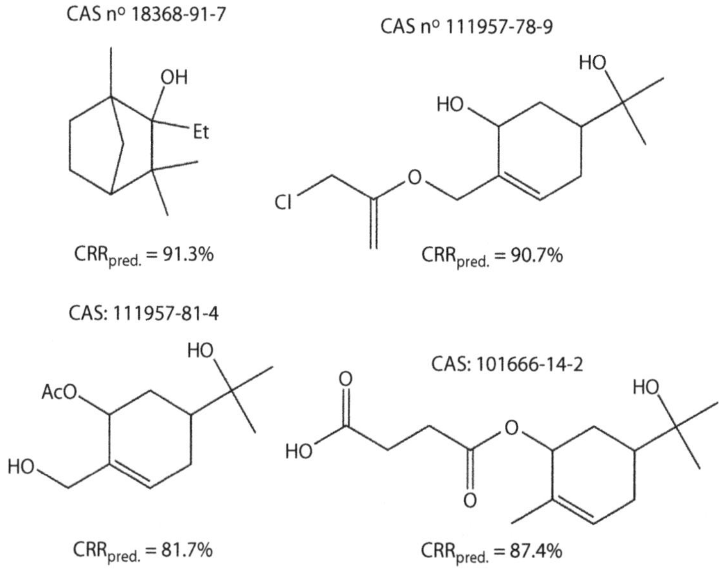

CAS n° 18368-91-7

$CRR_{pred.} = 91.3\%$

CAS n° 111957-78-9

$CRR_{pred.} = 90.7\%$

CAS: 111957-81-4

$CRR_{pred.} = 81.7\%$

CAS: 101666-14-2

$CRR_{pred.} = 87.4\%$

FIGURE 4.4 New compounds selected by MT, potentially active as repellents against *Ae. albopictus* mosquito.

previously as mosquito repellents. Compound CAS 101666-14-2 is known as antipyretic and could be a good candidate for experimental testing.

In 2009 Paluch et al. [63] described botanical sesquiterpenes as potential repellents against yellow fever mosquito. After isolating a class of 12 compounds of sesquiterpene derivatives, the authors assayed them for *Ae. aegypti* and used the data to build a QSAR model capable of identifying key properties of those derivatives. The most interesting aspect was the repellency mechanism of action lying in a group of topochemical and topological descriptors such as vapor pressure, LUMO energy, Mulliken population, and electrotopological state. Results from the study highlight the importance of electronic properties at the time to express repellency activity against mosquitoes.

Following the search for natural products active against mosquitoes, our research group developed various models using molecular topology jointly with MLRA for predicting the spatial repellency against the yellow fever mosquito, *Ae. aegypti*, from a group of natural sesquiterpenes [65]. Once the models were validated, molecular screening was performed for the election of new, potentially active sesquiterpenes (Figure 4.5). Some of the selected compounds also possessed antiparasitic activity (artemisinin as antimalarial, helenine as anthelmintic, and acaricidal or cedrol), thereby highlighting even more their potential as parasiticides.

An original study for the development of QSAR predictive models of *in vivo* antimalarial activity is the one by Srivastava et al. [45]. Here, the analyses of approximately 150 artemisinin analogs were conducted in order to find antimalarial activity. For the construction of the model, E-state, electronic, structural, topological quantum mechanics, and physicochemically based descriptors were used, such as the Balaban topological index and strain energy of the molecules.

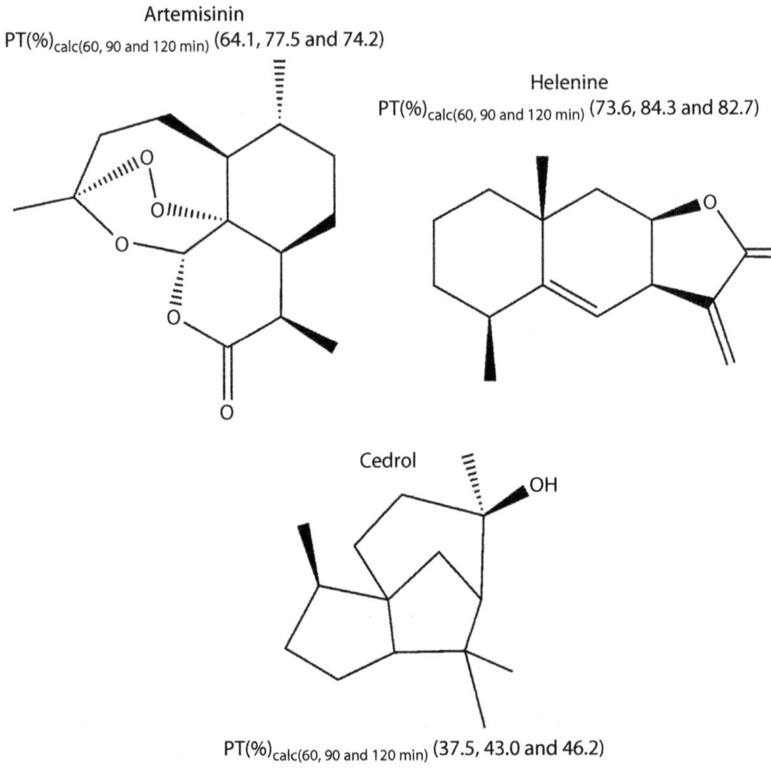

Artemisinin
$PT(\%)_{calc(60,\,90\text{ and }120\text{ min})}$ (64.1, 77.5 and 74.2)

Helenine
$PT(\%)_{calc(60,\,90\text{ and }120\text{ min})}$ (73.6, 84.3 and 82.7)

Cedrol
$PT(\%)_{calc(60,\,90\text{ and }120\text{ min})}$ (37.5, 43.0 and 46.2)

FIGURE 4.5 New compounds selected by MT, potentially active as repellents against *Ae. aegypti* mosquito.

The same year, our group [66] selected a topological-mathematical model based on MLRA to predict the mosquito repellent activity against *Ae. aegypti* (protection time, TP), for a group of 31 *N,N*-diethyl-*m*-toluamide derivatives (see Figure 4.6). The structural depiction was performed by using topological indices and a model with four variables for the prediction of log TP ($r^2 = 0.875$) was finally selected:

$$\log \mathrm{TP} = 8.40 - 0.91\, G_2 + 39.04\, J_4 - 7.80\frac{{}^3\chi_p}{{}^3\chi_p^v} + 6.24\left({}^4\chi_{pc} - {}^4\chi_{pc}^v\right) \quad (4.12)$$

$n = 31$, $r^2 = 0.875$, $SEE = 0.188$, $Q^2 = 0.818$, $F = 45$, $p < 0.0001$.

In the equation appear the topological charge indices G_2 and J_4, that evaluate the intramolecular charge transfer, and the branching descriptors ${}^4\chi_{pc}$ and ${}^4\chi_{pc}^v$. The model was validated by a cross-validation internal test and final results confirm its capability to predict the analyzed properties (with a prediction coefficient of $Q^2 = 0.818$).

Another proof of the effectiveness of topological descriptors lies in the results obtained by Deshpande et al. [46]. In this study, the antimalarial activity of

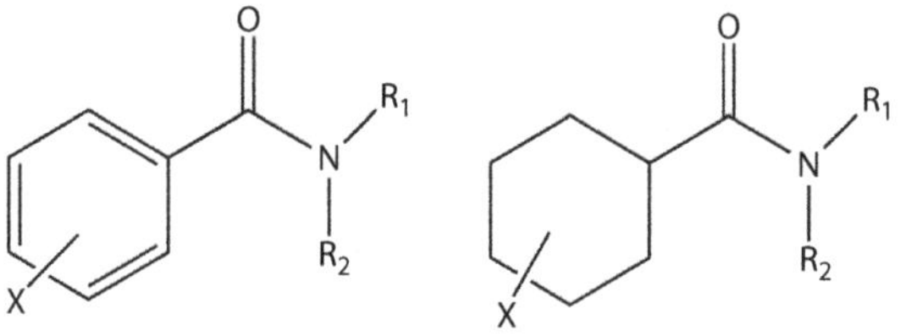

FIGURE 4.6 Structures of *N,N*-diethyl-*m*-toluamide derivatives.

N1-(7-chloro-4-quinolyl)-1,4-bis(3-aminopropyl) piperazine analogs was correlated to several descriptors, from constitutional to topological ones. The descriptors were calculated with DRAGON software and correlated by PLS analysis. Among the selected descriptors, "topological radial centric information," "number of double bonds," and "hydrophobicity" seem to be the most influential in predicting the antimalarial activity of the analogs. A summary of the most significant indices comprises ICR, GMTIV (Gutman molecular topological index by valence vertex degrees), SOK (Kier symmetry index), and IVDE (mean information content on the vertex degree equality) along with the 2D autocorrelation descriptors MATS6e and MATS8m.

In terms of novelty, a study by Gonzalez-Diaz et al. [47] about pharmacological activity on antimalarial properties is noteworthy. The authors presented a multitarget QSAR model constructed through the fusion of two software programs: MARCH-INSIDE and DRAGON. This novel technique enables the calculation of a new predictor for protein–ligand interactions. The best ANN model found is a multilayer perceptron (MLP) with profile MLP 21-31-1, which showed a sensitivity > 89% and a specificity >94%, corresponding to training accuracy over 92%.

In 2010 the same authors [48] applied Markov chain theory to calculate new multitarget spectral moments to get a QSAR model using spectral moments that predicted, for the first time, a mt-QSAR for 500 drugs tested in the literature against 16 parasite species, among which was *Plasmodium*, and another 207 drugs not tested in the literature. The goal was to identify those molecular fragments key for the activity of a given compound. As for the *Plasmodium*, the "aniline" fragment was the most significant one. The overall predictability performance of the model was 89.2%.

In 2011 Basak et al. [49] used two classes of graph-theoretic molecular descriptors—topological indices (TIs) and atom pairs (APs)—to derive high-quality, quantitative structure–activity relationships for inhibitors of dihydrofolate reductases (DHFRs) isolated from the wild and four mutant strains of *P. falciparum*. Through the application of principal components regression (PCR), PLS, and ridge regression (RR), they demonstrated that pairwise kappa values used in conjunction with critical macromolecules (enzymes or receptors) and collections of structurally broad sets of ligands for the respective biotarget(s) may be a useful tool in gauging the evolving mutual similarities/dissimilarities of mutating organisms from a purely computational chemistry point of view.

Roy et al. [50] published in 2011 an innovative study aimed at reducing the "noise" in terms of number of descriptors used to build regression equations, so as to focalize and improve their predictive performance.

The same year, our team applied MT and MLRA to find QSAR models for the prediction of the *in vitro* antiprotozoal activity against *P. falciparum* as well as the cytotoxicity of a group of pentamidine analogs [43]. After the models' validation, a molecular screening was carried out for searching out new compounds showing high activity and low cytotoxicity profiles.

The regression model selected was

$$\text{pIC}_{50} = 9.85 - 3.78\,G_3 + 1.22\,G_4^v + 44.41\,J_3^v + 0.60\,PR2 - 0.44\,PR3 \qquad (4.13)$$

$n = 35$, $r^2 = 0.8228$, $SEE = 0.294$, $Q^2 = 0.7442$, $F\,(5,29) = 26.9$, $p < 0.00001$.

Some of the selected compounds (Figure 4.7, comp. S9 and S13) exhibit an excellent activity profile, with $IC_{50} < 50$ nM (a value comparable to that of the reference drugs, pentamidine and chloroquine). If we consider that these molecules show also a suitable cytotoxicity profile (i.e., high SI values), then it turns these compounds into interesting antiprotozoal candidates.

The following year, Paliwal et al. [51] developed a QSAR study on 43 benzofuran derivatives with antimalarial activity using MLR and PLS. Satisfactory statistical values of $r^2 = 0.822$ and $r^2_{cv} = 0.783$ were obtained by MLR and a comparable $r^2_{cv} = 0.787$ by PLS. Among the most relevant descriptors were the ${}^4\chi^v_p$ Kier index and the Ipso atom E-state index.

A work by Pasquale et al. [67] led to the identification of chalcone derivatives as potential candidates for larvicidal activity. After calculating the descriptors

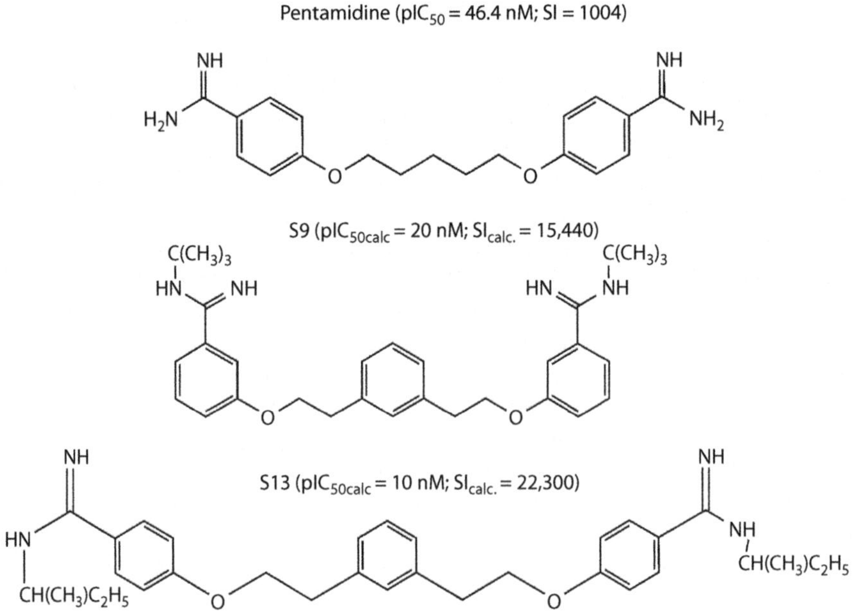

FIGURE 4.7 New pentamidine derivatives selected by MT, potentially active as antimalarics against *P. falciparum*.

with DRAGON software, using more than a thousand constitutional, topological, geometrical, and electronic molecular descriptors, the authors used a replacement method, an algorithm for generating multivariable linear regression QSAR models with minimized standard deviation (s), as the molecular descriptor selection approach.

The year 2013 represents one of the most prolific in the field of antiprotozoal QSAR investigation. Gupta [54] investigated the antimalarial activity and cytotoxicity of a quinoline family, through a combinatorial protocol in multiple linear regression (CP-MLR) approach. The regression resulted in the identification of various descriptors, using a three-stage classification. At the end of the search, 39 descriptors emerged as the most significant ones. A partial least squares analysis was carried out on these 39 descriptors to generate a "single window" structure–activity relationship. The PLS model elucidated 82.1% of predictive (LOO) as well as 85.6% of explained variance in the antimalarial activity. In the study of the CQ-sensitive 3D7 *Plasmodium* strain, the descriptors X3Av, MATS1p, GATS2p, BEHm4, and nNH2Ph were conducive to activity, while MATS6v and nCONHR were detrimental to activity. In the case of the CQ resistance strain, the descriptors IVDE, MATS8e, MATS8v, nNHR, nNH2Ph, and N-069 positively influenced the activity, whereas nCONHR showed negative influence on activity. These results reinforce the idea that topological indices play a role of great importance in the expression of antimalarial activity.

Another interesting study by Masand et al. [55] involved synthetic prodginines as potential antimalarial compounds against *P. falciparum*. The study involved multilinear regression and CoMFA analysis and demonstrated that the synergy between various techniques is essential to obtain fruitful results in assessing antimalarial activity. The best 3D-QSAR regression model showed an r^2 greater that 0.9 according to

$$\begin{aligned}\mathrm{pIC}_{50} = &-29.71 + 6.01\,ATS7v - 0.91\,{}^{2}\chi - 4.44\,GATS7p \\ &- 0.004Mor01m + 6.75\,ATS5v \end{aligned} \tag{4.14}$$

$n = 42$, $r^2 = 0.924$, $SEE = 0.372$, $Q^2 = 0.901$.

Topological indices such as GATS7p, ATS5v, and ATS7v are the most relevant.

The antimalarial activity of prodiginines against *P. falciparum* strains was also studied by Singh et al. [56]; the results are shown in Figure 4.8. The study indicated that the inertia moment at length 2, ${}^{6}\chi_{ch}$ Kier index, κ_3 kappa index, and the Wiener topological index plays an important role in antimalarial activity against the D6 strain, whereas descriptors as the inertia moment at length 2 and H-bond donors play important roles in antimalarial activity against the Dd2 strain. All developed QSAR models showed good correlation coefficients ($r^2 > 0.7$), higher F values ($F > 20$), and acceptable predictive power ($Q^2 > 0.6$).

A paper by Sharma et al. [57] presented a QSAR and pharmacophore study performed on 35 azaaurones (Figure 4.8) derivatives, in order to identify structural requirements for their antimalarial activity. The statistically significant 2D-QSAR model, with $r^2 = 0.906$ and $Q^2 = 0.815$, was developed by PLS and confirmed a

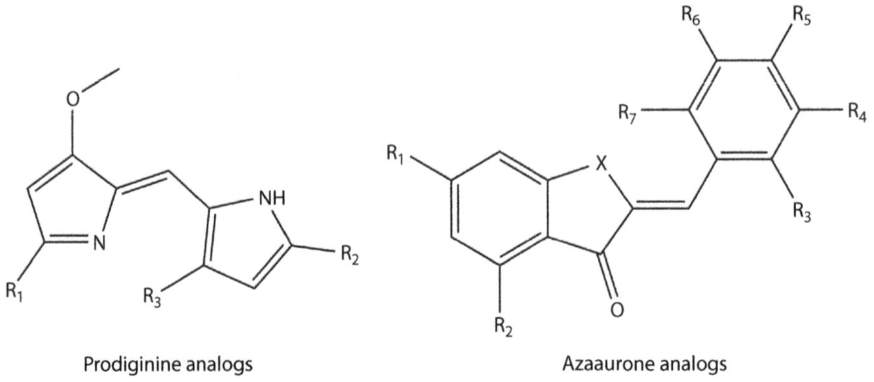

FIGURE 4.8 Structures of prodiginine and azaaurone analogs.

positive contribution of electrotopological indices, SsOHcount, and $SsCH_3$ index to antimalarial activity.

If 2013 was prolific in terms of research of novel antimalarial compounds; it was also centered in the search for new repellents. Oliferenko et al. [68] presented a paper in which they investigated the possibility to design novel repellents against *Ae. aegypti* using molecular field topology analysis, scaffold hopping, and molecular docking.

Even today, the research in the field of malaria treatment and plague control does not stop. A 2015 paper by Kurdekar and Jadhav [58] was based on the development of a python script that effectively uses descriptor and activity data in building and validating the best QSAR model. Topological descriptors like Wiener (W), Balaban (J), and Szeged (Sz) indices and quantum descriptors like energy of HOMO (EH) and energy of LUMO(EL) were selected as main pieces for the construction of the script. The models derived represented a potential opportunity for the prediction of antimalarial activity, among other pharmacological targets.

Finally, to conclude this bibliographic review, in 2015, our research group developed a multitarget QSAR able to simultaneously predict the activity of a given drug against three parasites with a high prevalence rate in humans: *P. falciparum, Trypanosoma brucei rhodesiense*, and *Leishmania donovani* [44]. The mt-QSAR study was conducted with a group of benzyl phenyl ether diamine derivatives following the approach by Speck-Planche et al. [70], applying statistical techniques such as MLRA and ANN.

The model selected showed a high-level prediction with variance values for the training, test, and external test of $r^2 = 0.964$, 0.924, and 0.900, respectively. Figure 4.9 shows the values of pCI_{50exp} versus pCI_{50calc} for all groups.

By analyzing the data obtained during the last 10 years, it is possible to distinguish two lines of research strongly marked. The first line aimed to research and develop new treatments to fight malaria, especially focused against the *Plasmodium* parasites. The second was centered in the effort to identify novel repellents effective against the *Ae. aegypti* vector.

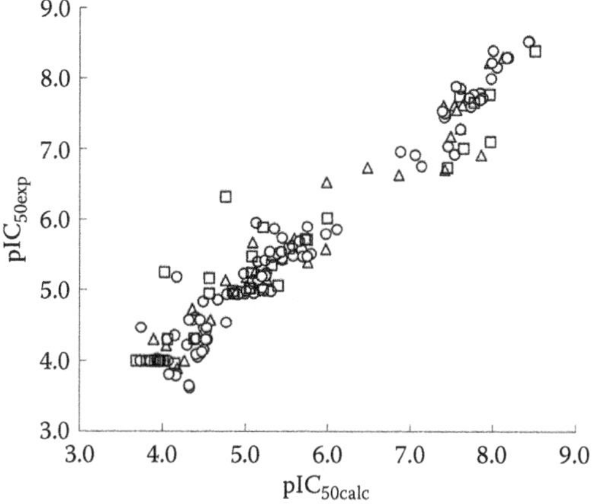

FIGURE 4.9 Plot of pCI_{50exp} versus pCI_{50calc} obtained with the mt-QSAR model selected by ANN (circles: training set; triangles: test set; squares: external test set).

4.2.3.2 Practical Example of Multitarget QSAR Study Applied to Repellent Activity

As already mentioned, multitarget QSAR became an avant-garde application in the QSAR field. The opportunity to develop models for the simultaneous prediction of different kinds of activities using the same statistical equations is undoubtedly useful.

A practical example of mt-QSAR application for the prediction of repellent activity of a group of amide trifluoromethylphenyl derivatives against *Ae. aegypti* and *An. albimanus* is shown in Table 4.4, which shows the MED (minimum effective dosage) values for each compound investigated. The MED is a measurement used to estimate the concentration level of repellent that fails to prevent mosquito bites equivalent to an ED_{99} [71].

In the absence of quantitative values of repellent activity for all compounds and vectors, linear discriminant analysis was used as the QSAR technique in order to classify each one in terms of repellent activity. To carry out the LDA analysis, it is necessary to prepare the data set since the repellent activity is different depending on the compound and the vector tested. The following steps are necessary:

- First, the 20 compounds analyzed were divided in two groups according to the vectors studied.
- Second, the output variable—namely, repellent activity, is a discriminant variable whose value is 1 (r) if MED < 25 μmol/cm^2 or 2 (nr) if MED > 25 μmol/cm^2.
- Third, to discriminate the activity of each compound as a function of the mosquito studied, it is necessary to introduce specific descriptors to identify each vector. This problem can be solved by inputting specific variables for each mosquito class by applying the approach of Speck-Planche et al.

TABLE 4.4
Structures and Experimental and Predicted Values of Repellent Activity against *Ae. aegypti* and *An. albimanus* for a Group of Trifluoromethylphenyl Amide Derivatives

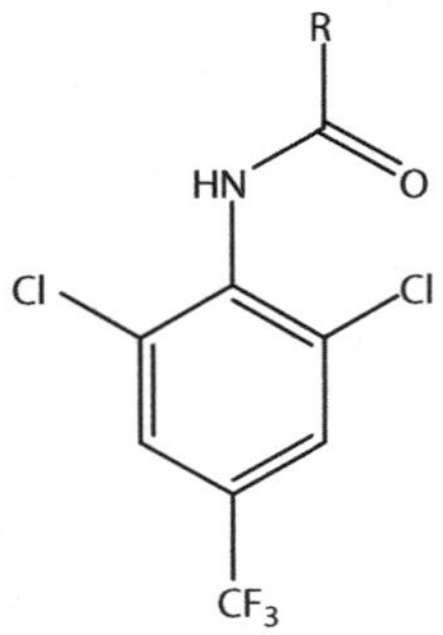

R, HN, O, Cl, Cl, CF_3

Structure 1

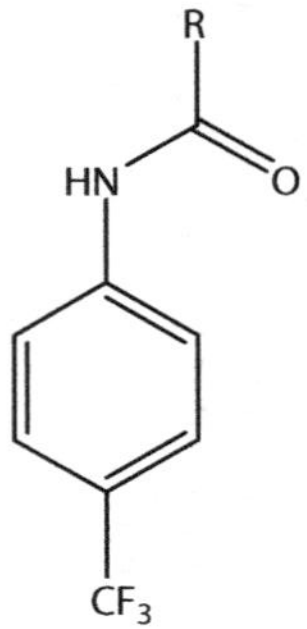

R

Series A
Series C: R-CF_3

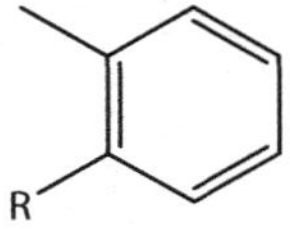

R, HN, O, CF_3

Structure 2

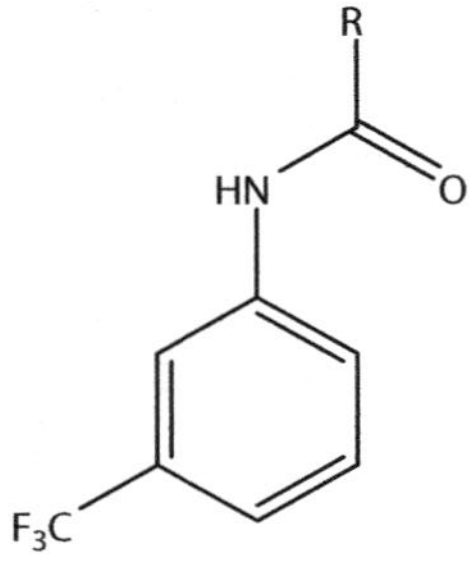

R

Series B

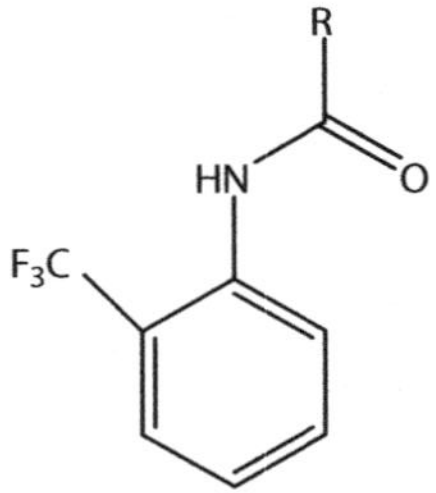

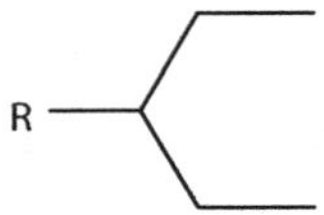

R, HN, O, F_3C

Structure 3

R

Series D

R, HN, O, F_3C

Structure 4

NO_2, R, NO_2

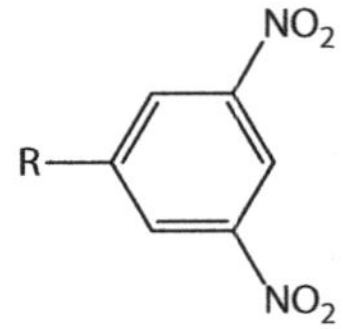

Series E

(Continued)

TABLE 4.4 (CONTINUED)
Structures and Experimental and Predicted Values of Repellent Activity against *Ae. aegypti* and *An. albimanus* for a Group of Trifluoromethylphenyl Amide Derivatives

	Aedes aegypti			*Anopheles albimanus*		
Comp.	**MED[a]**	**DF[b]**	**Prob(rep)[c]**	**MED[a]**	**DF[b]**	**Prob(rep)[c]**
1A	nr	–2.63	0.067	nr	–5.46	0.004
1B	nr	–6.11	0.002	nr	–8.93	0.000
1C	2.1	7.32	0.999	3.1	4.50	0.989
1D	nr	–5.91	0.003	nr	–8.73	0.000
1E[d]	1.9	–0.79	0.311[d]	nr	–3.62	0.026
2A	16.7	9.11	1.000	12.5	6.28	0.998
2B	nr	–8.41	0.000	nr	–11.23	0.000
2C[d]	9.2	6.82	0.999	nr	3.99	0.982[d]
2D	nr	–5.21	0.005	nr	–8.04	0.000
2E	nr	–6.58	0.001	nr	–9.40	0.000
3A	18.7	8.98	1.000	25	6.15	0.998
3B	nr	–8.49	0.000	nr	–11.31	0.000
3C	0.4	5.96	0.997	3.3	3.14	0.958
3D	nr	–5.32	0.005	nr	–8.14	0.000
3E	nr	–6.87	0.001	nr	–9.70	0.000
4A	0.1	11.71	1.000	3.1	8.89	1.000
4B	nr	–8.32	0.000	nr	–11.15	0.000
4C	0.04	10.53	1.000	1.4	7.70	1.000
4D*	0.5	–4.63	0.010*	nr	–7.45	0.001
4E	nr	–6.25	0.002	nr	–9.07	0.000
DEET	0.1			0.4		

[a] Minimum effective dosage (μmol/cm^2); nr = not repellent up to 25 μmol/cm^2. Source: M. Tsikolia et al., Pestic. Biochem. Physiol. 107 (2013), pp. 138–147.
[b] Values of the discriminant function obtained from Eq. 4.15.
[c] Values of the probability repellent activity.
[d] Values corresponding to predicted failures.

[68]. For doing that, the average values of each index, avgχ_i, avgG_i, were input for all compounds that were active (r) on the same mosquito. Also, the deviation of the χ_i and G_i index from the respective vector, desχ_i, desG_i (χ_i-avgχ_i and G_i-avgG_i), was calculated.

The selected discriminant function was

$$DF = 6.11 + 4.43\, des\, G_1^v - 93.44\, des \frac{{}^3\chi_p}{{}^3\chi_p^v} + 19.80\, des \frac{{}^3\chi_c}{{}^3\chi_c^v} \quad (4.15)$$

$n = 40, \lambda = 0.234, F(3,36) = 39.2, p < 0.00001.$

The statistical significance of the DF is high, $p < 0.00001$, and discriminatory, $\lambda = 0.234$. It involves the topological charge index G_1^V that evaluates intramolecular charge transfer, along with the connectivity indexes ${}^3\chi_c$ and ${}^3\chi_p$, which take into account the topological aspects of assembly and molecular branching. This function classifies correctly 87% of the repellent compounds and 96% of the nonrepellent compounds (in Table 4.4, observed failures are marked with *).

The quality of the discriminant function is supported by high values of AUC = 0.867 and Matthews correlation coefficient (MCC) = 0.839, as shown in the receiver operating characteristic (ROC) curve graph in Figure 4.10.

The domain of applicability of the selected model can be observed in the distribution diagram (see Figure 4.11). The highest probability of finding compounds with repellent activity is in the range of DF between 0 and 14.

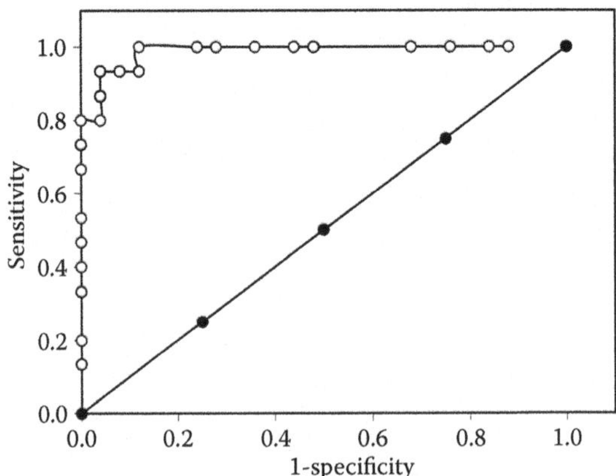

FIGURE 4.10 Receiver operator curve ROC for the LDA model. The diagonal line represents a model with no classification power in predicting binary outcomes.

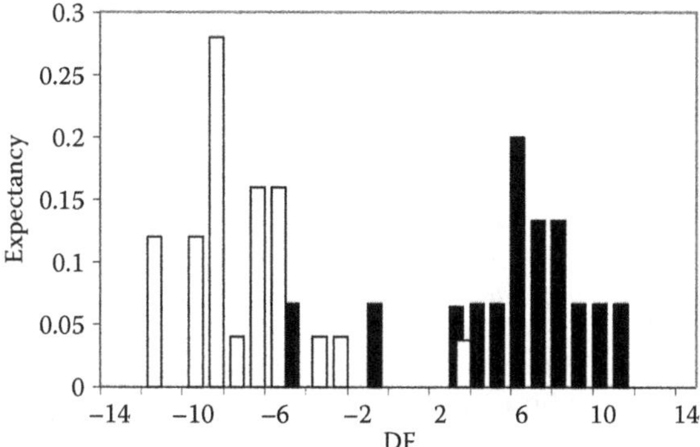

FIGURE 4.11 Pharmacological distribution diagram for repellent activity obtained using the discriminant function DF. (The black color represents the compounds with repellent activity and the white color the compounds without it.)

4.3 CONCLUSIONS

1. Molecular topology has proven to be an effective tool for the prediction of antimosquito activity of various organic compounds.
2. The use of new techniques, such as multitarget QSAR together with effective statistical tools, such as multilinear regression, discriminant analysis, or artificial neural networks, has proven essential to obtain the most reliable results in antimosquito activity.
3. The use of suitable molecular topological descriptors, among which are some introduced by our research group such as the topological charge indices, is also a basic condition.
4. Finally, based on what is stated in conclusions 2 and 3, topology mathematical models have been developed to allow a direct application in the search for new molecular candidates effective in controlling diseases caused by mosquitoes.

REFERENCES

[1] World Health Organization, *World Malaria Report, 2012. Geneva: WHO, 2012*, See http://www.who.int/malaria/publications/world_malaria_report_2013/wmr2013_country_profiles.pdf (2014).

[2] C.J. Murray, L.C. Rosenfeld, S.S. Lim, K.G. Andrews, K.J. Foreman, D. Haring, N. Fullman, M. Naghavi, R. Lozano, and A.D. Lopez, *Global malaria mortality between 1980 and 2010: A systematic analysis*, Lancet 379 (2012), pp. 413–431.

[3] L.C. Norris, B.J. Main, Y. Lee, T.C. Collier, A. Fofana, A.J. Cornel, and G.C. Lanzaro, *Adaptive introgression in an African malaria mosquito coincident with the increased usage of insecticide-treated bed nets*, Proc. Natl. Acad. Sci. U.S.A. 112 (2015), pp. 815–820.

[4] H.M. Patel, M.N. Noolvi, P. Sharma, V. Jaiswal, S. Bansal, S. Lohan, S.S. Kumar, V. Abbot, S. Dhiman, and V. Bhardwaj, *Quantitative structure–activity relationship (QSAR) studies as strategic approach in drug discovery*, Med. Chem. Res. 23 (2014), pp. 4991–5007.
[5] R. Natarajan, S.C. Basak, D. Mills, J.J. Kraker, and D.M. Hawkins, *Quantitative structure–activity relationship modeling of mosquito repellents using calculated descriptors*, Croat. Chem. Acta. 81 (2008), pp. 333–340.
[6] S.C. Basak, D.R. Mills, A.T. Balaban, and B.D. Gute, *Prediction of mutagenicity of aromatic and heteroaromatic amines from structure: A hierarchical QSAR approach*, J. Chem. Inf. Comput. Sci. 41 (2001), pp. 671–678.
[7] J. Devillers and A.T. Balaban, *Topological Indices and Related Descriptors in QSAR and QSPR*, Gordon & Breach, the Netherlands, 1999.
[8] L. Pogliani, *From molecular connectivity indices to semiempirical connectivity terms: Recent trends in graph theoretical descriptors*, Chem. Rev. 100 (2000), pp. 3827–3858.
[9] O. Ivanciuc, T. Ivanciuc, and A.T. Balaban, *Quantitative structure–property relationship study of normal boiling points for halogen-/oxygen-/sulfur-containing organic compounds using the CODESSA program*, Tetrahedron 54 (1998), pp. 9129–9142.
[10] H. Hosoya, M. Gotoh, M. Murakami, and S. Ikeda, *Topological index and thermodynamic properties. 5. How can we explain the topological dependency of thermodynamic properties of alkanes with the topology of graphs?* J. Chem. Inf. Comput. Sci. 39 (1999), pp. 192–196.
[11] R. García-Domenech, J. Gálvez, J.V. de Julián-Ortiz, and L. Pogliani, *Some new trends in chemical graph theory*, Chem. Rev. 108 (2008), pp. 1127–1169.
[12] S.C. Basak, D. Mills, B.D. Gute, and R. Natarajan, *Predicting pharmacological and toxicological activity of heterocyclic compounds using QSAR and molecular modeling*, in *QSAR and Molecular Modeling Studies in Heterocyclic Drugs I*, S.P. Gupta, ed., Springer, 2006, pp. 39–80.
[13] A.T. Balaban, I. Motoc, D. Bonchev, and O. Mekenyan, *Topological indices for structure-activity correlations*, in *Steric Effects in Drug Design*, Topic Drug Design 114 (1983), pp. 21–55.
[14] E. Estrada and E. Uriarte, *Recent advances on the role of topological indices in drug discovery research*, Curr. Med. Chem. 8 (2001), pp. 1573–1588.
[15] Y.M. Ponce, *Total and local (atom and atom type) molecular quadratic indices: Significance interpretation, comparison to other molecular descriptors, and QSPR/QSAR applications*, Bioorg. Med. Chem. 12 (2004), pp. 6351–6369.
[16] A.Z. Dudek, T. Arodz, and J. Galvez, *Computational methods in developing quantitative structure–activity relationships (QSAR): A review*, Comb. Chem. High Throughput Screen. 9 (2006), pp. 213–228.
[17] H. Wiener, *Structural determination of paraffin boiling points*, J. Am. Chem. Soc. 69 (1947), pp. 17–20.
[18] M. Randic, *Characterization of molecular branching*, J. Am. Chem. Soc. 97 (1975), pp. 6609–6615.
[19] L.B. Kier, W.J. Murray, M. Randić, and L.H. Hall, *Molecular connectivity V: Connectivity series concept applied to density*, J. Pharm. Sci. 65 (1976), pp. 1226–1230.
[20] A.T. Balaban, *Highly discriminating distance-based topological index*, Chem. Phys. Lett. 89 (1982), pp. 399–404.
[21] L.B. Kier, *Shape indexes of orders one and three from molecular graphs*, Quant. Struc.–Act. Relat. 5 (1986), pp. 1–7.
[22] L.B. Kier, *An index of molecular flexibility from kappa shape attributes*, Quant. Struct.–Act. Relat. 8 (1989), pp. 221–224.
[23] L.B. Kier and L.H. Hall, *An electrotopological-state index for atoms in molecules*, Pharm. Res. 7 (1990), pp. 801–807.

[24] E. Estrada, *Spectral moments of the edge adjacency matrix in molecular graphs. 1. Definition and applications to the prediction of physical properties of alkanes*, J. Chem. Inf. Comput. Sci. 36 (1996), pp. 844–849.
[25] J. Galvez, R. Garcia, M. Salabert, and R. Soler, *Charge indexes. New topological descriptors*, J. Chem. Inf. Comput. Sci. 34 (1994), pp. 520–525.
[26] L. EduSoft, *MolconnZ, version 4.05* (2003).
[27] R. Todeschini, V. Consonni, A. Mauri, and M. Pavan, *Dragon for windows (software for molecular descriptor calculations), version 5.4.* Talete srl: Milan, Italy (2006).
[28] S. Basak, D. Harriss, and V. Magnuson, *Polly 2.3*, Copyright of the University of Minnesota (1988).
[29] A. Katritzky, V. Lobanov, and M. Karelson, *CODESSA Software*, University of Florida, SemiChem, Shawnee, KS. 211 (1994).
[30] C.W. Yap, *PaDEL-descriptor: An open source software to calculate molecular descriptors and fingerprints*, J. Comput. Chem. 32 (2011), pp. 1466–1474.
[31] P.P. Roy, J.T. Leonard, and K. Roy, *Exploring the impact of size of training sets for the development of predictive QSAR models*, Chemom. Intel. Lab. Syst. 90 (2008), pp. 31–42.
[32] S. Wold, L. Eriksson, and S. Clementi, *Statistical validation of QSAR results*, Chemom. Meth. Molec. Design (1995), pp. 309–338.
[33] K. Roy and P.P. Roy, *Comparative chemometric modeling of cytochrome 3A4 inhibitory activity of structurally diverse compounds using stepwise MLR, FA-MLR, PLS, GFA, G/PLS and ANN techniques*, Eur. J. Med. Chem. 44 (2009), pp. 2913–2922.
[34] J. Devillers, *Strengths and weaknesses of the backpropagation neural network in QSAR and QSPR studies*, in *Neural Networks in QSAR and Drug Design*, J. Devillers, ed., Academic Press, London (1996), pp. 1–46.
[35] P. Liu and W. Long, *Current mathematical methods used in QSAR/QSPR studies*, Int. J. Molec. Sci. 10 (2009), pp. 1978–1998.
[36] K. Kourou, T.P. Exarchos, K.P. Exarchos, M.V. Karamouzis, and D.I. Fotiadis, *Machine learning applications in cancer prognosis and prediction*, Comput. Struct. Biotech. J. 13 (2015), pp. 8–17.
[37] Y.S. Prabhakar, *A combinatorial approach to the variable selection in multiple linear regression: Analysis of Selwood et al. data set—A case study*, QSAR Comb. Sci. 22 (2003), pp. 583–595.
[38] M.K. Gupta and Y.S. Prabhakar, *Topological descriptors in modeling the antimalarial activity of 4-(3′, 5′-disubstituted anilino) quinolines*, J. Chem. Inf. Model. 46 (2006), pp. 93–102.
[39] A.R. Katritzky, O.V. Kulshyn, I. Stoyanova-Slavova, D.A. Dobchev, M. Kuanar, D.C. Fara, and M. Karelson, *Antimalarial activity: A QSAR modeling using CODESSA PRO software*, Bioorg. Med. Chem. 14 (2006), pp. 2333–2357.
[40] N. Mahmoudi, J.V. de Julian-Ortiz, L. Ciceron, J. Galvez, D. Mazier, M. Danis, F. Derouin, and R. Garcia-Domenech, *Identification of new antimalarial drugs by linear discriminant analysis and topological virtual screening*, J. Antimicrob. Chemother. 57 (2006), pp. 489–497.
[41] N. Mahmoudi, R. Garcia-Domenech, J. Galvez, K. Farhati, J.F. Franetich, R. Sauerwein, L. Hannoun, F. Derouin, M. Danis, and D. Mazier, *New active drugs against liver stages of* Plasmodium *predicted by molecular topology*, Antimicrob. Agents Chemother. 52 (2008), pp. 1215–1220.
[42] R. García-Domenech, W. López-Peña, Y. Sanchez-Perdomo, J.R. Sanders, M.M. Sierra-Araujo, C. Zapata, and J. Gálvez, *Application of molecular topology to the prediction of the antimalarial activity of a group of uracil-based acyclic and deoxyuridine compounds*, Int. J. Pharm. 363 (2008), pp. 78–84.

[43] R. Garcia-Domenech, I. Denise Manhenje, Y. Monje, A. Lopez-Gonzalez, R. Marco, C. Tacho, and J. Galvez, *Search of QSAR models for predicting the antiprotozoal activity and cytotoxicity in vitro of a group of pentamidine analogous compounds*, Lett. Drug Design Discov. 8 (2011), pp. 172–180.
[44] R. Garcia-Domenech, R. Zanni, M. Galvez-Llompart, and J. Galvez, *Predicting antiprotozoal activity of benzyl phenyl ether diamine derivatives through QSAR multitarget and molecular topology*, Mol. Divers. 19 (2015), pp. 357–366.
[45] M. Srivastava, H. Singh, and P.K. Naik, *Quantitative structure–activity relationship (QSAR) of artemisinin: The development of predictive in vivo antimalarial activity models*, J. Chemom. 23 (2009), pp. 618–635.
[46] S. Deshpande, V.R. Solomon, S.B. Katti, and Y.S. Prabhakar, *Topological descriptors in modelling antimalarial activity: n 1-(7-chloro-4-quinolyl)-1, 4-bis (3-aminopropyl) piperazine as prototype*, J. Enzym. Inhib. Med. Chem. 24 (2009), pp. 94–104.
[47] F. Prado-Prado, X. García-Mera, M. Escobar, E. Sobarzo-Sánchez, M. Yañez, P. Riera-Fernandez, and H. González-Díaz, *2D MI-DRAGON: A new predictor for protein–ligands interactions and theoretic-experimental studies of US FDA drug-target network, oxoisoaporphine inhibitors for MAO-A and human parasite proteins*, Eur. J. Med. Chem. 46 (2011), pp. 5838–5851.
[48] F.J. Prado-Prado, X. García-Mera, and H. González-Díaz, *Multi-target spectral moment QSAR versus ANN for antiparasitic drugs against different parasite species*, Bioorg. Med. Chem. 18 (2010), pp. 2225–2231.
[49] S.C. Basak, D. Mills, and D.M. Hawkins, *Characterization of dihydrofolate reductases from multiple strains of* Plasmodium falciparum *using mathematical descriptors of their inhibitors*, Chem. Biodiv. 8 (2011), pp. 440–453.
[50] P.K. Ojha and K. Roy, *Comparative QSARs for antimalarial endochins: Importance of descriptor-thinning and noise reduction prior to feature selection*, Chemom. Intel. Lab. Syst. 109 (2011), pp. 146–161.
[51] S. Paliwal, J. Sharma, and S. Paliwal, *Quantitative structure–activity relationship analysis of bisbenzofuran cations as antimalarial agents employing multivariate statistical approach*, Indian J. Chem. B. 51 (2012), pp. 617–630.
[52] T. Qidwai, D. K Yadav, F. Khan, S. Dhawan, and R. S Bhakuni, *QSAR, docking and ADMET studies of artemisinin derivatives for antimalarial activity targeting plasmepsin II, a hemoglobin-degrading enzyme from* P. falciparum, Curr. Pharm. Des. 18 (2012), pp. 6133–6154.
[53] B.K. Sharma, S. Verma, and Y.S. Prabhakar, *Topological and physicochemical characteristics of 1,2,3,4-tetrahydroacridin-9(10H)-ones and their antimalarial profiles: A composite insight to the structure–activity relation*, Curr. Comput. Aid-Drug. Des. 9 (2013) pp. 317–35.
[54] M.K. Gupta, *CP-MLR/PLS-directed QSAR studies on the antimalarial activity and cytotoxicity of substituted 4-aminoquinolines*, Med. Chem. Res. 22 (2013), pp. 3497–3509.
[55] V.H. Masand, D.T. Mahajan, K.N. Patil, T.B. Hadda, M.H. Youssoufi, R.D. Jawarkar, and I.G. Shibi, *Optimization of antimalarial activity of synthetic prodiginines: QSAR, GUSAR, and CoMFA analyses*, Chem. Biol. Drug Des. 81 (2013), pp. 527–536.
[56] B. Singh, R. A Vishwakarma, and S. B Bharate, *QSAR and pharmacophore modeling of natural and synthetic antimalarial prodiginines*, Cur. Comput.-Aided Drug Des. 9 (2013), pp. 350–359.
[57] M.C. Sharma, S. Sharma, P. Sharma, and A. Kumar, *Pharmacophore and QSAR modeling of some structurally diverse azaaurones derivatives as anti-malarial activity*, Med. Chem. Res. 23 (2014), pp. 181–198.
[58] V. Kurdekar and H.R. Jadhav, *A new open source data analysis python script for QSAR study and its validation*, Med. Chem. Res. 24 (2015), pp. 1617–1625.

[59] J. Galvez, J. de Julian-Ortiz, and R. Garcia-Domenech, *Application of molecular topology to the prediction of potency and selection of novel insecticides active against malaria vectors*, J. Molec. Struct. Theochem. 727 (2005), pp. 107–113.

[60] A.R. Katritzky, D.A. Dobchev, I. Tulp, M. Karelson, and D.A. Carlson, *QSAR study of mosquito repellents using CODESSA pro*, Bioorg. Med. Chem. Lett. 16 (2006), pp. 2306–2311.

[61] R. Natarajan, S.C. Basak, and T.S. Neumann, *Novel approach for the numerical characterization of molecular chirality*, J. Chem. Inf. Model. 47 (2007), pp. 771–775.

[62] Z. Wang, J. Song, J. Chen, Z. Song, S. Shang, Z. Jiang, and Z. Han, *QSAR study of mosquito repellents from terpenoid with a six-member-ring*, Bioorg. Med. Chem. Lett. 18 (2008), pp. 2854–2859.

[63] G. Paluch, J. Grodnitzky, L. Bartholomay, and J. Coats, *Quantitative structure–activity relationship of botanical sesquiterpenes: Spatial and contact repellency to the yellow fever mosquito*, Aedes aegypti, J. Agric. Food Chem. 57 (2009), pp. 7618–7625.

[64] R. García-Domenech, J. Aguilera, A. El Moncef, S. Pocovi, and J. Gálvez, *Application of molecular topology to the prediction of mosquito repellents of a group of terpenoid compounds*, Mol. Divers. 14 (2010), pp. 321–329.

[65] R. Garcia-Domenech, P. Garcia-Mujica, U. Gil, C. Casanova, J. Mireilli Beltran, and J. Galvez, *Search of QSAR models for natural sesquiterpenes repellent activity against the yellow fever mosquito*, Aedes aegypti, AFINIDAD. 67 (2010), pp. 187–192.

[66] R. Garcia-Domenech, A. Bernués, S. Berrido, D. Diaz, and J. Gálvez, *Aplicación de la topología molecular en la predicción de la presión de vapor y la actividad repelente de mosquitos de un grupo de compuestos derivados de la n,n-dietil-m-toluamida*, AFINIDAD 66 (2009), pp. 439–444.

[67] G. Pasquale, G.P. Romanelli, J.C. Autino, J. García, E.V. Ortiz, and P.R. Duchowicz, *Quantitative structure–activity relationships of mosquito larvicidal chalcone derivatives*, J. Agric. Food Chem. 60 (2012), pp. 692–697.

[68] P.V. Oliferenko, A.A. Oliferenko, G.I. Poda, D.I. Osolodkin, G.G. Pillai, U.R. Bernier, M. Tsikolia, N.M. Agramonte, G.G. Clark, and K.J. Linthicum, *Promising* Aedes aegypti *repellent chemotypes identified through integrated QSAR, virtual screening, synthesis, and bioassay*, PloS ONE 8 (2013), pp. e64547.

[69] D. Mazier, N. Mahmoudi, K. Farhati, R. Garcia-Domenech, J. Galvez, F. Derouin, and M. Danis, *Compounds for preventing and treating* Plasmodium *infections*, PCT Int. Appl. (2009), WO 2009074649 A1 20090618, Eur. Pat. Appl. (2009), EP 2070522 A1 20090617.

[70] A. Speck-Planche, V.V. Kleandrova, and M.T. Scotti, *Fragment-based approach for the in silico discovery of multi-target insecticides*, Chemom. Intel. Lab. Syst. 111 (2012), pp. 39–45.

[71] M. Tsikolia, U.R. Bernier, M.R. Coy, K.C. Chalaire, J.J. Becnel, N.M. Agramonte, N. Tabanca, D.E. Wedge, G.G. Clark, and K.J. Linthicum, *Insecticidal, repellent and fungicidal properties of novel trifluoromethylphenyl amides*, Pestic. Biochem. Physiol. 107 (2013), pp. 138–147.

5 Pharmacophore Modeling Applied to Mosquito-Borne Diseases

Apurba K. Bhattacharjee

CONTENTS

ABSTRACT

Mosquito-borne diseases have an overwhelming impact on public health throughout the world. Tropical and subtropical countries are most affected by these diseases. The mosquito-borne diseases include malaria, dengue fever virus, Rift Valley fever virus, yellow fever, chikungunya virus, West Nile virus, lymphatic filariasis, Zika virus, and Japanese encephalitis. Despite several efforts by the WHO and the Gates Foundation to combat and eradicate these diseases, they still remain a big problem to the tropical world. Although mosquito repellents are effective and still good countermeasures against these infections, the repellents have several issues, including resistance and toxicity to limit their use. Thus, therapeutic interventions are necessary both for prophylactics and treatments from these infections. However, despite access to many available drugs for treatment, growing resistance to most of them has raised serious concerns and requires new countermeasures. The goal of this chapter is to illustrate the usefulness of *in silico* pharmacophore modeling as a tool for identification and design of new compounds for developing potential therapeutic agents against mosquito-borne diseases. The chapter will discuss earlier efforts of *in silico* research in this area through a comprehensive search of literature including efforts made in the author's own laboratory. The pharmacophore modeling presented here will be based on various computational strategies used for organizing molecular features to account for activity of potential therapeutics against these diseases. In addition, the chapter will illustrate how the models can be used as templates for virtual screening of compound databases and identify potential new candidates to counter the mosquito-borne diseases.

KEYWORDS

- *In silico* pharmacophore modeling
- Virtual screening
- Compound databases
- Stereoelectronic properties
- Drug design and discovery
- Potential therapeutics against mosquito borne diseases

5.1 INTRODUCTION

Mosquitoes are well known as the world's most notorious vectors for lethal human diseases that include malaria, dengue fever virus, Rift Valley fever virus, yellow fever, chikungunya virus, West Nile virus, lymphatic filariasis, Zika virus, and Japanese encephalitis [1–4]. Millions of people are infected annually by these infections, leading to hundreds of deaths, primarily transmitted by the *Aedes aegypti* species of mosquitoes. The impact of mosquito-borne diseases is enormous. A recent survey from the US National Center for Medical Intelligence indicated that infections from arthropod-borne diseases account for 28 of the top 40 endemic disease threats in the world [5]. In terms of disease transmission by arthropods, mosquitoes top the list and, since mosquitoes feed on blood, they cause more human suffering than any other organism.

The mosquito-borne infections not only kill but also debilitate millions of people. Often, the infected people cannot work to support themselves. Billions of dollars in productivity are lost in places where these infections run rampant. The deadliest of all mosquito-borne infections is malaria, a parasitic infection that mosquitoes directly inject into the bloodstream during a bite. The parasite species that causes a virulent infection from the mosquito bite is known as the *Plasmodium falciparum* [3]. According to the World Health Organization, it continues to be the cause of hundreds of deaths annually [1,2]. Once in the bloodstream, the parasites start traveling to the victim's liver, multiplying rapidly and continuing to produce more parasites while traveling farther in the bloodstream, destroying large number of red blood cells [3].

Although many important efforts have been made during the past decades, particularly by the WHO and the Gates Foundation [1,2], to combat and eradicate the mosquito borne diseases, the problem still remains a great headache to most governments of the tropical world. Except for yellow fever, there is no effective vaccine for the diseases transmitted by mosquitoes. Mosquitoes are not only limited to the previously named infections but also responsible for epidemic polyarthritis and several forms of encephalitis [3]. Increased international travel has enhanced this problem all over the world [2]. Research funding is directed more toward the study of parasite biology than the study of the mosquito vector. Even otherwise, vector control programs are limited only to the *Aedes* species of mosquitoes; these programs are increasingly threatened due to mosquitoes' resistance to insecticides. The growth of resistance to major insecticide classes has not been properly understood [5].

However, the focus of the present chapter will be on *in silico* pharmacophore modeling strategies, including additional *in silico* efforts that have been used to find potential new therapeutic agents and repellents to counter mosquito-borne diseases. Although pharmacophore modeling studies have been used for discovery of new drugs against many diseases like cancer, diabetes, and HIV, they have been applied far less to mosquito-borne diseases. In the past, combinatorial chemistry coupled with high-throughput screening shifted most of the drug discovery attention from structure-based approaches. However, in recent years, due to genome mappings and large-scale determination of protein structures, new chapters for drug discovery have opened up, starting from structure-based crystallographic protein structure active site analysis to design and identify new ligands as potential therapeutics. Unfortunately, the target structure of many proteins involved in mosquito-borne diseases is still unknown; for these target proteins, pharmacophore-based methods are found to be quite effective. These methods are not only useful for virtual screening and identification of new hits from compound databases but also provide useful insight for designing novel compounds as well as complementarity for binding at the active sites of unknown target structures. Thus, an integration of structure-based methods, pharmacophore modeling, virtual screening, and combinatorial chemistry could together be a better basis for more efficient drug discovery and design, reducing both cost and time.

5.1.1 Concept of Pharmacophore

The concept of pharmacophore is one of the most important steps toward understanding interactions between a receptor and its ligand. A literature survey reveals that Paul Ehrlich was probably the pioneer in offering a definition of pharmacophore in the early 1900s. He defined it as "a molecular framework that carries (*phoros*) the essential features responsible for a drug's (*pharmacon*) biological activity" [6,7]. This definition remained almost the same until Peter Gund, in 1977, provided a remarkably similar definition: "a set of structural features in a molecule that is recognized at a receptor site and is responsible for that molecule's biological activity" [8]. Gund is one of the pioneers in pattern searching methods based on functional features (pharmacophores) for screening of compound databases to identify new compounds sharing similar features. The technique led him to develop the first three-dimensional (3D) searching software, Molpad [8].

A more modern definition of a 3D pharmacophore was provided by the International Union of Pure and Applied Chemistry (IUPAC) in 1998. The official 1998 IUPAC definition of pharmacophore describes it as "an ensemble of steric and electronic features that are necessary for optimal interaction with a specific receptor target structure (a protein or an enzyme) to trigger or inhibit its biological response" [9]. It is an abstract concept but can be geometrically represented as a distribution of chemical features in a 3D space. The common chemical features are hydrogen bond acceptors and donors, aliphatic and aromatic hydrophobic sites, ring aromaticity, and ionizable sites that interact with complementary sites of a target structure. The features are either represented as vectors or point geometrical spherical objects. The vector directionality features—namely, those associated with the hydrogen bonding—are derived from symmetry, number of localized lone pairs, and the environment around the atom in a molecule (ligand atom) [10]. However, the important concept of pharmacophore for a medicinal chemist is that if two structurally different compounds share similar pharmacophore features, the compounds should have similar biological activity. For example, the illustration in (Figure 5.1 shows two structurally different

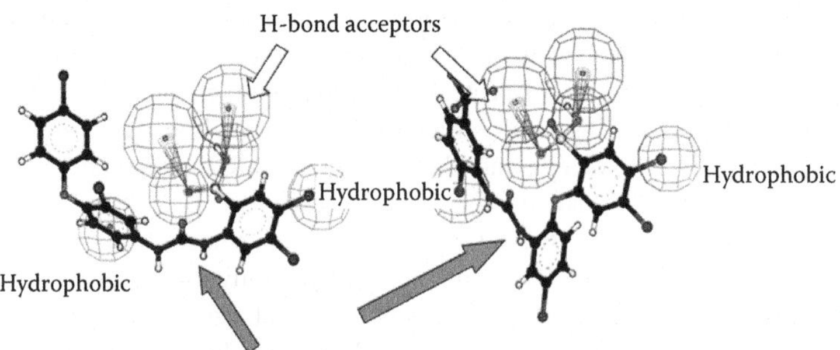
H-bond acceptors
Hydrophobic
Hydrophobic
Hydrophobic
Two structurally dissimilar molecules but fit to the same pharmacophore model (two H-bond acceptors and two hydrophobic sites) and both are potent antimalarials.

FIGURE 5.1 Example of 3D pharmacophore model: defining feature requirements in a molecular structure for antimalarial activity.

compounds sharing similar pharmacophore features containing two hydrogen bond acceptors and two hydrophobic sites. Since this pharmacophore was generated as a model for antimalarial activity, the compounds should have similar antimalarial activity. Indeed, as predicted, both the compounds show potent antimalarial activity [11].

However, the term pharmacophore is often misrepresented in medicinal chemistry [9]. Pharmacophore is essentially an abstract concept that provides an estimate of common molecular interaction capabilities of a group of bioactive compounds for its target receptor structure. It does not represent any molecule or functional group. The concept is very insightful for understanding the molecular recognition aspects of a target receptor shared by a set of bioactive compounds. Figure 5.1 shows two dissimilar compounds sharing common chemical features and also indicates that their interaction capacities are similar to the complementary site of the target structure. Pharmacophores are frequently used in medicinal chemistry for drug discovery projects for hit and lead identification as well as for lead optimization. Since pharmacophore transcends the chemical structural class and captures only the features responsible for activity, use of pharmacophore for virtual screening of compound databases has the advantage of identification of new potentially active compounds or chemical class (chemotypes) of compounds. The pharmacophores are also useful for *de novo* design of active compounds. However, there are a few implicit assumptions of pharmacophore models that are necessary to keep in mind. Structure of compounds used for model generation, validation, and database searches should be viewed along the following lines:

- The structures used in the model are responsible for the biological activity, not the metabolites.
- The conformation of a structure mapping the model is its bioactive conformation.
- The binding site is the same for all molecules in the proposed model.
- Biological activity is accounted only in terms of thermodynamic equilibrium, particularly by enthalpic energy considerations, assuming entropy for the molecules to be similar.
- Kinetics of the processes are ignored.
- Transport properties, diffusion, and solvent effects are largely avoided.

Nevertheless, the concept is very important and effective for computational medicinal chemists in drug discovery programs for both hit and lead identifications as well as for lead optimization. The generation of virtual libraries and virtual screening by pharmacophore has become a common technique in drug design and discovery research. The basic concept of mapping pharmacophore features on functional groups of screened compounds has not only allowed identification of a variety of compounds, but the models themselves are also useful as great visualization tools. For example, a rotatable 3D pharmacophore model mapped on a protein–ligand complex can provide useful insights on interaction possibilities and thereby facilitate interdisciplinary discussions with medicinal chemists, synthetic chemists, toxicologists, and pharmacologists for initiating interdisciplinary drug discovery programs [12]. However, experimental validation of the identified compounds obtained through pharmacophore-based virtual screening will be ultimately necessary to establish the reliability of the generated model.

5.1.2 Importance of Pharmacophore Modeling in Drug Discovery and Development

Discovery and development of new therapeutics are expensive and complex processes with ever changing technologies. On an average, it takes about 10 years and approximately five hundred million to a billion US dollars to bring a new drug or a new chemical entity (NCE) from the bench of discovery to the market for use [13,14]. Therefore, any technology that can improve the efficiency of the process is considered highly valuable to the pharmaceutical industry.

With the advent of high-speed modern computers, astronomical memory, and graphic tools, accomplishing computations and visualization of small to large biomolecules including proteins has become more efficient with greater precision. The graphic tools in modern computers have not only made possible visualization of three-dimensional structures of large protein molecules, but also allowed interactive virtual docking experiments between a potential drug molecule and the binding sites of a protein. Applications involving molecular modeling or *in silico* techniques have now become an integral part of basic science research that requires an understanding of molecular level information about environmental, biochemical, and biological processes. The current advances in these methodologies have direct applications ranging from accurate *ab initio* quantum chemical calculations of stereoelectronic properties, generation of three-dimensional pharmacophores, and performance of database searches to identify bioactive agents [15].

Increasing costs for pharmaceutical development have resulted in the emergence of many assorted approaches for improving the efficiency of the process in recent years. Among them, *in silico* screening or virtual screening of databases to identify potential new compounds has been quite successful [16,17]. Virtual screening is a process of intelligent use of computing to analyze large databases of chemical compounds to identify potential drug candidates. The process can serve as a complementary tool to high-throughput screening (HTS) for rapid and effective experimental assay for a large pool of compounds. Screening compounds by this method is essentially a knowledge-based approach and thus implicitly requires certain information about the nature of the receptor binding site or the nature of the ligand that is expected to bind effectively at the active site. However, the type of procedure followed in virtual screening for compound databases depends upon availability of information as input and requirement for the output [16,17].

Use of pharmacophore for virtual screening of databases is an intelligent knowledge-based, as well as efficient, approach for discovery of potential bioactive compounds that has an additional advantage for identification of new chemical classes or chemotypes. Pharmacophores may be derived in several ways—for example, by analogy to a natural substrate or known ligand, by inference from a series of dissimilar active analogs, or by direct analysis of the structure of a target protein [18–20]. However, a pharmacophore is used in two ways to identify new compounds that share its features, and thus may exhibit similar biologic response. In the first approach, *de novo* design can be performed by linking the disjointed parts of a pharmacophore together with chemical fragments to generate a hypothetical structure that is chemically reasonable but completely novel and can be synthesized. The second approach is virtual screening a compound database with the pharmacophore model to identify new potential hit compounds. One key advantage of virtual screening over *de novo* design is that it allows the identification of existing compounds that are either readily available or have a known synthetic procedure [21].

However, if three-dimensional structure of the target enzyme or the protein is available, small molecule docking procedures can be adopted to perform structure-based virtual screening followed by further short-listing through pharmacophore modeling and virtual screening of databases for identification of an ideal ligand. But if the three-dimensional structure of the target protein is unknown, feature-based pharmacophore modeling is best suited. This can be constructed from activity data of known compounds and the developed model template itself can be used for virtual screening to identify potential new hits. Pharmacophores may also be developed from other molecular properties, such as the absorption, distribution, metabolism, excretion (ADME) properties, toxicity data, lipophilicity, and drug-related properties. Identification of new, potentially active compounds using pharmacophore modeling has increasingly shown success in recent years [20,21]. In the past, too, development of pharmacophore models significantly contributed to the discovery of potent drugs [22–26] and in retrieving inhibitors for various drug targets. Pharmacophore models have led to the discovery of XEN2174 compounds, such as norepinephrine transporters (NETs) that are in phase II clinical trials as pain killers [23]. Pharmacophores have been developed for screening growth factor (ALK5) receptor inhibitors [24], and Checkpoint1 kinase inhibitors by retrieving potent molecules from databases, such as Specs, NCI, Maybridge, and CNPD [25]. For prediction of juvenile hormone antagonists, pharmacophore models have been generated from 3D quantitative structure–activity relationship (QSAR) studies [26].

5.2 METHODS FOR PHARMACOPHORE MODEL GENERATION

Pharmacophore models can be generated using three-dimensional QSAR pharmacophore generation methods. The two widely used pharmacophore modeling and screening software programs are LigandScout [27] and Discovery Studio [28]. These are used for generation, validation, and application for virtual screening of compound databases to identify new compounds. The 3D QSAR pharmacophore generation protocol utilizes the Discovery Studio [28] or an earlier version of the software known as CATALYST HypoGen algorithm [29] to derive a structure–activity relationship (SAR) hypothesis from a set of ligands with known activity values. Input ligand data should contain two properties: activity and uncertainty factors associated with reproducibility of the activities. Briefly, the methodology allows the use of known structure and activity data of a set of compounds (known as a training set) to create a hypothesis (pharmacophore), which characterizes activity of the compounds. The uncertainty factor represents a ratio range of uncertainty in activity value based on the expected statistical dispersion of biological data collection. Thus, an uncertainty value of 3 (or lower if the reproducibility of experimental data is more accurate) means that the actual activity of a particular compound is assumed to be situated somewhere in an interval ranging from one-third of to three times reported activity value of that compound. Lots of parameters may be tested. The pharmacophore model can be used as a template for database searches [19,21,30]. Conformational models of known structures are generated by creating a training set of compounds that emphasizes representative coverage within a range of the permissible Boltzman population with significant abundance (within 10.0 kcal/mol) of the calculated global minimum. This conformational model for pharmacophore generation that aims to identify the best three-dimensional arrangement of chemical features is usually selected for the modeling study. The pharmacophore features such as hydrogen bond donor/acceptor, hydrophobic sites, and positively and/or negatively ionizable sites distributed over a three-dimensional space explain the activity variations among the training set. The hydrogen bonding features are vectors, whereas all others are points.

The features in the chemical feature dictionary to be selected are usually as follows: hydrogen bond acceptor (HBA): 2 to 4; hydrogen bond donor (HBD): 0 to 2; hydrophobic (HY): 1 to 3; and ring aromatic (RA): 0 to 1. The minimum interfeature distance is 3, which specifies the minimum spacing in 3 Å between feature points. The mapping of a pharmacophore model onto an active compound is determined by its "fit score." The fit score or function is represented as

$$\text{Fit} = \Sigma w\left(1 - \Sigma(\text{Disp/Tol})^2\right)$$

where the equation variables are defined as w = adaptively determined weight of the hypothesis function.

The fit function not only checks if the function is mapped or not but also contains a distance term that measures the distance that separates the function on the molecule from the centroid of the hypothesis function. Both terms are used to calculate the geometric fit value.

The quality of pharmacophore model could be evaluated by a set of parameters as follows: correlation, total cost, null cost, fixed cost, and other statistical parameters. To get a reliable pharmacophore model, the total cost should be significantly different from null cost. For a pharmacophore hypothesis, a value of 40–60 bits between total cost and null cost (cost difference) might indicate that it had 75%–90% possibility of correlating the data. The greater the cost difference is, the less likely it is that the hypothesis reflects a chance correlation [28,29].

5.2.1 Pharmacophore Validation

Validation of a quantitative pharmacophore model is extremely important for reliability and usefulness of the model. It can be performed by determining whether the generated model was able to identify active structures and predict the activity accurately. In addition, a test set of active compounds may be constructed for determination of correlation (r^2) and compared with that of the training set. Furthermore,

Fischer's validation [28,29] and decoy set can be employed to validate the generated model. To achieve 95% confidence level of a model, the Fischer's randomization calculations should be conducted by scrambling the activity data of the training set molecules and assigning them new values, and then generating pharmacophore hypotheses using the same features and parameters originated for Hypo 1. The significance of the hypotheses is to be calculated using the following formula: $[1 - (1 + X)/Y] \times 100\%$, where X and Y are the number of training and test set molecules, respectively. The importance of scramble runs is that the model hypothesis (pharmacophore) was not generated by chance correlation but based on meaningful correlation [28,29]. Finally, for ultimate validation of the model, the identified compounds are to be tested first in *in vitro* assays and, subsequently, based on the promise of *in vitro* results, *in vivo* testing [30].

5.2.2 Virtual Screening of Compound Databases Using Generated Pharmacophore Model

With the aim to discover new compounds, the best generated pharmacophore model may be employed as a search query to retrieve molecules with novel and desired chemical features from different compound databases. However, it is important to note that all the compounds intended for virtual screening using a 3D pharmacophore from a database are to be formatted and stored in 3D multiconformer (within 20 kcal/mol) format. Mapping the model onto the most potent compound and converting it to a 3D shape based template may also be used for the virtual screening procedure [19,21,30]. A successful screening of a database of both trial and test set databases along with the "goodness of hits" (GH) may be identified by using the following equation:

$$GH = \{Ha(3A + Ht)/4HtA\} \times \{1 - Ht - Ha/D - A\}$$

where

Ht is the total number of trial compounds
Ha is number of known active compounds in the test database
A is the number of active compounds in the database
D is the total number of compounds in the database [10,28,29]

The identified compounds are to be mapped onto the pharmacophore model and the estimated (predicted) activities EC_{50} or IC_{50} values are to be noted. Estimated activities of less than 500 nM are usually considered for further study [30].

5.3 BIOLOGICAL ASSAY

It is important to note that although the pharmacophore-based directed approach for 3D compound database search is an efficient tool for extracting potential bioactive compounds, it is necessary to experimentally test the compounds and iteratively refine the model. Perfect mapping of any molecule to the pharmacophore model does not guarantee its experimental activity despite reflecting receptor complementarity. There may be several factors lacking, such as the perfect fit to the active site due to steric hindrance, electrostatics, overall lipophilicity, and other unforeseen parameters. Thus, development of appropriate biological assay is an important component of a pharmacophore-modeling-based drug discovery process. Compounds that map well onto the best generated pharmacophore model showing estimated activities EC_{50} or IC_{50} values less than 500 nM are usually considered for *in vitro* evaluation. Based on the promise of *in vitro* evaluations, favorable drug-like properties, and *in silico* ADME/toxicity assessments, the identified compounds are short-listed for *in vivo* animal studies [30].

5.4 PHARMACOPHORE MODELING FOR DISCOVERY OF POTENTIAL THERAPEUTICS AGAINST MOSQUITO-BORNE DISEASES

Review of pharmacophore modeling studies on mosquito-borne diseases is discussed next in the following order: malaria, dengue fever virus, lymphatic filariasis (LF), viral and Japanese encephalitis, West Nile virus, chikungunya virus, Rift Valley fever virus, and yellow fever.

5.4.1 Malaria

Malaria is the deadliest among all mosquito-borne infections. It is suffered by more than two billion people worldwide, mostly in the tropical countries. Currently, an estimated 300–500 million clinical cases are reported worldwide, from which approximately 600,000 deaths occur every year [2–5]. The deaths are mainly due to the most virulent species of malaria, the *Plasmodium falciparum* transmitted by the *Anopheles* mosquitoes [31,32]. President George W. Bush's malaria initiative in 2005, known as the President's Malaria Initiative (PMI) Program, has been the most effective project in combating malaria in the world [33]. According to the World Health Organization, "an estimated 4.3 million fewer malaria deaths occurred between 2001 and 2013, about 47% reduction in the number of deaths if malaria patterns of 2000 remained unchecked" [33]. The program has entered into its second phase to address drug and insecticide resistance to malaria, along with treatment monitoring and surveillance [33]. Although development of curative antimalarial agents is difficult due to various stages of manifestation of the parasite within the host, many antimalarial chemotherapeutics have been discovered over the past few decades. These include older drugs, such as chloroquine, mefloquine, and pyrimethamine/sulfadoxine, newer artemisinins (1994), artemether/lumefantrine (1999), atovaquone/proguanil (1999), chlorproguanil/dapsone (2003); and more recently the general ACT (artemisinin combination therapy) [34]. However, all of them have some issues ranging from growing resistance to the parasite to toxicity limiting their use [2]. In recent years, despite the completion of genome sequencing for *Plasmodium falciparum* and elimination of barriers to the use of state-of-the-art molecular and biological tools, no new antimalarial drugs have yet been discovered from genome-dependent experiments, though a wealth of new information has been produced for investigating the basic biology of the parasites [34]. Thus, continuing efforts are necessary for development of new effective and less toxic affordable drugs for treatment of malaria.

In silico pharmacophore modeling and, particularly, mapping of a validated antimalarial pharmacophore model have been recognized to be a powerful tool for rapid discovery of antimalarial agents including *in vivo* efficacy [35]. One of the earlier reports on the application of pharmacophore modeling to discovery of antimalarial agents was in 1997 by Grigorov et al. [36]. These researchers developed a 3D-QSAR pharmacophore model from bicyclic and tricyclic 1,2,4-trioxanes, elucidating the mode of action of these peroxidic antimalarials to facilitate further discovery. A few years later, Parenti et al. [37] reported a series of inhibitors of *Plasmodium falciparum* dihydrofolate reductase (*PfDHFR*) from a 3D pharmacophore model. The study not only predicted quantitatively the inhibition constants but also validated a target for discovery of antimalarial therapeutics. In addition, the study demonstrated an application of a combined pharmacophore screening and molecular docking approaches to discover new inhibitors for *PfDHFR* [38].

In another study [39], computational models were reported using the state-of-the-art machine learning algorithms. The models were evaluated based on multiple statistical criteria. The study found a random-forest-based approach to be better for accuracy as assessed from receiver operating characteristic (ROC) curve analysis. The study further evaluated the active molecules using a substructure-based approach to identify common substructures enriched in the active set and argued that the computational models generated could be effectively used to screen large molecular data sets to prioritize them for phenotypic screens, drastically reducing cost while improving the hit rate. However, no antimalarial drug has been reported so far by adopting the procedure.

In another study for a master's thesis [40], the author argued the necessity of exploration of the marine environment as a source for discovery of potential antimalarial therapeutics due to resistance of *Plasmodium falciparum* to the currently available antimalarial drugs. In this study, isonitrile, isocyanate, and isothiocyanate compounds were isolated from marine sponges and mollusks that were found to exhibit nanomolar antiplasmodial activities. Through QSAR studies and a literature precedent pseudoreceptor model, a pharmacophore for the design of novel antimalarial agents was proposed. Although the conventional idea suggests that the marine compounds exert their inhibitory action through interfering with the heme detoxification pathway in *P. falciparum*, this study demonstrated an inadequacy of the theory by going further on the mode of action using computational methods, formulating a series of modern computational procedures and validating and applying them to theoretical systems for investigation of the interaction between the marine compounds and the heme targets [40].

The validations of these algorithms, before their application to the marine compound-heme systems, were achieved through two case studies. The first was used to investigate the applicability of the statistical docking algorithm AutoDock [41] for the exploration of conformational space around the heme target. A theoretical *P. falciparum* 1-deoxy-D-xylulose-5-phosphate reductoisomerase (*Pf*DXR) enzyme model was subjected to rigorous docking simulations with over 30 different ligand molecules using the AutoDock algorithm [41]. This investigation facilitated a successful validation of a protein model and proposed that it can be used for rational design of new *Pf*DXR-inhibiting antiplasmodial compounds, as well as enabling an improvement of the docking algorithm for application to the heme systems.

The second case study was used to investigate the applicability of an *ab initio* molecular dynamics algorithm for simulation of bond-breaking and -forming events between the marine compounds and their heme targets. This validation involved the exploration of intermolecular interactions in a naturally occurring nonoligomeric zipper using the Car–Parrinello molecular dynamics (CPMD) method [42]. This study allowed proposal of a model for the intermolecular forces responsible for zipper self-assembly and showcased the CPMD method's abilities to simulate and predict bond-forming and -breaking events. Data from the computational analyses suggested that the interactions between marine isonitriles, isocyanates, and isothiocyanates occur through bondless electrostatic attractions rather than through formal intermolecular bonds as had been previously suggested. Accordingly, a simple bicyclic tertiary isonitrile was synthesized using Kitano and colleagues' relatively underutilized isonitrile synthetic method for the conversion of tertiary alcohols to their corresponding isonitriles [40]. This compound's potential for heme detoxification and inhibition was then explored *in vitro* via the pyridine-hemochrome assay. The assay data suggested that the synthesized isonitrile was capable of inhibiting heme polymerization in a similar manner to the known inhibitor chloroquine [40].

In another related study [43], a dynamic receptor-based pharmacophore model (DPM) of *Pf*SpdSyn complemented by a knowledge-based rational design strategy was adopted for antimalarial discovery. The development of a DPM allows incorporation of the protein to exhibit within the drug design process. The methodology resulted in a wealth of information of the chemical space for the active site and was incorporated in designing new inhibitors against *Pf*SpdSyn using a knowledge-based rational design strategy [43]. The active site of *Pf*SpdSyn was subdivided into four binding regions (DPM1–DPM4) to allow for the identification of fragments binding within these specific binding regions. DPM representatives of chemical characteristics of each binding region were constructed and subsequently screened against the drug-like subset from the ZINC database [44]. From the screens, a total of nine compounds were selected for *in vitro* assay, complementing each other in exploring specific active site binding characteristics. The test resulted in discovery of a new lead compound, *N*-(3-aminopropyl)-cyclohexylamine (NAC; Ki 2.8 μM) as inhibitor for *Pf*SpdSyn. NAC was specifically designed to bind both putrescine and decarboxylated adenosylmethionine cavities by chemically bridging the catalytic center that was confirmed by kinetic studies. The compound NAC shows great potential for lead optimization to increase its binding affinity. This study paved a new way for lead optimization and possibly for development of new antimalarial agents [43].

Since the type II fatty acid synthase pathway in the life cycle of malarial parasites has been well established and the enoyl acyl carrier protein (ACP) reductase of *Plasmodium falciparum* (*Pf*ENR) is known as the rate-determining step of the enzyme for its elongation, it has been a target for the development of antimalarial agents as well as vaccines. Toward this endeavor, a recent study [20] describes the identification of the rhodanine class of compounds as inhibitors of *Pf*ENR and generated 3D pharmacophore models for this enzyme for both rhodanine furans and phenyls. The pharmacophore model for rhodanine furan showed a hydrogen bond donor, two hydrogen bond acceptors, two metal ligators, three hydrophobic sites, and two aromatic ring features, whereas the pharmacophore model for the phenyl subclass showed two hydrogen bond donors, two hydrogen bond acceptors, one metal ligator, two hydrophobic sites, and two aromatic ring features. These models could be used for *in silico* screening of compound libraries for *Pf*ENR inhibitors. However, no new tested antimalarial compound from the models was reported [20].

Pharmacophore modeling efforts at the Walter Reed Army Institute of Research to discover antimalarial compounds in the past decade are summarized as follows. A part of the study in the form of a book chapter was reported earlier [45]. Our efforts on a pharmacophore modeling approach to discover potential new antimalarial compounds primarily stem from the fact that the crystal structure of many target proteins or enzymes for the antimalarial drugs was unknown. The strategy not only helped us to identify many new antimalarials but also provided many insights to possible mechanism of interactions with the unknown target receptors. One of our earliest reported studies in this area was virtual screening of an in-house compound database by developing a pharmacophore model from chloroquine (CQ)-resistant reversal agents [46,47]. First, we developed the model from known CQ reversal agents and then cross-validated the model by mapping it on a series of CQ reversal agents such as chloropheniramine, cyproheptadine, ketotefin, pizotyline, azatadine, loratadine, verapamil, and penfluridol. Mapping of the pharmacophore onto six well known CQ-resistance reversal agents showed excellent consistency. Next, we performed a database search using the pharmacophore for potential new CQ reversal agents from our in-house WRAIR—Chemical Information System database of over 290,000 compounds. The search resulted in identification of several 2,4-diamino-3*,4*-dichloro-6-quinazolinesulfonanilide analogs as promising candidates for CQ-resistant reversal studies. The identified lead compound was found to be a potent antimalarial in the mouse malaria presumptive causal prophylactic test as well as in an *in vivo* mouse malaria test [46].

Our next effort was to apply the methodology for proton pump inhibitors from four benzimidazoles—namely, omeprazole, lansoprazole, rabeprazole, and pantoprazole—that are clinically used as proton pump inhibitors [48]. The generated pharmacophore model was used in search of new compounds from our in-house database and identified 128 compounds that have similar features. Three of these compounds were observed to have antimalarial efficacy, not only *in vitro* but also in mouse malaria *in vivo* [48].

In continuation of our efforts for antimalarial drug discovery, we developed a pharmacophore model (Figure 5.2a) from a series of indolo[2,1-b]quinazoline-6,12-diones (tryptanthrins). These compounds exhibited remarkable *in vitro* antimalarial activity (below 100 ng/mL) and low cytotoxicity against sensitive and multidrug-resistant *Plasmodium falciparum* malaria [11]. Also, some of these compounds showed potency even where the parasite was 40% resistant to chloroquine and 20% resistant to mefloquine. However, despite possessing outstanding *in vitro* activity and well tolerated toxicity, these compounds failed to display *in vivo* activity, probably due to poor bioavailability and aqueous solubility. Nonetheless, the pharmacophore model that we developed from tryptanthrins was very useful for identification

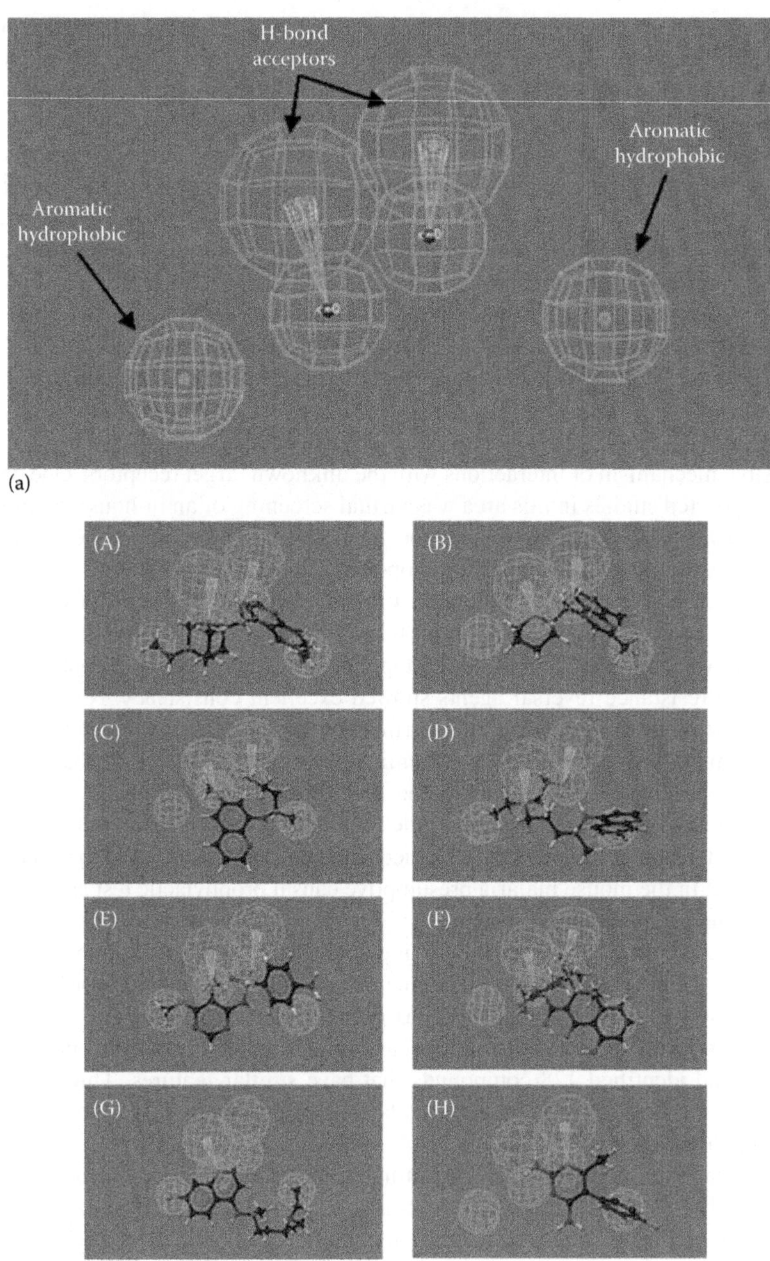

FIGURE 5.2 **(See color insert.)** (a) Pharmacophore for antimalarial activity of the tryptanthrins. (b) Mapping of the pharmacophore onto eight commonly used antimalarial drugs in the United States: (A) quinine, (B) mefloquine, (C) primaquine, (D) hydroxychloroquine, (E) sulfadoxine, (F) doxycycline, (G) chloroquine, and (H) pyrimethamine.

of many antimalarials and provided a fairly reliable foundation for 3D database searches. The model was found to map well on many antimalarial drugs, such as quinine, hydroxychloroquine, rhodamine dyes, and chalcones (Figure 5.2b) [11]. Interestingly, the observed mapping of the model onto quinine (Figure 5.2b—A) led us to hypothesize that, like quinine, the tryptanthrins should target heme polymerase in the *P. falciparum* tropozoites. Since the target structure for antimalarial activity of the tryptanthrins was unknown, we evaluated six substituted 4-azaindolo[2,1-*b*] quinazoline-6,12-dione analogs for hemin-binding affinity by ^{1}H NMR, x-ray crystallography, and *ab initio* quantum chemical methods [49]; we found the evidence for heme-tryptanthrin stacking organization in all these analogs. The observation was consistent with the proposed interactions with hemin determined separately by nuclear magnetic resonance (NMR) experiments [49].

Using the pharmacophore as a search template for virtual screening of an in-house database led us to identify five new aminoquinazoline derivatives showing potent activity in both *in vitro* and *in vivo* mouse malaria screening tests [11]. Thus, the model was not only useful for identification of a new class of antimalarials but also provided insight for a possible mechanism of action with the heme.

Our next effort was to apply structure-based drug design methodologies. Following the completion of the *P. falciparum* genome project, focus started shifting to specific proteins in parasites that are unique but critical for cellular growth and survival. With a direct role in the regulation of cellular proliferation, the cyclin-dependent protein kinases (CDKs) became attractive drug targets for discovery of new antimalarial chemotherapies and thus, toward this effort, we targeted the malarial CDK. Primarily, three plasmodial CDKs (*PfPK5*, *PfPK6*, and *Pfmrk*) were investigated. There are several inhibitors for CDKs that were reported to possess antiparasitic activity when assayed with the malarial parasites *in vitro* [50]. We developed a new pharmacophore model from known inhibitors targeting specifically the malarial *Pfmrk* [51] and used the model template for searching our in-house chemical database to identify new potential inhibitors. The procedure resulted in the discovery of 16 potent *Pfmrk* inhibitors [51]. The predicted inhibitory activity of some of these *Pfmrk* inhibitors from the molecular model agree exceptionally well with the experimental inhibitory values from the *in vitro* CDK assay [51]. Statistically, the most significant model obtained by us was found to contain two hydrogen-bond acceptor functions and two hydrophobic sites, including one aromatic-ring hydrophobic site [51]. Although the model was solely developed from structure–activity relationships and not from x-ray structural analysis of the known CDK2 structure, it was found to be consistent with the structure–functional requirements for binding of the CDK inhibitors in the ATP-binding pocket. Mapping of the pharmacophore on known CDK inhibitors that were tested in our *Pfmrk* assay, such as (a) indirubin, (b) staurosporine, (c) kenpaullone, (d) WR032428, and (e) WHI-P180, were observed to be consistent with the model [51]. Despite complexity of the malarial CDK activity, the predicted *Pfmrk* inhibition from the model was quite robust and could be useful for design of selective *Pfmrk* inhibitors, including assessment of the subtle differences in structure–function information between *Pfmrk* and other CDKs [50–52].

In another effort on structure-based drug discovery of antimalarials, we developed a phamacophore model for malarial *Pf*KASIII inhibitory activity and successfully utilized the model to identify new *Pf*KASIII inhibitors through virtual screening of databases [45,53]. From the electrostatic potential profile of only one known related inhibitor, thiolactomycin (TLM), we developed the pharmacophore model and utilized it iteratively to identify several new *Pf*KASIII inhibitors [45,53].

In another recent effort, we developed a pharmacophore model for chalcones from data of an in-house project and identified several new antimalarial agents through virtual screening of databases that were also helpful to design new antimalarial compounds [54–56]. Chalcones are known to rapidly metabolize by liver microsomes, but chalcone analogs with modified enone linkers were found to have significant improvement in metabolic stability. Our goal was to identify compounds that share the antimalarial properties of the chalcones, but lack the enone structure. However, despite an understanding the structural basis for antimalarial activity of the chalcones, its pharmacophore for activity remained unknown. We developed the first pharmacophore model for chalcones to obtain both structural and functional requirements for antimalarial activity [55]. The model enabled identification of several new antimalarial agents and facilitated the design of novel analogs [54,56]. The generated pharmacophore contained an aromatic and an aliphatic hydrophobic site, one hydrogen bond donor site, and a ring aromatic feature distributed over a three-dimensional space [55]. The activity of the compounds estimated by the pharmacophore was found to correlate well with those determined experimentally. Two of the identified compounds were found to be highly potent *in vitro* against all five strains of *P. falciparum* tested. Moreover, one compound showed significant potency in a malaria-infected mouse model [54]. The model was also reported to be useful for design of novel antimalarials [56]. The study therefore demonstrated how the chemical features of a set of diverse chalcones and chalcone-like compounds could be organized to develop a pharmacophore and be useful for discovery and design of novel antimalarial compounds.

In yet another recent study [57], we developed a pharmacophore model from the antimalarials 4(1*H*)-quinolones. These compounds are known to be highly effective in inhibiting the replication of *P. falciparum* along with synergism to the well-known antimalarial drug, atovaquone (Malarone). We not only developed a model for antimalarial activity of the quinolones comprising two aliphatic hydrophobic functions and one aromatic ring hydrophobic function, but also went beyond the pharmacophore to calculate mathematical descriptors directly related to the identified molecular structures and consistent with their predicted and experimental antimalarial activities [57].

5.4.2 Dengue Fever

Dengue fever and dengue hemorrhagic fever (DF/DHF) are mosquito-borne diseases of great public health concern that affect millions of people, particularly in tropical and subtropical parts of the world [58,59]. This infection causes a broad spectrum of clinical manifestations ranging from asymptomatic fever to potentially fatal hemorrhagic fever and dengue shock syndrome [60,61]. Among the affected tropical countries, the Philippines suffers the most from dengue, claiming the highest number of reported cases and deaths [58,62]. The main insect vector for the disease is the mosquito species *Ae. aegypti*. Despite its being considered one of the most significant mosquito-borne pathogens threatening global health, there are no approved antiviral drugs to combat the infection. Although pharmacophore modeling to discover antidengue drugs is rare, several virtual screening and docking studies have been reported in the literature, mainly on identification of promising targets for anti-dengue drug discovery. The reported studies are summarized next.

In a report, Wichapong et al. [63] identified NS2B/NS3 protease as an essential enzyme for viral replication and one of the promising targets in the search for drugs against dengue viral infection. In this study, virtual screening (VS) on four multi-conformational databases was carried out using multiple criteria. First, molecular dynamics simulations of the NS2B/NS3 protease with four known inhibitors were performed. The results demonstrated the importance of both electrostatic and van der Waals interactions in stabilizing the ligand–enzyme interaction, which accordingly led to generating three different pharmacophore models (structure-based, static, and dynamic models). These three models were employed for pharmacophore-based virtual screening for identification of potential antidengue compounds. Next, compounds passing the first criterion were further short-listed using Lipinski's rule of five to keep only compounds with drug-like properties. Following this, molecular docking calculations were performed to remove compounds that are unsuitable for ligand–enzyme interactions. Finally, free energy binding calculation of each compound was performed and compounds having better energy than known inhibitors were down-selected. This procedure led to short-listing 20 potential hits. However, no biological activity data on these 20 short-listed compounds were reported [63].

Wang et al. [64], in another paper on dengue virus, reported the entry of a small molecule inhibitor into the virus. They suggested that the entry of dengue virus into a host cell is mediated by its major envelope (E) protein. Since the crystal structure of the E protein is known to have a hydrophobic pocket that is presumably important for low-pH mediated membrane fusion, high-throughput docking with this hydrophobic pocket was performed in the study, and hits were evaluated in cell-based assays. One of the identified inhibitors showed an average 50% effective concentration of 119 nM against dengue virus serotype 2 in a human cell line. Performing mechanism-based olfaction studies, the authors demonstrated the compound's activity at an early stage of dengue virus infection, arresting the virus in vesicles that colocalize with endocytosed dextran and inhibit NS3 expression. The inhibitors described in this report may be considered molecular probes for further study for entry of flaviviruses into host cells [64].

Tomlinson et al. [65] summarized the importance of a structure-based drug discovery approach to address the shortcomings for discovery of antidengue drugs given the wealth of structural information available for dengue virus proteins. According to the study, most structure-based drug discovery paradigms compromise accuracy of free energy predictions for computational tractability, and thus are largely inefficient and ineffective in identifying drug leads. These researchers observed that a two-phased approach, virtual screening and molecular dynamic free energy of binding simulations, is necessary for discovering dengue drugs. The authors concluded that a combined approach provides perfect strength to the virtual screening and molecular dynamic free energy of binding simulations for identification of true antiviral drug leads without excessive false-positive predictions. According to the author's opinion, the approach can be adapted to combat any pathogen or disease with known virulence factor and structure. However, no antidengue compound having experimental validation was reported in the study [65].

In another study [66], design of an effective antidengue drug was reported using computer-aided drug designing techniques. Since computer-aided drug design (CADD) entails the use of biochemical information of ligand–receptor interaction sequentially to hypothesize the drug's refinements, docking of selected ligands having antidengue activity was performed with the active site of protein 2FOM (dengue virus NS2B/NS3 protease) to extract the biochemical information. Docking interactions were interpreted in terms of hydrogen bonding and hydrophobic and ionic interactions. On the basis of these interaction analyses and IC_{50} values, one ligand was recognized as "lead compound." Seven analogs of the lead compound were designed and docked with the active site of the protein. Interactions of the analogs in the active site of 2FOM protein were analyzed and, based on activity and high binding interactions, the compounds were recommended for clinical study and bulk synthesis [66].

In a patent application, Galindo et al. [67] reported several compounds obtained by using *in silico* methods that have pharmaceutical compositions for attenuating or inhibiting dengue virus infection. These compounds specifically interfere with multiple steps of the viral replication cycle that are associated with the arrival of the virus in the target cells and assembly of the progeny virions. A few compounds were listed as having antidengue activity. The invention also mentioned the use of these compounds for prophylactic and/or therapeutic treatment of infection caused by the four dengue virus serotypes as well as by other flaviviruses [67].

5.4.3 Lymphatic Filariasis (LF)

Lymphatic filariasis (LF) is a tropical disease caused by three parasites—namely, *Wuchereria bancrofti* (*Wb*), *Brugia malayi* (*Bm*), and *Brugia timori* (*Bt*)—found in a group of nematode worms belonging to the order Filariidae [68]. It is transmitted to humans by the infective bites of mosquitoes and a few other arthropods. Human filarial nematodes are endemic in tropical and subtropical regions around the world [68]. LF is one of the oldest and most debilitating diseases, affecting millions of people [69]. In India, *Wb* is mainly transmitted by the ubiquitous vector *Cx. quinquefasciatus*, which grows on dirty and polluted water, whereas *Bm* prevalence is restricted to the rural areas [70]. It is estimated that 700 million people are living in the endemic areas of the Southeast Asian region of the world [68,70]. Albendazole, DEC, and Ivermectin (Mectizan®) are the common drugs administered for treatment of this infection [68]. Several private pharmaceutical companies have played important roles in supporting treatments by donating billions of doses of these drugs to the endemic countries for elimination of LF under WHO [69,70].

However, for eradication of LF globally, new research strategies are needed. Discovery and development of more effective drugs and drug targets along with new methods for vector control, diagnostic tools, and techniques are necessary. A recent study [71] suggested statistical analysis of available data along with bioinformatics tools could be better suited for new filarial research. Bioinformatics-based proteomics and genomics approaches have offered promising tools to address the molecular level of information and their role in the infection. This has also provided an excellent opportunity to obtain insights about the proteins and genome for better understanding the inhibition process to validate a target for structure-based drug discovery. In the absence of the crystallographic structure of a validated target, Bhargavi et al. [71] made a homology model 3D structure of glutathione-S-transferase (GST), a target protein for LF, which was further explored by Azeez et al. [72] for discovery of a better drug against LF. In these reports, potential therapeutics against LF were identified using virtual screening of compound databases and testing *in vitro*. Other novel drug targets were also mentioned; however, the targets have not yet been validated. Several bioinformatics-based web resources and databases have been discussed in the study to enrich the literature for filarial research [68].

Continuing with the survey of literature reveals one pharmacophore modeling study used for identification of new compounds reported as potential drugs against filariasis disease [73]. The study suggests that, at the molecular level, a protein known as chitinase protein is responsible for the cause of filariasis. Since this protein has no known structure, using the information theory-based GOR (Garnier–Osguthorpe–Robson) method and SOPMA (self-optimized prediction method with alignment) *in silico* tools, a secondary structure for chitinase protein was predicted [73]. Using a protein structure prediction algorithm, a 3D structure was constructed and validated using modeler 9v8 and SAVS (Smart Access Vocabulary Server) algorithms [73]. Molecular properties were predicted and conformations of all the identified compounds were analyzed using different *in silico* protocols of Discovery Studio [28]. Down-selection of the identified compounds was performed using *in silico* ADME and toxicity descriptors of TOPKAT protocols in Discovery Studio [28]. Further down-selection was carried out by evaluating the pharmacokinetics and dynamics of the identified compounds using Discovery Studio [28]. After these analyses, three compounds—ZINC00001288, ZINCOO355669, and ZINC01165275 [44]—were predicted to be potent against filariasis. Docking these three compounds in the active site of chitinase protein, using ligand fit protocol by evaluating the dock score and relative energies, ZINC00001288 and ZINCOO355669 [44] were predicted to be efficient drugs for filariasis disease. However, the study was a theoretical one and no experimental validation of the target structure and the compounds was reported [73].

5.4.4 Viral Encephalitis

Viral encephalitis is a flavivirus commonly known as Japanese encephalitis (JE). It occurs predominantly in Asia affecting roughly 30 to 50 thousand people annually, most of them children [74]. It is a mosquito-borne infection transmitted by anthropophilic rice field-breeding mosquitoes of the *Culex* species [75]. Acute encephalitis is found to develop in 1–20 cases per 1,000 infections; 25% of these cases lead to death and 30% may produce serious neurological problems [74,75]. Although vaccines are available that have reduced the incidence of JE in some of these Asian countries, no antiviral therapy is currently available to treat the infection. Sampath et al. [76] recently reported a few molecular targets to facilitate drug discovery for this flavivirus. The targets are envelope glycoprotein, NS3 protease, NS3 helicase, NS5 methyltransferase, and NS5 RNA-dependent RNA polymerase. The NS3 protein of JE is a multifunctional protein combining protease, helicase, and nucleoside 50-triphosphatase (NTPase) activities. The crystal structure of the catalytic domain of this protein was recently solved. This enabled structure-based virtual screening for identification of novel inhibitors against JE NS3 helicase/NTPase. In a recent study [77], the natural ligand ATP and two known JE NS3 helicase/NTPase inhibitors were docked to their molecular target. The refined structure of the enzyme was used to construct a pharmacophore model for JE NS3 helicase/NTPase inhibitors. Through virtual screening of 1,161,000 compounds from the ZINC database [44], 15 compounds with high scores were selected and subjected to docking to the JE NS3 helicase/NTPase to examine their binding mode and verify screening results by consensus scoring procedure. No experimental validation of the 15 compounds was reported [77].

5.4.5 West Nile Virus

West Nile virus is a member of the *Flaviviridae* family and it is transmitted to humans through mosquito bites. Symptoms of the infection range from none to severe encephalitis (inflammation of the brain) or meningitis (inflammation of the lining of the brain and spinal cord). It has wide distribution across the world and causes serious health problems to many people. The virus actually causes a disease known as West Nile encephalitis (WNE). It is endemic in Asia, Africa, and the Middle East and periodically reported in many parts of the United States [78].

West Nile virus was discovered in 1937 in the West Nile district of Uganda. It emerged in the United States for the first time in the New York City area in August 1999. There were 62 reported human cases from which seven deaths occurred during the outbreak, creating a widespread panic. However, up till now no drug or vaccine is known to prevent the life cycle of this virus. There are some enzymes reported in the literature that are important for the virus replication, such as NS3 protease, RNA-dependent RNA-polymerase, and Helicase enzymes [79]. Since the West Nile virus NS3 protease is known to be responsible for the polyprotein cleavage during the life cycle of the virus, it is thought to be a target for inhibition to cause the virus death. Although no actual pharmacophore modeling study for identification of antiviral compounds against West Nile virus was reported in the literature, a QSAR-based pharmacophore model was reported with a training set of a few compounds showing potent activity against West Nile virus. These compounds contain 5-amino-1-(phenyl) sulfonyl-pyrazol-3-yl scaffold and target the NS3 protease [79]. The QSAR model was built from 10 active compounds showing a correlation coefficient ($R\hat{A}^2$) = 0.932. It was used for a virtual screening of databases to identify new compounds and to predict the activity of the unknown compounds. The researchers discovered four compounds showing high predictive activity against West Nile virus NS3 protease and validated its ability to discriminate the real active compounds through the QSAR-based virtual screening method [79]. The reported QSAR model showed high predictive power to predict compounds with expected potential activity. The use of molecular docking with the QSAR-based virtual screening process was successful in confirmation of the prediction. However, no experimental validation for activity of the identified compounds was reported [79].

5.4.6 Chikungunya Virus (CHIKV)

Chikungunya virus (CHIKV) is an arbovirus that is transmitted to humans by two species of mosquitoes, *Ae. aegypti* and *Ae. albopictus* [80]. Infection of CHIKV is often associated with fever, rash, and arthralgia. Currently, no vaccine or antiviral drug is available for prevention from or treatment of this infection. Although this virus is considered a tropical pathogen, adaptation of the virus to the mosquito species *Ae. albopictus* has resulted in recent outbreaks in Europe and the United States for being in common temperate zones [80]. The symptoms generally start within 4–7 days after the mosquito bite and acute infection lasts for 1–10 days with abrupt onset of fever, headache, fatigue, nausea, vomiting, rash, myalgia, and severe arthralgia [80]. However, due to lack of vaccines or effective antivirals against CHIKV, the treatment regimens are usually symptomatic and based on the clinical manifestations; therefore, new therapeutics are very much needed to counter CHIKV.

Recently Rashad et al. [81] presented a detailed review of the current medicinal chemistry scenario for discovery and development of chemotherapeutics targeting the CHIKV. The researchers used *in silico* virtual screening and docking studies along with structure-based design and identification of novel binding sites for discovery of inhibitors against the chikungunya virus envelope proteins. Availability of the crystal structures of several proteins of the CHIKV RNA genome and other related alphaviruses seems to be encouraging for the drug discovery process through target structure-based pharmacophore modeling, virtual library screening, and docking approaches. Bassettoa et al. [82] used a bioinformatics approach, considering a homology-based model for the CHIKV virus based on the crystal structure of nsP2 of the alphavirus VEE (Venezuelan equine encephalitis) virus [83], which is also a mosquito-borne viral pathogen, as a template. The study identified a few compounds that selectively inhibited CHIKV in a virus-cell-based CPE (cytopathic effect) reduction assay. However, further experimental studies have not been reported. Lucas-Hourani et al., on the other hand, used a phenotypic assay to identify new inhibitors targeting the same nonstructural nsP2 protein [84].

However, the study reported by Bassetto et al. [82] presents more details about the discovery of a novel series of compounds that inhibit CHIKV replication in the low micromolar ranges. Specifically, the study reports a virtual screening of 36 simulations of ~five million compounds on the viral protease CHIKV nsP2 and *in silico* investigation of structure–activity relationships (QSAR) of the identified hit compounds. Overall, a series of compounds, including the original hits, were evaluated in a virus-cell-based CPE reduction assay. The study concluded by reporting selective inhibitors for better understanding the CHIKV replication cycle as well as steps toward the development of a clinical candidate drug for treatment for the infection [82].

In another report [85], CHIKV virus replication and propagation were shown to be dependent on the protease activity of the viral nsP2 protein. It is reported to cleave the nsPlpolyprotein replication complex into a functional unit. The study investigates the nsP2 protease protein of CHIKV by adopting computational methods to find the best interaction site in the protease protein for identification of the best ligands inhibiting the nsP2 protein. Homology modeling was used to predict the 3D structure of the nsP2 protease protein. The 3D structure was validated using the SAVS algorithm, in which 85.2% of residues were present in the favored regions of the Ramachandran plot. Docking studies were carried out with various inhibitors and it was found that 4(*N*-butyl-9-[3,4-dipropoxy-5-(propoxymethyl)oxolan-2-yl] purin-6-amine) had the most stable interaction with the nsP2 CHIKV protease, indicating that this compound could be a promising inhibitor of the protein and a potential drug candidate against the CHIKV virus. However, no experimental data were reported [85].

In another study [86], a high-throughput virtual screening of commercially available databases, such as ZINC [44] and BindingDB [87], was carried out using the Openeye tools [88] and Schrodinger LLC software packages [89]. The Openeye Filter program [88] had been used to filter the database and the filtered outputs were docked using a high-throughput virtual screening (HTVS) protocol as implemented in the GLIDE package of Schrodinger LLC [89]. The top hit compounds were further used for enriching similar molecules from the database through a vROCS protocol [89] and a shape-based screening protocol as implemented in Openeye [88]. The adopted strategies provided different scaffolds as hits against CHIKV protease. The study selected three scaffolds: indole, pyrazole, and sulphone derivatives based on docking scores and synthetic feasibility. However, no experimental activity data were reported to support the model [86].

5.4.7 Rift Valley Fever

Rift Valley fever (RVF) is a viral disease spread primarily by mosquitoes that can affect both humans and animals [90]. It is generally found in regions of eastern and southern Africa where the sheep and cattle are raised, but the virus also exists in sub-Saharan Africa and Madagascar. RVF has never occurred in the United States [90]. The virus can cause several different disease symptoms and people typically have either no symptoms or a mild illness associated with fever and liver abnormalities. However, in some patients the illness can progress to hemorrhagic fever (which can lead to shock or hemorrhage), encephalitis (inflammation of the brain leading to headaches, coma, or seizures), or ocular disease (diseases affecting the eye). Approximately 1% of humans infected with RVF die from the disease. Case-fatality proportions are significantly higher for infected animals. Although no established course of treatment for patients infected with RVF virus is known, ribavirin, an antiviral drug, has shown promise in studies with monkeys and other animals. It could be useful for humans too, as mentioned in the report. In addition, a few studies suggest that interferon, immune modulators, and convalescent-phase plasma may also help in the treatment of patients with RVF [90,91].

However, no pharmacophore modeling study in aid for discovery of new therapeutics against RVF is known.

5.4.8 Yellow Fever

Yellow fever is one of the virus-infected diseases spread through mosquitoes; it kills more than 30 thousand people every year [92]. It is transmitted in urban areas by the bite of infective *Ae. aegypti* mosquitoes. It is an acute infectious viral disease of short duration with varying severity. In certain epidemic scenarios, the case fatality rate has been reported to exceed 50% among people. A preventive vaccine has been commercially available since the 1950s. In 2013, the World Health Organization concluded that "a single dose of vaccination is sufficient to confer lifelong immunity against yellow fever disease" [92]. Although no pharmacophore modeling study in aid of discovery of new therapeutics for yellow fever has been reported, a large number of compounds having potent activity against the disease has been published [92]. However, none of these candidate compounds has yet been approved for clinical use. Another recent study reports an efficient classification of a large data set (309 compounds) compiled from the ChEMBL database [93] using the Naïve Bayes method of KNIME-based classification models for yellow fever virus inhibition [94]. The best models obtained from this combined approach showed accuracy greater than 90% on the test set prediction (Matthew's correlation coefficients of >0.7). The authors suggested that all these models could be useful for virtual screening to identify inhibitors against yellow fever virus [94].

5.5 INSECT REPELLENTS

Numerous methods for controlling mosquito-borne diseases have been investigated over centuries, ranging from development of therapeutics to repellents and insecticides for treatment and prevention. However, the use of repellent still remains the most effective personal protection measure in reducing bites of mosquitoes and preventing vector-borne disease transmissions [95–97]. The most effective personal protection system used is a formulation of DEET (*N,N*-diethyl-1,3-toluamide) as a topical repellent that can defend against several types of biting arthropods [98]. However, as a repellent for human use, DEET has several issues. It is not equally effective against all insects and arthropod vectors of diseases. Moreover, many studies have shown that it can cause toxic encephalopathy, seizure, acute manic psychosis, cardiovascular problems, and dermatitis due to heavy or excessive use [99–101]. Efforts to find new repellents face numerous challenges and variables that affect the inherent repellency property of a chemical. Repellents do not all share a single mode of action, and surprisingly little is known about how repellents act on the target insects.

Thus, the quest to repel mosquitoes is not only an age-old problem but also a necessity for survival from mosquito-borne diseases and continues to be an important effort for scientific research. Discovery of new, environmentally safe repellents is an important goal for this effort. At the Walter Reed Army Institute, in pursuit of discovery for novel mosquito repellents and to better understand the mechanism of insect repellency, we performed a three-dimensional QSAR study and developed a pharmacophore model for potent repellent activity from a set of 11 known diverse insect repellents [102]. The generated model contained three hydrophobic sites and a hydrogen-bond acceptor site in specific locations around the 3D space of the known repellents. The pharmacophore not only showed an excellent correlation (correlation = 0.9) between the experimental protection time and the predicted protection time of the new repellents but the validity of the model also went beyond the list in the training set of compounds. It was found to map excellently onto a variety of other insect repellents, including a highly potent repellent compound that was extracted from the hair of the gaur, an animal frequently seen in Southeast Asia [102]. A US patent (#7,897,162) on the model, together with discovery of new arthropod repellents, was issued recently [97]. Details of these studies have already been reported [95,96].

In a recent study, Alcantara et al. [62] reported a three-dimensional structure of *Ae. aegypti* chorion peroxidase by homology modeling. A ModWeb server [103] was used that provided the most accurate model with QMEAN score [104] of 0.642. The protein model consists of 36.1% alpha-helices and 1% beta-strand. Ligand-binding sites in *Ae. aegypti* chorion peroxidase were identified using the SiteComp server [62]. *In silico* docking of a subset of the ZINC natural products database [44] was focused on the predicted binding site. Three ligands were found to be potential inhibitors of *Ae. aegypti* chorion peroxidase [62]. For future structure-based, as well as pharmacophore, modeling discovery of mosquito repellents, this model of *Ae. aegypti* chorion peroxidase could be useful.

5.6 FUTURE PERSPECTIVES

Since target 3D protein structure for biological activity of many drugs against mosquito-borne diseases is yet unknown and will continue to be a bottleneck for structure-based drug design, pharmacophore modeling from known active compounds and use of it for virtual screening of compound databases could be a reliable alternative approach for identification of new potential therapeutics. Even if the crystallographic structure of the 3D proteins is known or determined, certain specific features—such as the polarization and hydrogen-bonding effects arising from specific solvation patterns of water molecules in the common ATP-binding pocket and dynamic target–inhibitor interactions occurring intermittently to account for specificity of certain inhibitors—may not be available or detectible by crystallography. In order to account for such interactions, *in silico* methods, particularly quantum chemical methods, could be a better choice than most experimental methods. Furthermore, the prediction of protein- and ligand-binding affinity using 2D Bayesian categorization procedures may be adopted to build training sets of compounds obtained from docking of diverse drug-like molecules. Success of this approach has been reported [105]. However, for accomplishing a successful drug discovery and development program, the research team should have good synthetic chemistry support along with experienced molecular biologists and an efficient modeling group to facilitate lead optimization and a compound library of potential hits.

5.7 CONCLUDING REMARKS

Pharmacophore modeling studies clearly demonstrate a rational approach for discovery of potential therapeutic candidates solely from structure–activity relationship studies of known active compounds. However, testing for efficacies of the identified compounds—first, *in vitro* and, if found promising, then *in vivo* assays—is absolutely necessary. Stereoelectronic property analysis of a training set of active compounds can help to generate pharmacophore models with better reliability. Although the model can be built solely from structure–activity relationships, the model should be consistent with the observations of x-ray crystal structure if the target structure is known. For an unknown receptor structure, pharmacophore could be helpful to predict the complementary interaction capabilities in the active site of the target. However, it is important to note that despite perfect mapping onto the model, there may be many short-listed compounds that may show poor experimental activity. Thus, although the pharmacophore-directed 3D database search is an efficient approach for extracting potential bioactive compounds, it is necessary to experimentally test the identified compounds and iteratively refine the model. Perfect mapping of any molecule to the pharmacophore model does not guarantee its experimental activity despite reflecting receptor complementarity. There may be several factors lacking, such as the perfect fit to the active site due to steric hindrance, electrostatics, lipophilicity, and other unforeseen parameters. Thus, despite reasonably good mapping of the generated pharmacophore model, not all the compounds may be found to have similar efficacy. Nonetheless, pharmacophore models are useful tools for identification of new compounds. Finally, application of pharmacophore modeling for discovery of potential new therapeutics may be summarized in the following points:

- Pharmacophore models are very useful and efficient tools for drug discovery.
- It is important to note that there is no guarantee that all the compounds identified using pharmacophore models will show potent activity.
- Down-selected compounds are to be tested in *in vitro* assays first and subsequently in animals based on the promise of *in vitro* efficacy tests.
- Iterative use of 3D shape-based pharmacophore searching in conjunction with a large virtual compound library can be very effective in identifying new chemotypes with enhanced *in vitro* activity relative to the target molecule.
- *In silico* ADME/toxicity evaluations and docking methods could be useful for down-selection of compounds.
- In short, these *in silico* methods can maximize efficiency of therapeutic discovery.

ACKNOWLEDGMENTS

The author gratefully acknowledges valuable insights from numerous ex-colleagues in WRAIR for developing and improving the manuscript. Special thanks to Dr. Tulshi Saha of the National Institutes of Health (Bethesda, Maryland) for helping with the improvement of the chapter. The author also thanks the Bentham Science Publishers for permission to reproduce a few pages and the two figures in this chapter from my previous book chapter titled "Role of *In Silico* Stereo-electronic Properties and Pharmacophores in Aid of Discovery of Novel Antimalarials, Antileishmanials, and Insect Repellents" [45].

REFERENCES

[1] Business Insider, Health, 2014. http://www.businessinsider.com/r-gates-foundation-awards-notre-dame-23-million-for-malaria-dengue-studies-2014-14

[2] World Malaria Report (2009), World Health Organization: Geneva. Available at http://malaria.who.int/whosis/whostat/EN_WHS09_Full.pdf

[3] D. Brewste, *The story of mankind's deadliest foe*, Bio. Med. J. 323 (2001), pp. 289–290.

[4] B.F. Eldridge and J.D. Edman, *Medical Entomology: A Textbook on Public Health and Veterinary Problems Caused by Arthropods*, City Publishing Co., Chatsworth, CA, 2000.

[5] L.W. Kitchen, K.L. Lawrence, and R.E. Coleman, *The role of the United States military in the development of vector control products, including insect repellents, insecticides and bed nets*, J. Vector Ecol. 34 (2009), pp. 50–61.

[6] P. Ehrlich, *Über den Jetzigen Stand der Chemotherapie*, Ber. Deutsch. Chem. Ber. 42 (1909), pp. 17–47.

[7] E.J. Ariens, *Molecular pharmacology, a basis for drug design*, Forts. Arzneimittelforsch 10 (1966), pp. 429–529.

[8] P. Gund, *Three dimensional pharmacophore pattern searching*, Prog. Mol. Subcell. Biol. 5 (1977), pp. 117–143.

[9] A.R. Leach, V.J. Gillet, R.A. Lewis, and R. Taylor, *Three-dimensional pharmacophore methods in drug discovery*, J. Med. Chem. 53 (2010), pp. 539–558.

[10] O.F. Güner (ed.), *Pharmacophore, Perception, Development, and Use in Drug Design*, University International Line, IUL Biotechnology Series, San Diego, 2000.

[11] A.K. Bhattacharjee, M.G. Hartell, D.A. Nichols, R.P. Hicks, B. Stanton, J.E. van Hamont, and W.K. Milhous, *Structure–activity relationship study of antimalarial indolo [2,1-b]quinazoline-6,12-diones (tryptanthrins). Three dimensional pharmacophore modeling and identification of new antimalarial candidates*, Eur. J. Med. Chem. 39 (2004), pp. 59–67.

[12] G. Wolber, T. Seidel, F. Bendix, and T. Langer, *Molecule-pharmacophore superpositioning and pattern matching in computational drug design*, Drug Discov. Today 13 (2008), pp. 23–29.

[13] H. Kubinyi, *Success stories of computer-aided design*, in *Computer Applications in Pharmaceutical Research and Development*, S. Ekins and B. Wang, eds., Wiley-InterScience, New York, 2006, pp. 377–424.

[14] I.M. Kapetanovic, *Computer-aided drug discovery and development (CADDD): In silico-chemico-biological approach*, Chem. Biol. Interact. 171 (2008), pp. 165–176.

[15] B.L. Podlogar, I. Muegge, and L.J. Brice, *Computational methods to estimate drug development parameters*, Curr. Opinion Drug Discov. 12 (2001), pp. 102–109.

[16] W.P. Walters, M.T. Stahl, and M.A. Murko, *Virtual screening—An overview*, Drug Discov. Today 3 (1998), pp. 160–178.

[17] P.D. Lyne, *Structure-based virtual screening—A review*, Drug Discov. Today 7 (2002), pp. 1047–1055.

[18] A.K. Bhattacharjee, K. Kuča, K. Musilek, and R.K. Gordon, *In silico pharmacophore model for tabun-inhibited acetylcholinesterase (AChE) reactivators: A study of their stereoelectronic properties*, Chem. Res. Toxicol. 23 (2010), pp. 26–36.

[19] A.K. Bhattacharjee, J.A. Gordon, E. Marek, A. Campbell, and R.K. Gordon, *3D-QSAR studies of 2,2-diphenylpropionates to aid discovery of novel potent muscarinic antagonists*, Bioorg. Med. Chem. 17 (2009), pp. 3999–4012.

[20] G. Kumar, T. Banerjee, N. Kapoor, N. Surolia, and A. Surolia, *SAR and pharmacophore models for the rhodanine inhibitors of* Plasmodium falciparum *enoyl-acyl carrier protein reductase*, Life 62 (2010), pp. 204–213.

[21] A.K. Bhattacharjee, E. Marek, H.T. Le, and R.K. Gordon, *Discovery of non-oxime reactivators using an in silico pharmacophore model of oxime reactivators of OP-inhibited acetylcholinesterase*, Eur. J. Med. Chem. 49 (2012), pp. 229–238.

[22] P. Buchwald and N. Bodor, *Computer-aided drug design: The role of quantitative structure–property, structure–activity and structure–metabolism relationships (QSPR, QSAR, QSMR)*, Drug Future 27 (2002), pp. 577–588.

[23] A. Brust, E. Palant, D.E. Croker, B. Colless, R. Drinkwater, B. Patterson, C.I. Schroeder, D. Wilson, C.K. Nielson, M.T. Smith, D. Alewood, P.F. Alewood, and R.J. Lewis, *γ-Conopeptide pharmacophore development: Toward a novel class of norepinephrin transporter inhibitor (Xen2174) for pain*, J. Med. Chem. 52 (2009), pp. 6991–7002.

[24] J.X. Ren, L.L. Li, J. Zou, L. Yang, J.L. Yang, and S.Y. Yang, *Pharmacophore modeling and virtual screening for the discovery of new transforming growth factor b type I receptor (ALK5) inhibitors*, Eur. J. Med. Chem. 44 (2009), pp. 4259–4265.

[25] J.J. Chen, T.L. Liu, L.J. Yang, L.L. Li, Y.Q. Wei, and S.Y. Yang, *Pharmacophore modeling and virtual screening studies of Checkpoint kinase 1 inhibitors*, Chem. Pharm. Bull. 57 (2009), pp. 704–709.

[26] D. Liszekova, M. Palacovicova, M. Beno, and R. Farkas, *Molecular determinants of juvenile hormone action as revealed by 3D QSAR analysis in* Drosophila, PLoS ONE 4 (2009), pp. 1–15.

[27] V. Temml, T. Kaserer, Z. Kutil, P. Landa, T. Vanek, and D. Schuster, *Pharmacophore modeling for COX-1 and -2 inhibitors with LigandScout in comparison to Discovery Studio*, Future Med. Chem. 6 (2014), pp. 1869–1881.

[28] Discovery Studio, DS Version 2.5, Accelrys Inc., San Diego, CA, 2007, http://accelrys.com/products/discovery-studio/

[29] CATALYST version 4.10, Accelrys Inc., San Diego, CA, 2000, http://www.accelrys.com

[30] A.K. Bhattacharjee, E. Marek, H.T. Le, R. Ratcliffe, J.C. DeMar, D. Pervitsky, and R.K. Gordon, *Discovery of non-oxime reactivators using an in silico pharmacophore model of oxime reactivators of OP-inhibited acetylcholinesterase*, Eur. J. Med. Chem. 90 (2015), pp. 209–220.

[31] R.W. Snow, C.A. Guerra, A.M. Noor, H.Y. Myint, and S.I. Hay, *The global distribution of clinical episodes of* Plasmodium falciparum *malaria*, Nature 434 (2005), pp. 214–217.

[32] P.I. Trigg and A.V. Kondrachine, *The current global malaria situation*, in *Malaria Parasite Biology, Pathogenesis and Protection*, I.W. Sherman, ed., ASM Press: Washington, DC, 1998, pp. 11–22.

[33] S. Loewenberg, *Global role model. A successful malaria program enters its second phase*, Sci. Am. 5 (2015), pp. 18–19.

[34] T.N.C. Wells, P.L. Alonso, and W.E. Gutteridge, *New medicines to improve control and contribute to the eradication of malaria*, Nat. Rev. Drug Discov. 8 (2009), pp. 879–891.

[35] M.J. Dascombe, M.G.B. Drew, H. Morris, P. Wilairat, S. Auparakkitanon, W.A. Moule, S. Alizadeh-Shekalgourabi, P.G. Evans, M. Lloyd, A.M. Dyas, P. Carr, and F.M.D. Ismail, *Mapping antimalarial pharmacophores as a useful tool for the rapid discovery of drugs effective in vivo: Design, construction, characterization, and pharmacology of metaquine*, J. Med. Chem. 48 (2005), pp. 5423–5436.

[36] M. Grigorov, J. Weber, J.M.J. Tronchet, C.W. Jefford, W.K. Milhous, and D. Maric, *A QSAR study of the antimalarial activity of some synthetic 1,2,4-trioxanes*, J. Chem. Inf. Comput. Sci. 37 (1997), pp. 124–130.

[37] M.D. Parenti, S. Pacchioni, A.M. Ferrari, and G. Rastelli, *Three dimensional quantitative structure–activity relationship analysis of a set of* Plasmodium falciparum *dihydrofolate reductase inhibitors using a pharmacophore generation approach*, J. Med Chem. 47 (2004), pp. 4258–4267.

[38] G. Rastelli, S. Pacchioni, W. Sirawaraporn, R. Sirawaraporn, M.D. Parenti, and A. Ferrari, *Docking and database screening reveal new classes of* Plasmodium falciparum *dihydrofolate reductase inhibitors*, J. Med. Chem. 46 (2003), pp. 2834–2845.

[39] S. Jamal and V. Periwal, Open Source Drug Discovery Consortium, and V. Scaria, *Predictive modeling of anti-malarial molecules inhibiting apicoplast formation*, Bioinformatics 14 (2013), pp. 55, http://www.biomedcentral.com/1471-2105/14/55

[40] M.R. Adendorff, *Marine anti-malarial isonitriles: A synthetic and computational study*, M.S. diss., Rhodes University, South Africa, 2010.

[41] O. Trott and A.J. Olson, *AutoDock Vina: Improving the speed and accuracy of docking with a new scoring function, efficient optimization, and multithreading*, J. Comput. Chem. 31 (2010), pp. 455–461. http://autodock.scripps.edu

[42] R. Car and M. Parrinello, *Unified approach for molecular dynamics and density-functional theory*, Phys. Rev. Lett. 55 (1985), pp. 2471–2474.

[43] P.B. Burger, *Development of a dynamic receptor-based pharmacophore model of* Plasmodium falciparum *spermidine synthase for selective inhibitor identification*, Ph.D. thesis, University of Pretoria, Pretoria, South Africa, 2008.

[44] J.J. Irwin and B.K. Shoichet, *ZINC—A free database of commercially available compounds for virtual screening*, J. Chem. Inf. Model. 45 (2005), pp. 177–182.

[45] A.K. Bhattacharjee, *Role of in silico stereoelectronic properties and pharmacophores in aid of discovery of novel antimalarials, antileishmanials, and insect repellents*, in *Advances in Mathematical Chemistry and Applications*, S.C. Basak, G. Restrepo, and J.L. Villaveces, eds., Bentham Science Publishers, Amsterdam, 2014, pp. 273–305.

[46] A.K. Bhattacharjee, D.E. Kyle, J.L. Vennerstrom, and W.K. Milhous, *A 3D QSAR pharmacophore model and quantum chemical structure activity analysis of chloroquine (CQ)-resistance reversal*, J. Chem. Inf. Comput. Sci. 42 (2002), pp. 1212–1220.

[47] A.K. Bhattacharjee, D.E. Kyle, and J.L. Vennerstrom, *Structural analysis of chloroquine-resistance reversal by imipramine analogs*, Antimicrob. Agents Chemother. 45 (2001), pp. 2655–2657.

[48] M.A. Riel, D.E. Kyle, A.K. Bhattacharjee, and W.K. Milhous, *The efficacy of proton pump inhibitor drugs against* Plasmodium falciparum *in vitro and their probable pharmacophores*, Antimicrob. Agents Chemother. 46 (2002), pp. 2627–2632.

[49] R.P. Hicks, D.A. Nichols, C.A. DiTusa, D.J. Sullivan, M.G. Hartell, B.W. Koser, and A.K. Bhattacharjee, *Evaluation of 4-azaindolo[2,1-b]quinazoline-6,12-diones' interaction with hemin and hemozoin: A spectroscopic, x-ray crystallographic and molecular modeling study*, Internet Elec. J. Mol. Design 4 (2005), pp. 751–764.

[50] A.K. Bhattacharjee, *In silico 3D pharmacophores for aiding discovery of the Pfmrk (*Plasmodium *Cyclin-dependent protein kinases) specific inhibitors for therapeutic treatment of malaria*, Exp. Opin. Drug Discov. 2 (2007), pp. 1115–1127.

[51] A.K. Bhattacharjee, J.A. Geyer, C.L. Woodard, A.K. Kathcart, D.A. Nichols, S.T. Prigge, Z. Li, B.T. Mott, and N.C. Waters, *A three dimensional in silico pharmacophore model for Inhibition of* Plasmodium falciparum *cyclin dependent kinases and discovery of different classes of novel Pfmrk specific inhibitors*, J. Med. Chem. 47 (2004), pp. 5418–5426.

[52] A.K. Bhattacharjee, *Antimalarial drugs*, www.cen-online.org, 2010, July 19, pp. 3.

[53] P.J. Lee, J.B. Bhonsle, H.W. Gaona, D.P. Huddler, T.N. Heady, M. Kreishman-Deitrick, A.K. Bhattacharjee, W.F. McCalmont, L. Gerena, M. Lopez-Sanchez, N.E. Roncal, T.H. Hudson, J.D. Johnson, S.T. Prigge, and N.C. Waters, *Targeting the fatty acid biosynthesis enzyme, β-ketoacyl–acyl carrier protein synthase III (PfKASIII) in the identification of novel antimalarial agents*, J. Med. Chem. 52 (2009), pp. 952–963.

[54] C.E. Gutteridge, D.A. Nichols, S.M. Curtis, D.S. Thota, J.V. Vo, L. Gerena, G. Montip, C.O. Asher, D.S. Diaz, C.A. DiTusa, K.S. Smith, and A.K. Bhattacharjee, *In vitro and in vivo efficacy against* Plasmodium falciparum *and in vitro metabolism of 1-phenyl-3-aryl-2-propen-1-ones*, Bioorg. Med. Chem. Lett. 16 (2006), pp. 5682–5686.

[55] A.K. Bhattacharjee, D.A. Nichols, L. Gerena, N. Roncal, and C.E. Gutteridge, *An in silico 3D pharmacophore model of chalcones useful in the design of novel antimalarial agents*, Med. Chem. 3 (2007), pp. 317–326.

[56] C.E. Gutteridge, M.M. Hoffman, A.K. Bhattacharjee, W.K. Milhous, and L. Gerena, *In vitro efficacy of 7-benzylamino-1-isoquinolinamine against* Plasmodium falciparum *related to the efficacy of chalcone*, Bioorg. Med. Chem. Lett. 21 (2011), pp. 786–789.

[57] S.C. Basak, D. Mills, D.M. Hawkins, and A.K. Bhattacharjee, *Quantitative structure–activity relationship (QSAR) studies of antimalarial compounds from their calculated mathematical descriptors*, SAR and QSAR in Environ. Res. 21 (2010), pp. 103–125.

[58] World Health Organization, *Dengue fact sheet, 2009*. Available online: http://www.who.int/mediacentre/factsheets/fs117/en/(accessed Jan 2011).

[59] M.C. Kuo, P.L. Lu, J.M. Chang, M.Y. Lin, J.J. Tsai, Y.H. Chen, K. Chang, H.C. Chen, and S.J. Hwang, *Impact of renal failure on the outcome of dengue viral infection*, Clinical J. Amer. Soc. Nephrology 3 (2008), pp. 1350–1356.

[60] Y. Lei, Y. Huang, H. Zhang, L. Yu, M. Zhang, and A. Dayton, *Functional interaction between cellular p100 and the dengue virus 39 UTR*, J. Gen. Virol. 92 (2011), pp. 796–806.

[61] S.M. Tomlinson, R.D. Malmstrom, A. Russo, N. Mueller, Y.P. Pang, and S.J. Watowich, *Structure-based discovery of dengue virus protease inhibitors*, Antiviral Res. 82 (2009), pp. 110–114.

[62] E.P. Alcantara, *In silico identification of potential inhibitors of dengue mosquito,* Aedes aegypti *chorion peroxidase*, Comput. Biol. Bioinformatics 2 (2014), pp. 38–42.

[63] K. Wichapong, A. Nueangaudom, S. Pianwanit, W. Sippl, and S. Kokpol, *Identification of potential hit compounds for dengue virus NS2B/NS3 protease inhibitors by combining virtual screening and binding free energy calculations*, Trop. Biomed. 30 (2013), pp. 388–408.

[64] Q.Y. Wang, S.J. Patel, E. Vangrevelinghe, H.Y. Xu, R. Rao, D. Jaber, W. Schul, F. Gu, O. Heudi, N.L. Ma, M.K. Poh, W.Y. Phong, T.H. Keller, E. Jacoby, and S.G. Vasudevan, *A small-molecule dengue virus entry inhibitor*, Antimicrob. Agents Chemother. 53 (2009), pp. 1823–1831.

[65] S.M. Tomlinson, R.D. Malmstrom, and S.J. Watowich, *New approaches to structure-based discovery of dengue protease inhibitors*, Infect. Disorders—Drug Targets 9 (2009), pp. 1–17.

[66] S. Ruba, M. Arooj, and G. Naz, *In silico molecular docking studies and design of dengue virus inhibitors*, J. Pharm. Biol. Sci. 9 (2014), pp. 15–23.

[67] H. Galindo, V.G. Cruz, O.M. Lasa, A.M. Reyes, Y.F. Salazar, N.C. Santiago, and G.V.A. Roberto, *Chemical compounds having antiviral activity against dengue virus and other flaviviruses*, Intl patent application# W O 2009/106019 A3, (2010).

[68] E.M. Lipner, M.A. Law, E. Barnett, J.S. Keystone, F. von Sonnenburg, and L. Loutan, *Filariasis in travelers presenting to the GeoSentinel surveillance network*, PLoS Negl. Trop. Dis. 1 (2007), pp. e88.

[69] O.P. Sharma, Y. Vadlamudi, A.G. Kota, V.K. Sinha, and M.S. Kumar, *Drug targets for lymphatic filariasis: A bioinformatics approach*, J. Vector-Borne Dis. 50 (2013), pp. 155–162.
[70] DHR India Annual report 2010–2011.
[71] R. Bhargavi, S. Vishwakarma, and U.S. Murty, *Modeling analysis of GST (glutathione-s-transferases) from* Wuchereria bancrofti *and* Brugia malayi, Bioinform. 1 (2005), pp. 25–27.
[72] S. Azeez, R.O. Babu, R. Aykkal, and R. Narayanan, *Virtual screening and in vitro assay of potential drug like inhibitors from species against glutathione-s-transferase of filarial nematodes*, J. Mol. Model. 18 (2012), pp. 151–163.
[73] R.K. Saraf, S.K. Malhotra, R. Sharma, V.K. Mishra, N. Bora, and P.K. Yadav, *In silico inhibition of chitinase enzyme in filariasis and lead optimization by pharmacophore studies*, Online J. BioSci. Informat. 3 (2014), http://www.academia.edu/12119087/IN_SILICO_INHIBITION_OF_CHITINASE_ENZYME_IN_FILARIASIS_AND_LEAD_OPTIMIZATION_BY_PHARMACOPHORE_STUDIES.
[74] M. Diagana, P.M. Preux, and M. Dumas, *Japanese encephalitis revisited*, J. Neurol. Sci. 262 (2007), pp. 165–170.
[75] Y. Jackson, F. Chappuis, and L. Loutan, *Japanese encephalitis*, Rev. Med. Suisse 3 (2007), pp. 1233–1236.
[76] A. Sampath and R. Padmanabhan, *Molecular targets for flavivirus drug discovery*, Antiviral Res. 81 (2009), pp. 6–15.
[77] A. Kaczor and D. Matosiuk, *Structure-based virtual screening for novel inhibitors of Japanese encephalitis virus NS3 helicase/nucleoside triphosphatase*, Conference on Neglected Tropical Diseases, Cape Town, South Africa, 2009.
[78] West Nile Virus (WNV) Fact Sheet. http://www.cdc.gov/westnile/resources (accessed May 2015).
[79] M.A. Khedr, *Pharmacophore; QSAR-based virtual screening for computational prediction of West Nile virus ns3 protease inhibitors*, Academic J. 5 (2014), pp. 84.
[80] T.P. Monath (ed.), *The Arboviruses: Epidemiology and Ecology*, Vol. 5, CRC Press, Boca Raton, FL, 1989.
[81] A.A. Rashad and P.P. Keller, *Structure based design towards the identification of novel binding sites and inhibitors for the chikungunya virus envelope proteins*, J. Mol. Graph. Model. 44 (2013), pp. 241–252.
[82] M. Bassettoa, T.D. Burghgraeveb, L. Delangb, A. Massarottic, A. Colucciad, N. Zontaa, V. Gattid, G. Colombanoc, G. Sorbac, S. Romanod, G.C. Tronc, J. Neytsb, P. Leyssenb, and A. Brancalea, *Computer-aided identification, design and synthesis of a novel series of 4 compounds with selective antiviral activity against chikungunya virus*, Antiviral Res. 98 (2013), pp. 12–18.
[83] J. Esparza and A. Sánchez, *Multiplication of Venezuelan equine encephalitis (Mucambo) virus in cultured mosquito cells*, Arch. Virol. 49 (1975), pp. 273–280.
[84] M. Lucas-Hourani, A. Lupan, P. Desprès, S. Thoret, O. Pamlard, J. Dubois, C. Guillou, F. Tangy, P.-O. Vidalain, and H.M. Lehmann, *A phenotypic assay to identify chikungunya virus inhibitors targeting the nonstructural protein nsP2*, J. Biomol. Screen. 18 (2013), pp. 172–179.
[85] L. Bora, *Homology modeling and docking to potential novel inhibitor for chikungunya (37997) protein nsP2 protease*, J. Proteomics Bioinform. 5 (2012), pp. 54–59.
[86] S.S. Jadav, V. Jayaprakash, A. Basu, and B.N. Sinha, *Chikungunya protease domain—high throughput virtual screening*, Intern. Schol. Sci. Res. Innov. 6 (2012), pp. 38–39.
[87] T. Liu, Y. Lin, X. Wen, R.N. Jorrisen, and M.K. Gilson, *BindingDB: A web-accessible database of experimentally determined protein-ligand binding affinities*, Nucl. Acids Res. 35 (2007), D198–D201.
[88] OpenEye software, http://www.eyesopen.com/

[89] Schrodinger LLC software packages, http://www.schrodinger.com/

[90] CDC website. Rift Valley fever at http://www.cdc.gov/ncidod/dvrd/spb/mnpages/dispages/rvf.htm

[91] OIE, *Manual of Diagnostic Tests and Vaccines for Terrestrial Animals 2016*, Chapter 2.1.18, *Rift Valley fever (infection with Rift Valley fever virus)*, http://www.oie.int/en/international-standard-setting/terrestrial-manual/access-online/ (Accessed January 23, 2017).

[92] http://www.who.int/mediacentre/news/releases/2013/yellow_fever_20130517/en/

[93] N.Y. Mok and R. Brenk, *Mining the ChEMBL database: An efficient chemoinformatics workflow for assembling an ion channel-focused screening library*, J. Chem. Inf. Mod. 51 (2011), pp. 2449–2454.

[94] N.S.H.N. Moorthy and V. Poongavanam, *The KNIME based classification models for yellow fever virus inhibition*, RSC Adv. 5 (2015), pp. 14663–14669.

[95] A.K. Bhattacharjee, *In silico stereoelectronic profile and pharmacophore similarity analysis of juvenile hormone, juvenile hormone mimics (IGRs) and insect repellents may aid discovery and design of novel arthropod repellents*, in *Juvenile Hormones and Juvenoids. Modeling Biological Effects and Environmental Fate*, J. Devillers, ed., CRC Press, Boca Raton, FL, 2013, pp. 297–331.

[96] R.K. Gupta and A.K. Bhattacharjee, *Discovery and design of new arthropod/insect repellents by computer-aided molecular modeling*, in *Insect Repellents: Principles, Methods, & Use*, M. Debbon, S.P. Frances, and D. Strickman, eds., CRC Press, Boca Raton, FL, 2006, pp. 195–228.

[97] R.K. Gupta, A.K. Bhattacharjee, and D. Ma, US Patent: US7897162, 2011.

[98] B.F. Eldridge and J.D. Edman (eds), *Medical Entomology: A Textbook on Public Health and Veterinary Problems Caused by Arthropods*, Kluwer Academic Publisher, New York, 2000.

[99] G. Koren, D. Matsui, and B. Bailey, *DEET-based insect repellents: Safety implications for children and pregnant and lactating women*, Can. Med. Assoc. J. 169 (2003), pp. 209–212.

[100] G. Briassoulis, *Toxic encephalopathy associated with use of DEET insect repellents: A case analysis of its toxicity in children*, Hum. Exp. Toxicol. 20 (2001), pp. 8–14.

[101] M.S. Fradin, *Mosquitoes and mosquito repellents: A clinician's guide*, Ann. Intern. Med. 128 (1998), pp. 931–940.

[102] A.K. Bhattacharjee, W. Dheranetra, D.A. Nichols, and R.K. Gupta, *3D pharmacophore model for insect repellent activity and discovery of new repellent candidates*, QSAR Comb. Sci. 24 (2005), pp. 593–602.

[103] ModWeb server, https://modbase.compbio.ucsf.edu/modweb/.

[104] P. Benkert, S.C. Tosatto, and D. Schomburg, QMEAN: *A comprehensive scoring function for model quality assessment*, Proteins 71 (2008), pp. 261–277.

[105] A. Filiko, *Kinase-likeness as viewed by chemists and by kinases*, Cambridge Health Institute's Fifth Annual "Protein Kinase Targets: Drug Discovery and Design" meeting abstracts, June 4–6, 2007.

6 Comparative MIA-QSAR Proteochemometric Analysis of NS3 Protease Substrates in Dengue Virus

Mariene H. Duarte, Stephen J. Barigye, and Matheus P. Freitas

CONTENTS

ABSTRACT

The multivariate image analysis (MIA) applied to quantitative structure–activity relationship (QSAR) approach has been shown to be a promising tool in the modeling of bioactivities of chemical compounds. This chapter provides a comparative analysis of the traditional MIA-QSAR, aug-MIA-QSAR, and aug-MIA-QSARcolor approaches in the prediction of the Michaelis constant (K_m) and substrate cleavage rate (K_{cat}) of modified peptides against the dengue virus type 2 (DENV). The aug-MIA-QSARcolor method demonstrates superior predictive power compared to the traditional MIA-QSAR and aug-MIA-QSAR models, respectively, suggesting that the incorporation of color schemes to discriminate the different atom types in chemical structures and the modification of atomic sizes to achieve proportionality to the van der Waals radius improve the predictive ability of the MIA-QSAR method. Structural interpretation of the aug-MIA-QSARcolor models reveals that small, unbranched amino acids in the A1 and A4 positions are associated with greater substrate binding affinity (K_m) and a higher substrate cleavage rate (K_{cat}) when present in the positions A1 and A2, while acidic amino acids in the positions A2 and A4 are related with low substrate affinity and low substrate cleavage rates when in the A1 position. These findings provide relevant chemical information on the structural characteristics critical in the planning of novel target compounds against the DENV in a quicker and inexpensive way.

KEYWORDS

- MIA-QSAR
- aug-MIA-QSARcolor
- Partial least squares
- Multiple linear regression
- Protease NS3
- Dengue virus type 2

6.1 INTRODUCTION

Dengue is an acute febrile disease transmitted by a viral vector that affects tropical and subtropical regions around the world. Recent estimates indicate that 3.9 billion people in 128 countries are at risk of dengue infections [1]. In 2013, it was reported that 390 million dengue infections occurred in 2010 [2]. The geographical expansion and the increase in the number of cases and severity of the disease have caused dengue to become a serious public health problem worldwide [3,4].

Dengue transmission occurs primarily by female mosquitoes of the *Aedes aegypti* species and, to a lesser extent, by *Ae. albopictus* [4], whose etiologic agent is a virus of the *Flavivirus* genus (*Flaviviridae* family), which includes other disease-causing viruses such as the West Nile virus, Japanese encephalitis virus, Saint Louis encephalitis virus, and the yellow fever virus [5–8]. While vaccines for yellow fever and Japanese encephalitis are already available, there is no specific therapy or vaccine available against the Dengue virus (DENV), but rather palliative treatments to overcome the symptoms caused by the disease [9]. So, the prevention and reduction of dengue infections is entirely dependent on the control of mosquitoes or interruption of human–vector contact. These methods have not been shown to be satisfactory due to their dependence on social issues such as basic sanitation, collective community participation, and public health policies to contain mosquito-breeding sites or promote the use of nonspecific repellents. Thus, there is an urgent need to develop potent drugs and/or vaccines against the dengue infection.

Molecular modeling and virtual screening techniques offer valuable knowledge on the structural characteristics of peptides capable of inhibiting the DENV infectivity due to their affinity to the conformational intermediate in the fusion-promoting rearrangement. In this context, this chapter provides a comparative analysis of the multivariate image analysis applied to quantitative structure–activity relationship (MIA-QSAR) approaches in the prediction of the activity of modified peptides against the DENV using partial least squares (PLS) and multiple linear regression (MLR) methods. Additionally, an interpretation of the built models is performed with the aim of gaining greater insight on the structural features that favor the peptides' activity against the DENV, and thus contributing to the planning for the novel compounds effective against this infection.

6.2 ETIOLOGY AND STRUCTURAL CHARACTERISTICS OF THE DENV

The DENV comprises a spherical particle diameter of 40–50 nm, with a lipopolysaccharide envelope. The genome consists of a single strand of positive polarity RNA, with approximately 11,000 nucleotides forming a single open reading frame, which encodes a single polyprotein precursor sequence 5′-CprM(M)-E-NS1-NS2A-NS2B-NS3-NS4A-NS4B-NS5-3′. This viral polyprotein is cleaved by proteases in 10 gene products: three structural proteins (capsid, precursor membrane, and envelope) essential for entry, assembly, and viral maturation, and seven nonstructural proteins (i.e., NS1, NS2a, NS2B, NS3, NS4A, NS4B, NS5) responsible for acting in viral replication. There are four serotypes of the dengue virus—DENV-1, DENV-2, DENV-3 and DENV-4—characterized by different antigenic properties [10]. The four serotypes are responsible for dengue fever, dengue hemorrhagic fever, and dengue check syndrome [11] and are distributed in all endemic regions [12]. Infection by one of them gives permanent protection from the same serotype, but only temporary and partial immunity against the others [13,14]. These viral characteristics have hindered the development of effective drugs and vaccines to combat the disease. Moreover, the lack of an animal model for testing the effectiveness of vaccines has hindered *in vivo* experimental evaluations, since mice and monkey models show no pathology similar to humans when infected by the virus [15].

One of the most promising molecular targets studied for the development of specific therapies for dengue prevention is the multifunctional protein NS3, comprising 618 amino acids. This protein is vital for the post-translational proteolytic processing of the precursor polyprotein and essential for viral replication and virion maturation [16,17]. However, the NS3 requires the NS2B cofactor for its proteolytic activation and structural stability [18]. The NS3 protein contains two domains that possess multiple enzymatic activities: a trypsin-like serine protease domain located within the *N*-terminal amino acid residues [16,19,20] and an RNA helicase and NTPase in the *C*-terminal region [21,22].

6.3 THE MIA-QSAR METHOD IN MOLECULAR MODELING

6.3.1 Description of MIA-QSAR Method

The QSAR methods have been shown to be powerful tools in determining the structural profiles required to improve the drugs' performance, in the estimation of the bioactivity of drug candidates, and in the comprehension of the drugs' mechanisms of action [23,24]. These approaches are based on the understanding that the variation of the biological response for a series of molecules is related to the changes in their structural, chemical, and physical properties [25]. Modern QSAR modeling traces back to the 1960s with the seminal work of Hansch and Fujita; using the descriptors octanol/water partition coefficient (log P), the Hammett constant σ, and the lipophilicity parameter π, they achieved correlations with the biological activity of benzoic acids [26]. The Hansch–Fujita equation—together with the Free and Wilson model developed in the same year, based on additive substituent contributions [27]—reaffirmed the hypothesis that structural changes in a molecule alter its biological activity. Over the years, the QSAR methodology has progressed in many aspects, including the extension of the different methods to higher dimensions, incorporation of new computational approaches and statistical algorithms in the model building, and the definition of more chemically meaningful molecular descriptors that codify orthogonal chemical information [28]. Such advances have allowed for the application of QSAR studies in various areas, such as analytical chemistry, agriculture, environmental chemistry, toxicology and medicinal chemistry (particularly in the screening for new drugs).

In this perspective, the MIA-QSAR method [29], characterized as a two-dimensional (2D)-QSAR method, is considered as a promising tool in the screening for novel compounds of therapeutic interest. This is a simple technique, does not involve complex mathematical operations, and provides rapid characterization of chemical structures [30]. The MIA-QSAR approach is based on the premise that chemical structure images contain relevant information on the behavior of chemicals and may thus be useful in the description of their chemical, physicochemical, and biological properties. In this sense, the pixels for 2D chemical structure drawings are treated as descriptors for correlating the chemical structures with their activities [30]. Consequently, structural modifications or changes in the positions of the substituents in a series of molecules correspond to changes in the coordinates of the pixels of the image, and these changes explain the variance in the molecular properties [31].

An important preliminary step in the construction of MIA-QSAR models is the alignment of the molecules in a congeneric series of compounds with respect to a pixel common to all molecules. In pioneering MIA-QSAR studies, the models were constructed from images based on a binary pixel system for black-and-white images, where the pixel 765 (white) corresponds to blank spaces and the pixel 0 (black) to the atoms and bonds in a molecule. For this approach, the denomination **traditional MIA-QSAR** will be adopted. This method has been successfully used in the codifying of information on chemical and physicochemical phenomena, as well as a wide range of bioactivities [29,30,32–34] and its results have compared favorably with more sophisticated techniques such as comparative molecular field analysis (CoMFA) and comparative analysis of molecular similarity indices (CoMSIA) [30,35].

However, notwithstanding the satisfactory results obtained with the traditional MIA-QSAR approach, the use of a binary pixel system for the chemical images presents a drawback in the sense that some chemical structure details, such as the atomic radius and the nature of substituents are not considered, and this consequently precludes the possibility for the interpretation of the built models in terms of the atom

TABLE 6.1
Colors and Corresponding Pixels for Atoms according to the aug-MIA-QSAR and aug-MIA-QSARcolor Values (according to Pauling's Scale)

		aug-MIA-QSAR		aug-MIA-QSARcolor	
Atom	**E[a]**	**Colors**	**Pixel**	**Colors**	**Pixel**
H	2.1	Silver	612	Charcoal	210
C	2.5	Gray	426	Teal	250
N	3.0	Blue	279	Persian blue	300
O	3.5	Red	229	Scarlet	350
F	4.0	Electric blue	688	Turquoise	400
S	2.5	Gold	493	Olive	250
Cl	3.0	Green	289	Green	300
Br	2.8	Carmine	231	Maroon	280
I	2.5	Byzantium	294	Purple	250

[a] E: Pauling electronegativity.

types or functional groups responsible for the variation in the modeled properties, as a preliminary step in the rational design of novel structures. This limitation motivated the incorporation of color schemes to the MIA-QSAR method, yielding the **aug-MIA-QSAR** (augmented multivariate image analysis applied to QSAR) and the **aug-MIA-QSARcolor** schemes, respectively. In the latter, the atoms in the chemical structures are differentiated using randomly selected colors and the size of the atoms made to be proportional to the van der Waals radius [36–40] (see Table 6.1).

It is thus expected that biological properties that exhibit a strong relationship with the atomic or molecular bulk should correlate with the descriptors according to this approach [36–38]. As for aug-MIA-QSARcolor, the colors of atoms are made to be proportional to different atomic properties, such as electronegativity, atomic hydrophobicity, polarizability, atomic refractivity, covalent radius, etc. [40,41]. In this chapter, Pauling electronegativity is considered in the sense that the atom colors are chosen so that their pixel values perfectly correlate with Pauling's electronegativity for atoms [37,39,40]. For example, the colors for the carbon, oxygen, and fluorine atoms, with electronegativity values of 2.5, 3.5, and 4.0, respectively, are chosen to correspond to pixel values 250, 350, and 400, respectively (for details see Table 6.1). With the incorporation of the atomic radius and electronegativity, it is hoped that the aug-MIA-QSARcolor descriptors describe more accurately the structural characteristics of chemical compounds relative to the aug-MIA-QSAR and traditional MIA-QSAR methods.

6.3.2 Data Set of Antimicrobial Peptides for QSAR Modeling

A data set of 54 modified peptides known to be substrates for the NS3 protease of the DENV type 2 and whose Michaelis constant (K_m) and the kinetic parameter for the cleavage rate constant (K_{cat}) values have been experimentally determined was obtained from the literature (see Table 6.2) [42]. The black-and-white images of the modified peptides (i.e., for the traditional MIA-QSAR approach) were prepared using the ChemSketch [43] program and the colored images for the aug-MIA-QSAR and aug-MIA-QSARcolor approaches obtained using the GaussView program (Version 5.0.8) [44].

Each chemical structure was transferred to a work space in the Microsoft Windows Paint application of predefined dimension (i.e., 500 × 300 pixels for the traditional

TABLE 6.2
A1′–A4′ Sequence of Modified Peptides with General Structure Abz-RRRRXXXX-nY-NH$_2$[a] and Experimental Michaelis Constant (K_m) and Kinetic Parameter for Cleavage Rate Constant (K_{cat}) Values Obtained with DENV-2 Protease

Cpd	Peptide	K_m[b]	K_{cat}[b]	Cpd	Peptide	K_m[b]	K_{cat}[b]
1	APCN	0.73	−0.62	28	STPH	1.24	−0.03
2	HHGN	0.64	−1.1	29	GLGF	0.64	−0.05
3	DDGN	1.9	−1.15	30	DTSF	1.18	−1.22
4	SNSN	0.8	0.23	31	HPTF	1.09	−1.1
5	NWTN	0.78	−0.52	32	ADTF	1.43	−0.68
6	GTVN	0.81	−0.48	33	NSVF	1.04	−0.54
7	AGPN	0.62	−0.39	34	SAPF	0.73	0.3
8	SHCD	1.06	−0.13	35	ATCL	0.86	−0.66
9	APGD	1.29	−1.4	36	NGGL	0.86	−1
10	HWSD	1.17	−1.15	37	HDSL	1.27	−0.66
11	GGTD	1.11	−0.37	38	GSSL	1.02	−0.32
12	AAVD	1.26	−0.12	39	SLTL	0.7	0.25
13	NDVD	1.64	−1	40	DHTL	1.46	−1.15
14	NLPD	1.22	−0.66	41	DNPL	1.24	−1.15
15	HNCC	0.94	−0.85	42	HTG–	0.82	−1.1
16	NACC	0.51	−0.55	43	DLS–	1.29	−0.85
17	SWGC	0.46	−0.41	44	NHT–	0.78	−0.77
18	AHSC	0.75	−0.59	45	AWV–	0.35	−0.57
19	DSTC	1.11	−1.15	46	GPP–	0.71	−0.77
20	SGVC	0.63	−0.01	47	SAGM	0.54	0.27
21	GDPC	1.12	−0.74	48	SWPL	0.74	−0.2
22	DGCH	0.85	−1.05	49	AGVL	0.34	−0.47
23	ANGHH	0.85	−1.05	50	GTGN	0.48	−0.54
24	NPSH	0.71	−1.1	51	SAG–	0.56	0.4
25	GATH	0.67	−0.03	52	AAGM	0.39	−0.26
26	HLVH	1.15	−0.85	53	SAAM	0.52	0.19
27	HSPH	0.76	−1.15	54	SAGA	0.49	0.21

[a] Abz: *o*-aminobenzoic acid; nY: 3-nitrotyrosine.
[b] Expressed on logarithmic scale, original K_m and K_{cat} values in μM and min–1, respectively.

MIA-QSAR and 420 × 300 pixels for the aug-MIA-QSAR and aug-MIA-QSAR-color approaches) and each image was saved as a bitmaps file. The alignment of the molecules was carried out using a manual procedure in which a pixel sequence common to all molecules is set at a given coordinate, resulting in the overlapping of the molecules in a three-way arrangement of dimensions 54 × 500 × 300 (traditional MIA-QSAR) and 54 × 420 × 300 (aug-MIA-QSAR and aug-MIA-QSARcolor), respectively. After alignment, the three-way arrays were unfolded to give two-way matrices (see Figure 6.1 for an illustrative scheme for this procedure).

The obtained matrices were subsequently filtered for zero variance columns (white spaces common for all images and congruent parts of the chemical structures), yielding lower dimensionality matrices of size 54 × 4,723 (traditional MIA-QSAR), 54 × 22,756 (aug-MIA-QSAR), and 54 × 21,592 (aug-MIA-QSARcolor). The unfolding and dimensionality reduction procedure (elimination of zero variance columns) was performed using a free and user-friendly software denominated as the Chemoface program [45], downloadable via the Internet at http://ufla.br/chemoface/.

6.4 CHEMOMETRIC METHODS FOR MODELING AND VALIDATION

Given the large number of descriptors generated by MIA-QSAR approaches, the PLS technique was employed as the primary model-building method. In principle, this method is based on the transformation of the original variables into orthogonal projections (latent variables) that essentially encode all the information contained in the original data matrix [37]. For this regression analysis, the modified peptides data set was split into training and test series containing 43 (80%) and 11 (20%) compounds, respectively, in accordance with the literature [46], where a random procedure was performed. The quality of the PLS-based models was evaluated using the following parameters: determination coefficients for calibration (r^2) and leave-one-out cross-validation (r^2_{cv}), as well as the cross-validation root mean square error (RMSEcv). Additionally, the models were vigorously validated using the external validation (r^2_{pred}), and Y-randomization (statistically evaluated using RMSErand and r^2_{rand}) procedures to attest for the models' earnest predictive power and the absence of fortuitous correlation. Given the need of performing yet more vigorous validation procedures in QSAR modeling, additional parameters have been suggested to make the modeling results much more reliable [47]; these include the modified coefficient of determination (r^2_m) and penalized corrected coefficient of determination (${}^c r^2_p$) used to detect proximity between the observed and expected activity and to evaluate the difference of r^2 and r^2_{rand}, respectively [48–51]. These validation procedures were performed as well in the present study.

Although PLS is a preferred technique when dealing with data matrices with numerous variables relative to the instances, the fact that the original variables are converted into orthogonal projections makes the interpretation of the models in terms of the contribution of the atoms or atom types to the modeled activity complex because the original variables are "masked" in the latent variables. Therefore, in order to seek a more straightforward interpretation to the MIA-QSAR approaches, the MLR technique was also used. In MLR, the biological activity is estimated using

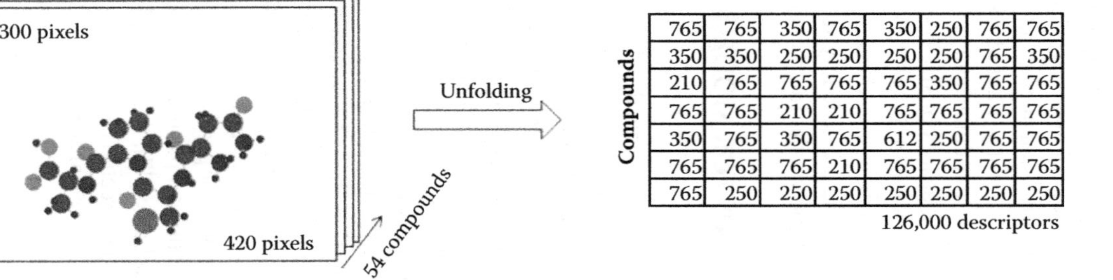

FIGURE 6.1 Illustration for the construction of the three-way multivariate image and unfolding procedures to yield a two-way data matrix for chemical data set.

a linear combination of independent variables contained in the matrix X. The key limitations of MLR are that it is more prone to over-fitting and chance correlation, in addition to the fact that exploring all possible combinations of the independent variables is tedious and time consuming. Topliss and Costello have proposed a rule consistent with Occam's razor for checking possible over-fitting, which stipulates that the ratio of chemicals to variables should always be equal to or higher than five [52]. To deal with the high dimensionality of the MIA-QSAR data matrices, a prefiltering technique based on Shannon's entropy (SE) was applied, where variables with less than 15% of the maximum entropy ($SE_{max} = Log_2 54$) were eliminated [53]. However, SE is a relevance-based filter and therefore does not deal with possible multicollinearity among the variables. Thus, the reduced matrix was further subjected to a prefiltering technique based on the correlation coefficient (x/x), where, for variables with correlation coefficients (x/x) ≥ 0.98, one variable is retained. Subsequently, MLR-based models were built using the filtered sets of variables and the models with the highest r^2 and r^2_{cv} values for each MIA-QSAR approach were retained. The obtained models were subsequently validated using the external validation and Y-randomization techniques and the corresponding parameters compared.

6.5 RESULTS AND DISCUSSION

6.5.1 PLS Analysis

The PLS models for the kinetic parameter K_m were built using three latent variables (LVs) for the traditional MIA-QSAR approach and two LVs for aug-MIA-QSAR and aug-MIA-QSARcolor approaches, respectively. For the kinetic parameter K_{cat}, six LVs were considered for the traditional MIA-QSAR and five LVs for aug-MIA-QSAR and aug-MIA-QSARcolor approaches, respectively. The number of LVs was determined based on the reduction of the leave-one-out cross-validation root mean square error (RMSEcv). The use of an excessive number of LVs may lead to over-fitting, notwithstanding the good correlation coefficients for calibration and internal validation [38]. Table 6.3 shows the statistical parameters for the PLS-based models obtained according to the traditional MIA-QSAR, aug-MIA-QSAR, and aug-MIA-QSARcolor approaches, respectively.

Generally, the built models showed good predictive capacity for the kinetic constant parameter of substrate cleavage (K_{cat}), with r^2 values > 0.7, and r^2_{cv}, r^2_{pred}, and r^2_m > 0.5 (see Table 6.3). In the comparison of the three MIA-QSAR approaches, the traditional MIA-QSAR method showed better results for calibration and external validation [$r^2 = 0.97$, $r^2_{pred} = 0.78$ ($r^2_m = 0.67$)] when compared to aug-MIA-QSAR [$r^2 = 0.89$, $r^2_{pred} = 0.70$ ($r^2_m = 0.58$)] and aug-MIA-QSARcolor [$r^2 = 0.90$, $r^2_{pred} = 0.59$ ($r^2_m = 0.57$)] approaches. However, the high r^2_{rand} value for the traditional MIA-QSAR (0.90) compared to the aug-MIA-QSARcolor (0.66) and aug-MIA-QSAR (0.62) approaches, suggests that the traditional MIA-QSAR model is more prone to chance correlation and may thus be considered unreliable. Moreover, the rather low ${}^{c}r^2_p$ value for the traditional MIA-QSAR (0.26), relative to the aug-MIA-QSAR (0.49) and aug-MIA-QSARcolor (0.46) methods, suggests rather low predictive power, although none of the approach attains the limit value of desirability (i.e., ${}^{c}r^2_p \geq 0.50$) [48].

TABLE 6.3
Statistics for PLS-Based QSAR Models Obtained for K_m and K_{cat} Parameters for Peptide Substrates against Protease NS3 DENV Type 2

	Traditional-MIA-QSAR		aug-MIA-QSAR		aug-MIA-QSARcolor	
Parameter	K_{cat}	K_m	K_{cat}	K_m	K_{cat}	K_m
LV	6	3	5	2	5	2
r^2_{cal}	0.97	0.66	0.89	0.44	0.90	0.44
r^2_{cv}	0.52	0.09	0.58	0.14	0.51	0.12
RMSEcv	0.08	0.21	0.16	0.22	0.15	0.35
r^2_{rand}	0.90	0.62	0.62	0.30	0.66	0.37
RMSErand	0.15	0.23	0.30	0.30	0.28	0.29
${}^cr^2_p$	0.26	0.16	0.49	0.25	0.46	0.17
r^2_{pred}	0.78	0.05	0.70	0.14	0.59	0.01
RMSEp	0.23	0.31	0.25	0.3	0.35	0.33
r^2_m	0.67	0.00	0.58	0.00	0.57	0.00

In this sense, the aug-MIA-QSAR-based model may be considered more predictive than the aug-MIA-QSARcolor and traditional MIA-QSAR approaches, as it shows better overall performance for the randomization and external validation procedures.

As for the PLS-based MIA-QSAR models for the Michaelis constant (K_m) of peptide substrates for the NS3 protease, it is observed that the traditional MIA-QSAR, aug-MIA-QSAR, and aug-MIA-QSARcolor approaches do not yield satisfactory results, since the values for r^2_{cv}, r^2_{pred}, ${}^cr^2_p$, and r^2_m were less than 0.5 and the r^2 less than 0.7 [48,49,51] (see Table 6.3). This result suggests that the PLS technique is unsuitable for modeling of the K_m parameter of the considered data set of modified peptides. Low performance of the PLS method for K_m was also observed in previous studies in the literature [42,46].

6.5.2 Multiple Linear Regression

Six variable MLR-based models were built for the K_m and K_{cat} parameters (Eqs. 6.1–6.6). The corresponding statistical parameters are also shown, using the traditional MIA-QSAR, aug-MIA-QSAR, and aug-MIA-QSARcolor approaches, respectively.

- Michaelis constant (K_m)
    - Traditional MIA-QSAR

$$K_m = 1.3702 - 0.0005\text{ X482} - 0.0004\text{ X874} - 0.0004\text{ X19020} - 0.0004\text{ X2243} - 0.0005\text{ X3869} - 0.0004\text{ X4451} \quad (6.1)$$

n = 54 (43 training and 11 test), $r^2_{\text{cal}} = 0.77$, $r^2_{\text{cv}} = 0.68$, RMSEcv = 0.18, $r^2_{\text{rand}} = 0.18$, ${}^{c}r^2_{\text{p}} = 0.67$, $r^2_{\text{pred}} = 0.24$, $r^2_{\text{m}} = 0.00$.

- aug-MIA-QSAR

$$K_{\text{m}} = 2.1299 - 0.0021\ \text{X}1377 + 0.0007\ \text{X}1704 - 0.0008\ \text{X}4256 - 0.017\ \text{X}6809 + 0.0025\ \text{X}18185 - 0.0008\ \text{X}19371 \quad (6.2)$$

n = 54 (43 training and 11 test), $r^2_{\text{cal}} = 0.64$, $r^2_{\text{cv}} = 0.49$, RMSEcv = 0.26, $r^2_{\text{rand}} = 0.12$, ${}^{c}r^2_{\text{p}} = 0.58$, $r^2_{\text{pred}} = 0.58$, $r^2_{\text{m}} = 0.53$.

- aug-MIA-QSARcolor

$$K_{\text{m}} = 0.7346 + 0.0013\ \text{X}1377 - 0.0016\ \text{X}1704 + 0.0010\ \text{X}4256 - 0.005\ \text{X}6809 + 0.0006\ \text{X}18185 - 0.0010\ \text{X}19371 \quad (6.3)$$

n = 54 (43 training and 11 test), $r^2_{\text{cal}} = 0.71$, $r^2_{\text{cv}} = 0.60$, RMSEcv = 0.21, $r^2_{\text{rand}} = 0.13$, ${}^{c}r^2_{\text{p}} = 0.64$, $r^2_{\text{pred}} = 0.64$, $r^2_{\text{m}} = 0.5$.

- Cleavage rate constant (K_{cat})
  - Traditional MIA-QSAR

$$K_{\text{cat}} = 0.3369 - 0.0015\ \text{X}482 + 0.0008\ \text{X}874 - 0.0004\ \text{X}19020 + 0.0006\ \text{X}224 - 0.0002\ \text{X}3869 - 0.0005\ \text{X}4451 \quad (6.4)$$

n = 54 (43 training and 11 test), $r^2_{\text{cal}} = 0.90$, $r^2_{\text{cv}} = 0.87$, RMSEcv = 0.18, $r^2_{\text{rand}} = 0.13$, ${}^{c}r^2_{\text{p}} = 0.83$, $r^2_{\text{pred}} = 0.66$, $r^2_{\text{m}} = 0.66$.

  - aug-MIA-QSAR

$$K_{\text{cat}} = -1.2412 + 0.0015\ \text{X}803 - 0.0011\ \text{X}1843 + 0.0011\ \text{X}2988 - 0.0020\ \text{X}7062 + 0.0023\ \text{X}7655 - 0.0006\ \text{X}11236 \quad (6.5)$$

n = 54 (43 training and 11 test), $r^2_{\text{cal}} = 0.87$, $r^2_{\text{cv}} = 0.82$, RMSEcv = 0.21, $r^2_{\text{rand}} = 0.11$, ${}^{c}r^2_{\text{p}} = 0.81$, $r^2_{\text{pred}} = 0.76$, $r^2_{\text{m}} = 0.60$.

  - aug-MIA-QSARcolor

$$K_{\text{cat}} = -1.7105 + 0.0009\ \text{X}309 + 0.0013\ \text{X}2090 - 0.012\ \text{X}2126 + 0.0005\ \text{X}2498 + 0.0006\ \text{X}3892 - 0.0008\ \text{X}7320 \quad (6.6)$$

n = 54 (43 training and 11 test), $r^2_{\text{cal}} = 0.88$, $r^2_{\text{cv}} = 0.81$, RMSEcv = 0.21, $r^2_{\text{rand}} = 0.11$, ${}^{c}r^2_{\text{p}} = 0.82$, $r^2_{\text{pred}} = 0.92$, $r^2_{\text{m}} = 0.80$.

As may be observed in these equations, the models for the K_m and K_{cat} parameters generally show good behavior demonstrated by the quality of the respective statistics. In the comparison of the performance of the K_m models for the different MIA-QSARs approaches (see Eqs. 6.1–6.3), it is observed that despite the superior calibration and cross-validation parameters for the traditional MIA-QSAR approach [r^2 (0.77), r^2_{cv} (0.68), ${}^c r^2_p$ (0.67)] relative to the aug-MIA-QSAR [r^2 (0.64), r^2_{cv} (0.49), ${}^c r^2_p$ (0.58)] and aug-MIA-QSARcolor [r^2 (0.71), r^2_{cv} (0.60), ${}^c r^2_p$ (0.64)] approaches, the aug-MIA-QSARcolor-based model possesses greater robustness and predictive power, since it yields higher values for external validation parameters [i.e., r^2_{pred} (0.64) and r^2_m (0.55)] compared to the traditional MIA-QSAR [r^2_{pred} (0.24), r^2_m (0.00)] and aug-MIA-QSAR [r^2_{pred} (0.58), r^2_m (0.53)] approaches.

As for the substrate K_{cat} models (Eqs. 6.4–6.6), it is observed that while comparable performance is observed for the three approaches when the calibration and internal validation procedures are compared—that is, traditional MIA-QSAR [r^2_{cal} (0.90), r^2_{cv} (0.87)], aug-MIA-QSAR [r^2_{cal} (0.87), r^2_{cv} (0.82)], and aug-MIA-QSARcolor [r^2_{cal} (0.88), r^2_{cv} (0.83)], superior performance is observed for the aug-MIA-QSARcolor approach for the external validation procedure [i.e., r^2_{pred} (0.92), r^2_m (0.80)] relative to the traditional MIA-QSAR [r^2_{pred} (0.66), r^2_m (0.66)] and aug-MIA-QSAR [r^2_{pred} (0.76), r^2_m (0.60)] approaches.

The superior predictive power for the aug-MIA-QSARcolor approach suggests that the incorporation of more meaningful chemical information to the images—particularly, the modification of the atomic sizes to achieve proportionality to the van der Waals radius and the use of color schemes carefully defined to correlate with the atomic electronegativity—improves the performance of the MIA-QSAR method. Also, in addition to contributing to the improvement of the predictive power, the color schemes open the way to the chemical interpretation of the built models, since the atoms are identified by their pixel values in the variables. In the next section, aug-MIA-QSARcolor models for the K_{cat} and K_m parameters are analyzed to interpret the chemical information coded thereof, which consequently contributes to the development of more promising substrates with greater affinity for the NS3 protein of the DENV.

6.5.3 Structural Interpretation of the aug-MIA-QSARcolor Models

An analysis of the variables contained in the aug-MIA-QSARcolor model for the K_m parameter (see Table 6.4) reveals that the variables X1377 and X1704 correspond to pixel coordinates for the A1 amino acid side chains, the variables X4256 and X6809 codify information on A2 amino acid chains, and the variables X18185 and X19371 encode information on A4 amino acid R-groups (note that each variable corresponds to a single pixel coordinate in the multivariate image). The absence of variables corresponding to pixel coordinates for the A3 side chains suggests minimal contribution for amino acids in this position to the kinetic parameter K_m, corroborating the results obtained in a previous study by Prusis et al. [42].

A simple inductive analysis reveals that blank space pixel values (765) for the variables X1377 and X1704 are indicative of small-size R-groups for the A1 amino acid (e.g., forming part of alanine, serine, and glycine). When concurrent with

TABLE 6.4
Variables Contained in aug-MIA-QSARcolor Model for K_m Parameter

Cpd	X1377	X1704	X4256	X6809	X18185	X19371
1[a]	765	765	765	765	210	210
2[a]	300	250	612	612	210	210
3	765	250	765	350	210	210
4	765	765	765	350	210	210
5	300	250	250	250	210	210
6	765	765	765	210	765	765
7	765	765	765	765	210	210
8	765	765	612	612	765	210
9	765	765	765	765	765	210
10	300	250	250	250	765	210
11	765	765	765	765	765	210
12	765	765	765	765	765	210
13	300	250	765	350	765	210
14	300	250	765	612	210	210
15	300	250	765	350	765	765
16	300	250	765	765	765	765
17	765	612	250	250	765	765
18	765	765	765	612	765	765
19[a]	765	250	765	765	765	765
20	765	765	765	765	765	765
21	765	765	765	350	765	765
22[a]	300	250	765	765	765	765
23	765	765	765	350	612	765
24[a]	300	250	765	765	612	765
25	765	765	765	765	612	765
26	300	250	765	612	612	765
27	300	250	765	210	612	765
28	765	612	765	210	612	765
29	765	765	765	612	765	612
30	350	250	765	210	765	612
31	300	250	765	765	765	612
32	765	765	765	350	765	612
33	300	250	765	210	765	612
34	765	612	765	765	765	612
35[a]	765	765	765	210	765	765
36	300	250	765	765	765	765
37	300	250	765	350	765	765
38[a]	765	765	765	210	765	765
39	765	765	765	612	765	765
40	765	250	612	612	765	765
41	350	250	765	350	765	765
42	300	250	765	210	765	765

(Continued)

TABLE 6.4 (CONTINUED)
Variables Contained in aug-MIA-QSARcolor Model for K_m Parameter

Cpd	X1377	X1704	X4256	X6809	X18185	X19371
43	765	250	765	612	765	765
44	300	250	612	612	765	765
45[a]	765	765	250	250	765	765
46	765	765	765	765	765	765
47	765	765	765	765	765	765
48[a]	765	612	250	250	765	765
49[a]	765	765	765	765	765	765
50	765	765	765	210	765	765
51[a]	765	765	765	765	765	765
52	765	765	765	765	765	765
53	765	765	765	765	765	765
54	765	765	765	765	765	765

[a] Test compounds.

similar size R-groups in A4 (identified by blank spaces [765] for substituents X18185 and X19371; e.g., forming part of glycine, alanine, leucine, and methionine), low K_m values are observed (i.e., high activity). In contrast, when small-size R-groups in A1 are concurrent with a carboxyl group (identified by pixels 765 [blank space] and 210 [H] for X18185 and X19371, respectively, i.e., hydrogen forming part of the carboxyl group) containing side chains in A4, an opposite effect is achieved (i.e., is related to high K_m values). Even more interesting, irrespective of the R-groups in A1 and A4, when the A2 side chain comprises a carboxyl group (identified by pixels 765 [blank space] and 350 [O] for variables or pixel positions X4256 and X6809, respectively), the K_m value increases. Note that the pixel positions X4256 and X6809 (or X18185 and X19371) dictate that the carboxyl group be separated by approximately one carbon distance from the peptide chain, as is in aspartic acid. This result is consistent with previous reports in the literature where low substrate affinity was attributed to the presence of aspartic acid units in the peptide substrates. Conversely, a bicyclic system R-group for the A2 amino acid (identified by carbon [250] atoms at pixel positions X4256 and X6809) is related to low K_m values (high substrate affinity), with a synergic effect if the A1 and A4 side chains also favor low K_m values. This is an interesting finding in line with an earlier study in which tryptophan (indole derivate) was cited as the most favorable amino acid for this position for high substrate affinity [42].

A painstaking assessment of the variables constituting the MLR-based aug-MIA-QSARcolor model (see Table 6.5) for the substrate cleavage constant indicates that the variables X309, X2090, X2126, and X2498 correspond to pixel positions for the A1 amino acid R-groups, while the variables X6660 and X7320 encode information on the A2 amino acid side chains. Given that no variable in the aug-MIA-QSARcolor model corresponds to R-groups in the A3 and A4 positions, it may be inferred that

TABLE 6.5
Variables Contained in aug-MIA-QSARcolor Model for K_{cat} Parameter

Cpd	X309	X2090	X2126	X2498	X6660	X7320
1[a]	765	765	765	765	765	765
2[a]	250	250	250	250	612	250
3	210	350	250	250	350	250
4	765	765	350	765	350	250
5	765	350	250	250	250	250
6	765	765	765	765	765	350
7	765	765	210	210	765	765
8	765	765	350	765	612	250
9	765	765	765	210	765	765
10	250	250	250	250	250	250
11	765	765	765	765	765	765
12	765	765	765	765	765	210
13	612	350	250	250	350	250
14	765	350	250	250	765	250
15	250	612	250	250	350	250
16	612	350	250	250	765	210
17	765	765	350	765	250	250
18	765	765	765	210	612	250
19[a]	210	350	250	250	765	350
20	765	765	350	765	765	765
21	765	765	765	765	350	250
22[a]	612	350	250	250	765	765
23	765	765	765	210	350	250
24[a]	612	350	250	250	765	765
25	765	765	765	765	765	210
26	250	250	250	765	765	250
27	250	250	250	250	765	350
28	765	765	350	765	765	350
29	765	765	765	765	765	250
30	210	350	250	250	765	350
31	250	250	250	250	765	765
32	765	765	765	765	350	250
33	612	350	250	250	765	350
34	765	765	350	765	765	210
35[a]	765	765	765	210	765	350
36	612	350	250	250	765	765
37	250	612	250	250	350	250
38[a]	765	765	765	765	765	350
39	765	765	350	765	765	250
40	210	350	250	250	612	250
41	210	350	250	250	350	250
42	250	250	250	250	765	350

(Continued)

TABLE 6.5 (CONTINUED)
Variables Contained in aug-MIA-QSARcolor Model for K_{cat} Parameter

Cpd	X309	X2090	X2126	X2498	X6660	X7320
43	210	350	250	250	765	250
44	612	350	250	250	612	250
45[a]	765	765	210	210	250	250
46	765	765	765	765	765	765
47	765	765	350	765	765	210
48[a]	765	765	350	765	250	250
49[a]	765	765	765	765	765	765
50	765	765	765	765	765	350
51[a]	765	765	350	765	765	210
52	765	765	765	765	765	210
53	765	765	350	765	765	210
54	765	765	350	765	765	210

[a] Test compounds.

these have no influence on the substrate cleavage constant, and thus these positions may be important for the modification of other properties, such as the physicochemical and pharmacokinetic factors, without significantly altering the cleavage constant. This finding is consistent with a previous study on tetrapeptide aldehyde inhibitors [54], where it was reported that A2 side chain interactions are more important than those of A1, A3, and A4 side chains and that modifications in A3 and A4 positions are tolerated without significant reduction in cleavage constant substitutions, with A4 depicting little contribution to the binding of the enzyme.

Small-size side chains in A1—identified by the pixel values 765 (X309), 765 (X2090), 350 (X2126), and 765 (X2498), corresponding to a hydroxyl group separated by a one-carbon distance from the peptide backbone as in serine and blank space pixel values (765) for all the four pixel positions, indicating much smaller R-groups in this position, such as hydrogen and a methyl group as in alanine and glycine—are associated with high K_{cat} values. These findings are consistent with previous studies where it was reported that small, unbranched amino acids in the A1 position contribute to the increase in the cleavage rate of the substrates as these favor the enzyme–substrate interactions [54–57]. Note that when the tetrapeptide substrate contains small, unbranched amino acids in A1 and a methyl group variant for the A2 amino acid (identified by pixel values 765 [blank space] and 210 [H] for the variables X6660 and X7320, respectively), greater cleavage constant values are obtained. The opposite is observed when the A2 amino acid contains a cyclic side chain forming part of the backbone (identified by blank spaces for the variables X6660 and X7320), as in proline. The rest of the R-group substituents for A2 do not show a particular trend as they are spread all over the K_{cat} range of values. This suggests that A2 allows for a diverse range of amino acids, corroborating the findings of a previous study by Li et al. [54]. On the other hand, amide—identified by the pixel values 612 (X309),

350 (X2090), 250 (X2126), and 250 (X2498); imidazole—represented by the carbon atom pixel (250) for the four variables; or carboxyl—identified by the pixel values 210 (X309), 350 (X2090), 250 (X2126), and 250 (X2498)—groups for the A1 amino acid side chains are related to low K_{cat} values. Thus, asparagine, histidine, and aspartic acid are deemed unsuitable for this position. Altogether, these findings provide valuable information on the structural characteristics key in the development of novel tetrapeptides with greater activity against the DENV.

6.6 CONCLUSION

Since the inception of the MIA-QSAR approach in 2005, it has undergone changes that seek to incorporate more meaningful chemical information to the structural images. The use of color schemes to discriminate different atom types and the modification of atomic sizes according to the van der Waals radius have improved the performance of the MIA-QSAR models and opened the way to their structural interpretation. In the present chapter, the aug-MIA-QSARcolor approach yielded more predictive MLR-based models for the K_m and K_{cat} kinetic parameters than the traditional MIA-QSAR and aug-MIA-QSAR approaches, respectively. The variables contained in the aug-MIA-QSARcolor model highlighted the importance of small-size unbranched amino acids in the A1 and A4 positions in increasing the substrate affinity to the NS3 protein while in the A1 and A2 positions, increasing the rate of substrate cleavage. On the other hand, acidic amino acids in the A2 and A4 positions were found to be related with low substrate affinity to the NS3 protein and low substrate cleavage values in the A1 position. These findings provide relevant chemical information useful in the planning of target compounds against the DENV in a quicker and inexpensive way.

ACKNOWLEDGMENTS

The authors are thankful to FAPEMIG for the financial support of this research and for the studentship (M.H.D.), and to CNPq for fellowships (to S.J.B. and M.P.F.).

REFERENCES

[1] O.J. Brady, P.W. Gething, S. Bhatt, J.P. Messina, J.S. Brownstein, A.G. Hoen, C.L. Moyes, A.W. Farlow, T.W. Scott, and S.I. Hay, *Refining the global spatial limits of dengue virus transmission by evidence-based consensus*, PLoS Negl. Trop. Dis. 6 (2012), pp. e1760.

[2] S. Bhatt, P.W. Gething, O.J. Brady, J.P. Messina, A.W. Farlow, C.L. Moyes, J.M. Drake, J.S. Brownstein, A.G. Hoen, and O. Sankoh, *The global distribution and burden of dengue*, Nature 496 (2013), pp. 504–507.

[3] M.G. Guzman and E. Harris, *Dengue*, Lancet 385 (2014), pp. 453–465.

[4] W.H.O. WHO, *Dengue and severe dengue*, (2015), http://www.who.int/mediacentre/factsheets/fs117/en/

[5] M. Bessaud, B.A. Pastorino, C.N. Peyrefitte, D. Rolland, M. Grandadam, and H.J. Tolou, *Functional characterization of the NS2B/NS3 protease complex from seven viruses belonging to different groups inside the genus Flavivirus*, Virus Res. 120 (2006), pp. 79–90.

[6] J.S. Mackenzie, D.J. Gubler, and L.R. Petersen, *Emerging flaviviruses: The spread and resurgence of Japanese encephalitis, West Nile and dengue viruses*, Nat. Med. 10 (2004), pp. S98–S109.

[7] J.M. Medlock, K.M. Hansford, F. Schaffner, V. Versteirt, G. Hendrickx, H. Zeller, and W.V. Bortel, *A review of the invasive mosquitoes in Europe: Ecology, public health risks, and control options*, Vector Borne Zoonotic Dis. 12 (2012), pp. 435–447.

[8] T. Solomon and M. Mallewa, *Dengue and other emerging flaviviruses*, J. Infect. 42 (2001), pp. 104–115.

[9] S. Kalayanarooj, *Clinical manifestations and management of dengue/DHF/DSS*, Trop. Med. Health 39 (2011), pp. 83.

[10] P.W. Mason, P.C. McAda, T.L. Mason, and M.J. Fournier, *Sequence of the dengue-1 virus genome in the region encoding the three structural proteins and the major non-structural protein NS1*, Virology 161 (1987), pp. 262–267.

[11] R.S. Lanciotti, J.G. Lewis, D.J. Gubler, and D.W. Trent, *Molecular evolution and epidemiology of dengue-3 viruses*, J. Gen. Virol. 75 (1994), pp. 65–76.

[12] S. Dinu, I.R. Pănculescu-Gătej, S.A. Florescu, C.P. Popescu, A. Sîrbu, G. Oprişan, D. Bădescu, L. Franco, and C.S. Ceianu, *Molecular epidemiology of dengue fever cases imported into Romania between 2008 and 2013*, Travel Med. Infect. Dis. 13 (2015), pp. 69–73.

[13] N.T. Darwish, Y.B. Alias, and S.M. Khor, *An introduction to dengue—Disease diagnostics*, Trend Anal. Chem. 67 (2015), pp. 45–55.

[14] E.A. Ashley, *Dengue fever*, Trends Anaesth. Crit. Care 1 (2011), pp. 39–41.

[15] R.V. Gibbons and D.W. Vaughn, *Dengue: An escalating problem*, Brit. Med. J. 324 (2002), pp. 1563–1566.

[16] B. Falgout, M. Pethel, Y. Zhang, and C. Lai, *Both nonstructural proteins NS2B and NS3 are required for the proteolytic processing of dengue virus nonstructural proteins*, J. Virol. 65 (1991), pp. 2467–2475.

[17] D. Luo, T. Xu, R.P. Watson, D. Scherer-Becker, A. Sampath, W. Jahnke, S.S. Yeong, C.H. Wang, S.P. Lim, and A. Strongin, *Insights into RNA unwinding and ATP hydrolysis by the flavivirus NS3 protein*, EMBO J. 27 (2008), pp. 3209–3219.

[18] J. Lescar, D. Luo, T. Xu, A. Sampath, S.P. Lim, B. Canard, and S.G. Vasudevan, *Towards the design of antiviral inhibitors against flaviviruses: The case for the multifunctional NS3 protein from dengue virus as a target*, Antiviral Res. 80 (2008), pp. 94–101.

[19] T.J. Chambers, C.S. Hahn, R. Galler, and C.M. Rice, *Flavivirus genome organization, expression, and replication*, Annu. Rev. Microbiol. 44 (1990), pp. 649–688.

[20] J.F. Bazan and R.J. Fletterick, *Detection of a trypsin-like serine protease domain in flaviviruses and pestviruses*, Virology 171 (1989), pp. 637–639.

[21] R. Assenberg, E. Mastrangelo, T.S. Walter, A. Verma, M. Milani, R.J. Owens, D.I. Stuart, J.M. Grimes, and E.J. Mancini, *Crystal structure of a novel conformational state of the flavivirus NS3 protein: Implications for polyprotein processing and viral replication*, J. Virol. 83 (2009), pp. 12895–12906.

[22] A.E. Gorbalenya, E.V. Koonin, A.P. Donchenko, and V.M. Blinov, *Two related superfamilies of putative helicases involved in replication, recombination, repair and expression of DNA and RNA genomes*, Nucleic Acids Res. 17 (1989), pp. 4713–4730.

[23] R.D. Brown and Y.C. Martin, *The information content of 2D and 3D structural descriptors relevant to ligand–receptor binding*, J. Chem. Inf. Comput. Sci. 37 (1997), pp. 1–9.

[24] E. Estrada, E. Molina, and I. Perdomo-López, *Can 3D structural parameters be predicted from 2D (topological) molecular descriptors?*, J. Chem. Inf. Comput. Sci. 41 (2001), pp. 1015–1021.

[25] Z. Garkani-Nejad and B. Ahmadi-Roudi, *Modeling the antileishmanial activity screening of 5-nitro-2-heterocyclic benzylidene hydrazides using different chemometrics methods*, Eur. J. Med. Chem. 45 (2010), pp. 719–726.

[26] T. Hansch and J. Fujita, *ρ–σ–π analysis. A method for the correlation of biological activity and chemical structure*, J. Am. Chem. Soc. 86 (1964), pp. 1616–1626.

[27] S.M. Free and J.W. Wilson, *A mathematical contribution to structure–activity studies*, J. Med. Chem. 7 (1964), pp. 395–399.

[28] R. Todeschini and V. Consonni, *Molecular Descriptors for Chemoinformatics*, vol. 41, Wiley, Weinheim, 2009.

[29] M.P. Freitas, S.D. Brown, and J.A. Martins, *MIA-QSAR: A simple 2D image-based approach for quantitative structure–activity relationship analysis*, J. Mol. Struct. 738 (2005), pp. 149–154.

[30] M.P. Freitas, *MIA-QSAR modelling of anti-HIV-1 activities of some 2-amino-6-arylsulfonylbenzonitriles and their thio and sulfinyl congeners*, Org. Biomol. Chem. 4 (2006), pp. 1154–1159.

[31] R.A. Cormanich, C.A. Nunes, and M.P. Freitas, *Chemical drawings correlate to biological properties: MIA-QSAR*, Quím. Nova 35 (2012), pp. 1157–1163.

[32] R.A. Cormanich, M.P. Freitas, and R. Rittner, *2D chemical drawings correlate to bioactivities: MIA-QSAR modelling of antimalarial activities of 2,5-diaminobenzophenone derivatives*, J. Braz. Chem. Soc. 22 (2011), pp. 637–642.

[33] M. Goodarzi and M.P. Freitas, *On the use of PLS and N-PLS in MIA-QSAR: Azole antifungals*, Chemom. Intell. Lab. Sys. 96 (2009), pp. 59–62.

[34] M.P. Freitas, *Multivariate image analysis applied to QSAR: Evaluation to a series of potential anxiolytic agents*, Chemom. Intell. Lab. Sys. 91 (2008), pp. 173–176.

[35] R.A. Cormanich, M. Goodarzi, and M.P. Freitas, *Improvement of multivariate image analysis applied to quantitative structure–activity relationship (QSAR) analysis by using wavelet-principal component analysis ranking variable selection and least-squares support vector machine regression: QSAR study of checkpoint kinase WEE1 inhibitors*, Chem. Biol. Drug Des. 73 (2009), pp. 244–252.

[36] M. Duarte, S. Barigye, E. da Mota, and M. Freitas, *Computational modelling of the antischistosomal activity for neolignan derivatives based on the MIA-SAR approach*, SAR QSAR Environ. Res. (2015), pp. 1–12.

[37] M. Duarte, S. Barigye, and M. Freitas, *Exploring MIA-QSARs' for antimalarial quinolon-4 (1H)-imines*, Comb. Chem. High Through. Scr. (2014), pp. 208–216.

[38] M.P. Freitas and M.H. Duarte, *Evolution of multivariate image analysis in QSAR: The case for neglected diseases*, in *Quantitative Structure–Activity Relationships in Drug Design, Predictive Toxicology, and Risk Assessment*, K. Roy, ed., 2015, IGI Global, Hershey, PA, pp. 84.

[39] M.C. Guimarães, E.G. Mota, D.G. Silva, and M.P. Freitas, *aug-MIA-QSPR modelling of the toxicities of anilines and phenols to* Vibrio fischeri *and* Pseudokirchneriella subcapitata, Chemom. Intell. Lab. Sys. 134 (2014), pp. 53–57.

[40] C.A. Nunes and M.P. Freitas, *Introducing new dimensions in MIA-QSAR: A case for chemokine receptor inhibitors*, Eur. J. Med. Chem. 62 (2013), pp. 297–300.

[41] M.R. Freitas, S.J. Barigye, and M.P. Freitas, *Coloured chemical image-based models for the prediction of soil sorption of herbicides*, RSC Adv. 5 (2015), pp. 7547–7553.

[42] P. Prusis, M. Lapins, S. Yahorava, R. Petrovska, P. Niyomrattanakit, G. Katzenmeier, and J.E. Wikberg, *Proteochemometrics analysis of substrate interactions with dengue virus NS3 proteases*, Bioorg. Med. Chem. 16 (2008), pp. 9369–9377.

[43] ACD/ChemSketch 2012. Toronto, Canada.

[44] R. Dennington, II, T.A. Keith, and J.M. Millam, *GaussView 5.0. 8, Gaussian*, Inc., Wallingford, CT (2008).

[45] C.A. Nunes, M.P. Freitas, A.C.M. Pinheiro, and S.C. Bastos, *Chemoface: A novel free user-friendly interface for chemometrics*, J. Braz. Chem. Soc. 23 (2012), pp. 2003–2010.

[46] J.M. Silla, C.A. Nunes, R.A. Cormanich, M.C. Guerreiro, T.C. Ramalho, and M.P. Freitas, *MIA-QSPR and effect of variable selection on the modeling of kinetic parameters related to activities of modified peptides against dengue type 2*, Chemom. Intell. Lab. Sys. 108 (2011), pp. 146–149.
[47] N. Chirico and P. Gramatica, *Real external predictivity of QSAR models. Part 2. New intercomparable thresholds for different validation criteria and the need for scatter plot inspection*, J. Chem. Inf. Model. 52 (2012), pp. 2044–2058.
[48] I. Mitra, A. Saha, and K. Roy, *Exploring quantitative structure–activity relationship studies of antioxidant phenolic compounds obtained from traditional Chinese medicinal plants*, Mol. Simulat. 36 (2010), pp. 1067–1079.
[49] P.K. Ojha, I. Mitra, R.N. Das, and K. Roy, *Further exploring r_m^2 metrics for validation of QSPR models*, Chemom. Intell. Lab. Sys. 107 (2011), pp. 194–205.
[50] K. Roy, P. Chakraborty, I. Mitra, P.K. Ojha, S. Kar, and R.N. Das, *Some case studies on application of "r_m^2" metrics for judging quality of quantitative structure–activity relationship predictions: Emphasis on scaling of response data*, J. Comput. Chem. 34 (2013), pp. 1071–1082.
[51] P.P. Roy and K. Roy, *On some aspects of variable selection for partial least squares regression models*, QSAR Comb. Sci. 27 (2008), pp. 302–313.
[52] J.G. Topliss and R.J. Costello, *Chance correlations in structure–activity studies using multiple regression analysis*, J. Med. Chem. 15 (1972), pp. 1066–1068.
[53] R.W.P. Urias, S.J. Barigye, Y. Marrero-Ponce, C.R. García-Jacas, J.R. Valdes-Martiní, and F. Perez-Gimenez, *IMMAN: Free software for information theory-based chemometric analysis*, Molec. Divers. 19 (2015), pp. 305–319.
[54] J. Li, S.P. Lim, D. Beer, V. Patel, D. Wen, C. Tumanut, D.C. Tully, J.A. Williams, J. Jiricek, and J.P. Priestle, *Functional profiling of recombinant NS3 proteases from all four serotypes of dengue virus using tetrapeptide and octapeptide substrate libraries*, J. Biol. Chem. 280 (2005), pp. 28766–28774.
[55] T.J. Chambers, R.C. Weir, A. Grakoui, D.W. McCourt, J.F. Bazan, R.J. Fletterick, and C.M. Rice, *Evidence that the N-terminal domain of nonstructural protein NS3 from yellow fever virus is a serine protease responsible for site-specific cleavages in the viral polyprotein*, Proc. Natl. Acad. Sci. USA 87 (1990), pp. 8898–8902.
[56] Z. Yin, S.J. Patel, W.-L. Wang, W.-L. Chan, K.R. Rao, G. Wang, X. Ngew, V. Patel, D. Beer, and J.E. Knox, *Peptide inhibitors of dengue virus NS3 protease. Part 2: SAR study of tetrapeptide aldehyde inhibitors*, Bioorg. Med. Chem. Lett. 16 (2006), pp. 40–43.
[57] S. Chanprapaph, P. Saparpakorn, C. Sangma, P. Niyomrattanakit, S. Hannongbua, C. Angsuthanasombat, and G. Katzenmeier, *Competitive inhibition of the dengue virus NS3 serine protease by synthetic peptides representing polyprotein cleavage sites*, Biochem. Biophys. Res. Commun. 330 (2005), pp. 1237–1246.

7 Mosquito-Active Cry δ-Endotoxins from *Bacillus thuringiensis* subsp. *israelensis*

Structural Insights into Toxin-Induced Pore Architecture

Chalermpol Kanchanawarin and Chanan Angsuthanasombat

CONTENTS

ABSTRACT

Computational and theoretical methods have recently played crucial parts in helping us to comprehend toxic mechanisms at the molecular level of two closely related dipteran-specific toxins: Cry4Aa and Cry4Ba. These mosquito-larvicidal proteins produced by *Bacillus thuringiensis* subsp. *israelensis* have been exploited as a safe bioinsecticide for controlling mosquitoes. We have been employing various computational techniques, such as protein-structure alignments, protein bioinformatics, homology-based modeling, protein–protein docking, and molecular dynamics (MD) simulations, to study Cry4Aa and Cry4Ba toxins. Nevertheless, if we desire to engineer these protein-based insecticides to be a better mosquito-active toxin, it is necessary to understand details of their toxic mechanisms. To this end, computational methods have been employed together with biochemical and biophysical experiments to explore structure–function relationships and dynamics of these two Cry toxins. Their structural basis of membrane-pore formation has allowed us to model and design further experimentation. Overall, this chapter provides the essential fundamentals of *in silico* techniques used to explore more insights into the Cry toxin-induced pore architecture and their applications for redesigning a better modified active toxin.

KEYWORDS

- *Bacillus thuringiensis* subsp. *israelensis*
- Cry4Aa toxin
- Cry4Ba toxin
- Homology modeling
- Molecular dynamics
- Larvicide

7.1 INTRODUCTION TO MOSQUITO-ACTIVE Cry TOXINS: PROTEIN-BASED BIOINSECTICIDES

Bacillus thuringiensis (*Bt*) is a Gram-positive soil bacterium that has been used widely as a bioinsecticide in agriculture and for vector control due to its remarkable toxicity to various kinds of insect larvae. It produces parasporal crystalline inclusions that contain two families of insecticidal proteins known as δ-endotoxins—that is, cytolytic (Cyt) and crystal (Cry) toxins that are variously toxic to insect larvae and nematodes [1]. Unlike chemical insecticides, *Bt* toxins are specifically toxic only to certain insect larvae and are harmless to the other organisms and the environment, so they are more suitable to be used as a safe and effective biopesticide. In particular, two closely related Cry toxins from *Bt* subsp. *israelensis* (*Bti*) (i.e., Cry4Ba and Cry4Aa) are specifically active against mosquito larvae of the genera *Aedes* and *Anopheles*, which are vectors of several serious human diseases including dengue hemorrhagic fever, chikungunya, yellow fever, and malaria. Thus far, these mosquito-active Cry toxins have been produced commercially and used as bioinsecticides for controlling mosquitoes and diseases and are much safer than chemical insecticides.

Our current understanding of how Cry toxins function is as follows. After ingestion of these bacteria by a susceptible mosquito larva, the released Cry toxins have been thought to form lytic pores in the larval midgut epithelial cell membrane according to their common mechanism [2] (shown in Figure 7.1). This pore-forming mechanism consists roughly of six key steps (i.e., inclusion solubilization, proteolytic activation, receptor binding, oligomerization, membrane insertion, and pore formation). During the first step of solubilization, the 130 kDa inactive Cry protoxins in the crystalline inclusions are solubilized in the larval midgut lumen under alkaline conditions. Subsequently, in the second step of activation, the protoxins are cleaved at specific sites by larval gut proteases to yield active toxins of ~65 kDa [3]. In the third step of receptor binding, these activated Cry toxin monomers bind to specific receptors located on the midgut epithelial cell membrane, facilitating their getting

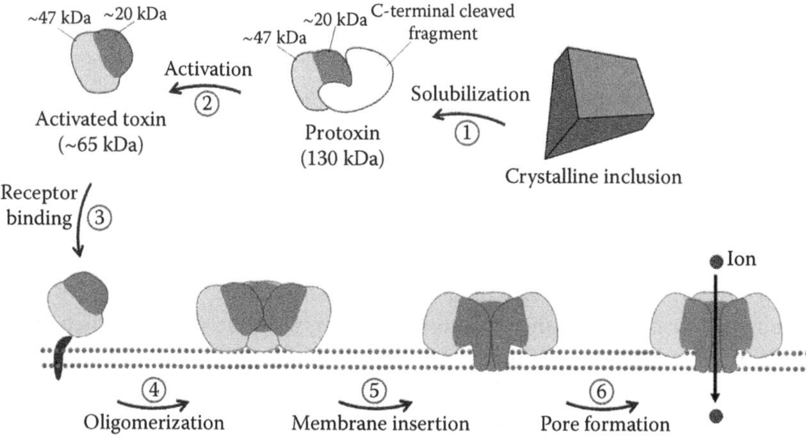

FIGURE 7.1 Proposed model of pore-forming mechanism of mosquito-active Cry toxins (modified from P. Ounjai, PhD thesis, Institute of Molecular Biosciences, Mahidol University, Bangkok, Thailand, 2007, pp. 85). These toxins are produced in the form of crystalline protoxin inclusions in *Bt* bacteria. Upon ingestion by a mosquito larva, the released protoxin inclusions are solubilized in the alkaline midgut lumen and subsequently activated by gut proteases to yield a ~65 kDa active toxin comprising two noncovalently associated fragments of ~47 kDa and ~20 kDa (C. Angsuthanasombat et al., J. Biochem. Mol. Biol. 37 (2004), pp. 304–313). Binding of the activated toxin monomer to its specific receptors would promote toxin–membrane association. Consequently, conformational changes of membrane-bound monomers would lead to molecular assembly of a prepore trimer, which is capable of inserting into the membrane to generate an ion-leakage pore (W. Sriwimol et al., J. Biol. Chem. 290 (2015), pp. 20793–20803).

interacted with lipid molecules in the target cell membrane and possibly forming a membrane-bound complex [4]. Then in the fourth step of oligomerization, these Cry toxin–lipid complexes get together and oligomerize into a trimeric prepore on the membrane [4,5]. In the last two steps, the prepore trimer of Cry toxin inserts its pore-forming parts through the membrane to form ion-leakage pores. This results in osmotic cell lysis and eventual death of the target larvae [6–8].

Specific binding to their individual receptors is believed to be the critical determinant in target insect specificity. Currently, a variety of membrane-bound proteins have been identified as functional receptors for several Cry mosquito-active toxins, including *Anopheles* and *Aedes* cadherin-like proteins for Cry4Ba and Cry11Aa, respectively [9,10]. Two different glycosylphosphatidylinositol (GPI)-anchored proteins—that is, GPI-anchored alkaline phosphatase (ALP) and GPI-anchored aminopeptidase-N (APN)—have also been identified as potential receptors mediating toxicity of individual mosquito-active toxins in different target larvae, including *Anopheles* ALP for Cry11Ba [11], *Aedes* ALPs for Cry4Ba and Cry11Aa [12–14], *Anopheles* APN for Cry11Ba [15,16], and *Aedes* APNs for Cry4Ba and Cry11Aa [17–19]. Interestingly, Cyt1Aa and Cyt2Aa toxins from *Bt* subsp. *israelensis* and *darmstadiensis*, respectively, have also been reported as functional alternative receptors, which play an important role in synergistic interactions with Cry4Ba or Cry11Aa [20–22]. However, the structural basis for such toxin–receptor interactions is still unclear and needs to be explored.

7.2 STRUCTURE–FUNCTION RELATIONSHIPS OF Cry4Ba AND Cry4Aa TOXINS AND THEIR MUTANTS

In 2004, before the three-dimensional (3D) structure of Cry4Ba was available, we used homology-based modeling to construct structural models of Cry4Ba and Cry4Aa toxins [23]. This allowed us to see the overall molecular structures of these two mosquito-active toxins and to make use of them to design mutation experiments. In particular, the loop linking α_4 and α_5 in the pore-forming domain (DI) of Cry4Aa was found to possess a unique disulfide bond and a Pro-rich region. Moreover, the receptor-binding domain (DII) of both toxins was shown to be quite different, reflecting their specific binding to the receptors on the target cell membrane. Then, in 2005 and 2006, when the 3D structures of Cry4Ba and Cry4Aa toxins were solved by x-ray crystallography (PDB ID 1W99 and 2C9K, respectively) [24,25], the observed structural features were confirmed. Both x-ray structures of Cry4Ba and Cry4Aa toxins were very similar to those of other Cry toxins and also to our homology-based modeled structures [23].

Dissimilar to the single α/β domain-Cyt1Aa structure [26], all known Cry toxin structures, including the Cry4Ba structure as shown in Figure 7.2(a), share the same overall fold features, thereby suggesting their general mechanism of action that is conceivably somewhat different from the toxic action of the Cyt toxin family [1,27]. The Cry structures display a wedge-shaped appearance (estimated dimensions of 55 × 65 × 75 Å) with three distinctive structural and functional domains running from the N- to C-terminus—that is, an α-helical bundle (DI), a β-sheet prism (DII), and a β-sheet sandwich (DIII) [1,28]. The first domain (DI) is a membrane pore-forming domain consisting of a group of eight helices—α_1, α_{2a}, α_{2b}, α_3, α_4, α_5, α_6, and α_7, initially assigned by the Ellar group [29]—in which the most hydrophobic helix (α_5) is surrounded by seven outer helices. Note that α_5 is actually not entirely hydrophobic, but rather displays an amphipathic character, as all of its polar or charged side chains in the interhelical space are engaged in hydrogen (H) bonds or salt bridges [24,25,29,30]. The β-prism domain (DII) consists of several β-sheets forming three sides of a prism. It functions as a receptor-binding domain using various loops connecting these β-sheets [21,31–34]. DIII is the C-terminal domain whose function is not yet clear but it has been shown to be able to bind sugar molecules [35]. This domain consists of two layers of β-sheets forming a sandwich.

In our earlier studies, the functional role of DI and DII of the two closely related Cry4 mosquito-active toxins (Cry4Aa and Cry4Ba) has been intensively investigated. We have provided direct evidence for liposomal membrane-permeating activity of the α_4-loop-α_5 fragment purified from the engineered Cry4Ba toxin [36]. Similar studies by other researchers using synthetic peptides corresponding to α_4, α_5, and α_4-loop-α_5 of Cry1Ac showed that the loop linking α_4 and α_5 is required for efficient penetration of these two transmembrane helices into the lipid bilayers to induce membrane permeabilization [37]. This notion was supported by our findings that revealed a structural requirement for larvicidal activity of one highly conserved aromatic side chain found within the α_4–α_5 loop of both Cry4Aa (Tyr^{202}) and Cry4Ba (Tyr^{170}), conceivably for stabilizing the membrane-associated toxin pore complex [38,39]. Likewise, two other conserved aromatic side chains (i.e., Tyr^{249} and Phe^{264}), which are oriented

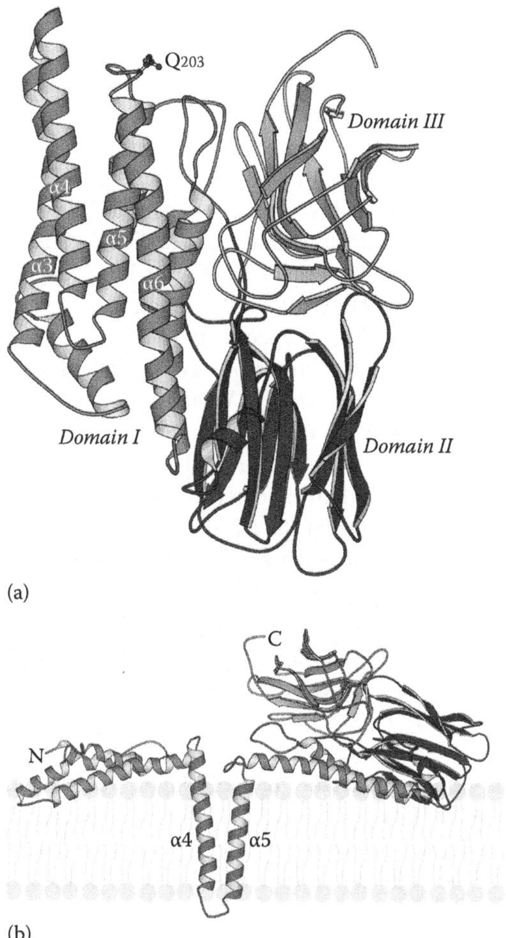

FIGURE 7.2 Complete structure of the Cry4Ba-R203Q mutant. (a) R203Q mutant structure obtained by combination of new x-ray structure (N. Thamwiriyasati et al., Acta Crystallogr. Sect. F 66 (2010), pp. 721–724, recent unpublished data) and homology modeling of the missing α_1 and α_2 based on the Cry4Aa structure (W. Sriwimol et al., J. Biol. Chem. 290 (2015), pp. 20793–20803; P. Boonserm et al., J. Bacteriol. 188 (2006), pp. 3391–3401). (b) Umbrella-like model for Cry toxins suggests that the α_4–α_5 hairpin is inserted into the membrane while the rest of the toxin lies on the membrane surface (E. Gazit et al., PNAS 95 (1998), pp. 12289–12294).

on the same side of Cry4Ba-α_7, were also found to be crucial for toxicity [40]. We have further shown that this highly conserved α_7 could undergo a conformational change to transform into a membrane-inserted β-hairpin, which may perhaps act as a lipid anchor needed for membrane penetration by the pore-forming α_4-loop-α_5 [41]. Recently, we have validated the functional importance of intrinsic stability of the unique α_4–α_5 loop contributed by the Pro^{193}Pro^{194}_Pro^{196} cluster within Cry4Aa-DI

[42]. We have also shown that the polarity of the α_4–α_5 loop-residue—Asn^{166}—plays a key role in ion permeation through the Cry4Ba toxin-induced pore [43]. For studies of DII, it seems that most researchers have confined themselves to exploring a possible involvement in receptor binding of only three surface-exposed loops formed at the apex of this domain (i.e., loops β_2–β_3, β_6–β_7, and β_{10}–β_{11}). However, we have clearly demonstrated that two other β-hairpin loops of β_4–β_5 and β_8–β_9 within the receptor-binding domain also participate in receptor recognition [21,31]. More recently, we have shown that charge-reversal substitutions at Asp^{454} (D454R and D454K) placed within the receptor-binding β_{10}–β_{11} loop of Cry4Ba strikingly increase the toxin activity against less susceptible *Culex* mosquito larvae, suggesting a key role of the charged side chain in determining target insect specificity [32].

To be able to perform MD simulations of Cry4Aa and Cry4Ba toxins, the missing parts in their structures require fixing. Various computational techniques were used to fix all these missing parts in the x-ray structures of Cry4Ba and Cry4Aa toxins. In the previous Cry4Ba structure, α_1 and α_2 were found to be missing [24]. Moreover, α_3 seems to be longer than expected, about two helical twists too long. On the other hand, the x-ray structure of the Cry4Aa toxin is more complete with only a few missing short loops and is much easier to fix by loop modeling [44,45] and MD simulations [46]. Since 2010, we have tried to improve our crystallization techniques to resolve a more complete structure of the Cry4Ba-R203Q mutant toxin (one trypsin-cleavage site at Arg^{203} was removed, thus producing a 65 kDa active fragment upon trypsin digestion and retaining high larval toxicity) [47]. The obtained new structure (PDB ID 4MOA) has α_1 and α_2 but in very disordered forms and still has long α_3 (reference 47 and unpublished data). We suspected that this was possibly due to α_3 being still too long, therefore interrupting α_1 and α_2. We thus modeled the missing α_1 and α_2 of the Cry4Ba toxin by performing homology-based modeling using the Cry4Aa structure as a template since these two closely related mosquito-active toxins display the highest amino acid sequence similarity of 32.2% [23]. We subsequently integrated it to the 2010 x-ray structure and used MD simulations to obtain an equilibrated stable structure of the Cry4Ba-R203Q mutant toxin as shown in Figure 7.2(a). Moreover, we have recently used this complete structure to set up a structural simulating system of Cry4Ba monomer upon interacting with micelles of OG (octyl-β-D-glucoside) [4], as an OG molecule was modeled via a ParamChem server [48] and a spherical micelle consisting of 80 OG molecules was constructed via a VMD program [49]. This simulated Cry4Ba atomic structure interacting with an OG micelle from 100 ns MD simulations has been used to fit into a 3D reconstructed model of the toxin trimer obtained by negative-stain electron microscopy (EM) in combination with single-particle reconstruction [4].

7.3 STRUCTURAL MODELING AND SIMULATIONS OF Cry TOXIN PREPORE TRIMERS

For molecular mechanisms of oligomerization-mediated pore formation of Cry toxins, one of the proposed models is that the active Cry toxin monomers would rearrange upon receptor binding, thus bringing on their conformational changes for forming an oligomeric prepore, which is capable of penetrating the target membrane to generate a membrane-lytic pore [50,51]. Alternatively, the activated Cry toxins could individually insert into the target membrane where they afterward assemble into an oligomeric pore complex, causing cell lysis [52,53]. Nevertheless, the exact description of either proposed model still remains to be critically verified. In our earlier studies, we have made use of detergent dialysis-driven two-dimensional (2D) crystallization to directly demonstrate for the first time a symmetrical trimeric structure of a membrane-associated Cry4Ba toxin complex, although this structure was still inadequate to provide detailed insights into molecular organization of the pore architecture [5].

Despite the fact that the trimeric form can be considered conceivably correct as the Cry toxin's functional oligomericity supported by other studies [54,55], the tetrameric pore complex has also been reported via other different techniques [52,56]. Further attempts have therefore been made to provide more critical insights into our trimeric pore's model by using two direct rendering techniques—that is, single-particle EM and HS-AFM (high-speed atomic force microscopy)—for envisioning 3D structure and trimeric assembly of the membrane-associated Cry4Ba toxin. Particularly, HS-AFM was successfully employed to provide for the first time a real-time trimeric arrangement through interactions of individual toxin monomers associated with lipid membrane bilayers, suggesting a need of membrane-induced conformational changes for facilitating the trimer assembly of such mosquitocidal pore-forming proteins [4].

As shown in Figure 7.3(a), both modeled structures give an idea about relative positioning of individual domains in the context of the trimeric prepore complex. The trimeric Cry4Ba pore complex looks like a three-blade propeller with all three

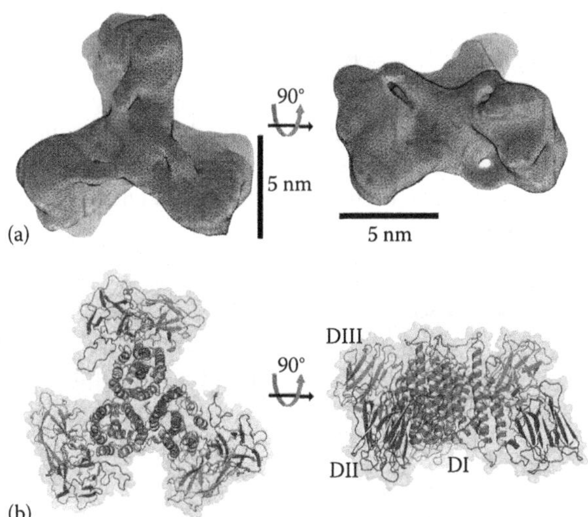

FIGURE 7.3 (See color insert.) Electron density map of Cry4Ba trimer compared with the Cry4Aa trimeric model structure. (a) Top and side views of an EM image of Cry4Ba trimer (figures were taken and modified from W. Sriwimol et al., J. Biol. Chem. 290 (2015), pp. 20793–20803); (b) top and side views of structural model of Cry4Aa trimer. Proteins are rendered as ribbons and colored by domain (DI = red, DII = blue, and DIII = green). vdW surface is also shown.

copies of the pore-forming DI making contact at the center and the three copies of the receptor-binding DII forming three legs (more details follow). This trimeric notion has led us to further construct a full atomic model of the Cry4Aa trimer using computational methods. We performed protein structural alignment of the Cry4Aa monomeric structure on the trimeric Cry4Ba structure to obtain the full atomic structure of Cry4Aa trimer as shown in Figure 7.3(b). We used MD simulations to equilibrate the structure in solution and then used the methods of molecular mechanics combined with Poisson–Boltzmann and surface area calculations to compute binding free energy and showed that the obtained trimeric Cry4Aa structure in aqueous solution is quite stable [46].

It is noteworthy that the trimeric structure of Cry4Aa, which is very much related to Cry4Ba, would be stable in solution only if the binding free energy is negative, i.e., the free energy of the Cry toxin trimer is lower than that of three separated Cry toxin monomers in solution. This can be shown by computing the binding free energy, ΔG_{bind}, from the following equation:

$$\Delta G_{bind} = G_{tri} - 3 \times G_{mon} \tag{7.1}$$

Here, G_{tri} and G_{mon} are the free energies of Cry toxin trimeric and monomeric structures obtained from two MD simulations of Cry toxin monomer and trimer in solution, respectively. The free energies of the Cry toxin can be determined from the sum of the molecular mechanical potential energies, E_{MM} and the solvation free energy, G_{sol}, and subtracted by the entropic energies, TS, as follows:

$$G = E_{MM} + G_{sol} - TS \tag{7.2}$$

The molecular mechanical potential is computed from the sum of three bonded-interaction potentials (bond stretching, $E_{stretch}$; bond bending, E_{bend}; and bond rotationing, E_{rot}) and two nonbonded interaction potentials (electrostatic, E_{elec}, and van der Waals, E_{vdW}) as follows:

$$E_{MM} = (E_{stretch} + E_{bend} + E_{rot}) + (E_{elec} + E_{vdW}) \tag{7.3}$$

The solvation free energy, G_{sol}, is calculated as the sum of its polar and nonpolar parts:

$$G_{sol} = G_{sol}^{pol} + G_{sol}^{non-pol} \tag{7.4}$$

The entropic energy, TS, is computed as the sum of the entropies due to translational motion, (S_{trans}), rotational motion, (S_{rot}), and vibrational motion, (S_{vib}), of the Cry toxin:

$$TS = T(S_{trans} + S_{rot} + S_{vib}) \tag{7.5}$$

As can be inferred from Figure 7.4, the obtained propeller-shaped structure of the Cry4Aa trimer has striking similarity to the 3D-reconstructed EM trimer model of the Cry4Ba toxin. The full atomic structure of the Cry4Aa trimer revealed that its monomers form the trimer by packing their α_3 and α_4 together at the central interface. Moreover, combined molecular mechanics (MM) and electrostatic interaction calculations revealed that this trimeric structure has lower free energy than that of three separated monomers and is therefore comparatively more stable.

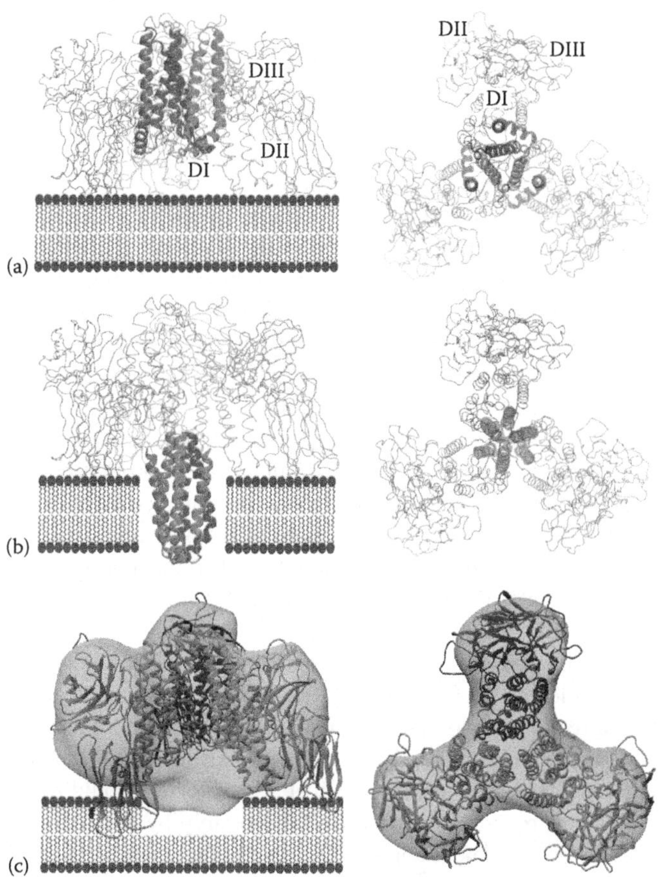

FIGURE 7.4 **(See color insert.)** Pore formation of the prepore Cry4Aa trimer compared with the EM density map of Cry4Ba trimer. Side and top views of prepore Cry4Aa trimer (lines in red, blue, and green) before (a) and after (b) the insertion of three α_4–α_5 hairpins (ribbons in red, blue, and green) into the membrane (T. Taveecharoenkool et al., PMC Biophys. 3 (2010), pp. 10); (c) side and top views of the EM density map of Cry4Ba trimer (gray surface) in its membrane-bound state docked with three copies of Cry4Ba monomers (ribbons in red, blue, and green) for comparison (W. Sriwimol et al., J. Biol. Chem. 290 (2015), pp. 20793–20803).

7.4 INSIGHTS INTO MEMBRANE-ASSOCIATED Cry TOXIN–PORE COMPLEX

Until now, the step at which the Cry oligomerization pathway may happen, either in aqueous solution [51,57] or in a membrane-associated state [52,53], remains to be undoubtedly elucidated. However, our recent studies have clearly revealed that the 65-kDa Cry4Ba monomer, unlike its OG micelle-associated monomer, was unable to self-assemble to form an oligomeric complex in aqueous solution since no defined oligomeric complex was observed for the purified trypsin-activated Cry4Ba-R203Q toxin in the carbonate-based solution (toxin-solubilization buffer, pH 9.0) as analyzed via a seminative PAGE (polyacrylamide gel electrophoresis) system (containing a detergent—SDS—only in gel and running buffers and no boiling of samples) [4]. Thus, this could imply requirements of a particular driving force in promoting a molecular recognition between Cry4Ba toxin monomers needed for the trimer assembly. In other words, a proper conformational transition of the toxin monomer could be an important prerequisite for toxin-pore oligomerization. It has been suggested that multiple polar interactions, especially a network of H-bonds and ionic interactions, appeared to exhibit a key role in the stability of oligomer assemblies [58]. As mentioned before, we have shown via combined MD and continuum solvent studies that the 65 kDa Cry4Aa toxin could form a stable trimer in aqueous solution as mostly attributed by the intersubunit interactions via certain polar-uncharged and charged side chains in the pore-forming DI [46]. Moreover, we have demonstrated via mutagenesis studies that one highly conserved polar side chain—Asn^{183}—located in the DI-α_5 plays an important role in Cry4Ba trimer formation and hence larval toxicity [53]. Accordingly, a major driving force in promoting the molecular recognitions between OG micelle-induced monomers needed for the Cry4Ba trimer assembly could possibly be a set of connections of H-bonds and ionic interactions.

In line with several other works and the established mechanistic models for the membrane-inserted state of Cry toxins, knowing the structure of Cry4Aa prepore trimer has allowed us to anticipate how *Bt* Cry toxins can form the lytic pore by inserting their three α_4–α_5 transmembrane hairpins into the lipid bilayers. Since the pore-forming DI is located at the center of the prepore trimer, the three copies of α_4–α_5 hairpins are already in positions to form a pore as shown in Figure 7.4(a). To form a functional transmembrane pore, the three α_4–α_5 hairpins have to be fully inserted into the membrane as shown in Figure 7.4(b). This process would require the separation of individual α_4–α_5 hairpins from DI, possibly by the umbrella model in Figure 7.2(b). However, as mentioned earlier, by the fact that the hydrophobic faces of the outer amphipathic helices of DI face inward, these activated Cry toxins must undergo conformational changes mostly within DI so as to convert this pore-forming domain into a transmembrane pore in which the hydrophobic surfaces would be in close contact with the lipids [23,28]. Thus, the individual protomeric subunits within the propeller-shaped trimer that displays an apparent protrusion from DI, possibly either the α_4–α_5 hairpin [59] or lipid-sensing α_7 [41], would be expected to have an overall conformation different from the water-soluble monomer.

As illustrated in Figure 7.4(c), our 3D reconstructed EM structure of the propeller-shaped Cry4Ba trimer associated with the membrane reveals that during the beginning of the membrane-inserting step, the three transmembrane hairpins are lowered such that the individual loops connecting α_4 and α_5 would be interacting with the lipid head groups. When the three α_4–α_5 hairpins are inserted into the membrane, they would also have to rearrange such that they could form a pore with all key polar residues lining the pore interior. Based on this depiction, we have decided to construct a pore model of the Cry4Ba toxin by learning how α_4 could be docked with α_5 using the Hex program [60]. Hex is a protein–protein docking program that uses Fourier series expansion of spherical harmonics functions to represent the protein surface to help accelerate the search for low-energy protein–protein docking conformations. In our studies, we extracted the structures of α_4 and α_5 from the Cry4Ba crystal structure (Figure 7.5a) and then docked them together by using the Hex program to find two possible packings between them. One was the internal packing between α_4 and α_5 within an α_4–α_5 hairpin (*see* Figure 7.5b) and the other was the external packing between two other nearby α_4–α_5 hairpins. With this obtained α_4–α_5 packing among hairpins, we would be able to construct a trimeric pore model of selected Cry toxins by arranging three α_4–α_5 hairpins on an equilateral triangle as shown in Figure 7.5(c). Subsequently, individual docked pairs were moved and rotated such that (1) each pair of docked helices (at the interface between two hairpins) is packed together according to the acquired α_4–α_5 docking result, and (2) all three Arg^{158} side chains on each α_4 are positioned in the pore interior, as can be seen in Figure 7.5(d).

Moreover, we have made use of this trimeric pore model to investigate certain key residues on the Cry4Ba α_4–α_5 loop (e.g., Asn^{166}), as suggested by mutagenesis experiments that its polarity is crucially involved in ion permeation through the toxin-induced pore [43]. We have performed MD simulations in a fully hydrated DMPC (1,2 dimyristoyl-*sn*-glycero-3-phosphocholine) system constructed by using a membrane builder module from the CHARMM GUI server [61]. In this context, structural simulations of the wild-type pore and its mutants (N166A and N166I) revealed that Asn^{166} could directly interact with the polar head groups of DMPC lipids, leading to pore opening as shown in Figure 7.6. Possibly, the partial negative charge of oxygen

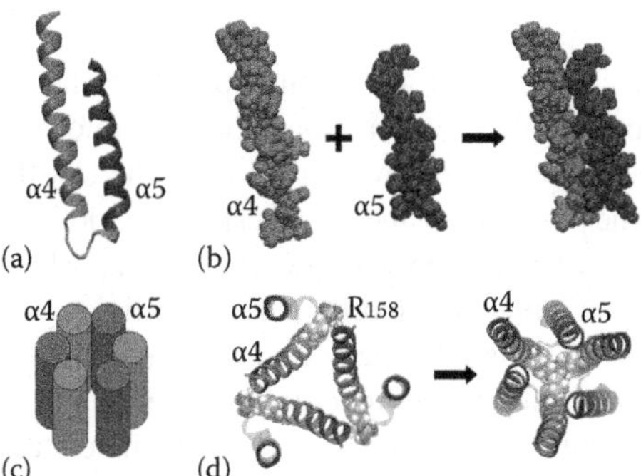

FIGURE 7.5 **(See color insert.)** Trimeric pore modeling of Cry4Ba toxin. (a) An α_4–α_5 hairpin was extracted from the x-ray structure of Cry4Ba toxin monomer; (b) α_4 and α_5 were separated and then docked together by Hex program to find two possible packings between them. (c, d) Three α_4–α_5 hairpins were arranged on an equilateral triangle and then they were moved and rotated such that each pair of α_4 and α_5 was packed according to the results from (b).

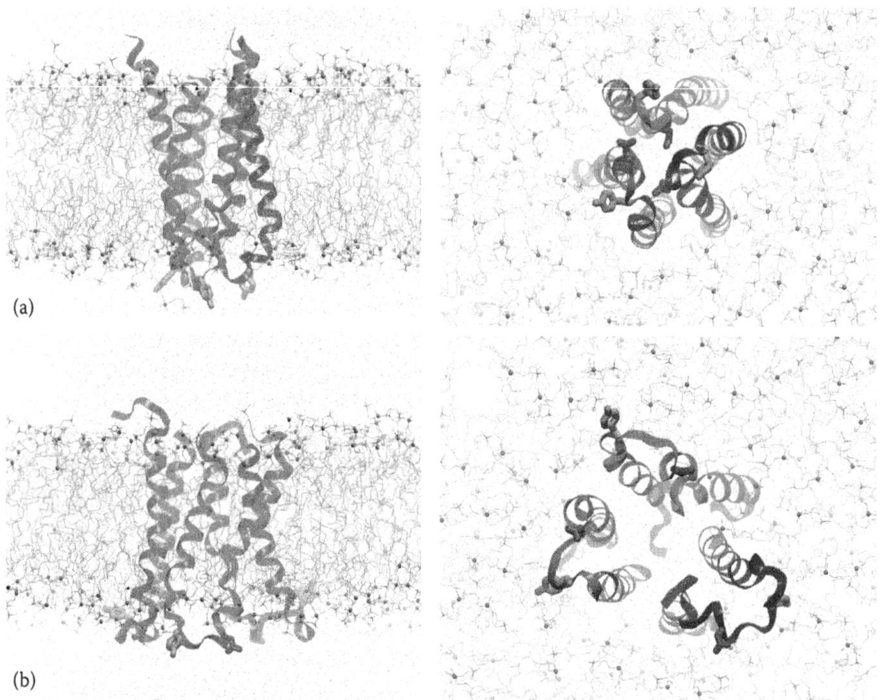

FIGURE 7.6 **(See color insert.)** Opening of trimeric Cry4Ba pore in DMPC membrane during 60-ns MD simulations. MD snapshots at time $t = 0$ ns (a), and $t = 60$ ns (b) from the side and bottom views. The three α_4–α_5 hairpins were rendered as ribbons and colored in red, green, and blue. Asn^{166} and Tyr^{170}, which directly interacted with lipid head groups, were drawn as capsules and colored in purple and orange, respectively. DMPC lipids were shown as lines and colored in cyan (carbon), blue (nitrogen), and red (oxygen) with phosphorus atoms rendered as small gold spheres. Only oxygen atoms of water were shown as red points.

atoms on Asn^{166} side chains could attract the positively charged choline moiety of DMPC lipids. Such protein–lipid interactions could therefore trigger the opening of the toxin-induced pore. As can be also inferred from the Cry4Ba pore model, Asn^{166} would basically face the pore lumen and the interfacial regions of the lipid membrane. Thus, replacing Asn^{166} with a nonpolar side chain (e.g., Ala or Ile) could directly influence its interactions with surrounding water and lipids, thus having an adverse consequence on conformational changes of such a trimeric pore. Nevertheless, further studies on toxin-induced pore architecture and precise mechanism of ion passage through the pore would be of great interest. A detailed understanding of the structural basis for ion permeable-pore formation by these two closely related mosquito-active toxins (Cry4Aa and Cry4Ba) is very important because such comprehension would definitely bolster the development of engineered insecticidal pore-forming proteins from *Bt* biopesticides for the control of human-disease vectors.

ACKNOWLEDGMENTS

We are grateful to Dr. Chompounoot Imtong for her kind assistance in thorough reading and making graphics. This work was supported in part by grants from Faculty of Science, Kasetsart University to C.K. and the Thailand Research Fund (BRG-58-8-0002) to C.A.

REFERENCES

[1] E. Schnepf, N. Crickmore, J. Van Rie, D. Lereclus, J. Baum, J. Feitelson, D.R. Zeigler, and D.H. Dean, Bacillus thuringiensis *and its pesticidal crystal proteins*, Microbiol. Mol. Biol. Rev. 62 (1998), pp. 775–806.

[2] P. Ounjai, *Molecular biophysical study of the* Bacillus thuringiensis *Cry4Ba toxin pore structure*, PhD thesis, Institute of Molecular Biosciences, Mahidol University, Bangkok, Thailand, 2007, pp. 85.

[3] H. Höfte and H.R. Whiteley, *Insecticidal crystal proteins of* Bacillus thuringiensis, Microbiol. Rev. 53 (1989), pp. 242–255.

[4] W. Sriwimol, A. Aroonkesorn, S. Sakdee, C. Kanchanawarin, T. Uchihashi, T. Ando, and C. Angsuthanasombat, *Potential prepore trimer formation by the* Bacillus thuringiensis *mosquito-specific toxin: Molecular insights into a critical prerequisite of membrane-bound monomers*, J. Biol. Chem. 290 (2015), pp. 20793–20803.

[5] P. Ounjai, V.M. Unger, F.J. Sigworth, and C. Angsuthanasombat, *Two conformational states of the membrane-associated* Bacillus thuringiensis *Cry4Ba δ-endotoxin complex revealed by electron crystallography: Implications for toxin-pore formation*, Biochem. Biophys. Res. Commun. 361 (2007), pp. 890–895.

[6] B.H. Knowles, ed. *Mechanism of action of* Bacillus thuringiensis *insecticidal δ-endotoxins*, in *Advances in Insect Physiology*, P.D. Evans, ed., Academic Press, Cambridge, MA, 1994, pp. 275–308.

[7] B.H. Knowles and D.J. Ellar, *Colloid-osmotic lysis is a general feature of the mechanism of action of* Bacillus thuringiensis *δ-endotoxins with different insect specificity*, Biochim. Biophys. Acta—Gen. Subj. 924 (1987), pp. 509–518.

[8] C.R. Pigott and D.J. Ellar, *Role of receptors in* Bacillus thuringiensis *crystal toxin activity*, Microbiol. Mol. Biol. Rev. 71 (2007), pp. 255–281.

[9] G. Hua, R. Zhang, M.A.F. Abdullah, and M.J. Adang, Anopheles gambiae *cadherin AgCad1 binds the Cry4Ba toxin of* Bacillus thuringiensis israelensis *and a fragment of AgCad1 synergizes toxicity*, Biochemistry 47 (2008), pp. 5101–5110.

[10] J. Chen, K.G. Aimanova, L.E. Fernandez, A. Bravo, M. Soberon, and S.S. Gill, Aedes aegypti *cadherin serves as a putative receptor of the Cry11Aa toxin from* Bacillus thuringiensis *subsp.* israelensis, Biochem. J. 424 (2009), pp. 191–200.

[11] G. Hua, R. Zhang, K. Bayyareddy, and M.J. Adang, Anopheles gambiae *alkaline phosphatase is a functional receptor of* Bacillus thuringiensis jegathesan *Cry11Ba toxin*, Biochemistry 48 (2009), pp. 9785–9793.

[12] M. Dechklar, K. Tiewsiri, C. Angsuthanasombat, and K. Pootanakit, *Functional expression in insect cells of glycosylphosphatidylinositol-linked alkaline phosphatase from* Aedes aegypti *larval midgut: A* Bacillus thuringiensis *Cry4Ba toxin receptor*, Insect Biochem. Mol. Biol. 41 (2011), pp. 159–166.

[13] L.E. Fernandez, K.G. Aimanova, S.S. Gill, A. Bravo, and M. Soberón, *A GPI-anchored alkaline phosphatase is a functional midgut receptor of Cry11Aa toxin in* Aedes aegypti *larvae*, Biochem. J. 394 (2006), pp. 77–84.

[14] L.E. Fernandez, C. Martinez-Anaya, E. Lira, J. Chen, A. Evans, S. Hernández-Martínez, H. Lanz-Mendoza, A. Bravo, S.S. Gill, and M. Soberón, *Cloning and epitope mapping of Cry11Aa-binding sites in the Cry11Aa-receptor alkaline phosphatase from* Aedes aegypti, Biochemistry 48 (2009), pp. 8899–8907.

[15] M.A. Abdullah, A.P. Valaitis, and D.H. Dean, *Identification of a* Bacillus thuringiensis *Cry11Ba toxin-binding aminopeptidase from the mosquito,* Anopheles quadrimaculatus, BMC Biochem. 7 (2006), pp. 1–6.

[16] R. Zhang, G. Hua, T.M. Andacht, and M.J. Adang, *A 106-kDa aminopeptidase is a putative receptor for* Bacillus thuringiensis *Cry11Ba toxin in the mosquito* Anopheles gambiae, Biochemistry 47 (2008), pp. 11263–11272.

[17] A. Aroonkesorn, K. Pootanakit, G. Katzenmeier, and C. Angsuthanasombat, *Two specific membrane-bound aminopeptidase N isoforms from Aedes aegypti larvae serve as functional receptors for the* Bacillus thuringiensis *Cry4Ba toxin implicating counterpart specificity*, Biochem. Biophys. Res. Commun. 461 (2015), pp. 300–306.

[18] J. Chen, K.G. Aimanova, S. Pan, and S.S. Gill, *Identification and characterization of* Aedes aegypti *aminopeptidase N as a putative receptor of* Bacillus thuringiensis *Cry11A toxin*, Insect Biochem. Mol. Biol. 39 (2009), pp. 688–696.

[19] S. Saengwiman, A. Aroonkesorn, P. Dedvisitsakul, S. Sakdee, S. Leetachewa, C. Angsuthanasombat, and K. Pootanakit, *In vivo identification of* Bacillus thuringiensis *Cry4Ba toxin receptors by RNA interference knockdown of glycosylphosphatidylinositol-linked aminopeptidase N transcripts in* Aedes aegypti *larvae*, Biochem. Biophys. Res. Commun. 407 (2011), pp. 708–713.

[20] P.E. Cantón, E.Z. Reyes, I.R. de Escudero, A. Bravo, and M. Soberón, *Binding of* Bacillus thuringiensis *subsp.* israelensis *Cry4Ba to Cyt1Aa has an important role in synergism*, Peptides 32 (2011), pp. 595–600.

[21] C. Lailak, T. Khaokhiew, C. Promptmas, B. Promdonkoy, K. Pootanakit, and C. Angsuthanasombat, Bacillus thuringiensis *Cry4Ba toxin employs two receptor-binding loops for synergistic interactions with Cyt2Aa2*, Biochem. Biophys. Res. Commun. 435 (2013), pp. 216–221.

[22] C. Pérez, L.E. Fernandez, J. Sun, J.L. Folch, S.S. Gill, M. Soberón, and A. Bravo, Bacillus thuringiensis *subsp.* israelensis *Cyt1Aa synergizes Cry11Aa toxin by functioning as a membrane-bound receptor*, Proc. Natl. Acad. Sci. USA 102 (2005), pp. 18303–18308.

[23] C. Angsuthanasombat, P. Uawithya, S. Leetacheewa, W. Pornwiroon, P. Ounjai, T. Kerdcharoen, G. Katzenmeier, and S. Panyim, Bacillus thuringiensis *Cry4A and Cry4B mosquito-larvicidal proteins: Homology-based 3D model and implications for toxin activity*, J. Biochem. Mol. Biol. 37 (2004), pp. 304–313.

[24] P. Boonserm, P. Davis, D.J. Ellar, and J. Li, *Crystal structure of the mosquito-larvicidal toxin Cry4Ba and its biological implications*, J. Mol. Biol. 348 (2005), pp. 363–382.

[25] P. Boonserm, M. Mo, C. Angsuthanasombat, and J. Lescar, *Structure of the functional form of the mosquito larvicidal Cry4Aa toxin from* Bacillus thuringiensis *at a 2.8-angstrom resolution*, J. Bacteriol. 188 (2006), pp. 3391–3401.

[26] S. Cohen, S. Albeck, E. Ben-Dov, R. Cahan, M. Firer, A. Zaritsky, and O. Dym, *Cyt1Aa toxin: Crystal structure reveals implications for its membrane-perforating function*, J. Mol. Biol. 413 (2011), pp. 804–814.

[27] B.A. Federici, H.W. Park, and D.K. Bideshi, *Overview of the basic biology of* Bacillus thuringiensis *with emphasis on genetic engineering of bacterial larvicides for mosquito control*, Open Toxinol. J. 3 (2010), pp. 83–100.

[28] C. Angsuthanasombat, *Structural basis of pore formation by mosquito-larvicidal proteins from* Bacillus thuringiensis, Open Toxinol. J. 3 (2010), pp. 119–125.

[29] J. Li, J. Carroll and D.J. Ellar, *Crystal structure of insecticidal δ-endotoxin from* Bacillus thuringiensis *at 2.5 Å resolution*, Nature 353 (1991), pp. 815–821.

[30] P. Grochulski, L. Masson, S. Borisova, M. Pusztai-Carey, J.L. Schwartz, R. Brousseau, and M. Cygler, Bacillus thuringiensis *CryIA(a) insecticidal toxin: Crystal structure and channel formation*, J. Mol. Biol. 254 (1995), pp. 447–464.

[31] T. Khaokhiew, C. Angsuthanasombat, and C. Promptmas, *Correlative effect on the toxicity of three surface-exposed loops in the receptor-binding domain of the* Bacillus thuringiensis *Cry4Ba toxin*, FEMS Microbiol. Lett. 300 (2009), pp. 139–145.

[32] S. Visitsattapongse, S. Sakdee, S. Leetacheewa, and C. Angsuthanasombat, *Single-reversal charge in the β10–β11 receptor-binding loop of* Bacillus thuringiensis *Cry4Aa and Cry4Ba toxins reflects their different toxicity against* Culex *spp. larvae*, Biochem. Biophys. Res. Commun. 450 (2014), pp. 948–952.

[33] S. Likitvivatanavong, K.G. Aimanova, and S.S. Gill, *Loop residues of the receptor binding domain of* Bacillus thuringiensis *Cry11Ba toxin are important for mosquitocidal activity*, FEBS Lett. 583 (2009), pp. 2021–2030.

[34] S. Pacheco, I. Gómez, I. Arenas, G. Saab-Rincon, C. Rodríguez-Almazán, S.S. Gill, A. Bravo, and M. Soberón, *Domain II loop 3 of* Bacillus thuringiensis *Cry1Ab toxin is involved in a "Ping Pong" binding mechanism with* Manduca sexta *aminopeptidase-N and cadherin receptors*, J. Biol. Chem. 284 (2009), pp. 32750–32757.

[35] J.L. Jenkins, M.K. Lee, A.P. Valaitis, A. Curtiss, and D.H. Dean, *Bivalent sequential binding model of a* Bacillus thuringiensis *toxin to gypsy moth aminopeptidase N receptor*, J. Biol. Chem. 275 (2000), pp. 14423–14431.

[36] S. Leetachewa, G. Katzenmeier, and C. Angsuthanasombat, *Novel preparation and characterization of the α4-loop-α5 membrane-perturbing peptide from the* Bacillus thuringiensis *Cry4Ba δ-endotoxin*, J. Biochem. Mol. Biol. 39 (2006), pp. 270–277.

[37] D. Gerber and Y. Shai, *Insertion and organization within membranes of the δ-endotoxin pore-forming domain, helix 4-loop-helix 5, and inhibition of its activity by a mutant helix 4 peptide*, J. Biol. Chem. 275 (2000), pp. 23602–23607.

[38] Y. Kanintronkul, I. Sramala, G. Katzenmeier, S. Panyim, and C. Angsuthanasombat, *Specific mutations within the α4–α5 loop of the* Bacillus thuringiensis *Cry4B toxin reveal a crucial role for Asn-166 and Tyr-170*, Mol. Biotechnol. 24 (2003), pp. 11–19.

[39] W. Pornwiroon, G. Katzenmeier, S. Panyim, and C. Angsuthanasombat, *Aromaticity of Tyr-202 in the α4–α5 loop is essential for toxicity of the* Bacillus thuringiensis *Cry4A toxin*, J. Biochem. Mol. Biol. 37 (2004), pp. 292–297.

[40] K. Tiewsiri and C. Angsuthanasombat, *Structurally conserved aromaticity of Tyr*^{249} *and Phe*^{264} *in helix 7 is important for toxicity of the* Bacillus thuringiensis *Cry4Ba toxin*, J. Biochem. Mol. Biol. 40 (2007), pp. 163–171.

[41] K. Tiewsiri, W.B. Fischer, and C. Angsuthanasombat, *Lipid-induced conformation of helix 7 from the pore-forming domain of the* Bacillus thuringiensis *Cry4Ba toxin: Implications for toxicity mechanism*, Arch. Biochem. Biophys. 482 (2009), pp. 17–24.

[42] C. Imtong, C. Kanchanawarin, G. Katzenmeier, and C. Angsuthanasombat, Bacillus thuringiensis *Cry4Aa insecticidal protein: Functional importance of the intrinsic stability of the unique α4–α5 loop comprising the Pro-rich sequence*, Biochim. Biophys. Acta—Proteins and Proteomics 1844 (2014), pp. 1111–1118.

[43] T. Juntadech, Y. Kanintronkul, C. Kanchanawarin, G. Katzenmeier, and C. Angsuthanasombat, *Importance of polarity of the α4–α5 loop residue—Asn*^{166} *in the pore-forming domain of the* Bacillus thuringiensis *Cry4Ba toxin: Implications for ion permeation and pore opening*, Biochim. Biophys. Acta—Biomembranes 1838 (2014), pp. 319–327.

[44] C.S. Soto, M. Fasnacht, J. Zhu, L. Forrest, and B. Honig, *Loop modeling: Sampling, filtering, and scoring*, Proteins 70 (2008), pp. 834–843.

[45] Z. Xiang, C.S. Soto, and B. Honig, *Evaluating conformational free energies: The colony energy and its application to the problem of loop prediction*, Proc. Natl. Acad. Sci. USA 99 (2002), pp. 7432–7437.

[46] T. Taveecharoenkool, C. Angsuthanasombat, and C. Kanchanawarin, *Combined molecular dynamics and continuum solvent studies of the pre-pore Cry4Aa trimer suggest its stability in solution and how it may form pore*, PMC Biophys. 3 (2010), pp. 10.

[47] N. Thamwiriyasati, S. Sakdee, P. Chuankhayan, G. Katzenmeier, C.J. Chen, and C. Angsuthanasombat, *Crystallization and preliminary x-ray crystallographic analysis of a full-length active form of the Cry4Ba toxin from* Bacillus thuringiensis, Acta Crystallogr. Sect. F 66 (2010), pp. 721–724.

[48] K. Vanommeslaeghe, E. Hatcher, C. Acharya, S. Kundu, S. Zhong, J. Shim, E. Darian, O. Guvench, P. Lopes, I. Vorobyov, and A.D. Mackerell, *CHARMM general force field: A force field for drug-like molecules compatible with the CHARMM all-atom additive biological force fields*, J. Comput. Chem. 31 (2010), pp. 671–690.

[49] W. Humphrey, A. Dalke, and K. Schulten, *VMD: Visual molecular dynamics*, J. Mol. Graph. 14 (1996), pp. 33–38.

[50] A. Bravo, I. Gómez, J. Conde, C. Muñoz-Garay, J. Sánchez, R. Miranda, M. Zhuang, S.S. Gill, and M. Soberón, *Oligomerization triggers binding of a* Bacillus thuringiensis *Cry1Ab pore-forming toxin to aminopeptidase N receptor leading to insertion into membrane microdomains*, Biochim. Biophys. Acta—Biomembranes 1667 (2004), pp. 38–46.

[51] I. Gómez, J. Sánchez, R. Miranda, A. Bravo, and M. Soberón, *Cadherin-like receptor binding facilitates proteolytic cleavage of helix α-1 in domain I and oligomer pre-pore formation of* Bacillus thuringiensis *Cry1Ab toxin*, FEBS Lett. 513 (2002), pp. 242–246.

[52] N. Groulx, H. McGuire, R. Laprade, J.L. Schwartz, and R. Blunck, *Single molecule fluorescence study of the* Bacillus thuringiensis *toxin Cry1Aa reveals tetramerization*, J. Biol. Chem. 286 (2011), pp. 42274–42282.

[53] S. Likitvivatanavong, G. Katzenmeier, and C. Angsuthanasombat, *Asn^{183} in α5 is essential for oligomerisation and toxicity of the* Bacillus thuringiensis *Cry4Ba toxin*, Arch. Biochem. Biophys. 445 (2006), pp. 46–55.

[54] I. Gómez, J. Sánchez, C. Muñoz-Garay, V. Matus, S.S. Gill, M. Soberón, and A. Bravo, Bacillus thuringiensis *Cry1A toxins are versatile proteins with multiple modes of action: Two distinct pre-pores are involved in toxicity*, Biochem. J. 459 (2014), pp. 383–396.

[55] C. Muñóz-Garay, L. Portugal, L. Pardo-López, N. Jiménez-Juárez, I. Arenas, I. Gómez, R. Sánchez-López, R. Arroyo, A. Holzenburg, C.G. Savva, M. Soberón, and A. Bravo, *Characterization of the mechanism of action of the genetically modified Cry1AbMod toxin that is active against Cry1Ab-resistant insects*, Biochim. Biophys. Acta—Biomembranes 1788 (2009), pp. 2229–2237.

[56] X. Lin, K. Parthasarathy, W. Surya, T. Zhang, Y. Mu, and J. Torres, *A conserved tetrameric interaction of Cry toxin helix α3 suggests a functional role for toxin oligomerization*, Biochim. Biophys. Acta—Biomembranes 1838 (2014), pp. 1777–1784.

[57] S. Guo, Y. Zhang, F. Song, J. Zhang, and D. Huang, *Protease-resistant core form of* Bacillus thuringiensis *Cry1Ie: Monomeric and oligomeric forms in solution*, Biotechnol. Lett. 31 (2009), pp. 1769–1774.

[58] E.M. Hotze, A.P. Heuck, D.M. Czajkowsky, Z. Shao, A.E. Johnson, and R.K. Tweten, *Monomer–monomer interactions drive the prepore to pore conversion of a β-barrel-forming cholesterol-dependent cytolysin*, J. Biol. Chem. 277 (2002), pp. 11597–11605.

[59] E. Gazit, P.L. Rocca, M.S.P. Sansom, and Y. Shai, *The structure and organization within the membrane of the helices composing the pore-forming domain of* Bacillus thuringiensis *δ-endotoxin are consistent with an "umbrella-like" structure of the pore*, Proc. Natl. Acad. Sci. USA 95 (1998), pp. 12289–12294.

[60] D.W. Ritchie, D. Kozakov, and S. Vajda, *Accelerating and focusing protein–protein docking correlations using multi-dimensional rotational FFT generating functions*, Bioinformatics 24 (2008), pp. 1865–1873.

[61] S. Jo, J.B. Lim, J.B. Klauda, and W. Im, *CHARMM-GUI membrane builder for mixed bilayers and its application to yeast membranes*, Biophys. J. 97 (2009), pp. 50–58.

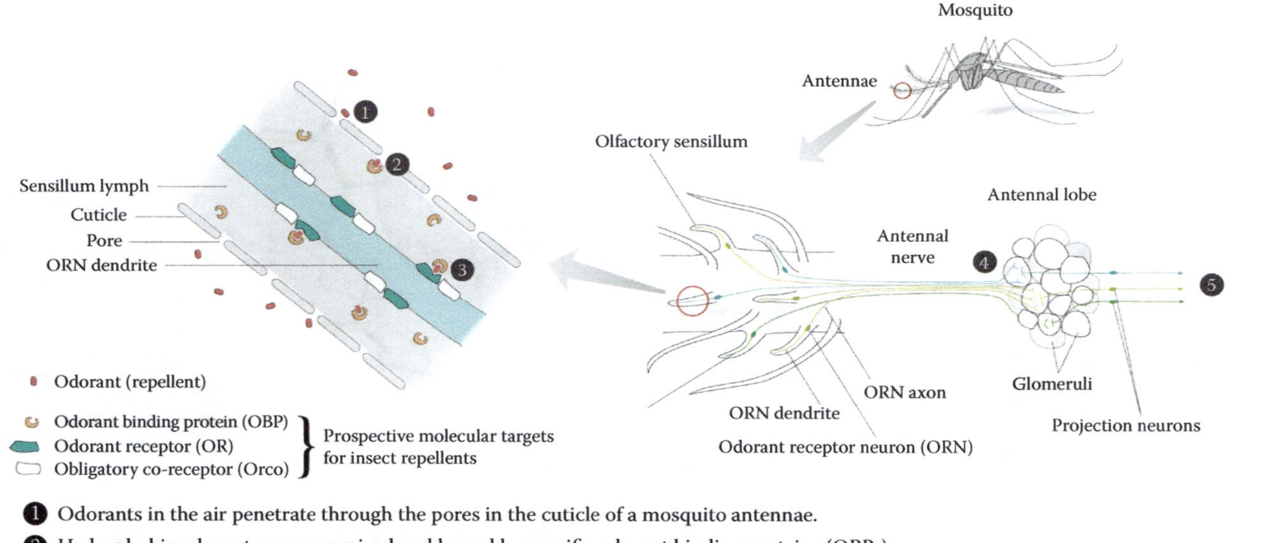

FIGURE 2.1 Schematic illustration of odor reception in mosquitoes. (Modified from C.I. Bargmann, Nature 444 (2006), pp. 295–301; E. Jacquin-Joly and C. Merlin, J. Chem. Ecol. 30 (2004), pp. 2359–2397; A.F. Carey and J.R. Carlson, PNAS 108 (2011), pp. 12987–12995.)

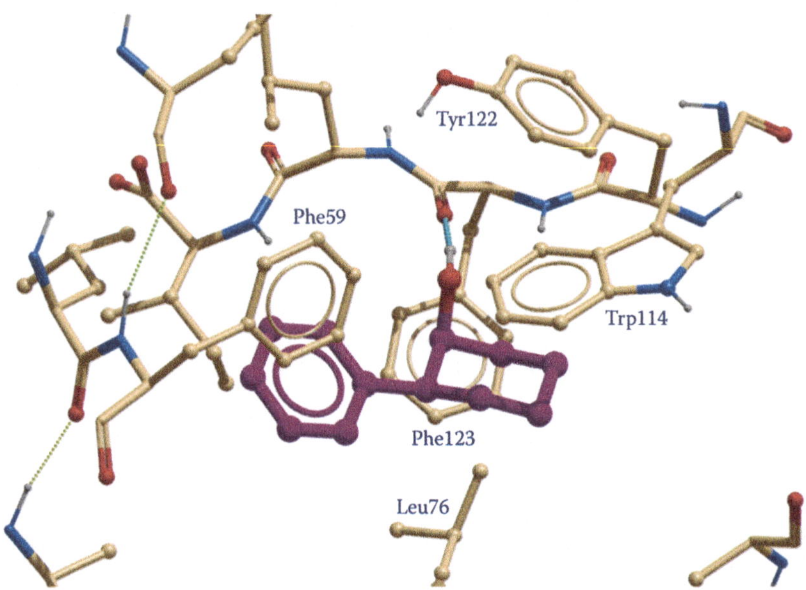

FIGURE 2.4 Atomic details on how **YF24** binds to OBP1, as depicted by ICM Browser (MolSoft, www.molsoft.com). An anchoring interaction that defines the position and orientation of the ligand is the hydrogen bond between the hydroxyl group of **YF24** and the backbone carbonyl group of Phe123. The rest of the interaction is driven by a set of aromatic and hydrophobic residues, Phe59, Leu76, Trp114, Tyr122, and Phe123, that accommodates the cyclohexylbenzene core. Only proximate residues making contact with **YF24** are shown.

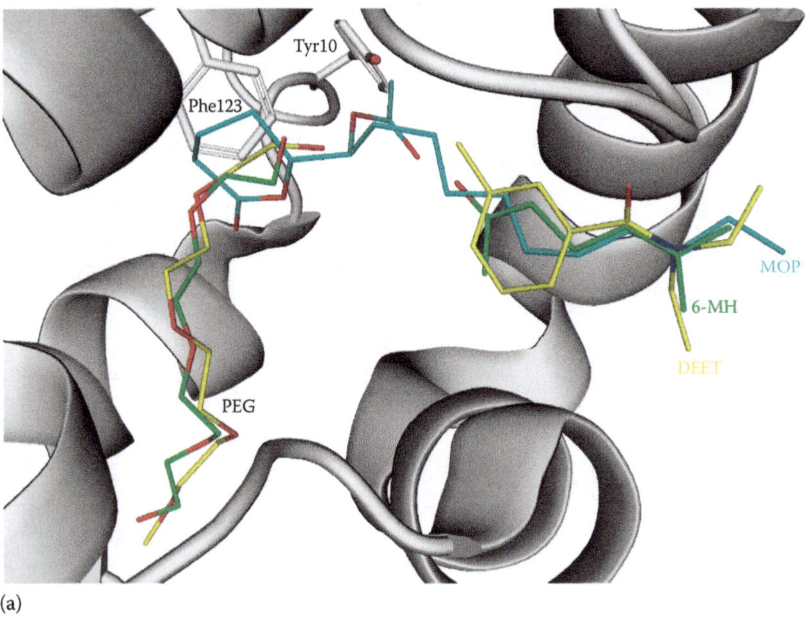

FIGURE 3.5 (a) Schematic presentation illustrating the binding mode of DEET and PEG (yellow, PDB id 3N7H), 6-MH, and PEG (green, PDB id 4FQT), and MOP (cyan, PDB id 3OGN) to the binding site of AgamOBP1 and CquiOBP1, respectively. *(Continued)*

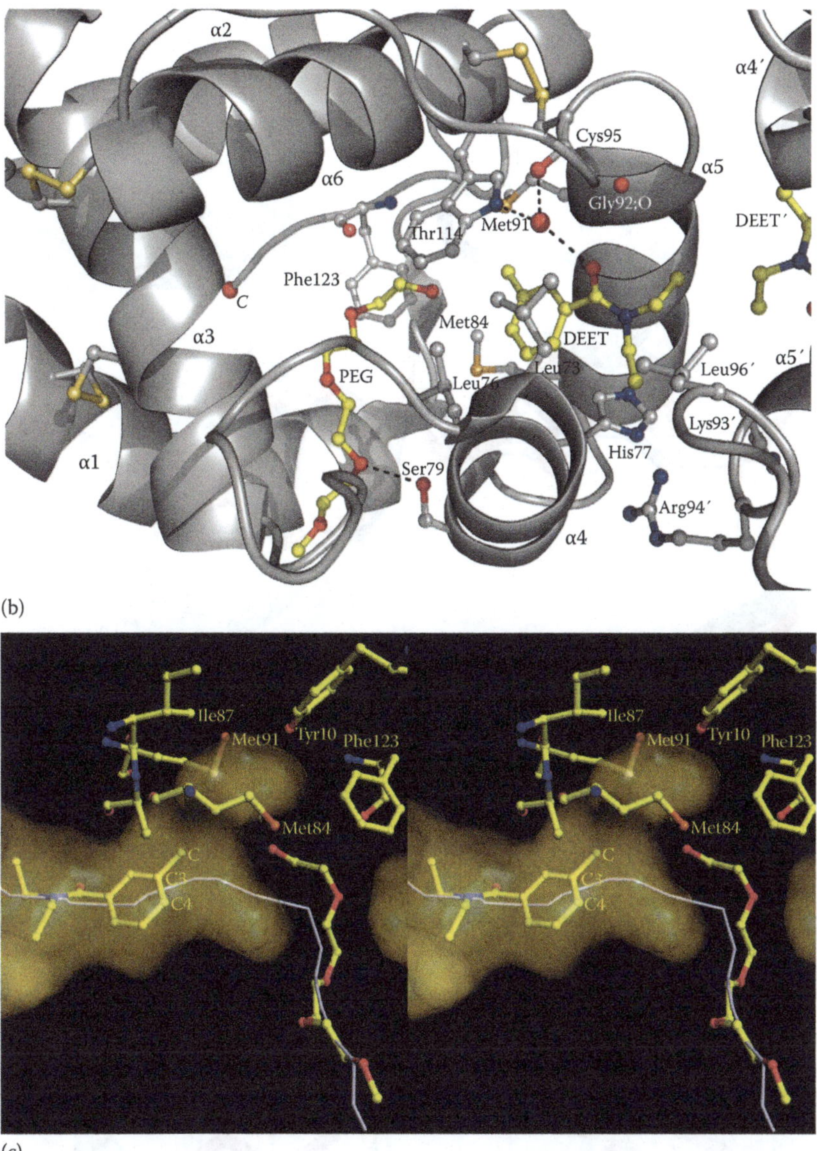

FIGURE 3.5 (CONTINUED) (b) DEET and PEG (with carbon atoms colored in yellow) form one hydrogen bond each, with a water molecule and the side chain of Ser79, respectively. Hydrogen bonds are shown as dashed lines. The contacts between ligands and AgamOBP1 are dominated by nonpolar vdW interactions. In particular, DEET makes vdW interactions with residues of α4, α5 helices and residues from the other monomer as well as with the neighboring DEET's molecule. The three disulfide bonds are represented in light orange sticks (K.E. Tsitsanou et al., Cell. Mol. Life Sci. 69 (2012), pp. 283–297). (With kind permission from Springer Science and Business Media.) (c) A stereodiagram of the Connolly molecular surface (yellow) in the vicinity of DEET's binding site. The corresponding path of the binding tunnel is represented as a white line. The existence of an empty pocket that diverges from the tunnel direction and emerges above the 3-methyl group of DEET is apparent. The shape of this cavity suggests that substitution at 3 (*meta-*) and 4 (*para-*) positions of the benzene ring could be exploited for the design of DEET analogs with improved affinity for AgamOBP1 (K.E. Tsitsanou et al., Cell. Mol. Life Sci. 69 (2012), pp. 283–297). (With kind permission from Springer Science and Business Media.)

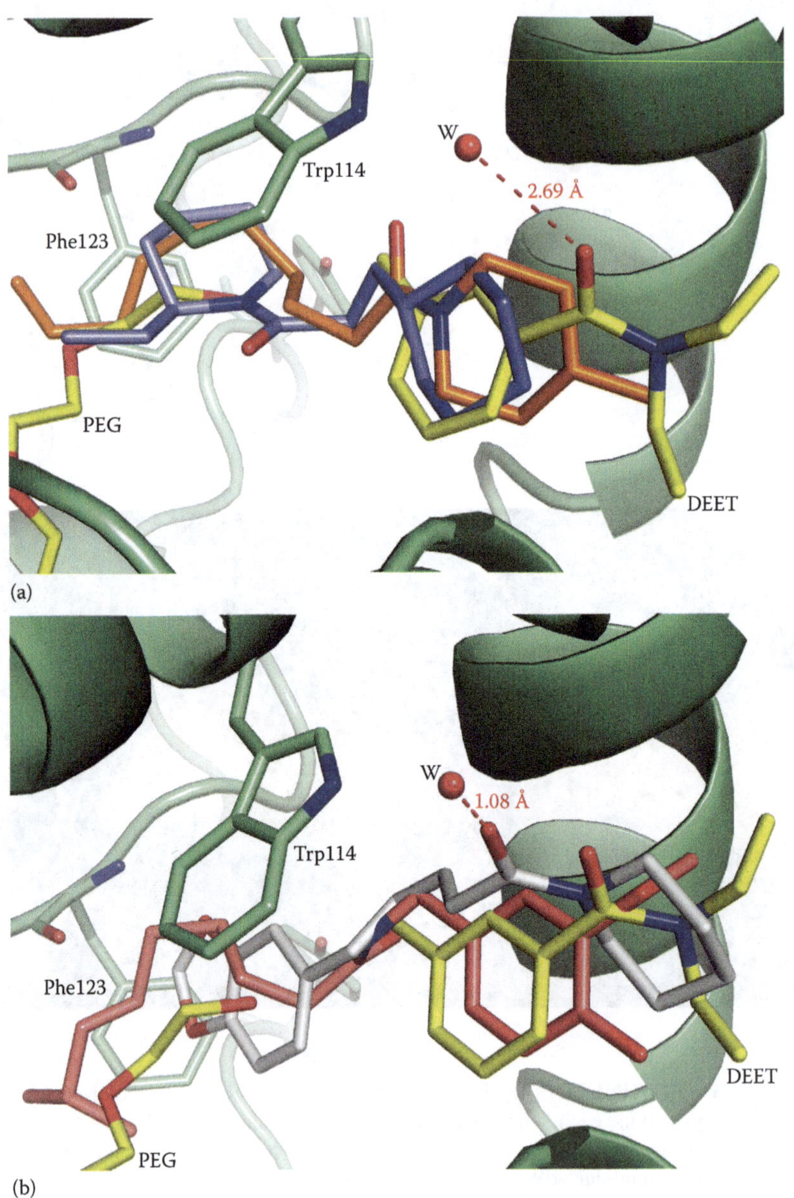

FIGURE 3.7 Representation of the AgamOBP1 binding site with DEET and PEG (yellow) superimposed with the tested compounds, showing the predicted binding modes of (a) the active compounds 4-methyl-1-(1-oxodecyl)piperidine (orange) and 1-(3-cyclohexyl-1-oxopropyl)-2-ethylpiperidine (blue) as well as of (b) piperine (gray) and capsaicin (raspberry).

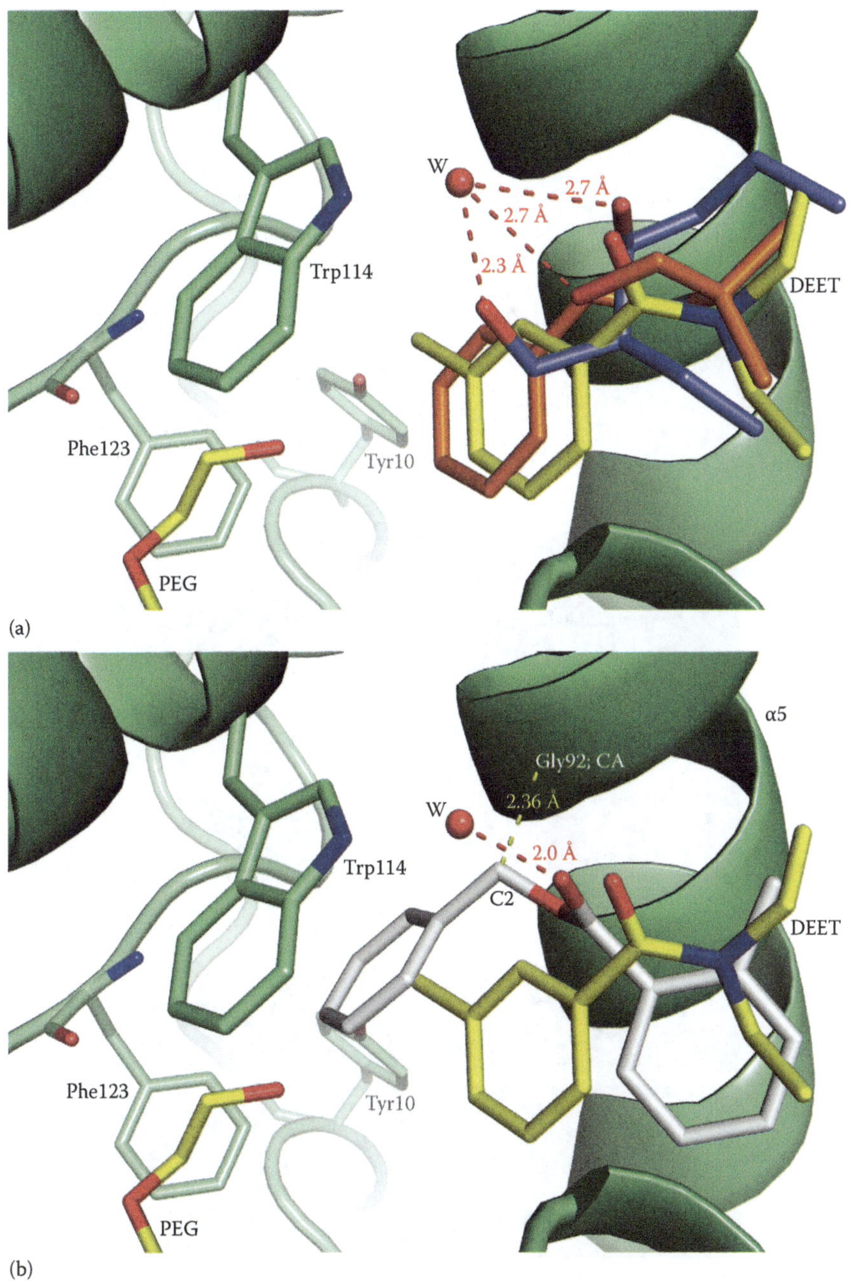

FIGURE 3.11 Representation of the AgamOBP1 binding site with DEET and PEG (yellow) superimposed with tested compounds. (a) The predicted binding modes of the active compounds oxazolidine,4,4-dimethyl-2-phenyl (orange) and Rutgers 612 (2-ethyl-1,3-hexane diol) (blue); (b) the predicted binding mode of *o*-toluic acid benzyl ester, a methyl derivative of benzyl benzoate (gray).

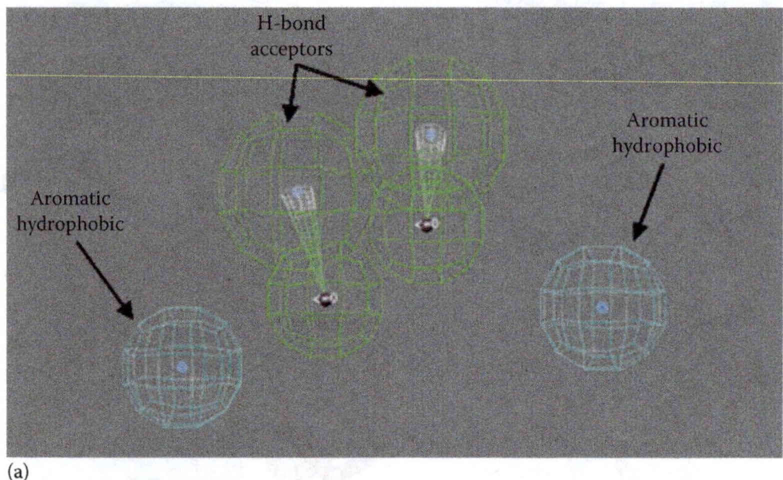

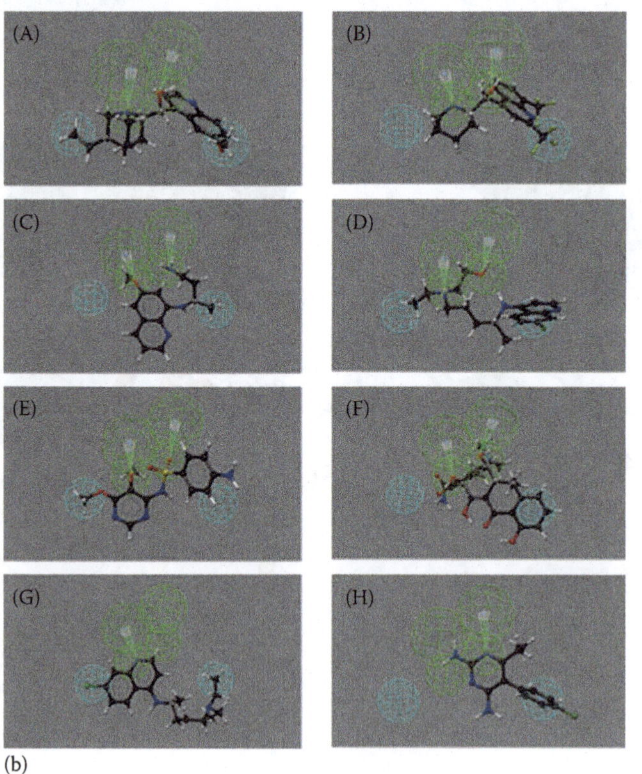

FIGURE 5.2 (a) Pharmacophore for antimalarial activity of the tryptanthrins. (b) Mapping of the pharmacophore onto eight commonly used antimalarial drugs in the United States: (A) quinine, (B) mefloquine, (C) primaquine, (D) hydroxychloroquine, (E) sulfadoxine, (F) doxycycline, (G) chloroquine, and (H) pyrimethamine.

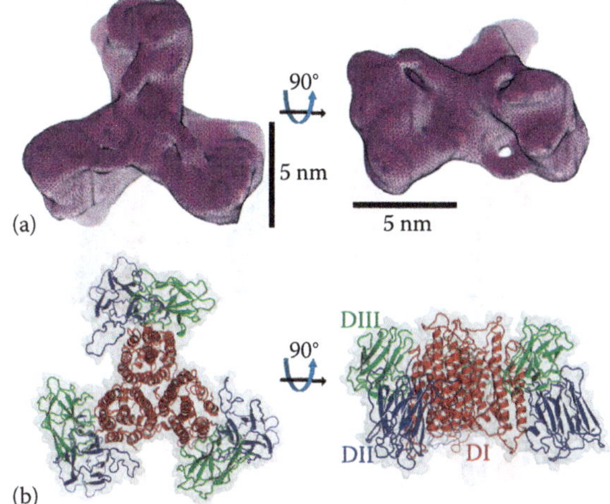

FIGURE 7.3 Electron density map of Cry4Ba trimer compared with the Cry4Aa trimeric model structure. (a) Top and side views of an EM image of Cry4Ba trimer (figures were taken and modified from W. Sriwimol et al., J. Biol. Chem. 290 (2015), pp. 20793–20803); (b) top and side views of structural model of Cry4Aa trimer. Proteins are rendered as ribbons and colored by domain (DI = red, DII = blue, and DIII = green). vdW surface is also shown.

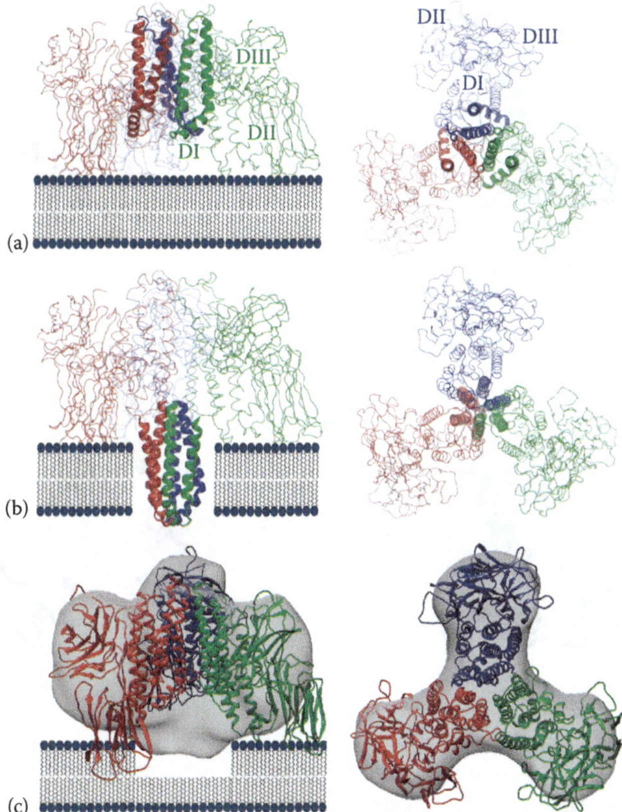

FIGURE 7.4 Pore formation of the prepore Cry4Aa trimer compared with the EM density map of Cry4Ba trimer. Side and top views of prepore Cry4Aa trimer (lines in red, blue, and green) before (a) and after (b) the insertion of three α_4–α_5 hairpins (ribbons in red, blue, and green) into the membrane (T. Taveecharoenkool et al., PMC Biophys. 3 (2010), pp. 10); (c) side and top views of the EM density map of Cry4Ba trimer (gray surface) in its membrane-bound state docked with three copies of Cry4Ba monomers (ribbons in red, blue, and green) for comparison (W. Sriwimol et al., J. Biol. Chem. 290 (2015), pp. 20793–20803).

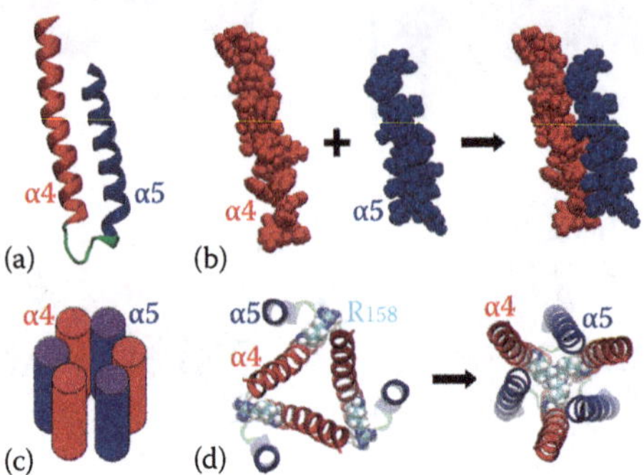

FIGURE 7.5 Trimeric pore modeling of Cry4Ba toxin. (a) An α_4–α_5 hairpin was extracted from the x-ray structure of Cry4Ba toxin monomer; (b) α_4 and α_5 were separated and then docked together by Hex program to find two possible packings between them. (c, d) Three α_4–α_5 hairpins were arranged on an equilateral triangle and then they were moved and rotated such that each pair of α_4 and α_5 was packed according to the results from (b).

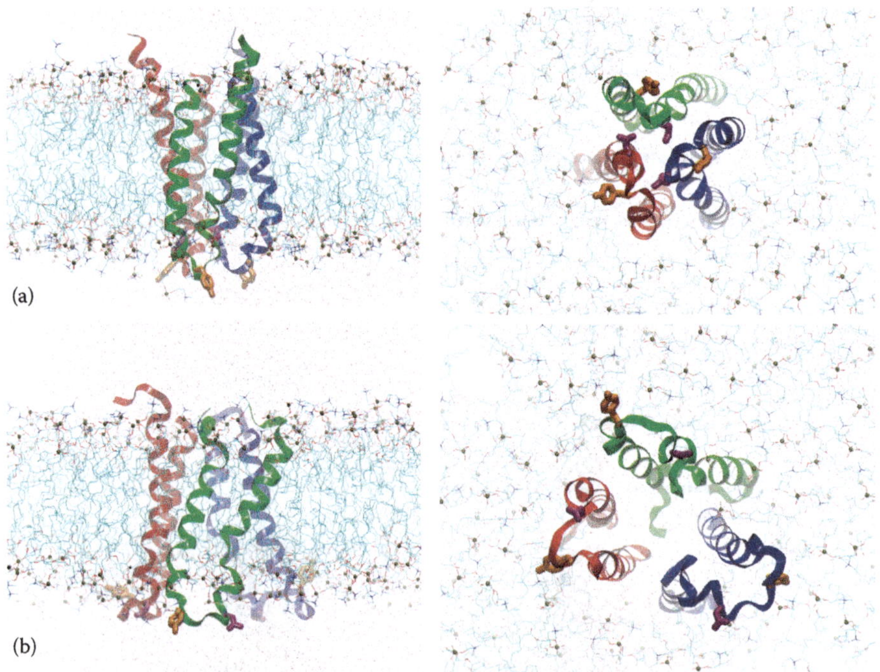

FIGURE 7.6 Opening of trimeric Cry4Ba pore in DMPC membrane during 60-ns MD simulations. MD snapshots at time $t = 0$ ns (a), and $t = 60$ ns (b) from the side and bottom views. The three α_4–α_5 hairpins were rendered as ribbons and colored in red, green, and blue. Asn^{166} and Tyr^{170}, which directly interacted with lipid head groups, were drawn as capsules and colored in purple and orange, respectively. DMPC lipids were shown as lines and colored in cyan (carbon), blue (nitrogen), and red (oxygen) with phosphorus atoms rendered as small gold spheres. Only oxygen atoms of water were shown as red points.

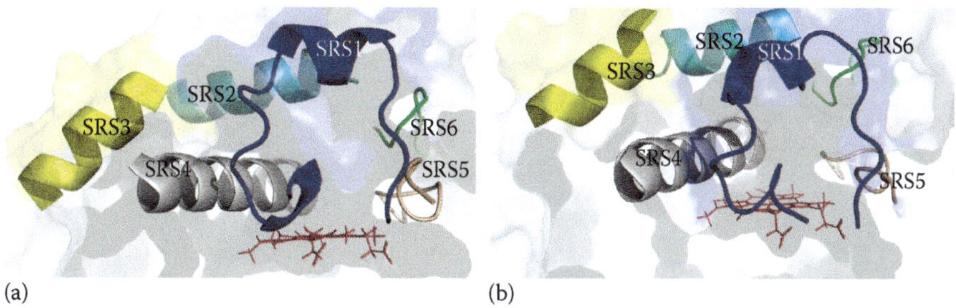

FIGURE 12.3 Active sites and predicted substrate recognition sites (SRSs) designated 1-6 of CYP6AA3 (a) and CYP6P7 (b) homology models (P. Lertkiatmongkol, E. Jenwitheesuk, and P. Rongnoparut, BMC Res. Notes 4 (2011), pp. 321). SRSs are depicted in cartoon. The heme group is shown at the lower part of the structure.

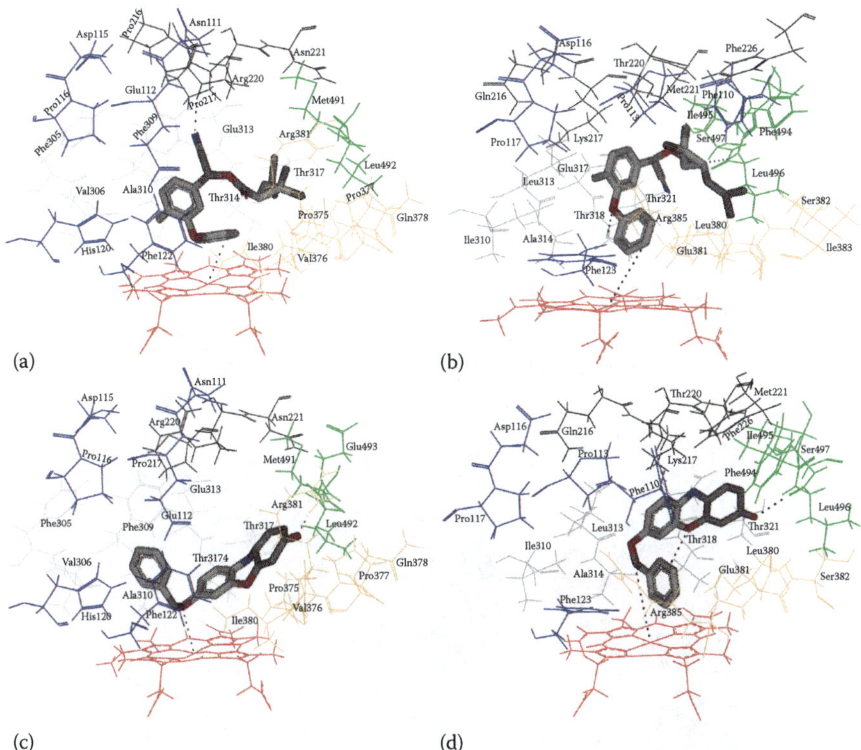

FIGURE 12.4 Docking simulation of cypermethrin (a, b) and benzyloxyresorufin (c, d) into CYP6AA3 (a, c) and CYP6P7 (b, d). The heme group of P450s is shown at the lower part of each figure. Amino acid residues interacted with ligands are presented.

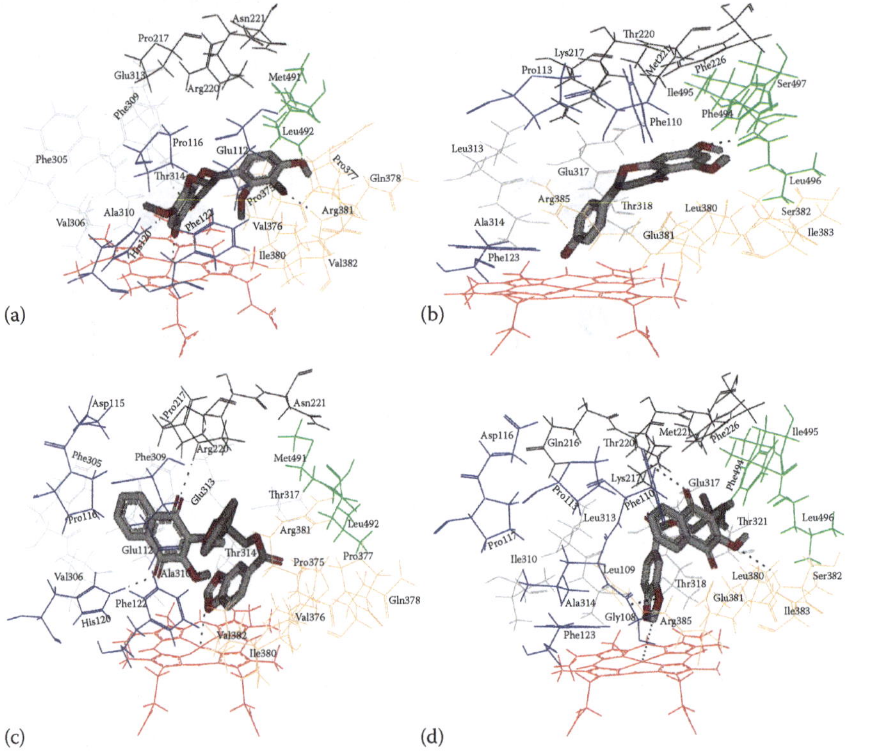

FIGURE 12.5 Docking simulation of 5,4′-dihydroxy-7,8,2′3′-tetramethoxyflavone (a), 5,7,4′-trihydroxyflavone (apigenin, b), and rhinacanthin-D (c, d) into CYP6AA3 (a, c) and CYP6P7 (b, d). Other details are the same as in the caption of Figure 12.4.

FIGURE 15.1 Indoor residual spraying in Mayotte, France.

8 Structural Scaffolding for New Mosquito Larvicides

Kunal Roy and Rahul B. Aher

CONTENTS

ABSTRACT

Mosquito-borne diseases like malaria, filariasis, dengue, chikungunya, yellow fever, Japanese encephalitis, etc. can be fatal, as they account for millions of severe infectious cases and deaths every year. They have caused serious health challenges in most of the Asian and African countries due to the tropical and subtropical climatic conditions in these regions. These diseases are transmitted by different species of mosquito vectors, such as *Anopheles* (malaria), *Aedes* (chikungunya, dengue, and yellow fever), and *Culex* (filariasis and Japanese encephalitis), while the causative organisms are protozoa (malaria) and viruses (other diseases). The control of mosquito-borne diseases has become difficult due to the widespread resistance of the malaria parasites to the chemotherapeutic agents, undesirable effects of the therapeutic drugs on nontarget sites, environmental hazards, nonavailability of specific treatments for mosquito-borne viral infections, etc. The existing vector control measures are not reliable for long-term mosquito control. The best way to prevent outbreaks of mosquito-borne diseases is to target the mosquito vector at the larval stage, which is an attractive and potential site due to mosquitoes' stationary nature in their breeding habitats. In the present chapter, we have made an attempt to focus on the structure–activity relationship (SAR) studies performed on derivatives of different scaffolds such as monoterpenes, *para*-benzoquinones, eugenols, 2,4-dienoates, eudesmanolides, triptamine amides, chalcones, furanochalcone and furanoflavonoid analogs to the karanjin moiety, benzophenone-linked indole analogs, dihydroguaiaretic acids (DGAs), tetrahydroquinolines (THQs), halogenated coumarins, thiadiazolotriazin-4-ones, phenylethylamines, phosphorylated/thiophosphorylated benzimidazoles and benzothiazoles, etc. for the larvicidal activity to different mosquito species.

KEYWORDS

- Antimalarial
- Mosquito larvicidal
- Structure–activity relationship
- Vector control

8.1 INTRODUCTION

Mosquitoes are one of the most menacing insects and the major source of vector-borne diseases (VBD) in both humans and domestic animals of tropical countries. Although a few species of mosquitoes are harmless or even useful to the humans, the females of most species (*Anopheles*, *Aedes*, etc.) are ectoparasites feeding on the blood of various kinds of the hosts and giving rise to a variety of deadly vector diseases. Millions of people become sick or die annually due to epidemics of vector-borne diseases such as malaria, dengue, filariasis, Japanese encephalitis, chikungunya, and yellow fever [1,2].

Malaria is transmitted by the bite of an infected female *Anopheles* mosquito [3], while dengue fever, also known as breakbone fever, is caused by the mosquito species of genus *Aedes*, particularly *Aedes aegypti* [4]. The other related diseases due to *Aedes* species are yellow fever and chikungunya. Yellow fever is caused by yellow fever virus (YFV) belonging to the genus *Flavivirus* and is spread by the bite of female mosquitoes of the *Ae. aegypti* species [5]. Chikungunya is an infection caused by the alpha chikungunya virus, which is passed to humans by the mosquito species of genus *Ae. albopictus* and *Ae. aegypti* [6]. Lymphatic filariasis, commonly known as elephantiasis, is a parasitic disease caused by the bites of *Culex quinquefasciatus* mosquitoes [7], while Japanese encephalitis (JE) is a disease caused by an arbovirus that is mainly transmitted by the bite of infected *Cx. tritaeniorhynchus* mosquitoes [8].

8.2 VECTOR CONTROL MEASURES AND CHALLENGES

The most widely used and recommended method to control epidemics of vector-borne diseases is the use of insecticides and larvicides in houses, surroundings, and larval habitats. There are only four classes of chemical insecticides—namely, organochlorines (DDT, dieldrin, heptachlor, and heptachlor), organophosphates (reldan, temephos, fenthion, chlorpyrifos, pirimiphos-methyl), carbamates (aldicarb, carbofuran, carbaryl, ethienocarb, etc.), and pyrethroids (allethrin, permethrin, cyfluthrin, etc.)—used as a part of vector control programs [9,10]. The structures of some of the commercially available insecticides are given in Figure 8.1. A prolonged and incessant use of these insecticides on a larger scale has led to the widespread development of resistance in many vector populations [11]. Also, the usage of these chemicals on a larger scale is not eco-friendly, since prolonged exposure leads to serious health concerns to plant, human, and animal life. This has created a need to search for an alternative method or discovery of stage-specific (larval) new molecules, which should be safe, effective, and environmentally friendly. The larval stage of mosquito is one of the important target sites of vector control agents because of the stationary nature of mosquito larvae in their breeding habitats [12]. In this chapter, we review all the recently identified scaffolds and the structure–activity relationship (SAR) studies performed for the design of potent larvicidal agents.

8.3 MOSQUITOCIDAL SCAFFOLDS AVAILABLE FOR FURTHER RESEARCH

The resistance of different mosquito species to multiple insecticides has accelerated the research for identification of novel mosquitocidal agents. In order to circumvent the existing problems of currently available insecticides, various new chemical scaffolds have been discovered using chemical synthesis and plant-based isolation of compounds. The identified different scaffolds are valuable resources of information for the discovery of novel, safe, eco-friendly, and resistance-free vector-control agents. The chemical structures of different scaffolds identified against the mosquito vectors *Aedes* (*Ae. albopictus*, *Ae. aegypti*), *Culex* (*Cx. quinquefasciatus*, *Cx. pipiens pallens*), and *Anopheles* (*An. stephensi*, *An. albimanus*, *An. arabiensis*) are given in Tables 8.1 through 8.3, respectively.

8.4 SAR STUDIES FOR MOSQUITO LARVICIDAL ACTIVITY

8.4.1 SAR Studies on Larvicidal Activity for *Aedes aegypti* Vector

Santos et al. [4] reported larvicidal activity of naturally available monoterpenes and some synthesized monoterpenes against the *Ae. aegypti* L. (Diptera: Culicidae). The larvicidal activity was determined at the third-instar larval stage of *Ae. aegypti* species. The study was performed to identify the structural modifications required for improving the larvicidal activity in monoterpenes. The following modifications were required to be considered in monoterpene derivatives (Figure 8.2) for improving the larvicidal potency:

1. Limonene showed the highest larvicidal activity, while neoisopulegol showed the lowest potency. The unsaturated cyclic hydrocarbons possessing endo- and exo-double bonds in *R*- and *S*-limonene (LC_{50} = 27 and 30 ppm,

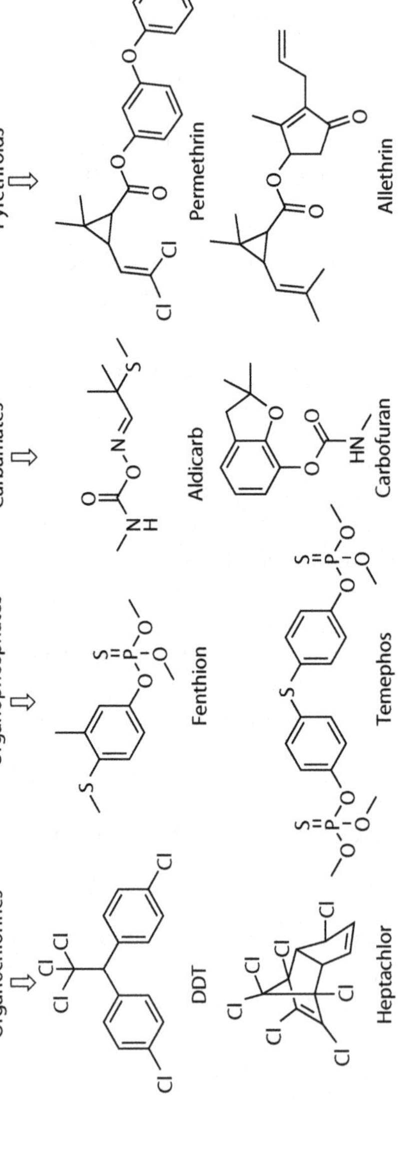

FIGURE 8.1 Structures of some commercial chemical insecticides.

TABLE 8.1
Scaffolds Discovered against *Aedes aegypti*[a] and *Aedes albopictus*[b] (Vector of Dengue, Chikungunya, and Yellow Fever)

No.	Name	Scaffolds
1	Hydrazones[a] [13]	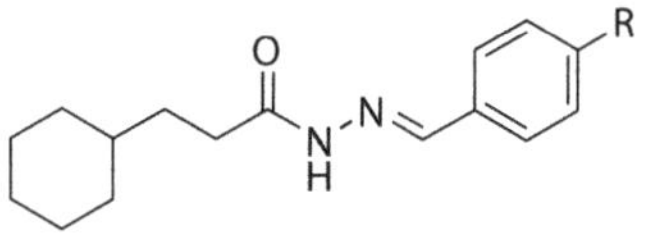
2	Thiourea linked triazole moiety[a] [14]	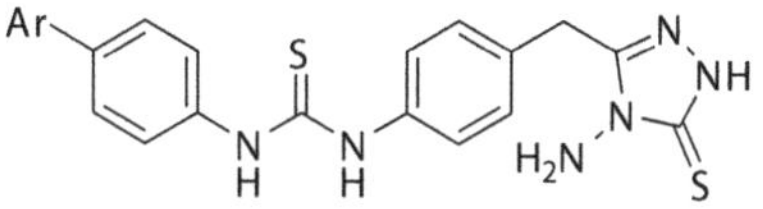
3	Phosphorylated benzimidazoles[b] [9]	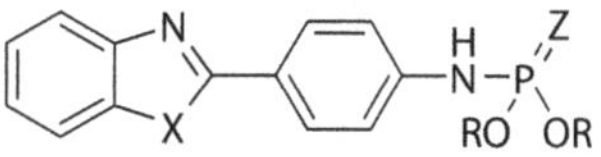
4	Ethyl-ether dilapiole[a] [15]	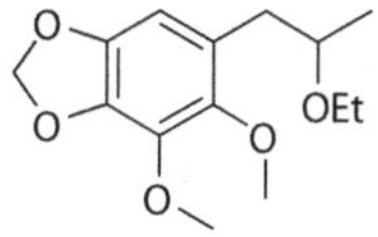
5	Monoterpenes[a] [16]	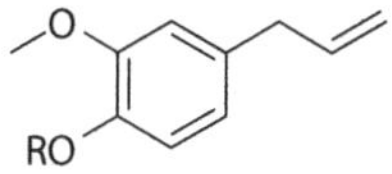
6	Triorganotin *para*-substituted benzoates[a] [17]	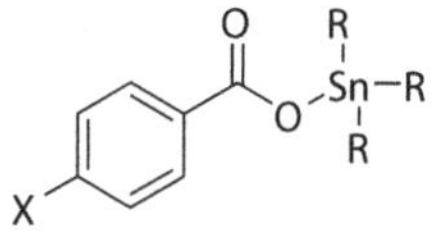
7	Triorganotin chrysanthemumates[a] [18]	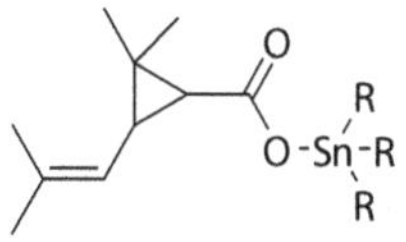
8	Alantolactone[a] [19]	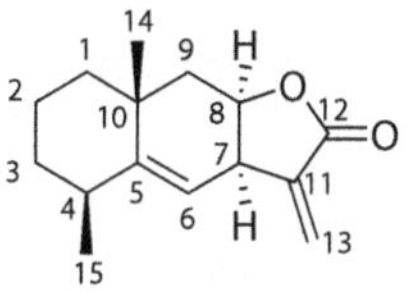
9	Hydrazide-hydrazones[a] [20]	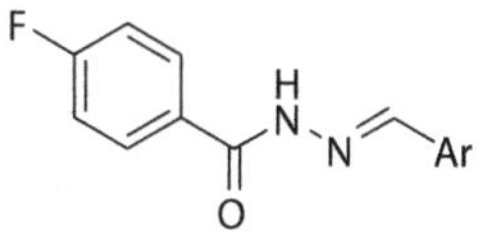
10	Eugenol[a] [12]	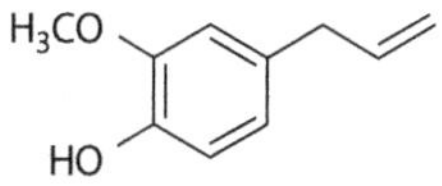

(*Continued*)

TABLE 8.1 (CONTINUED)
Scaffolds Discovered against *Aedes aegypti*[a] and *Aedes albopictus*[b] (Vector of Dengue, Chikungunya, and Yellow Fever)

No.	Name	Scaffolds
11	Phenylethylmines[a] [1]	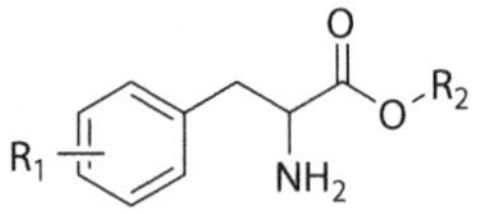
12	*R*-Limonene[a] [4]	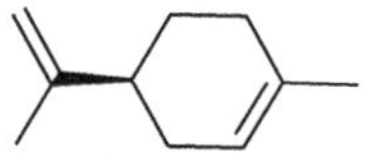
13	Triorganotin 2-(*p*-chlorophenyl)-3-methylbutyrates[a] [21]	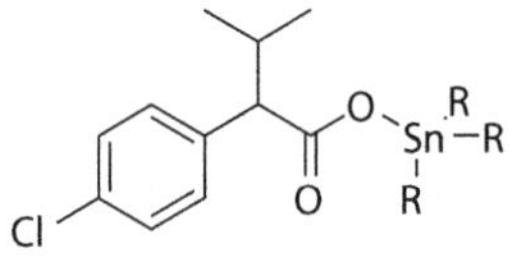
14	*para*-Benzoquinones[a] [22]	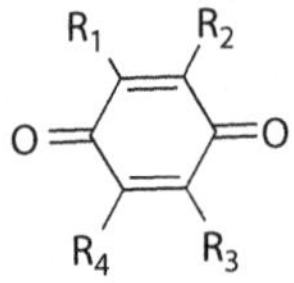
15	2,4-Dienoates[a] [23]	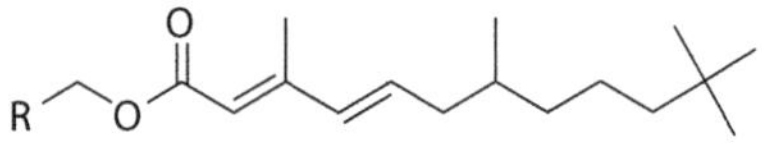
16	Triorganotin 2,2,3,3-tetramethylcyclo propanecarboxylates[a] [24]	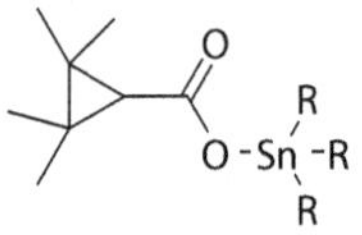
17	Isoalantolactone[a] [19]	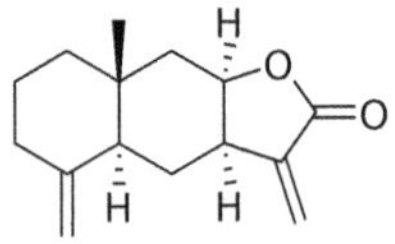
18	3-Acetyl-2,5-disubstituted-2,3-dihydro-1,3,4-oxadiazole[a] [20]	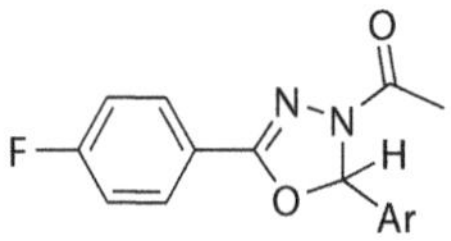
19	Triptamine amides [25]	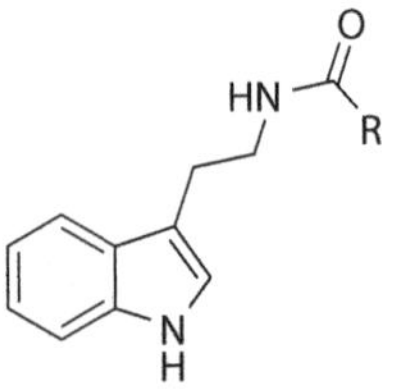

TABLE 8.2
Scaffolds Discovered against *Culex quinquefasciatus*[a] and *Culex pipiens pallens*[b] (Vector of Lymphatic Filariasis)

No.	Name	Scaffolds
1	Linear furano-chalcones[a] [2]	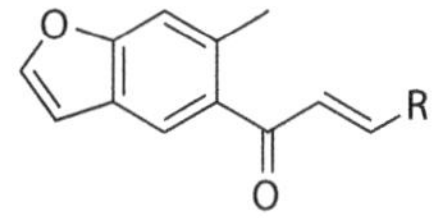
2	Dihydroguaiaretic acids[b] [26]	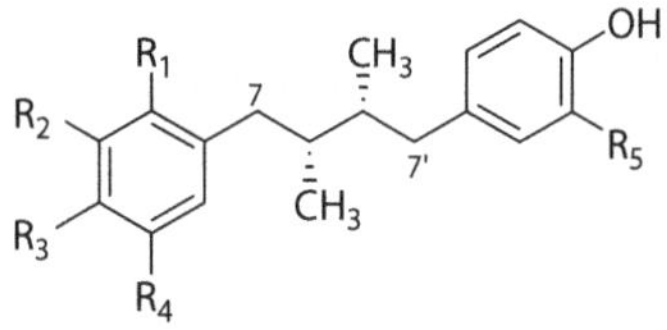
3	Phenylethylamines[a] [1]	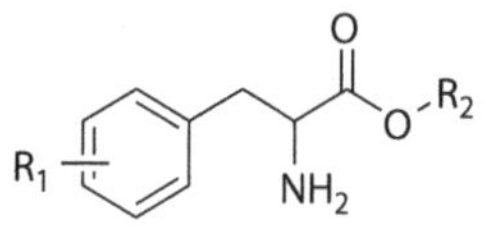
4	Chalcone derivatives[a] [7]	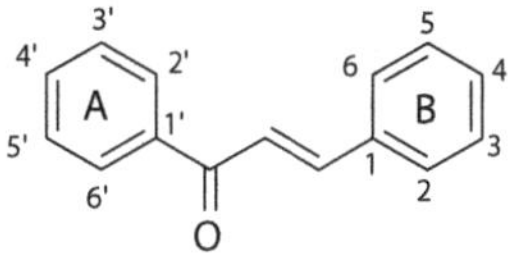
5	Phosphorylated benzimidazoles[a] [9]	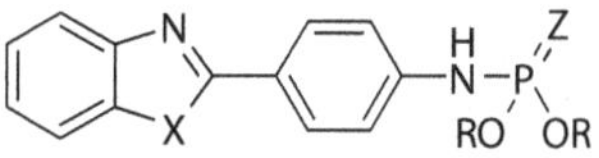
6	Benzophenone linked indoles[a] [27]	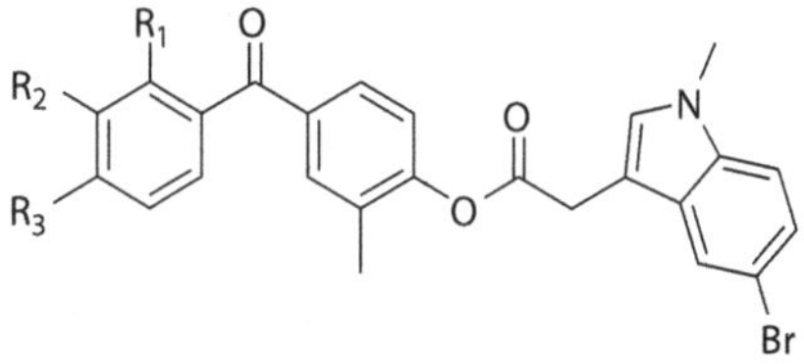
7	Triorganotin 2-(*p*-chlorophenyl)-3-methylbutyrates[a] [21]	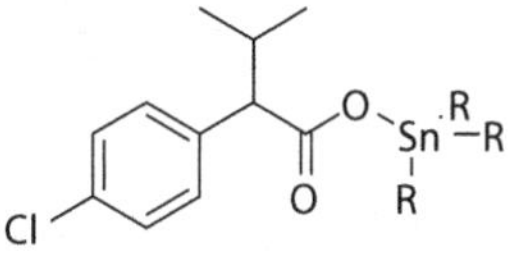
8	Isoxazolyl pyrimidoquinolines[a] [28]	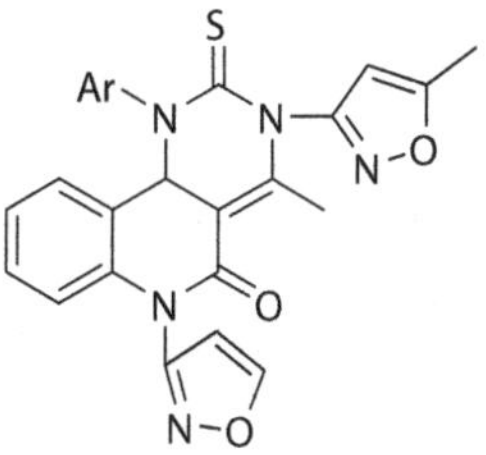

(*Continued*)

TABLE 8.2 (CONTINUED)
Scaffolds Discovered against *Culex quinquefasciatus*[a] and *Culex pipiens pallens*[b] (Vector of Lymphatic Filariasis)

No.	Name	Scaffolds
9	Fluorosubstituted benzoylphenylureas[b] [29]	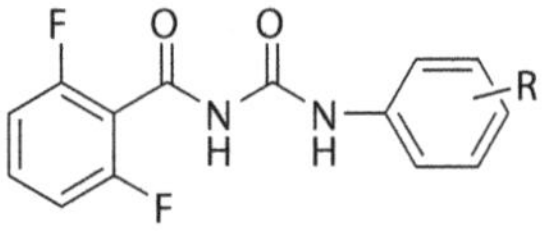
10	Triorganotin 2,2,3,3-tetramethylcyclopropanecarboxylates[a] [24]	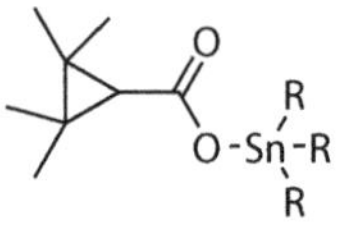
11	Angular furanochalcones[a] [2]	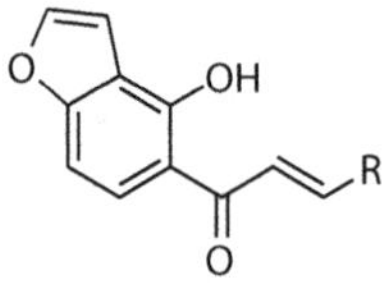
12	Tetrahydroquinolines[b] [30]	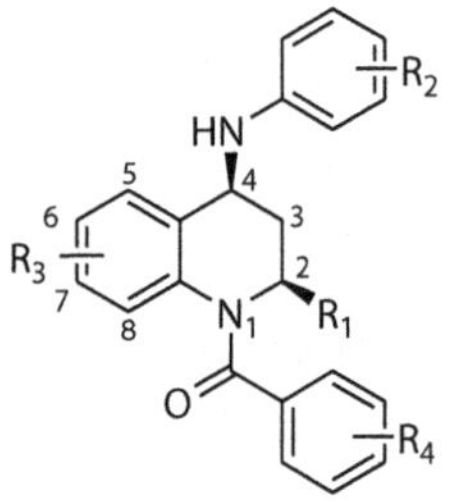
13	Benzoylpyridazyl ureas[b] [31]	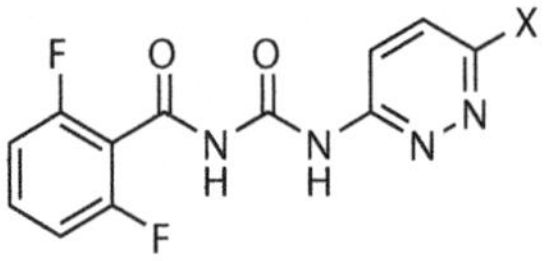
14	*N*-benzoyl-*N*′-phenyl-*N*′-sulfenylureas[b] [32]	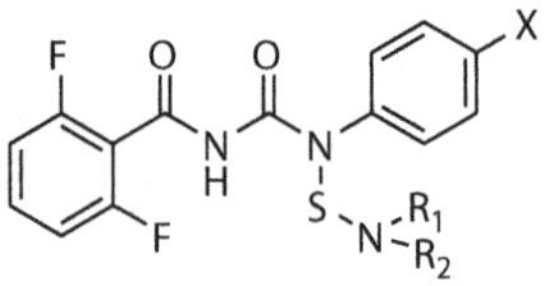
15	Benzoylphenylureas[b] [33]	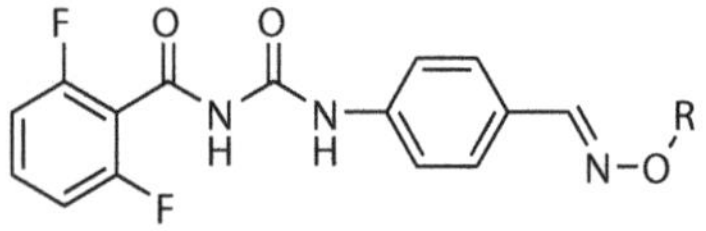
16	Benzoheterocyclic diacylhydrazines[b] [34]	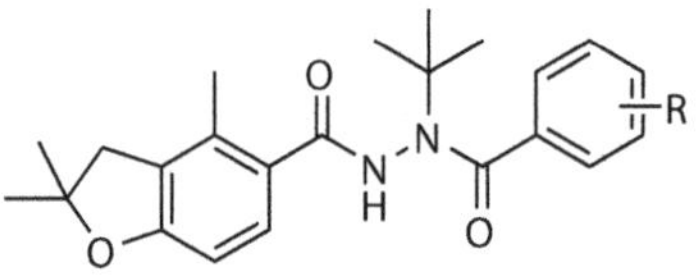

(Continued)

TABLE 8.2 (CONTINUED)
Scaffolds Discovered against *Culex quinquefasciatus*[a] and *Culex pipiens pallens*[b] (Vector of Lymphatic Filariasis)

No.	Name	Scaffolds
17	Benzoylphenylureas-linked oxazoles[b] [35]	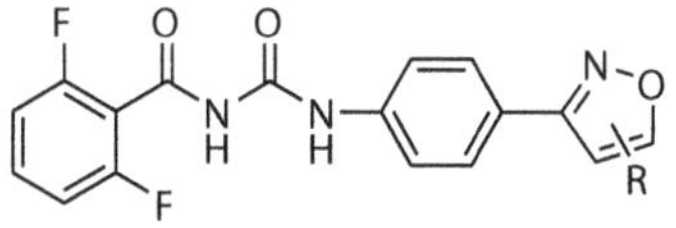
18	Podophyllotoxin[b] [36]	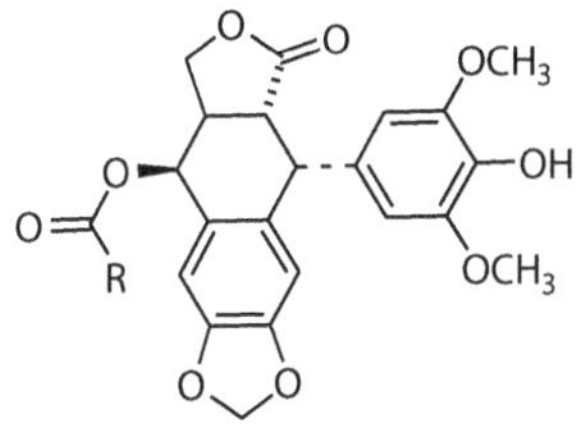
19	Triorganotin chrysanthemumates[a] [18]	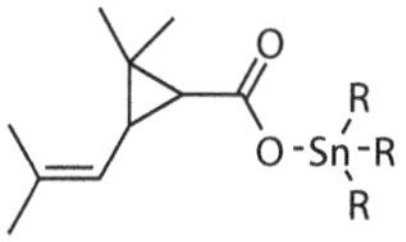

respectively) were the most potent compounds. Changing the position of the exo-double bond in limonene has resulted in a twofold decrease in the larvicidal potency (γ-terpinene; LC_{50} = 56 ppm). A fivefold decrease in potency was observed when there was an absence of the exo-double bond in limonene and presence of a three-member ring (bicyclic compound 3-carene; LC_{50} = 150 ppm).

2. An addition of heteroatoms in the cyclic hydrocarbon structure of limonene reduced the larvicidal potency, which might also be related to the decrease in the lipophilicity.
3. The number of conjugated double bonds increased the larvicidal potency. For example, thymol, carvacrol, and eugenol showed higher larvicidal potencies than carvone.
4. Presence of a hydroxyl group in the cyclic structure and replacement of double bonds by epoxides decreased the larvicidal potency.

Sousa et al. [22] reported a comparative study of assessment of larvicidal activity of six *para*-benzoquinones derivatives against the third-instar mosquito larvae of *Ae. aegypti*. The following structural changes for improving the activity were to be considered while optimizing the benzoquinones derivatives (Figure 8.3):

1. The presence of an alkyl group was crucial for retaining the larvicidal activity as evident from the lowest potency of the unsubstituted *para*-benzoquinone (compound 1a; LC_{50}: 90 ppm). The *para*-benzoquinone with two methyl

TABLE 8.3
Scaffolds Discovered against *Anopheles stephensi, Anopheles albimanus,* and *Anopheles arabiensis*

Mosquito-Borne Disease	Mosquito Vector	Scaffolds: Name	Scaffolds: Structure
Malaria	*Anopheles stephensi*	Thiadiazolotriazin-4-ones [37]	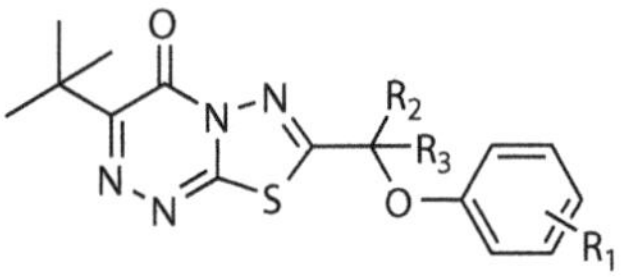
		Triorganotin *para*-substituted benzoates [17]	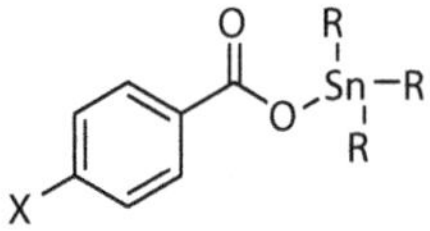
		Triorganotin 2-(*p*-chlorophenyl)-3-methylbutyrates [21]	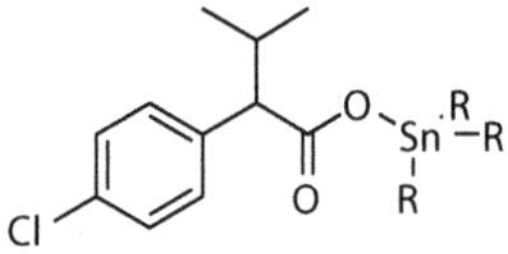
		Triorganotin chrysanthemumates [18]	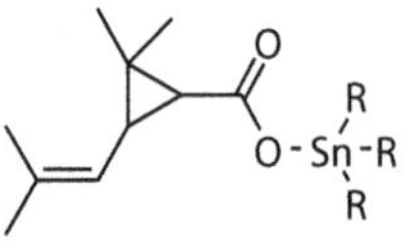
		Triorganotin 2,2,3,3-tetramethylcyclo propanecarboxylates [24]	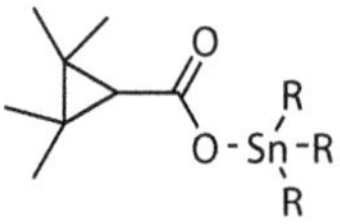
	Anopheles albimanus	Phenylethylamines [1]	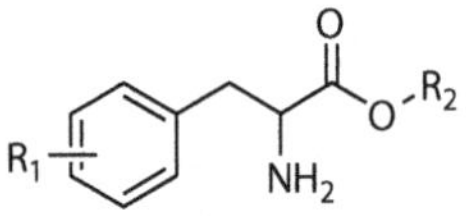
	Anopheles arabiensis	6-Halogenated coumarins [10]	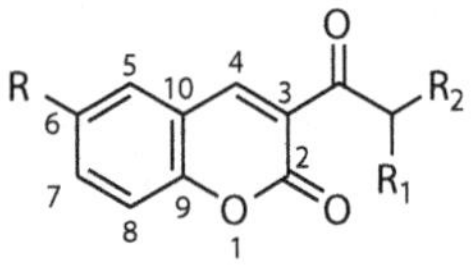
		2,6-Substituted benzo[d] thiazole [3]	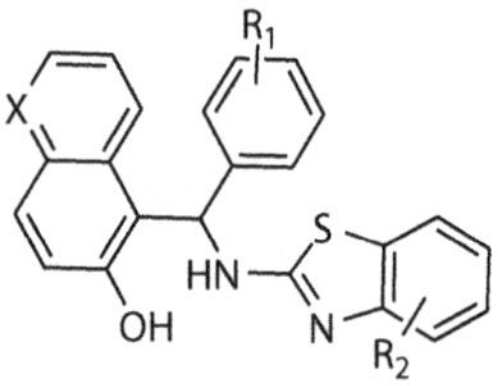

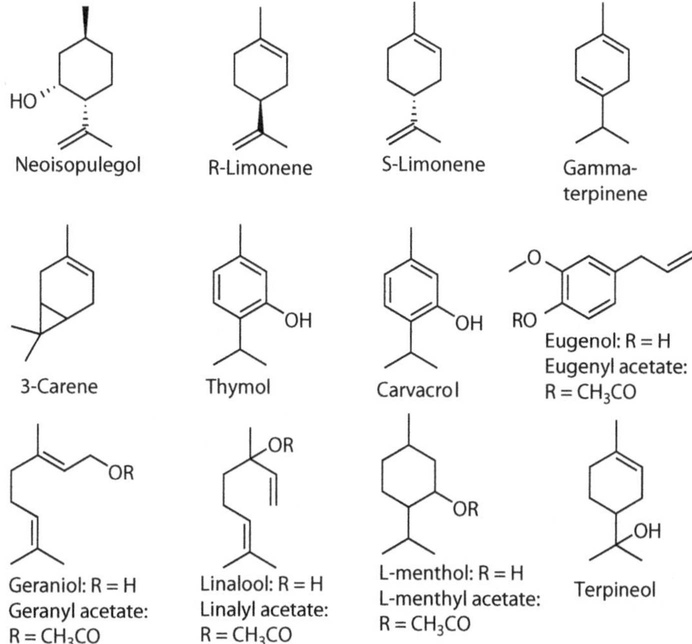

FIGURE 8.2 Structures of different monoterpenes.

R_4, R_1, R_3, R_2, O, O

1a: $R_1 = R_2 = R_3 = R_4 = H$
1b: $R_1 = CH_3$; $R_2 = R_3 = R_4 = H$
1d: $R_1 = CH_3$; $R_2 = R_3 = H$; $R_4 = CH_3$
1e: $R_1 = CH_3$; $R_2 = R_4 = H$; $R_3 = CH_3$
1f: R_1 = isopropyl; $R_2 = R_4 = H$; $R_3 = CH_3$

1c, LC_{50}: 33 ppm (most active)

Tectoquinone

FIGURE 8.3 General structure of *para*-benzoquinone derivative.

groups (compound 1d; LC_{50}: 42 ppm) was more active compared to the monomethyl derivative. Subsequently, changing the position of the methyl group also influenced the larvicidal activity as observed when comparing compound 1d with compound 1e (LC_{50}: 57 ppm). These modifications of the position of the methyl group on *para*-benzoquinone were in agreement with the literature data, which showed that anthraquinone skeleton with a methyl group at C-2 position (tectoquinone) exhibited the strongest larvicidal activity.

2. When the methyl group of compound 1e was replaced by a bulkier group (isopropyl group, compound 1f; LC_{50}: 48 ppm), the activity was slightly enhanced.

3. The activity differences among compounds 1b, 1c, and 1f suggested that monosubstituted *para*-benzoquinones with bulky groups had higher larvicidal activity than disubstituted quinones with methyl and isopropyl groups. The presented knowledge of SAR is important to develop novel larvicidal compounds potentially suitable to control the dengue mosquito.

Pandey et al. [16] reported the larvicidal activity of monoterpenes and their acetyl derivatives against the fourth-instar larvae of *Ae. aegypti*, and the structure–activity relationship (SAR) study was investigated in order to identify the structural features necessary for the mosquitocidal activity. The following SAR features were revealed from the investigation: The derivatization of monoterpenes (eugenol, geraniol, linalool, L-menthol, and terpeniole) by the acetylation of a hydroxyl group resulted in significant enhancement of the larvicidal activity (Figure 8.2). Eugenol (LC_{50} = 82.8 ppm) was found to be the most active among the other monoterpenes, while its acetylated derivative (eugenyl acetate; LC_{50} = 50.2 ppm) was identified to be one of most potent analogs among all the derivatives tested. The activity of the acetate derivatives was in following order:

Eugenyl acetate > linalyl acetate > terpinyl acetate > menthyl acetate > geranyl acetate

Modification of the side chain of eugenol leads to a considerable decrease in the larvicidal activity to 1,415.1 ppm [10]. The plausible reason for this change is that an increased number of hydroxyl groups might prevent the compound from penetrating the larvae cuticle. Both linalool (LC_{50} = 242.6 ppm) and geraniol (LC_{50} = 415.0 ppm) are acyclic oxygenated monoterpenes, but the larger activity of linalool was observed due to the presence of a tertiary alcohol group, while geraniol contains only a primary alcohol group. The greater activity of terpineol (LC_{50} = 331.7 ppm) compared to the L-menthol (LC_{50} = 365.8 ppm) was observed due to the fact that terpineol contains a tertiary hydroxyl group and a double bond, while L-menthol is saturated and has no double bond or hydroxyl group attached to the cyclohexane ring.

Barbosa et al. [12] studied the structure–activity relationship of eugenol derivatives and their mosquitocidal activity, which was determined by testing against the third-instar larvae of *Ae. aegypti*. Different derivatives of eugenol (compound 2a: LC_{50} = 93.3 ppm; parent compound) were synthesized and the larvicidal activity of other derivatives (Figure 8.4) was compared in order to elucidate the structural factors governing the activity. Substitution of the hydroxyl group in compound 2a by bulky groups such as benzoyl, *tert*-butyl silyl, etc. resulted in eight- and threefold decreases (compound 2b: LC_{50} = 706.8 ppm; compound 2c: LC_{50} = 278.9 ppm) in larvicidal activity. Similarly, replacement of the hydroxyl group (compound 2e: LC_{50} = 295.9 ppm) of compound 2a by a carboxylic acid group led to a decrease in activity by three orders of magnitude. Changing the side chain of eugenol (LC_{50} = 93.3 ppm) with other groups as in compound 2d led to a drastic reduction in the activity (compound 2d: LC_{50} = 1,415.1 ppm). The explanation for this major change in activity is that considerable increase in the number of hydroxyl groups might prevent penetration of the larvae cuticle.

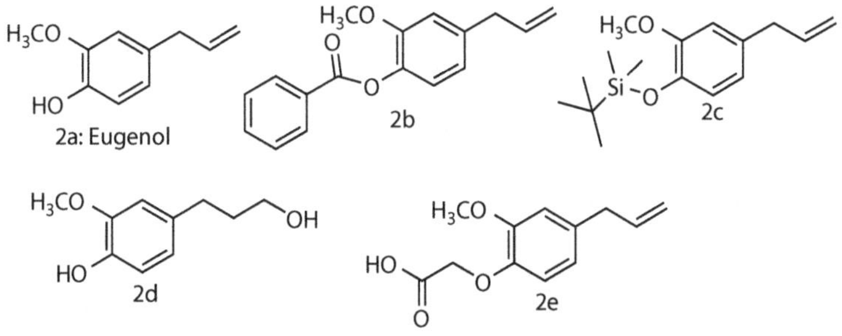

FIGURE 8.4 Structures of eugenol derivatives.

Devillers et al. [23] utilized 188 chemicals (mostly 2,4-dienoate derivatives) and their larvicidal activity against the *Ae. aegypti* vector for deriving SAR as well as for the development of structure–activity models. These compounds were tested previously against the younger larval and last larval instars of *Ae. aegypti* species. They divided the 2,4-dienoate derivatives into three parts for correlating the structures with the larvicidal activity (Figure 8.5). Part A consisted of different functional groups linked to the carbonyl group adjacent to the diene group; part B comprised a middle carbon chain that consisted of double bonds and different substituents (mostly methyl or ethyl groups) and part C was the terminal end of the chain opposite to the carbonyl group representing the different atoms or functional groups linked to the dimethyl carbon that terminated the middle carbon chain.

The following changes were suggested by Devillers et al. [23] for designing a molecule that can inhibit the larval development of *Ae. aegypti*:

1. If part A is with an oxygen atom or a sulfur atom linked to the carbonyl group, part B is 2,4-diene group substituted by a methyl group in positions 3 and 7, and part C is a hydrogen atom, then the activity increases, while presence of a nitrogen atom or carbon atom in part A reduces the activity.
2. The length of part A also influences the activity. Changing the length of part A from methyl to ethyl group (without changing part B and part C), when the atom linked to the carbonyl group is either sulfur or oxygen, is favorable for the activity.
3. The nature and location of substituents in part B (parts A and C unchanged) also change the activity of the molecules.

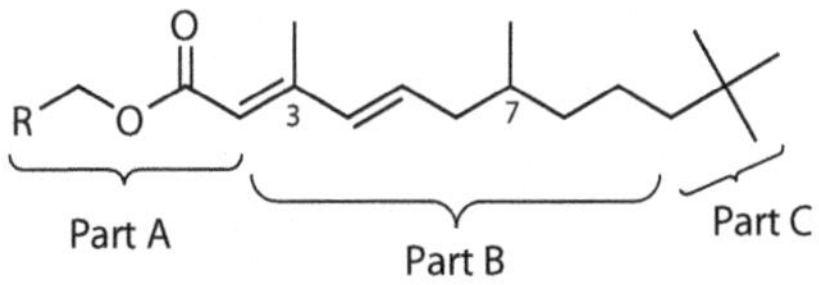

FIGURE 8.5 General structure of 2,4-dienoate derivatives.

Cantrell et al. [19] performed structural modifications on two eudesmanolides (namely, alantolactone and isoalantolactone) by epoxidations, reductions, catalytic hydrogenation, and Michael additions to the α,β-unsaturated lactones, and the corresponding modified derivatives (Figure 8.6) were screened against the *Ae. aegypti* vector in order to understand the functional groups necessary for maintaining and/or increasing the activity. These compounds were tested against the first-instar larval stage of the *Ae. aegypti* species. The presence of α,β-unsaturated lactone moiety in eudesmanolides is important for maintaining the activity, which was justified by the absence of a double bond in 11,13-dihydroalantolactone (3b) and 11, 13-dihydroisoalantolactone (4b). The epoxy synthetic isomers of alantolactone (3a), isoalantolactone (4a), 11,13-dihydroalantolactone, and 11,13-dihydroisoalantolactone were devoid of larvicidal activity. The saturation of the double bond between C-5 and C-6 (compound 3c) or diol derivatives (compound 3d) of compound 3b showed no activity. The diethyl analog (compound 6a) was slightly less active (LC_{50} = 14.4 μg/mL) than its parent compound 4a (LC_{50} = 10 μg/mL), while the diethyl analog compound 5a of compound 3a (LC_{50} = 36.2 μg/mL) showed a LC_{50} value > 125 μg/mL. The piperidine analogs (compound 5b: LC_{50} = 12.4 μg/mL; compound 6b, LC_{50} = 55.1 μg/mL) of both compound 3a and compound 4a had activity comparable to their parent compounds respectively.

Oliveira et al. [25] synthesized some triptamine amide derivatives and evaluated them against the mosquito vector of *Ae. aegypti* in order to identify the activity affecting groups of the larvicidal activity. These compounds were tested against the third-instar larval stage of *Ae. aegypti.* The structure–activity relationship revealed that the chlorinated derivatives of triptamine amides (compound 7a, 7b [most active]) exhibited up to 10-fold higher potency than the nonchlorinated compounds

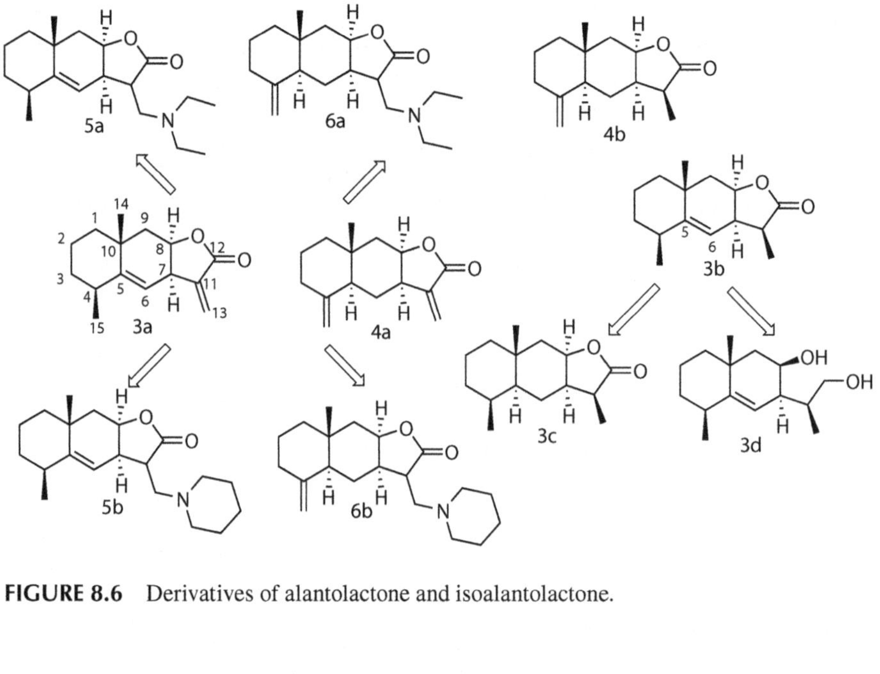

FIGURE 8.6 Derivatives of alantolactone and isoalantolactone.

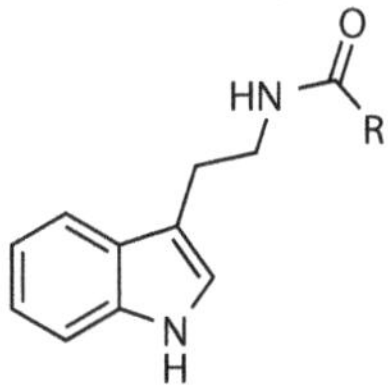

7a: R = CH_2Cl, LC_{50}: 134.9 ppm
7b: R = $C(Cl)_3$, LC_{50}: 48.6 ppm
7c: R = CH_2CH_3, LC_{50}: 443.2 ppm
7d: R = $(CH_2)_2CH_3$, LC_{50}: 271.4 ppm
7e: R = $(CH_2)_3CH_3$, LC_{50}: 222.0 ppm
7f: R = $C(CH_3)_3$, LC_{50}: 123.3 ppm

FIGURE 8.7 Derivatives of triptamine amide.

(compound 7c–7f, Figure 8.7). The larvicidal activity also was enhanced as the number of methylene group increased (compound 7c–7e) in the amide hydrocarbon chain, and this observation suggested that the hydrophobicity was an important factor for the observed enhancement of the larvicidal activity. This observation was in agreement with the previous studies of SAR of eudesmanolides [19].

8.4.2 SAR Studies on Larvicidal Activity for *Culex* Vector

8.4.2.1 Studies for *Cx. quinquefasciatus* Species

Begum et al. [7] reported synthesis and larvicidal activity of chalcone derivatives against the mosquito vector *Cx. quinquefasciatus*. These derivatives were tested against the third-instar larval stage of *Cx. quinquefasciatus*.

The SAR analysis (Figure 8.8) revealed the following points:

1. Chalcones having electron releasing groups on either ring A or on ring B showed higher activity.
2. Replacement of the phenyl ring (ring B) by a furan ring exhibited higher larvicidal activity.
3. Hydroxylated chalcones showed potent activity, especially when there is an –OH group at the 2′-position on ring A.
4. The larvicidal activity was significantly reduced when the ring B was substituted with an electron withdrawing group (3-NO_2, 4-NO_2). Some compounds bearing –NO_2 group on ring B, and one –OH group on the 4′-position of ring A surprisingly showed enhanced activity. But the presence of a –NO_2 group (ring B) and at the same time an –OH group at the 2′-position (ring A) exhibited low activity.
5. Substitution of chlorine at the 4-position of ring B also enhanced the activity.

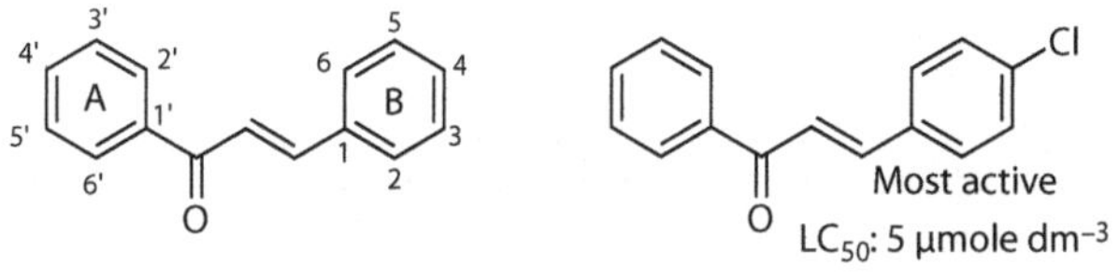

Most active
LC_{50}: 5 μmole dm^{-3}

FIGURE 8.8 General structure of chalcone derivatives.

6. Presence of an electron releasing group (4-OH, 3-OCH_3) on both rings A and B decreased the activity abruptly.
7. The activity was also reduced when ring A was replaced by a methyl group or C_6H_5-CH=CH– group, or ring B was replaced by a C_6H_5-CH=CH– group.
8. Extending the conjugation or blocking the α,β-unsaturated ketone moiety by derivatization of chalcone drastically decreased the larvicidal activity.

Satyavani et al. [2] reported the synthesis and biological testing of a series of furanoflavonoids (analogous to karanjin) and their precursor furanochalcones against the mosquito larvae of *Cx. quinquefasciatus* strain (Figure 8.9). These compounds were tested against the fourth-instar larvae of larval *Cx. quinquefasciatus* strain. The synthesized angular furanochalcones exhibited better activity compared to their linear counterparts. Also, the angular furanochalcones possessing no substituent on the aromatic ring and its corresponding angular furanoflavonoid showed the maximum toxicity to the *Culex* larvae strain. The pyridyl group in the furanochalcones exhibited an excellent larvicidal activity. Moreover, the angular furanochalcones showed better activity than their corresponding angular furanoflavonoids. The hydroxyl phenyl group in furanoflavonoid exhibited better toxicity than their corresponding precursor.

Ranganatha et al. [27] synthesized benzophenone integrated indole analogs and evaluated them against the late third- or early fourth-instar larvae of filariasis vector mosquitoes, *Cx. quinquefasciatus*.

The highest larvicidal activity was observed due to the following substitutions in the benzophenone integrated indole scaffold (Figure 8.10):

1. Presence of a methoxy group at the *para* position of the benzoyl ring
2. A methyl group at the *para* position of the benzoyl ring
3. No substitution in the benzoyl ring of the benzophenone nucleus

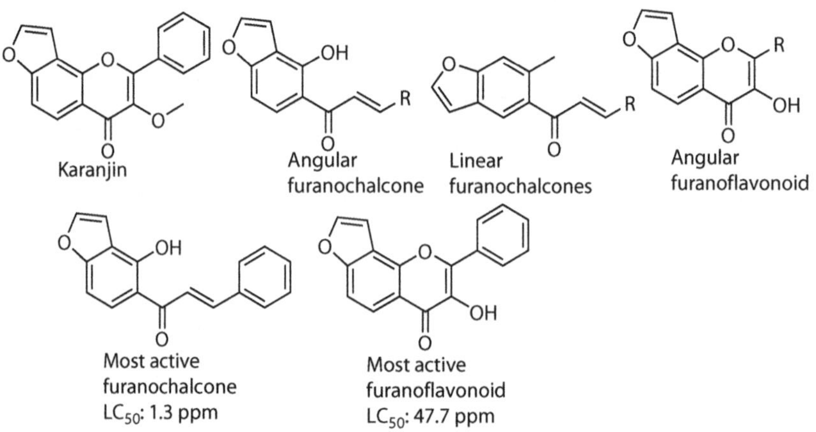

FIGURE 8.9 Structure of furanoflavonoids and their precursor.

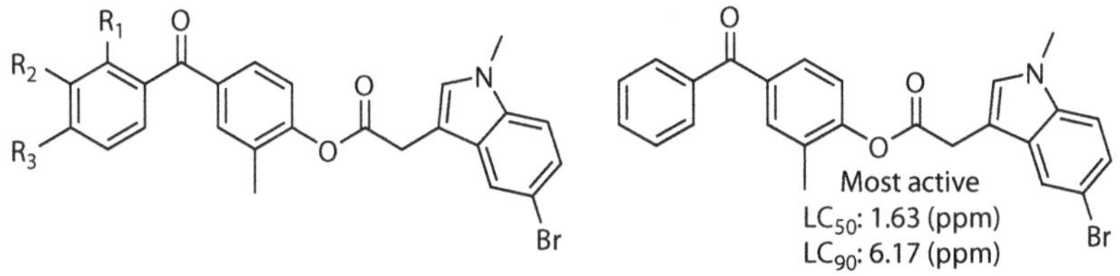

FIGURE 8.10 General structure of benzophenone comprise indole analogs.

These compounds could be further optimized for the development of a powerful armory for the control of the mosquito population in the endemic areas of insecticide resistance.

8.4.2.2 Studies for *Cx. pipiens* Species

Nishiwaki et al. [26] reported the synthesis and larvicidal activity of lignan analogs (–)-dihydroguaiaretic acid (DGA)/secoisolariciresinol (SECO) derivatives against the mosquito vector *Cx. pipiens* in order to elucidate the structure–activity relationship between the chemical structure and the larvicidal activity. These derivatives were tested against the third-instar larval stage of the *Cx. pipiens* strain. The comparison of the larvicidal activity of stereoisomers of DGA and SECO revealed that the DGA isomers were more active [(–)-DGA: potent isomer], while SECO isomers exhibited low larvicidal activity.

(–)-DGA (natural compound) was one of the most active molecules (LC_{50}: 3.52 [10^{-5} molar]); therefore, different substitutions were tried on DGA in order get the most active larvicidal compound. Compounds having no substitution on the 7-phenyl ring (compound 8a) showed equal potency to that of (–)-DGA (Figure 8.11). An introduction of a hydroxyl group on the 7-phenyl ring (other than the *ortho* position) had no change on the larvicidal activity, while substitutions with a hydroxyl group at the *ortho* position (compound 8b) decreased the activity by 13-fold when compared with (–)-DGA. The two di-hydroxylated (compound 8c, 7d), tri-hydroxylated (compound 8e),

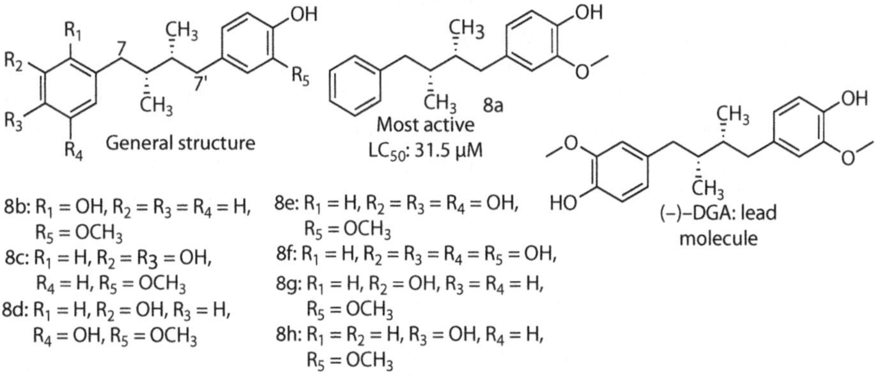

FIGURE 8.11 General structure and potent derivatives of dihydroguaiaretic acid.

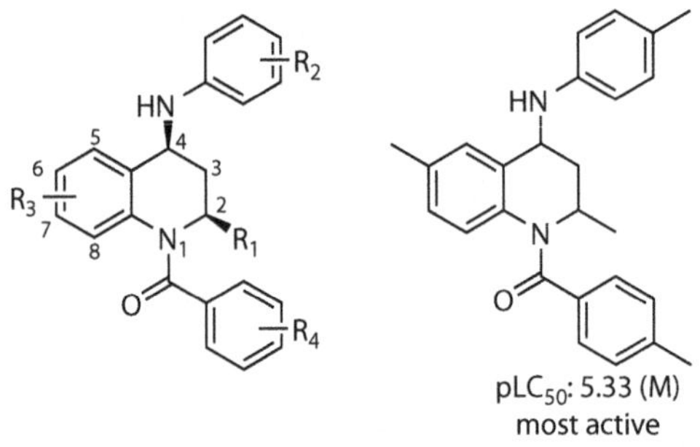

pLC$_{50}$: 5.33 (M)
most active

FIGURE 8.12 General structure of tetrahydroquinoline.

and penta-ol (compound 8f) derivatives exhibited much weaker activity compared to the monohydroxylated derivative (compound 8g, 8h). Compounds bearing more hydroxyl substitutions were hydrophilic compounds, which in turn suggested that hydrophobicity was associated with the larvicidal activity.

DGA derivatives possessing 3-OH (compound 8g) and 4-OH (compound 8h) substitutions induced paralysis and tremors in the larvae; thus, they were thought to exert their action through either the neuronal or respiratory system in mosquitoes. All the key modifications should be helpful for developing novel vector control agents.

Kitamura et al. [30] synthesized four types of tetrahydroquinoline (THQ) compounds having *cis* configuration and evaluated them against the second-instar larvae of mosquito *Culex pipiens pallens*. These compounds have ecdysone agonistic activity. Ecdysone is a steroidal prohormone of the major insect molting hormone 20-hydroxyecdysone, which is secreted from the prothoracic glands. Tetrahydroquinolines regulate the insect molting and metamorphosis processes and are known as insect growth disruptors (IGDs). They are specific to dipterans (insects having two wings), particularly to mosquito ecdysone receptors (EcRs).

The *cis*-(2*R*,4*S*) enantiomer (most active THQ analog) showed 55 times higher larvicidal activity than the *cis*-(2*S*,4*R*) enantiomer (Figure 8.12).

The R1 group should have *2R* (methyl/hydrogen) conformation for the ecdysone agonist activity, while the *2S* methyl substituent lowers the activity (Figure 8.12). This may be because the *2S* methyl group contributes to the steric collision with the receptor, whereas the *2R* methyl group or hydrogen does not. The presence of an aniline moiety and *4S* chiral conformation is important for the activity. The larvicidal activity can be improved by approximately threefold by changing the 4-chloro to 4-bromo at the R4 position. This SAR information is important for the design of novel ecdysone agonists.

8.4.3 SAR Studies on Larvicidal Activity for *Anopheles* Vector

8.4.3.1 Study for *An. arabiensis* Species

Narayanaswamy et al. [10] reported the synthesis and screening of halogenated coumarins (bromo/chloro substituted analogs) for larvicidal property against

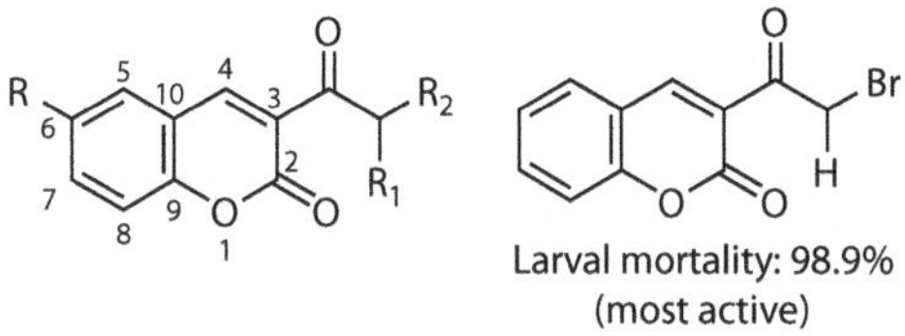

Larval mortality: 98.9%
(most active)

FIGURE 8.13 General structure of coumarin scaffold.

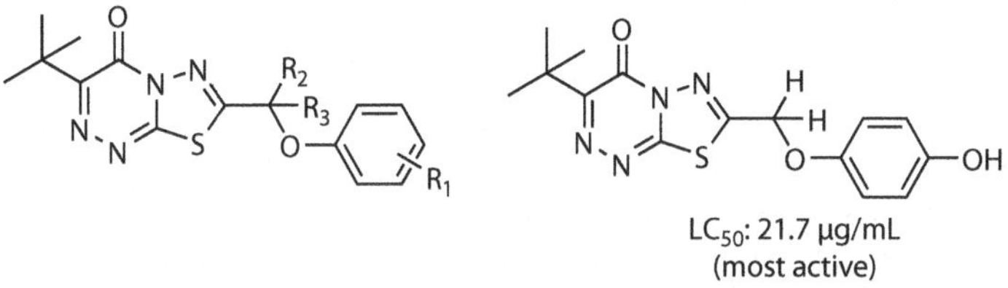

LC_{50}: 21.7 μg/mL
(most active)

FIGURE 8.14 General structure of 3-*tert*-butyl-[1,3,4]thiadiazolo[2,3-c][1,2,4]triazin-4-ones.

An. arabiensis, one of the dominant vectors of malaria in Africa. These compounds were tested against the third-instar larval stage of *An. arabiensis* species.

The presence of electron-withdrawing halogens (bromine and chlorine) on acetyl groups (Figure 8.13) at the third and sixth positions of the coumarin nucleus increases the larvicidal activity, which is comparable to the activity of the positive control temephos (positive control; larval mortality = 100%). A bromine group at the C-5 and C-8 positions of 4-methyl-7-hydroxy coumarin derivative shows both larvicidal and ovicidal activity against the vectors of *Ae. aegypti* and *Cx. quinquefasciatus.*

8.4.3.2 Study for *An. stephensi* Species

Castelino et al. [37] synthesized a series of novel 3-*tert*-butyl-[1,3,4]thiadiazolo[2,3-c][1,2,4]triazin-4-ones and tested them for larvicidal activity against the malaria vector *An. stephensi.* The larvicidal activity was determined at the late third- or early fourth-instar larval stage of the *An. stephensi.* The SAR study showed the significance of the presence of the 4-hydroxyl group (most active compound; Figure 8.14) on a phenyl group for the toxicity against mosquito larvae. Similarly to the hydroxyl group, the presence of other electron-releasing groups such as methyl, methoxy, acetyl, etc. on a phenyl ring of thiadiazolotriazin-4-one scaffold also contributed to the larvicidal activity.

8.4.4 SAR Studies on Larvicidal Activity for Multiple Vectors

8.4.4.1 Studies for *An. albimanus, Ae. aegypti,* and *Cx. quinquefasciatus* Species

Quevedo et al. [1] reported the larvicidal activity of phenylethylamine derivatives against larvae (late third- and early fourth-instar stages) of the mosquito species

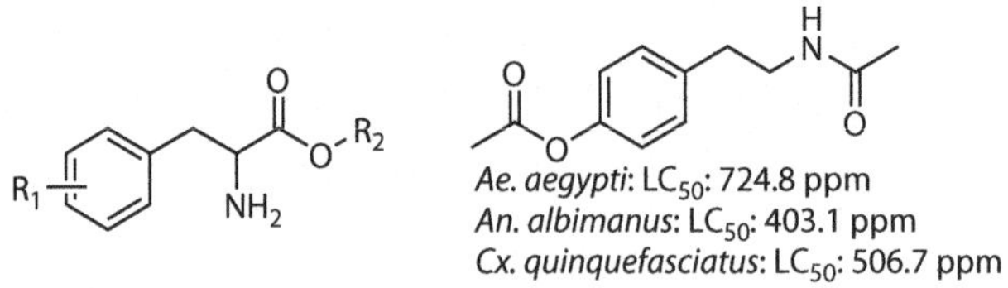

Ae. aegypti: LC_{50}: 724.8 ppm
An. albimanus: LC_{50}: 403.1 ppm
Cx. quinquefasciatus: LC_{50}: 506.7 ppm

FIGURE 8.15 Structure of general scaffold and most active larvicidal compound against mosquito species.

An. albimanus, *Ae. aegypti*, and *Cx. quinquefasciatus*. The structure–larvicidal relationship shows that the following features affect larvicidal activity (Figure 8.15):

1. The phenylethylamine nucleus present in different organic compounds possesses moderate larvicidal activity in itself.
2. The phenolic hydroxyl group is not decisive in the larvicidal activity.
3. Macrocyclic amino acids have activity comparable to or less than that observed for simple phenylethylamines.
4. The hydrocarbon chain on the carboxyl affects larvicidal activity.
5. The intermolecular interactions are crucial for the larvicidal activity of phenylethylamines, which is the new facet to be considered while designing alternative environmentally friendly insecticides.

8.4.4.2 Studies for *Ae. albopictus* and *Cx. quinquefasciatus* Species

Bandyopadhyay et al. [9] reported the synthesis and larvicidal activity of a novel series of benzimidazole- and benzothiazole-linked phosphoramidates and phosphoramidothioates and benzimidazole-linked phenylphosphoramidates and phenylphosphoramidothioates against the late third- to early fourth-instar larvae of *Ae. albopictus* and *Cx. quinquefasciatus* vectors, respectively. The compound with the highest toxicity (LC_{50}) to *Ae. albopictus* showed LC_{50} of 6.42 and 5.25 mg/L at 24 and 48 h, respectively, while it was 7.01 and 3.88 mg/L against *Cx. quinquefasciatus* (Figure 8.16). The compound containing P=S groups showed more larvicidal activity compared to the compound possessing C=O groups (Figure 8.16). Also, the compounds bearing the [NH-P(=S)$(OEt)_2$] group exhibited a better activity profile (toxicity) than the other compounds. These findings of SAR for mosquitocidal activity suggested that structural factors such as nature of substitution in the ring and water solubility of the target compounds could further enhance their activity.

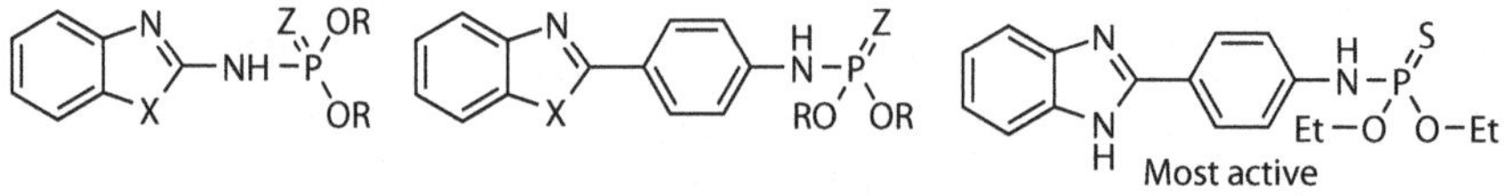

Most active

FIGURE 8.16 General structure of thiophosphorylated benzimidazoles and benzothiazoles.

8.5 OVERVIEW AND CONCLUSION

The available classes of insecticides and larvicides are not environmentally safe, since prolonged and injudicious application of these chemicals leads to contamination of the ecosystem by concentrating on the nontarget sites. Also, insecticide resistance has resulted in the significant loss of potency to the commonly used insecticides. These two major problems of available insecticidal/larvicidal agents have encouraged us to bring together the information of available larvicidal scaffolds and the structural modification studies performed against the different species of mosquitoes. The information provided in this chapter has concatenated the mosquitocidal scaffolds (larval stage) for further research as well as their SAR/QSAR studies of larvicidal activity against the mosquito vector of *Culex*, *Aedes*, and *Anopheles*, respectively. The larval stage of mosquito is an attractive target for insecticide because of their restricted distribution within their breeding habitat and availability of diverse chemical templates against the larval stage.

The structure–activity relationship of monoterpenes and their acetylated derivative provided the indication that structural variations such as shape, size, degree of unsaturation, type and position of functional groups, and type of derivatization in monoterpenoids are quite essential for the enhanced larvicidal activity against the *Ae. aegypti* species. The acetylated derivatives of monoterpenes (eugenyl acetate, linalyl acetate, terpinyl acetate, menthyl acetate, geranyl acetate) have higher activity than their corresponding monoterpenes. Among benzoquinone compounds, the 2-isopropyl derivative of *para*-benzoquinone is the most potent larvicidal derivative, whereas the unsubstituted *para*-benzoquinone is the less potent derivative against the *Ae. aegypti* species. The synthetic isomers of isoalantolactone obtained by structural variations such as epoxidation, reduction, catalytic hydrogenation, and Michael additions are less active than the isoalantolactone itself against *Ae. aegypti*, while many of the corresponding isomers of alantolactone are more active than the parent scaffold. The chlorinated derivatives of triptamine amide showed higher potency against the *Ae. aegypti* species than the nonchlorinated derivatives. Also, the increase in the number of methylene carbons in triptamine amides resulted in an increase in the larvicidal potency. This indicates the significance of hydrophobicity in enhancing the larvicidal activity.

Chalcone derivatives having electron-releasing groups on either ring A or ring B show increased toxicity against the larvae of *Cx. quinquefasciatus*, while those with electron-withdrawing groups, particularly on ring B, show decreased toxicity to the larvae. The angular furanochalcones exhibited better larvicidal activity than their corresponding angular furanoflavonoids against the *Cx. quinquefasciatus* species. The benzophenone tagged indole derivatives having electron-releasing substituents such as methyl, methoxy have higher potency against the *Cx. quinquefasciatus* species than the other derivatives bearing electron-withdrawing halogens (chloro, bromo, and fluoro) on the benzoyl moiety. The hydroxylated derivatives of dihydroguaiaretic acid (DGA) show lower activity than the natural compound (DGA), which signals that the hydrophobicity property would probably be an important factor for the larvicidal activity against *Cx. pipiens* species. The (*2R,4S*)-enantiomer of tetrahydroquinoline derivative exhibited 55 times higher mosquito larvicidal activity than the (*2S,4R*)-enantiomer against *Cx. pipiens* species, which suggests the significance of specific chiral conformation for the larvicidal activity.

The hydroxyl moiety on the phenyl ring of thiadiazolotriazin-4-one scaffold is the necessary substituent for having the maximum toxicity against the *An. stephensi* strain. The chloro and bromo substituted (third and sixth position) coumarin derivatives exhibited the larvicidal activity comparable to the positive control temephos against the malaria vector of *An. arabiensis.* The phenylethylamine nucleus possesses a moderate larvicidal activity in itself. The different derivatives of phenylethylamine showed larvicidal activity against the multiple mosquito species of *An. albimanus*, *Ae. aegypti*, and *Cx. quinquefasciatus*. The triorganotin derivatives possess a broad spectrum of larvicidal activity against the mosquito vectors *Ae. aegypti*, *An. stephensi*, and *Cx. quinquefasciatus*, respectively. Also, this class of compounds is known to biodegrade into nontoxic species in the environment, and there are no known reports of resistance to these agents. Therefore, the triorganotin class of compounds needs to be researched extensively for the next-generation larvicides.

The new strategies for mosquito larval control may involve designing potent larvicidal agents or using one larvicide (like triorganotins) along with another, so as to have synergistic effects, and thereby reducing the amount of chemicals needed for the task. The second way that may prove fruitful in combating resistance is to target the multiple stages of the mosquito vector. The third way is to explore the nature-inspired strategy for enhancing the efficacy and potency of available larvicides. Thus, the information presented here covers the maximum chemical domain (scaffolds) of mosquito larvicidal agents, which should be useful for the development of safe and effective larvicidal agents for vector control.

ACKNOWLEDGMENTS

R.B.A. is thankful to the Indian Council of Medical Research, New Delhi, for providing financial assistance in the form of a senior research fellowship (SRF).

REFERENCES

[1] R. Quevedo, N. Nunez-Dallos, and M.L. Quinones, *Larvicidal activity of single and macrocyclic tyrosine derivatives against three important vector mosquitoes*, Res. Chem. Intermed. 41 (2015), pp. 5283–5292.

[2] S.R. Satyavani, S. Kanjilal, M.S. Rao, R.B. Prasad, and U.S. Murthy, *Synthesis and mosquito larvicidal activity of furanochalcones and furanoflavonoids analogous to karanjin*, Med. Chem. Res. 24 (2015), pp. 842–850.

[3] K.N. Venugopala, M. Krishnappa, S.K. Nayak, B.K. Subrahmanya, J.P. Vaderapura, R.K. Chalannavar, M.G. Raquel, and B. Odhav, *Synthesis and antimosquito properties of 2,6-substituted benzo [d]thiazole and 2,4-substituted benzo [d] thiazole analogues against* Anopheles arabiensis, Eur. J. Med. Chem. 65 (2013), pp. 295–303.

[4] S.R. Santos, M.A. Melo, A.V. Cardoso, R.L. Santos, D.P. de Sousa, and S.C. Cavalcanti, *Structure–activity relationships of larvicidal monoterpenes and derivatives against* Aedes aegypti *Linn*, Chemosphere 84 (2011), pp. 150–153.

[5] http://www.who.int/mediacentre/factsheets/fs100/en/ (accessed on May 5, 2015).

[6] http://www.who.int/mediacentre/factsheets/fs327/en/ (accessed on May 5, 2015).

[7] N.A. Begum, N. Roy, R.A. Laskar, and K. Roy, *Mosquito larvicidal studies of some chalcone analogues and their derived products: Structure–activity relationship analysis*, Med. Chem. Res. 20 (2011), pp. 184–191.

[8] http://www.who.int/mediacentre/factsheets/fs386/en/ (accessed on May 5, 2015).

[9] P. Bandyopadhyay, M. Sathe, S.N. Tikar, R. Yadav, P. Sharma, A. Kumar, and M.P. Kaushik, *Synthesis of some novel phosphorylated and thiophosphorylated benzimidazoles and benzothiazoles and their evaluation for larvicidal potential to* Aedes albopictus *and* Culex quinquefasciatus, Bioorg. Med. Chem. 24 (2014), pp. 2934–2939.

[10] V.K. Narayanaswamy, R.M. Gleiser, K. Kasumbwe, B.E. Aldhubiab, M.V. Attimarad, and B. Odhav, *Evaluation of halogenated coumarins for antimosquito properties*, Sci. World J. (2014), pp. 1–6.

[11] S. Chakraborty, S. Singha, and G. Chandra, *Mosquito larvicidal effect of orthophosporic acid and lactic acid individually or their combined form on* Aedes aegypti, Asian Pac. J. Trop. Dis. 3 (2010), pp. 954–956.

[12] J.D. Barbosa, V.B. Silva, P.B. Alves, G. Gumina, R.L. Santos, D.P. Sousa, and S.C. Cavalcanti, *Structure–activity relationships of eugenol derivatives against* Aedes aegypti *(Diptera: Culicidae) larvae*, Pest. Manag. Sci. 68 (2012), pp. 1478–1483.

[13] N. Tabanca, D.E. Wedge, A. Ali, I.A. Khan, Z.A. Kaplancikli, and M.D. Altintop, *Antifungal, mosquito deterrent, and larvicidal activity of N-(benzylidene)-3-cyclohexylpropionic acid hydrazide derivatives*, Med. Chem. Res. 22 (2013), pp. 2602–2609.

[14] B. Kocyigit-Kaymakcioglu, A.O. Celen, N. Tabanca, A. Ali, S.I. Khan, I.A. Khan, and D.E. Wedge, *Synthesis and biological activity of substituted urea and thiourea derivatives containing 1,2,4-triazole moieties*, Molecules 18 (2013), pp. 3562–3576.

[15] P.R.C. Domingos, A.C. da Silva Pinto, J.M.M. dos Santos, and M.S. Rafael, *Insecticidal and genotoxic potential of two semi-synthetic derivatives of dillapiole for the control of* Aedes (Stegomyia) aegypti *(Diptera: Culicidae)*, Mutat. Res.-Gen. Toxicol. Environ. 772 (2014), pp. 42–54.

[16] S.K. Pandey, S. Tandon, A. Ahmad, A.K. Singh, and A.K. Tripathi, *Structure–activity relationships of monoterpenes and acetyl derivatives against* Aedes aegypti *(Diptera: Culicidae) larvae*, Pest Manag. Sci. 69 (2013), pp. 1235–1238.

[17] Q. Duong, X. Song, E. Mitrojorgji, S. Gordon, and G. Eng, *Larvicidal and structural studies of some triphenyl- and tricyclohexyltin para-substituted benzoates,* J. Organomet. Chem. 691 (2006), pp. 1775–1779.

[18] A. Zapata, D.P. Mclean, J.H. Delao Hernández, A.C. de Dios, X. Song, and G. Eng, *Synthesis, structural and larvicidal studies of a series of triorganotin chrysanthemumates*, Appl. Organomet. Chem. 25 (2011), pp. 777–782.

[19] C.L. Cantrell, J.W. Pridgeon, F.R. Fronczek, and J.J. Becnel, *Structure–activity relationship studies on derivatives of eudesmanolides from* Inula helenium *as toxicants against* Aedes aegypti *larvae and adults*, Chem. Biodivers. 7 (2010), pp. 1681–1697.

[20] N. Tabanca, A. Ali, U.R. Bernier, I.A. Khan, B. Kocyigit-Kaymakcioglu, E.E. Oruç-Emre, S. Unsalan, and S. Rollas, *Biting deterrence and insecticidal activity of hydrazide-hydrazones and their corresponding 3-acetyl-2,5-disubstituted-2,3-dihydro-1,3,4-oxadiazoles against* Aedes aegypti, Pest. Manag. Sci. 69 (2013), pp. 703–708.

[21] G. Eng, X. Song, A. Zapata, A.C. de Dios, L. Casabianca, and R.D. Pike, *Synthesis, structural and larvicidal studies of some triorganotin 2-(p-chlorophenyl)-3-methylbutyrates*, J. Organomet. Chem. 692 (2007), pp. 1398–1404.

[22] D.P. De Sousa, Y.W. Vieira, M.P. Uliana, M.A. Melo, T.J. Brocksom, and S.C. Cavalcanti, *Larvicidal activity of para-benzoquinones*, Parasitol. Res. 107 (2010), pp. 741–745.

[23] J. Devillers, A. Doucet-Panaye, and J.P. Doucet, *Structure–activity relationship (SAR) modelling of mosquito larvicides*, SAR QSAR Environ. Res. 26 (2015), pp. 263–278.

[24] X. Song, A. Zapata, J. Hoerner, A.C. de Dios, L. Casabianca, and G. Eng, *Synthesis, larvicidal, QSAR and structural studies of some triorganotin 2,2,3,3-tetramethylcyclopropanecarboxylates*, Appl. Organomet. Chem. 21 (2007), pp. 545–550.

[25] R.R.B. Oliveira, T.B. Brito, A. Nepel, E.V. Costa, A. Barison, R.S. Nunes, R.L.C. Santos, and S.C.H. Cavalcanti, *Synthesis, activity, and QSAR studies of tryptamine derivatives on third-instar larvae of* Aedes aegypti *Linn*, Med. Chem. 10 (2014), pp. 580–587.

[26] H. Nishiwaki, A. Hasebe, Y. Kawaguchi, M. Akamatsu, Y. Shuto, and S. Yamauchi, *Larvicidal activity of (−)-dihydroguaiaretic acid derivatives against* Culex pipiens, Biosci. Biotechnol. Biochem. 75 (2011), pp. 1735–1739.

[27] V.L. Ranganatha, A.B. Begum, T. Prashanth, H.D. Gurupadaswamy, S.K. Madhu, S. Shivakumar, and S.A. Khanum, *Synthesis and larvicidal properties of benzophenone comprise indole analogues against* Culex quinquefasciatus, Drug Invent. Today 5 (2013), pp. 275–280.

[28] E. Rajanarendar, M.N. Reddy, K.R. Murthy, K. Govardhan Reddy, S. Raju, M. Srinivas, B. Praveen, and M. Srinivasa Rao, *Synthesis, antimicrobial, and mosquito larvicidal activity of 1-aryl-4-methyl-3, 6-bis-(5-methylisoxazol-3-yl)-2-thioxo-2,3,6,10b-tetrahydro-1H-pyrimido [5,4-c] quinolin-5-ones*, Bioorg. Med. Chem. Lett. 20 (2010), pp. 6052–6055.

[29] R. Sun, Y. Liu, Y. Zhang, L. Xiong, and Q. Wang, *Design and synthesis of benzoylphenylureas with fluorinated substituents on the aniline ring as insect growth regulators*, J. Agric. Food Chem. 59 (2011), pp. 2471–2477.

[30] S. Kitamura, T. Harada, H. Hiramatsu, R. Shimizu, H. Miyagawa, and Y. Nakagawa, *Structural requirement and stereospecificity of tetrahydroquinolines as potent ecdysone agonists*, Bioorg. Med. Chem. Lett. 24 (2014), pp. 1715–1718.

[31] R. Sun, Y. Zhang, F. Bi, and Q. Wang, *Design, synthesis, and bioactivity study of novel benzoylpyridazyl ureas*, J. Agric. Food Chem. 57 (2009), pp. 6356–6361.

[32] R. Sun, Y. Zhang, L. Chen, Y. Li, Q. Li, H. Song, H. Runqiu, Bi. Fuchun, and Q. Wang, *Design, synthesis, and insecticidal activities of new N-benzoyl-N′-phenyl-N′-sulfenylureas*, J. Agric. Food Chem. 57 (2009), pp. 3661–3668.

[33] R. Sun, M. Lü, L. Chen, Q. Li, H. Song, Bi. Fuchun, H. Runqiu, and Q. Wang, *Design, synthesis, bioactivity, and structure–activity relationship (SAR) studies of novel benzoylphenylureas containing oxime ether group*, J. Agric. Food Chem. 56 (2008), pp. 11376–11391.

[34] Z. Huang, Q. Cui, L. Xiong, Z. Wang, K. Wang, Q. Zhao, Bi. Fuchun, and Q. Wang, *Synthesis and insecticidal activities and SAR studies of novel benzoheterocyclic diacylhydrazine derivatives*, J. Agric. Food Chem. 57 (2009), pp. 2447–2456.

[35] R. Sun, Y. Li, L. Xiong, Y. Liu, and Q. Wang, *Design, synthesis, and insecticidal evaluation of new benzoylureas containing isoxazoline and isoxazole group*, J. Agric. Food Chem. 59 (2011), pp. 4851–4859.

[36] X. Di, Y. Liu, Y. Liu, X. Yu, H. Xiao, X. Tian, and R. Gao, *Synthesis and insecticidal activities of pyridine ring derivatives of podophyllotoxin*, Pestic. Biochem. Physiol. 89 (2007), pp. 81–87.

[37] P.A. Castelino, P. Naik, J.P. Dasappa, R.S. Sujayraj, K.S. Chandra, K. Chaluvaiah, R. Nair, M.V.S. Kumari, G. Kalthur, and S.K. Adiga, *Synthesis of novel thiadiazolotriazin-4-ones and study of their mosquito-larvicidal and antibacterial properties*, Eur. J. Med. Chem. 84 (2014), pp. 194–199.

9 SAR and QSAR Modeling of Structurally Diverse Juvenoids Active on Mosquito Larvae

James Devillers, Annick Doucet-Panaye, and Jean-Pierre Doucet

CONTENTS

ABSTRACT

Juvenoids are chemicals that mimic the activity of the juvenile hormone secreted by the *corpora allata*. These man-made chemicals selectively target and disrupt the endocrine system of insects, derailing their development irreversibly. Consequently, they are particularly suited as larvicides for the control of mosquitoes. In this context, an attempt was made to review all the structure–activity relationship (SAR) and quantitative structure–activity relationship (QSAR) models aiming at predicting the juvenoid activity of chemicals against mosquitoes. Analysis of these models shows that, to be correctly modeled, the juvenoid activity, even for structurally related molecules, needs to be encoded by different categories of descriptors, which are very often related to some structural characteristics of the molecules and to steric effects. The 1-octanol/water partition coefficient appears less significant except when it also encodes size and branching information such as the autocorrelation vector of lipophilicity. Last, this review reveals that the supervised nonlinear methods outperform the linear regression techniques in the design of powerful predictive QSAR models for juvenoids.

KEYWORDS

- SAR
- QSAR
- Juvenile hormone mimics (JHMs)
- Linear regression
- Nonlinear techniques

9.1 INTRODUCTION

The most efficient way to control mosquitoes is to eliminate the standing water where the females lay their eggs and their larvae live and grow. There is a wide variety of sites in urban, suburban, and rural areas for breeding, depending on the ecology of the mosquito species. Thus, studies have shown that *Aedes aegypti* (L.) prefer containers holding drinking water, especially those retained near or inside homes within urban areas. In contrast, *Culex pipiens* L. and *Cx. quinquefasciatus* (Say) need larger water bodies with higher organic matter and the two species are often found in abandoned pools, catch basins, and sewers [1]. *Ae. albopictus* (Skuse) is more eclectic in its breeding sites. In urban and suburban areas, this includes catch basins and plant saucers in homes or cemeteries, while in rural areas, this can consist in buckets and drums in vegetable gardens [1,2]. The elimination of large larval sites such as swamps, ditches, and flooded areas is usually a task handled by professionals in mosquito control programs. This includes draining and drying the areas of water, establishing ditches or canals and controlling the aquatic weeds, and so on. In urban environments, because mosquito females deposit their eggs in a wide assortment of man-made containers, the extensive and repetitive collaboration of homeowners and inhabitants is needed for optimizing the reduction of the breeding sites.

The second efficient method used to control mosquitoes is larviciding. This utilizes the application of insecticides targeted at immature mosquitoes—namely, the larvae or pupae. These are applied to bodies of water harboring the larvae. Larvicides differ from adulticides in that they are directed at a limited targeted area, which is the body of water and only that area where the larvae grow and mature. The use of larvicides optimizes the efficiency of insecticide control because mosquito larvae cannot escape from the treated water [3]. There are different categories of larvicides. Some are biopesticides such as the *Bacillus thuringiensis israelensis* (*Bti*) while most of them are chemical pesticides having different mechanisms of action. Among the latter category, those mimicking the activity of the juvenile hormone in insects, which are termed juvenoids, have been the subject of intensive research. Historically, the synthesis of juvenoids with insect control potential was preceded by the chemical identification of naturally occurring insect juvenile hormones (JHs), which, at least as a group of closely related homologs, turned out to be rather ubiquitous in insects. These JHs were found to be essential regulators of insect development, particularly with respect to metamorphosis and reproduction. Application of juvenoids proved to be capable of derailing the development process thoroughly and irreversibly, leading ultimately to insect population collapse [4,5]. The search for newly active juvenoids has benefited from the use of *in silico* methods, especially the structure–activity relationship (SAR) and quantitative structure–activity relationship (QSAR) modeling approaches [6,7].

Consequently, the aim of this study was to review the main structure–activity models dedicated to the search of new juvenoids potentially active on mosquitoes. Rather than a compilation of all the existing models, an attempt was made to highlight the diversity of the methods used to describe the molecules as well as the interest in testing different statistical methods on the same data set for optimizing the performances of the models.

9.2 STRUCTURE–ACTIVITY RULES FOR JUVENILE HORMONE ACTIVITY IN MOSQUITOES

Iwamura and colleagues developed different series of molecules for estimating their juvenile hormone activity (JHA) potential against *Cx. pipiens*. These chemicals were undecen-2-one oxime *O*-ethers and undecen-2-yl carbamates [8], (4-phenoxyphenoxy)- and (4-benzylphenoxy)alkanaldoxime *O*-ethers, their ether and hydroxylamine congeners [9,10], (4-alkoxyphenoxy)- and (4-alkylphenoxy)alkanaldoxime *O*-ethers, and their congeners in which the oxime group was replaced by the function ether, ester, amide, carbamate, urea, or aromatic (benzene or pyridine) [11,12]. Attempts were made to relate the structure of these molecules to their JHA against the common mosquito. The results showed that the most important features for JHA were the overall length of the chain molecule, the dimensions of the molecular ends, and the position of the functional group incorporated at one end of the molecule [11,13]. Other authors already pinpointed the importance of these kinds of structural characteristics for explaining the JHA of molecules (see, for example, references [14–17]), but to our knowledge, Iwamura and colleagues were the first to have proposed molecular descriptors for encoding these structural features in their studied series of molecules. For *Cx. pipiens*, the optimum length *D* is about 21–22 Å, and the position-specific interaction site of the functional groups in the molecules with optimum activity is about 4.6 Å distant from one end of the molecule [18]. Electrostatic potentials were calculated for a variety of functional groups found in active insect JHMs studied by Iwamura's group. Quantum chemical calculations showed that the contours of the electrostatic potentials of these functions have negative peak in the plane that perpendicularly bisects their skeletal plane [13]. Hayashi et al. [18] showed that the conditions for highly potent JHA—namely, an optimum molecular length and the position of the functional group—were the same for *Musca domestica* (Diptera, Muscidae) and *Spodoptera litura* (Lepidoptera, Noctuidae).

9.3 TWO- AND THREE-DIMENSIONAL QSAR MODELS ON *CULEX PIPIENS*

Niwa et al. [19] used a data set of 48 (4-phenoxyphenoxy)- and (4-benzylphenoxy) alkanaldoxime *O*-ethers (Figure 9.1) for which JHA data on *Cx. pipiens* were

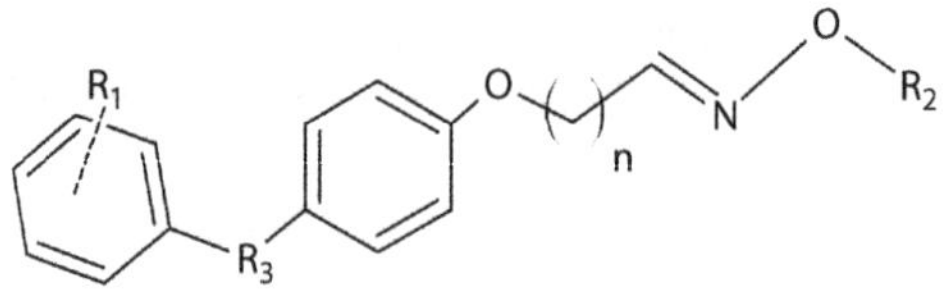

FIGURE 9.1 Structures of (4-phenoxyphenoxy)- and (4-benzylphenoxy)-alkanaldoxime *O*-ethers.

available. JHA was expressed in terms of pI_{50} (M), the logarithm of the reciprocal of the concentration at which 50% inhibition of mosquito metamorphosis was observed. To express the steric features of the molecules, the authors defined the D-axis that passes through the alkoxy oxygen atom and the oxygen atom β to the oxime carbon atom in the fully extended conformation of the acetaldoxime O-ethers. By definition, the angle between the D-axis and the bond that connects the β-oxygen atom to the oxime moiety is 40.02°. For the propionaldoxime O-ethers, the D-axis was drawn so as to pass on the oxygen atom β to the oxime carbon and satisfy this angle condition (Figure 9.2). D_1, the length along the D-axis of the terminal benzyl and phenoxy moieties, and D_2, the length of the alkoxy end, were also defined. Notice that $D = D_1 + D_2$. To express the steric dimensions of the two terminal moieties in more detail, the authors further defined other steric parameters. W_1 was defined as the width of the benzene end in the direction perpendicular to its connecting axis to the rest of the molecule. W_1° corresponded to the minimum width of the substituents. T_2 was the thickness of the alkoxy moiety in vertical direction to the zigzag backbone plane. All the steric parameters were expressed in angstroms [19]. The authors also experienced the usefulness of the

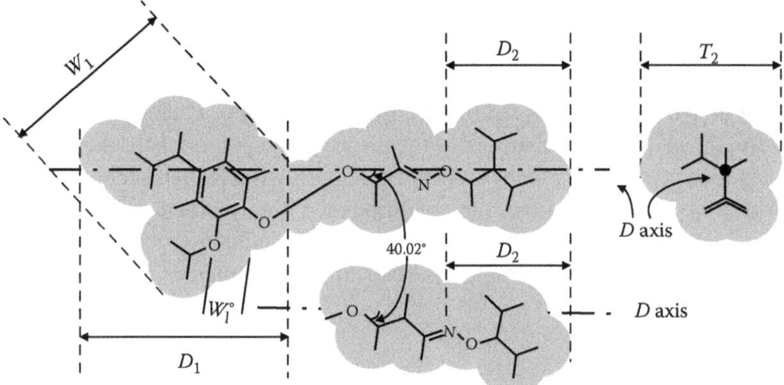

FIGURE 9.2 Schematic representation of the steric parameters. The ends of the bars of the structure represent hydrogen atoms. (Redrawn from A. Niwa et al., J. Agric. Food Chem. 36 (1988), pp. 378–384.)

1-octanol/water partition coefficient (log *P*) in the design of the QSAR models. Their best model was the following:

$$\mathrm{pI}_{50} = 2.91(1.62)D - 0.12(0.06)D^2 + 11.47(3.17)W_1 - 0.78(0.22)W_1^2 - 1.92(0.59)W_1^{\circ} + 4.67(1.94)T_2 - 0.48(0.21)T_2^2 + 1.30(0.28)Ip - 59.81(14.20) \tag{9.1}$$

$n = 48$, $s = 0.34$, $r = 0.95$, $F = 43.14$.

Ip is an indicator variable having a value of 1 for the propionaldoxime series and 0 for acetaldoximes. In Eq. (9.1), *n* is the number of chemicals, *s* is the standard deviation, *r* is the correlation coefficient, *F* is the Fisher statistics, and the figures in parentheses are the 95% confidence intervals.

The squared descriptor parameters in Eq. (9.1) let us suppose that some degrees of nonlinearity exist between the structure of the studied molecules and their JHA. Consequently, an attempt was made to use a purely nonlinear statistical tool for modeling the data [6]. Thus, a three-layer perceptron (TLP) [20], with the five initial descriptors as input layer neurons, was experienced. Three neurons were selected for constituting the hidden layer and one neuron, the pI_{50} values, constituted the output layer. Calculations were made with the data mining module of Statistica™ software (StatSoft, France). Niwa et al. [19] only tried to find a model that fitted at best the whole data set without verifying the prediction performances of Eq. (9.1) although, according to the Organization for Economic Cooperation and Development (OECD) principles for validation of the (Q)SAR models [21], the prediction performances of the models must always be estimated, preferentially from an external test set. This is even more important for the TLPs because they are well known to suffer from problems of over-fitting if they are not correctly used [20]. To avoid this problem, the data set was randomly split into a learning set (LS) to train the TLP, a test set (TS) for monitoring the learning phase via the estimation of the performances of the network while it was under training, and a validation set (VS) to determine how well the network predicted "new" data that were used neither to train the model nor to test its performance when being trained. The LS, TS, and VS included 34, 7, and 7 chemicals, respectively. A 5/3/1 TLP with a bias connected to the hidden and output layers and using the BFGS (Broyden–Fletcher–Goldfarb–Shanno) second-order training algorithm allowed obtaining the best prediction results. The hidden and output activation functions were both a negative exponential function.

The convergence was obtained after only 40 cycles. With such a configuration, the correlation coefficients for the LS, TS, and VS were equal to 0.92, 0.94, and 0.93, respectively. An overall correlation coefficient of 0.92 was calculated from the prediction results obtained with the three sets. It is noteworthy that other random selections of the three sets, only keeping the same proportions between them, led to broadly the same prediction performances. Because the previously selected TLP model included a rather high number of connections, attempts were made to reduce the number of connections in the network. A 5/2/1 TLP with the same configuration, after 26 cycles, led to correlation coefficients of 0.90, 0.95, and 0.92 for the LS,

TS, and VS, respectively. An overall correlation coefficient of 0.90 was calculated. Devillers [6] also tried to use a support vector regression (SVR) analysis [22,23] for modeling the data from the five initial descriptors. The data set was randomly split into a learning set of 36 chemicals and an external test set of 12 chemicals. An SVR type 1 with a radial basis function as kernel provided the best prediction results with C, ε, and γ set to 27, 0.1, and 0.60, respectively. With such a configuration, the correlation coefficients for the learning and external test sets were equal to 0.96 and 0.92, respectively. An overall correlation coefficient of 0.95 was calculated from the prediction results obtained with the two sets. The simulation performances of the SVR model are quite satisfying and we assume that they outperform those obtained with Eq. (9.1), despite the same correlation coefficient of 0.95, because the selection of this configuration has been made on the basis of its performances on a randomly selected learning set and external test set at the opposite of Eq. (9.1), where attempts were only made to fit at best the whole data set. Notice that other selections of learning and external test sets showed similar performances [6].

Hayashi et al. [24] synthesized a series of 30 (4-alkoxyphenoxy)-alkanaldoxime *O*-ethers and 14 (4-alkylphenoxy)-alkanaldoxime *O*-ethers (Figure 9.3), which were then tested on larvae of *Cx. pipiens*. The activity was expressed in terms of pI_{50} values (*M*), the logarithm of the reciprocal of the concentration at which 50% inhibition of metamorphosis was observed. Different steric parameters were defined for describing the molecules. The *D*-axis was defined as passing through the phenoxy and oxime oxygen atoms in the fully extended conformation of the acetaldoxime *O*-ethers (Figure 9.4). The angle between the *D*-axis and the bond that connects the phenoxy oxygen atom to the oxime moiety equals 40.02°. For the propionaldoxime

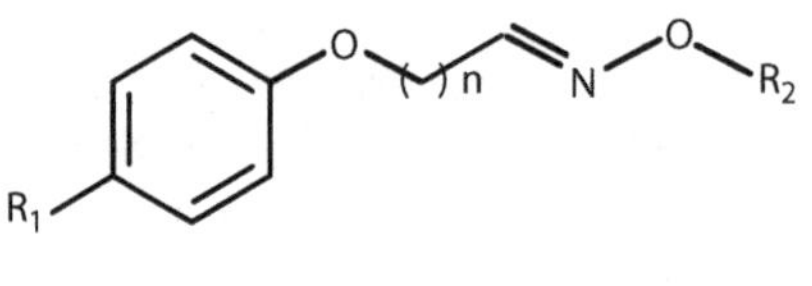

FIGURE 9.3 Generic structure of (4-alkoxyphenoxy)alkanaldoxime *O*-ethers and (4-alkylphenoxy)-alkanaldoxime *O*-ethers (R_1 = alkoxy, alkyl; R_2 = Et, *n*-Pr, *i*-Pr; n = 1, 2).

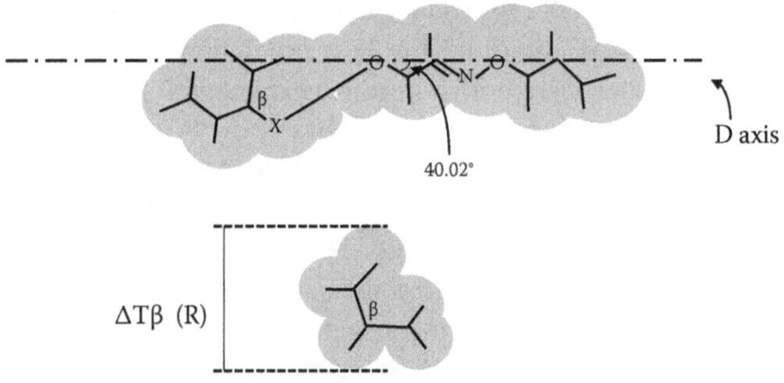

FIGURE 9.4 Definition of the steric parameters. (Redrawn from T. Hayashi et al., J. Agric. Food Chem. 37 (1989), pp. 467–472.)

O-ethers, the *D*-axis was drawn so as to pass through the phenoxy oxygen atom, satisfying this angle condition. The length of the whole molecule was measured along the *D*-axis and named *D*. To express the bulkiness arising from the β-branch of the 4-substituents, the authors [24] defined Tβ as the thickness at the β-position measured in the direction vertical to the zigzag plane (Figure 9.4). In the QSAR analysis, the value relative to the unbranched chemicals (ΔTβ) was used.

The regression model obtained for the alkoxy chemicals was the following:

$$\begin{aligned} \mathrm{pI}_{50} &= 3.40(1.51)D - 0.08(0.03)D^2 + 1.12(0.44)Ip \\ &\quad + 1.59(0.49)I_{br}(\mathrm{OR}) - 31.06(16.94) \end{aligned} \tag{9.2}$$

$n = 30$, $r = 0.93$, $s = 0.46$, $F = 41.65$.

The presence of D and D^2 in Eq. (9.2) reflects the existence of an optimum length for the activity of the molecules. *Ip* is an indicator variable that takes the value of 1 for the propionaldoximes and 0 for the acetaldoximes. I_{br}(OR) is an indicator variable that takes 1 for the compounds having a branch at β of the 4-alkoxy substituents and 0 for the others.

The regression model obtained for the alkoxy and the alkyl chemicals was the following:

$$\begin{aligned} \mathrm{pI}_{50} &= 3.24(1.35)D - 0.07(0.03)D^2 + 1.18(0.35)Ip + 1.55(0.42)I_{br}(\mathrm{OR}) \\ &\quad + 4.34(0.96)\Delta\mathrm{T}\beta(\mathrm{R}) - 1.37(0.39)\Delta\mathrm{T}\beta(\mathrm{R})^2 - 29.24(15.10) \end{aligned} \tag{9.3}$$

$n = 44$, $r = 0.96$, $s = 0.46$, $F = 72.92$.

The letter R in parentheses after the ΔTβ means that the term is for the 4-alkyl compounds. Higher activity is showed by 4-alkyl chemicals with an appropriate β-branch than by those with no β-branch.

An attempt was made by Basak et al. [25] to derive QSAR models from the data generated by Fujita's team. The models were derived from 304 pI_{50} (logarithm of the reciprocal of the concentration (*M*) at which 50% of metamorphosis was observed) obtained on *Cx. pipiens* larvae. The molecules were described by means of topostructural indices (e.g., Wiener number, Randic indices), topochemical indices (e.g., valence and bond connectivity indices, E-state indices), the triplet indices and Balaban's *J* indices, geometric or three-dimensional (3D) indices (kappa shape indices), and atom pairs [26]. Briefly, an atom pair was defined as a substructure consisting of two nonhydrogen atoms, *i* and *j*, and their interatomic separation: {atom descriptor$_i$} - {separation} - {atom descriptor$_j$} where {atom descriptor} contains information regarding atom type, number of nonhydrogen neighbors, and the number of π electrons. Thus, for example, C1X2-6-O0X2 represents a carbon atom with one π electron and two nonhydrogen neighbors (atom no. 1) and an oxygen atom with no π electrons and two nonhydrogen neighbors (atom no. 2), with an interatomic separation (including both the atoms) of six [6]. Molecules were described by a large number of diverse descriptors including 268 global molecular descriptors (topostructural, topochemical, and geometrical), 13 quantum chemical descriptors, and 915 atom pairs (substructural counts). The descriptors were grouped in three sets—namely, the atom pairs (APs), the global descriptors (all the others), and the whole set. Ridge regression (RR) [27], partial least squares regression (PLS) [28], and principal component regression (PCR) [29] were used as statistical methods.

Three different predictor-thinning methods—namely, a modified Gram–Schmidt algorithm, a marginal soft thresholding algorithm, and Lasso (least absolute shrinkage and selection operator) [30]—were utilized to reduce the number of descriptors prior to developing QSAR models. The initial data set was split into five calibration data sets of random sample sizes 60/110/160/210/260 and the remaining 244/194/144/94/44 chemicals were used for validations. For each of these five divisions of data, the three predictor-thinning methods were used. Subsets of 50, 100, 150, 200, and 250 predictors were chosen by each method (though this was not always possible, depending on the size of the calibration data). Ridge regression was used to compute the models that were evaluated for the calculation of the 10-fold cross-validated q^2 values. Lasso was not found to be a very effective method in handling a large set of molecular descriptors because the number of predictors retained could not exceed the number of observations. The results revealed that the modified Gram–Schmidt algorithm was suited to trim the number of predictors in the global molecular descriptor set where collinearity of the descriptors was the major concern [25]. On the other hand, the soft thresholding approach was found to be an effective tool in subset selection from a diverse set of descriptors having both sparsity and multicollinearity, as in the case of the combined set of atom pairs and global molecular descriptors. After the evaluation of the applicability of descriptor trimming, Basak et al. [25] decided to fit the JHA of all 304 chemicals using 250 predictors selected from the combined set of predictors by the marginal soft threshold method, and to use the ridge regression as a statistical tool for deriving the model. The final model was derived on 244 descriptors and led to a 10-fold q^2 equal to 0.60. Among the 16 predictors with the highest t-values, a majority of them were atom pairs and a few were triplet indices. The electrotopological state of oxygen (–O–) was found to be the most important factor that affects JHA. The E-state index of an atom is a measure of the intrinsic state that is perturbed by every other atom in the molecule, accounting for the valence electronegativity of the atom and its local chemical environment. Four of the atom pairs that contain oxygen also appeared as important moieties affecting the JHA of the studied chemicals. The presence in the final model of triplet indices derived from both adjacency and distance matrices also indicated the importance of shape of the ligand. Another interesting observation is that none of the semiempirical quantum chemical parameters, such as E_{HOMO}, E_{LUMO}, or $E_{HOMO} - E_{LUMO}$, were selected in the final model [25].

A subset of 143 molecules was used by Latha et al. [31] to derive 3D QSAR models based on comparative molecular field analysis (CoMFA) and comparative molecular similarity analysis (CoMSIA) [32]. A test set of 14 chemicals was selected to estimate the prediction performance of the models. Each chemical was inserted into a 3D lattice with grid points separated by 2.0 Å in x, y, and z directions and an extension of 4 Å beyond the aligned molecules in all directions. The surroundings of each molecule were mapped by calculating the interactions between probe atoms and each molecule at each grid point. Three different probes were used: a carbon atom (the C_3 probe), a water molecule (the H_2O probe), and a plus-two charged calcium ion (the Ca^{2+} probe), reflecting the steric field, hydrogen bonding field, and the electrostatic field, respectively [31]. The steric (van der Waals) and electrostatic (columbic) field descriptors were calculated on all the lattice points by adding the individual interaction energy between each atom of the chemical with an sp3 carbon probe atom with a +1 charge and 1.52 Å van der Waals radius. A distant dependent dielectric constant was used to treat the electrostatic interaction. A default cutoff equal to 30 kcal/mol was used to truncate the steric and electrostatic field energies. An arbitrary cutoff value of 0.2 was applied to reduce the data set to a limited number of variables [31]. One of the molecules with the highest activity and the lowest internal energy was selected as template. The CoMFA model included 11 PLS components. The cross-validated (q^2), coefficient of determination (r^2), standard error of estimate (*SEE*), and *F* test values equaled 0.568, 0.944, 0.313, and 178.799, respectively. The steric field contributed to 51.8% and the electrostatic field to 48.2%. The CoMSIA model included only six PLS components but the statistics were of lower quality. Indeed, the q^2, r^2, *SEE*, and *F* values equaled 0.505, 0.773, 0.616, and 69.244, respectively. The steric, electrostatic, hydrophobic, H-bond donor, and H-bond acceptor fields contributed to 12.1%, 23.5%, 17.2%, 21.3%, and 25.9%, respectively. While an $r^2_{\text{pred}} = 0.70$ was obtained with the CoMFA model, a very bad correlation was obtained with the CoMSIA model due to the presence of strong outliers. No explanation was given by the authors [31] and no other trial results based on different training and test sets were performed.

9.4 TWO-DIMENSIONAL SAR AND QSAR MODELS ON *AEDES AEGYPTI*

Juvenile hormone activity values on last larval instars of *Ae. aegypti*, originally reported by Henrick et al. [33] for about 100 structurally diverse (2*E*,4*E*)-3,7,11-trimethyl-2,4-dodecadienoates, were used by Nakayama et al. [34] for computing different QSAR models. The JHA values originally reported by Henrick et al. [33] in parts per million were converted into log $1/IC_{50} = pI_{50}$ (mmol). Figure 9.5 shows the common skeleton of the 2,4-dodecadienones in which X expresses the substituents at the carbonyl C1 atom of the dodecadienone skeleton and Y is the longest of the C11 substituents in terms of length Ly along the bond axis. The molecules were described by different steric and hydrophobic parameters and used to compute step by step various QSAR models based on different functional groups. Eq. (9.4) represents the final QSAR model on *Ae. aegypti* selected by Nakayama et al. [34].

$$\begin{aligned} pI_{50} = {} & 3.65(1.26)L_x - 0.35(0.11)L_x^2 + 1.08(1.12)D - 0.06(0.06)D^2 + 1.90(1.13)\log P \\ & - 0.14(0.09)\log P^2 + 0.57(0.25)B_x - 0.71(0.41)I_N \\ & + 0.86(0.35)I_{OR} - 1.39(0.65)I_{br} - 0.65(0.37)I_{(-)} - 16.35(5.21) \end{aligned} \tag{9.4}$$

$n = 85$, $r = 0.89$, $s = 0.53$.

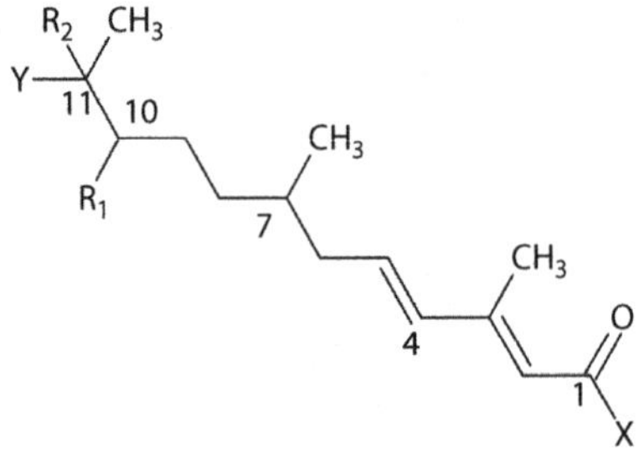

FIGURE 9.5 Structure of 2,4-dodecadienone derivatives with X = OR, SR, NHR, NR2, alkyl; R_1 = H, OR, SEt, 10-ene, 11-ene, 10-epoxy, oxo; Y = OR, SR, OCOR, Me, Et; R_2 = H, Me, Cl.

In Eq. (9.4), L_x is the STERIMOL length of the X end along the bond axis (C1-X). D is the maximum length of the whole molecule. B_x expresses the bulkiness toward the carbonyl group of α-substituents in the alcohol moiety of ester derivatives and/or thiol ester derivatives. I_N is an indicator variable that takes the value of 1 for the amides and 0 for the other chemicals. I_{OR} is an indicator variable that takes the value of 1 for the compounds whose Y moiety is an alkoxy or a hydroxy group and otherwise 0. I_{br} is an indicator variable for the chemicals having a branch at any position in the X moiety of ketone derivatives. $I_{(-)}$ is an indicator variable coding enantiomeric effects. It takes the value of 1 for the *(R)*-(–)-isomer and zero for the (±) mixture. The hydrophobicity of the whole molecule was estimated by log *P* [34,35].

Eq. (9.4) is quite complex, including a lot of parameters with their square terms while leading to rather modest results. Hansch and Leo [36] have tried to obtain a simpler model than Eq. (9.4) by deriving a bilinear model [37] from the descriptors calculated by Nakayama et al. [34]. To do so, the L_x, L_x^2, D, and D^2 descriptors were omitted and 10 chemicals were removed without justification. This led to Eq. (9.5).

$$\begin{aligned}\mathrm{pI}_{50} = {} & 1.16(0.71)\log P - 1.04(0.81)\log(\beta 10^{\log P} + 1) + 0.42(0.25)B_x - 0.57(0.41)I_N \\ & + 0.96(0.27)I_{OR} - 1.18(0.63)I_{br} - 0.91(0.31)I_{(-)} - 1.78(3.20)\end{aligned} \tag{9.5}$$

$n = 75$, $r = 0.86$, $s = 0.52$.

According to Hansch and Leo [36], the relatively high standard deviations and low correlation coefficients of Eqs. (9.4) and (9.5) clearly called for further analyses. Consequently, recently, Devillers [6] has tried to compute a model presenting better performances while including fewer numbers of descriptors. To do so, a TLP [20] was used as statistical engine to optimize the search for complex relationships between the pI_{50} values recorded on *Ae. aegypti* and the molecular descriptors selected by Nakayama et al. [34] but without accounting for the squared terms.

The original data set was randomly split into a learning set (LS), test set (TS), and validation set (VS) of 69, 8, and 8 chemicals, respectively. Use of L_x, D, B_x, log *P*, I_N, I_{OR}, and $I_{(-)}$ as input neurons in the TLP allowed us to obtain a model with good performances. Because with the Statistica™ software (StatSoft, France) a Boolean descriptor is encoded as two input neurons, a 10/4/1 TLP was designed. Such a TLP with a bias connected to the hidden and output layers and the use of the BFGS second-order training algorithm led to satisfying prediction results. The hidden and output activation functions were a hyperbolic tangent function and a negative exponential function, respectively. The convergence was obtained after 138 cycles. With such a configuration, the correlation coefficients for the LS, TS, and VS were equal to 0.92, 0.94, and 0.95, respectively. An overall correlation coefficient of 0.92 was calculated from the prediction results obtained with the three sets. The TLP model has been more robustly derived than Eq. (9.4) and it shows better predictive performances. Thus, for example, inspection of the prediction results (not shown) revealed that 13 chemicals (15%) have residuals ≥ 0.8 (in absolute values) when Eq. (9.4) is used to estimate the pI_{50} values of the 85 studied 2,4-dodecadienones, while the use of the TLP model leads to more than twice fewer chemicals with a residual value ≥ 0.8 (7%). Other trials with random selections of 80% (LS), 10% (TS), and 10% (VS) of the data set and with the same TLP architecture led to broadly the same prediction results. In the same way, use of five or six neurons on the hidden layer increased the performances of the models. It is noteworthy that with such TLP architecture, most of the time, good performances were obtained on the TS and VS despite the high number of connections within the networks [6].

Basak et al. [38] also used the biological data reported by Henrick et al. [33] to derive various QSAR models, focusing on the selection of the most informative descriptors as well as the most efficient statistical method. They included a higher number of chemicals in their modeling process than Nakayama et al. [34], Hansch and Leo [36], and Devillers [6]. Thus, they considered 143 JHA values obtained on last larval instars of yellow fever mosquitoes and 35 censored data reported to be less than a given datum.

The molecules were described by means of topostructural and topochemical indices, triplet indices and Balaban's J indices, geometric or 3D indices, and atom pairs. Initially, 1,173 descriptors were computed, including 915 atom pairs. Ridge regression (RR), principal component regression (PCR), and partial least squares regression were used as statistical tools, and the modified Gram–Schmidt orthogonalization was employed to trim the 258 global molecular descriptors to a size of 100. The leave-one-out cross-validated q^2 values obtained with the RR, PCR, and PLS methods equaled 0.361, 0.053, and 0.327, respectively. RR gave the best results, so it was used to compute the models from atom pairs and from the whole set of descriptors after their trimming by means of the soft threshold method. On the basis of q^2 values, atom pairs provided the best results ($q^2 = 0.408$ with 200 APs).

Basak et al. [38] suspected that the use of a nonlinear method could improve the prediction performances. Thus, recursive partitioning [39] was used to capture any nonlinear relation between the JHA measured on *Ae. aegypti* and the global descriptors. The r^2 values obtained for the dendrograms and predictors that led to significant splits in recursive partitioning equaled 0.54. While the different results obtained by Basak et al. [38] are interesting, the performances of their numerous models remain rather modest and the models obtained on *Ae. aegypti* do not outperform those obtained by Nakayama et al. [34], Hansch and Leo [36], and Devillers [6].

Devillers et al. [40] also tried to obtain a QSAR model with a large domain of application from the data of Henrick et al. [33] completed by others [41–43]. A database of 188 chemicals with their pI_{50} values (millimolar) was used. The chemicals were geometry optimized with the AM1 semiempirical method in HyperChem (Hypercube, Gainesville, FL) and exported to MOPAC. The structures were then introduced into the CODESSA (comprehensive descriptors for structural and statistical analysis) Pro software (http://www.codessa-pro.com) to compute 242 topological, geometrical, and quantum-chemical descriptors. In addition, the 1-octanol/water partition coefficient (log *P*) of the molecules was also calculated from Hyperchem. After exclusion of the descriptors that had a too high number of null values or constant and those with pairwise correlations >0.80, this led to 87 descriptors. A backward feature selection procedure was then used to select sets of 20 descriptors considered to be suited to encode the structural and functional characteristics of the studied molecules.

The 188 chemicals were also described by means of the modified autocorrelation method [44]. From the fragmental constants of Rekker and Mannhold [45], an autocorrelation vector L, representing lipophilicity, was derived. Second, an autocorrelation vector MR, encoding molar refractivity, was designed from the fragmental constants of Hansch and Leo [46] or directly from the Lorentz–Lorenz equation. Last, autocorrelation vectors encoding the H-bonding acceptor ability (HBA) and H-bonding donor ability (HBD) of the molecules were also calculated by means of Boolean contributions (i.e., 0/1). The autocorrelation vectors were calculated by means of AUTOCOR™ 2.4 (CTIS, France) using SMILES (simplified molecular input line entry system) notation as inputs. For the autocorrelation vectors L and MR, a truncation of the vectors was performed in order to obtain 17 autocorrelation components (distances 0–16 in the graphs of the molecules). For the HBA autocorrelation vector, the first three components were calculated (distances 0 to 2), while for the HBD autocorrelation vector, only the first component was selected. A feature selection by forward and backward procedures and also with the genetic algorithm/neural net procedure of the Statistica™ software was used to retrieve the most discriminating autocorrelation descriptors. CODESSA and autocorrelation descriptors were treated separately to derive models.

PLS [28], classification and regression tree (CART) [47], random forest [48] and boosting regression tree analyses [49], Kohonen self-organizing maps [50], linear artificial neural networks, TLPs, radial basis function artificial neural networks [20], and support vector machines (SVMs) [23] were tested as statistical tools. Numerous trials were performed with each method by optimizing their intrinsic parameters. The data set was randomly split into training and test sets of 80% and 20%, respectively.

Due to the high number of censored data (i.e., 37/188), all the attempts to derive QSAR models from the chemicals described by various subsets of CODESSA descriptors or autocorrelation descriptors and by using the linear or nonlinear methods selected in this study were not satisfying. The obtained equations showed rather modest performances with a significant number of outliers. As a result, SAR models were derived by using $pI_{50} = 4$ as threshold value leading to 124 inactive molecules and 64 active molecules.

About 75% of good predictions were repeatedly obtained on training and test sets by using a type 1 SVM with an RBF kernel and the following set of autocorrelation descriptors as input: L_2, L_6, L_8, L_{11}, L_{12}, L_{16}, MR_0, MR_7, MR_{12}, and MR_{15} (the figures in subscript represent the distances in the molecular graph). It is worth noting that penalty scores were always used to counterbalance the difference of size between the active and inactive molecules. Broadly, the same percentages of prediction were obtained with the CODESSA descriptors. Use of random forest and boosting tree analyses with the 10 previous autocorrelation descriptors as inputs provided broadly the same results as those with the SVM but the models required a significant number of trees and they were more difficult to derive.

Among the artificial neural networks tested in the study, the TLP led to the best results with possibly at least 90% of good predictions on both sets and, in addition, fewer numbers of descriptors. Thus, for example, an 8/5/2 TLP allowed us to obtained 90.73% of good predictions for the learning set and 94.59% of well classified chemicals for the test set. The eight neurons of the input layer were L_2, L_6, L_8, L_{11}, MR_0, MR_7, MR_{12}, and MR_{15}. It is noteworthy that with the Data Miner module of the Statistica™ software, a Boolean activity is encoded by means of two output neurons. Biases were also included in the network architecture. In the selected configuration, the BFGS algorithm was used during the learning phase and the convergence was obtained only after 186 cycles. The hidden and output activation functions were both tanh (hyperbolic tangent).

Even if the selected 8/5/2 TLP model performed quite well, an attempt was made to refine the model by optimizing the predictions of very active molecules. To do so, a new two-class TLP model was designed from the same set of descriptors as input neurons but with a threshold value selected at $pI_{50} = 5$. This led to considering a class of 166 inactive molecules versus a class of 22 active molecules in the modeling process. Numerous case studies and theoretical works have shown [51–53] that a TLP was able to correctly model very unbalanced data sets provided that some precautions were taken. Thus, the test set was extended to 100 molecules in order to have the possibility to deeply estimate the predictive performances of the models. The inactive molecules were randomly split into learning and the external test sets. Conversely, due to the limited number of active molecules, for each series of trials, six active molecules were manually allocated to the external test set in order to better account for the representation of their chemical structures. An 8/4/2 TLP model trained in 68 cycles by a BFGS algorithm was selected. The hidden and output activation functions were both exponential. It allowed us to obtain 96.60% and 97% of good predictions for the learning set and the external test set, respectively. The six active chemicals belonging to the external test set were all correctly predicted by the selected 8/4/2 TLP model. Among the 16 active chemicals belonging to the learning set, only two chemicals were misclassified by the model. The TLP models derived in this study were used to propose new chemical structures potentially active as juvenoids against *Ae. aegypti* larvae.

9.5 CONCLUSIONS

More than 30 years ago, juvenoids gained much interest due to their high developmental toxicity against the larvae of insect pests and, conversely, their low toxicity vis-à-vis mammals. Consequently, a huge number of molecules was synthesized and tested on larvae of insects including mosquitoes. This enthusiasm was dampened when it was realized that these molecules were also active on nontarget insects. Indeed, due to their mechanism of action, the juvenoids belong to the category of endocrine disruptors [54]. Nevertheless, the search for new active juvenoids against mosquitoes has to be encouraged for at least the three following reasons:

- The problems of resistance of the mosquitoes to insecticides are so important [55,56] that an effort has to be continuously maintained for the discovery of new active molecules to use alone or in combination.
- Among the arthropods, there are differences of sensitivity between the taxa for the same juvenoid. This explains the poor interspecies correlations [6] even if the most active juvenoids generally show large activity spectra.
- The high difference of sensibility between the larvae and adults has been exploited in the autodissemination method [57] based on the use of the pyriproxyfen, which is highly toxic for the mosquito larvae but not lethal or repellent to adults.

The search for new juvenoids has benefited from the use of SAR and QSAR modeling for cost reduction and time saving, but more important, by the rational approach offered by this paradigm, which allows one to gain insights into the mechanism of action of molecules.

Analysis of the SAR and QSAR models derived on *Cx. pipiens* and *Ae. aegypti* shows that to be correctly modeled, the juvenoid activity, even for structurally related molecules, needs to be encoded by different categories of descriptors, which are very often related to some structural characteristics of the molecules and/or to steric effects. The 1-octanol/water partition coefficient appears less significant except when it also encodes size and branching information such as the autocorrelation vector of lipophilicity.

Last, this review reveals that the supervised nonlinear methods outperform the linear regression techniques in the design of powerful predictive QSAR models for juvenoids. This finding is not surprising because these nonlinear tools have aroused great interest in finding complex relationships between the activity or the property of structurally diverse sets of molecules and their structure [58–60; see also Chapters 10 and 11].

ACKNOWLEDGMENT

The financial support from the French Agency for Food, Environmental and Occupational Health & Safety (Anses) is gratefully acknowledged (contract #EST-2012/2/64).

REFERENCES

[1] I. Unlu, A. Farajollahi, D. Strickman, and D.M. Fonseca, *Crouching tiger, hidden trouble: Urban sources of* Aedes albopictus *(Diptera: Culicidae) refractory to source-reduction*, PLoS ONE 8 (2013), pp. e77999.

[2] F. Baldacchino, B. Caputo, F. Chandre, A. Drago, A. della Torre, F. Montarsi, and A. Rizzoli, *Control methods against invasive* Aedes *mosquitoes in Europe: A review*, Pest Manag. Sci. 71 (2015), pp. 1471–1485.

[3] C. Duchet, M. Larroque, T. Caquet, E. Franquet, C. Lagneau, and L. Lagadic, *Effects of spinosad and* Bacillus thuringiensis israelensis *on a natural population of* Daphnia pulex *in field microcosms*, Chemosphere 74 (2008), pp. 70–77.

[4] G.B. Staal, *Insect growth regulators*, in *Proceedings of the National Conference on Urban Entomology*, P.A. Zungoli, ed., Department of Entomology, Clemson University, 1986, pp. 59–64.

[5] J. Devillers, *Juvenile hormones and juvenoids: A historical survey*, in *Juvenile Hormones and Juvenoids. Modeling Biological Effects and Environmental Fate*, J. Devillers, ed., CRC Press, Boca Raton, FL, 2013, pp. 1–14.

[6] J. Devillers, *SAR and QSAR modeling of juvenile hormone mimics*, in *Juvenile Hormones and Juvenoids. Modeling Biological Effects and Environmental Fate*, J. Devillers, ed., CRC Press, Boca Raton, FL, 2013, pp. 145–174.

[7] J. Devillers, C. Lagneau, A. Lattes, J.C. Garrigues, M.M. Clémenté, and A. Yébakima, *In silico models for predicting vector control chemicals targeting* Aedes aegypti, SAR QSAR Environ. Res. 25 (2014), pp. 805–835.

[8] A. Nakayama, H. Iwamura, A. Niwa, Y. Nakagawa, and T. Fujita, *Development of insect juvenile hormone active oxime O-ethers and carbamates*, J. Agric. Food Chem. 33 (1985), pp. 1034–1041.

[9] A. Niwa, H. Iwamura, Y. Nakagawa, and T. Fujita, *Development of (phenoxyphenoxy)- and (benzylphenoxy)propyl ethers as potent insect juvenile hormone mimetics*, J. Agric. Food Chem. 37 (1989), pp. 462–467.

[10] A. Niwa, H. Iwamura, Y. Nakagawa, and T. Fujita, *Development of N,O-disubstituted hydroxylamines and N,N-disubstituted amines as insect juvenile hormone mimetics and the role of the nitrogenous function for activity*, J. Agric. Food Chem. 38 (1990), pp. 514–520.

[11] T. Hayashi, H. Iwamura, and T. Fujita, *Insect juvenile hormone mimetic activity of (4-substituted)phenoxyalkyl compounds with various nitrogenous and oxygenous functions and its relationship to their electrostatic and stereochemical properties*, J. Agric. Food Chem. 39 (1991), pp. 2029–2038.

[12] T. Hayashi, H. Iwamura, and T. Fujita, *Development of 4-alkylphenyl aralkyl ethers and related compounds as potent insect juvenile hormone mimetics and structural aspects of their activity*, J. Agric. Food Chem. 38 (1990), pp. 1965–1971.

[13] T. Hayashi, H. Iwamura, and T. Fujita, *Electrostatic and stereochemical aspects of insect juvenile hormone active compounds: A clue for high activity*, J. Agric. Food Chem. 38 (1990), pp. 1972–1977.

[14] V.B. Wigglesworth, *Chemical structure and juvenile hormone activity: Comparative tests on* Rhodnius prolixus, J. Insect Physiol. 15 (1969), pp. 73–94.

[15] J.W. Patterson and M. Schwarz, *Chemical structure, juvenile hormone activity and persistence within the insect of juvenile hormone mimics for* Rhodnius prolixus, J. Insect Physiol. 23 (1977), pp. 121–129.

[16] H.A. Schneiderman, A. Krishnakumaran, V.G. Kulkarni, and L. Friedman, *Juvenile hormone activity of structurally unrelated compounds*, J. Insect Physiol. 11 (1965), pp. 1641–1649.

[17] G. Brieger, *Juvenile hormone mimics: Structure–activity relationships for* Oncopeltus fasciatus, J. Insect Physiol. 17 (1971), pp. 2085–2093.

[18] T. Hayashi, H. Iwamura, T. Fujita, N. Takakusa, and T. Yamada, *Structural requirements for activity of juvenile hormone mimetic compounds against various insects*, J. Agric. Food Chem. 39 (1991), pp. 2039–2045.

[19] A. Niwa, H. Iwamura, Y. Nakagawa, and T. Fujita, *Development of (phenoxyphenoxy)- and (benzylphenoxy)alkanaldoxime o-ethers as potent insect juvenile hormone mimics and their quantitative structure–activity relationship*, J. Agric. Food Chem. 36 (1988), pp. 378–384.

[20] J. Devillers, *Neural Networks in QSAR and Drug Design*, Academic Press, London, 1996.

[21] Anonymous, *The Principles for Establishing the Status of Development and Validation of (Quantitative) Structure–Activity Relationships (Q)SARs,* OECD document, ENV/JM/TG(2004)27.

[22] V.N. Vapnik, *The Nature of Statistical Learning Theory*, Springer, Berlin, 1995.

[23] N. Cristianini and J. Shawe-Taylor, *An Introduction to Support Vector Machines and Other Kernel-Based Learning Methods*, Cambridge University Press, 2000.

[24] T. Hayashi, H. Iwamura, Y. Nakagawa, and T. Fujita, *Development of (4-alkoxyphenoxy)- and (4-alkylphenoxy)-alkanaldoxime o-ethers as potent insect juvenile hormone mimics and their structure–activity relationships*, J. Agric. Food Chem. 37 (1989), pp. 467–472.

[25] S.C. Basak, R. Natarajan, D. Mills, D.M. Hawkins, and J.J. Kraker, *Quantitative structure–activity relationship modeling of juvenile hormone mimetic compounds for* Culex pipiens *larvae, with a discussion of descriptor-thinning methods*, J. Chem. Inf. Model. 46 (2006), pp. 65–77.

[26] J. Devillers and A.T. Balaban, *Topological Indices and Related Descriptors in QSAR and QSPR,* Gordon and Breach Science Publishers, Amsterdam, the Netherlands, 1999.

[27] A.E. Hoerl and R.W. Kennard, *Ridge regression: Applications to nonorthogonal problems*, Technometrics 12 (1970), pp. 69–82.

[28] S. Wold, M. Sjöström, and L. Eriksson, *PLS-regression: A basic tool of chemometrics*, Chemom. Intell. Lab. Syst. 58 (2001), pp. 109–130.

[29] J. Devillers, D. Zakarya, M. Chastrette, and J.C. Doré, *The stochastic regression analysis as a tool in ecotoxicological QSAR studies*, Biomed. Environ. Sci. 2 (1989), pp. 385–393.

[30] R. Tibshirani, *Regression shrinkage and selection via the Lasso*, J. Royal Stat. Soc. B 58 (1996), pp. 267–288.

[31] R.S. Latha, R. Vijayaraj, J. Padmanabhan, E.R.A. Singam, K. Chitra, and V. Subramanian, *3D-QSAR studies on the biological activity of juvenile hormone mimetic compounds for* Culex pipiens *larvae*, Med. Chem. Res. 22 (2013), pp. 5948–5960.

[32] J.P. Doucet and A. Panaye, *Three Dimensional QSAR: Applications in Pharmacology and Toxicology*, CRC Press, Boca Raton, FL, 2010.

[33] C.A. Henrick, W.E. Willy, and G.B. Staal, *Insect juvenile hormone activity of alkyl (2E,4E)-3,7,11-trimethyl-2,4-dodecadienoates. Variations in the ester function and in the carbon chain*, J. Agric. Food Chem. 24 (1976), pp. 207–218.

[34] A. Nakayama, H. Iwamura, and T. Fujita, *Quantitative structure–activity relationship of insect juvenile hormone mimetic compounds*, J. Med. Chem. 27 (1984), pp. 1493–1502.

[35] H. Iwamura, K. Nishimura, and T. Fujita, *Quantitative structure–activity relationships of insecticides and plant growth regulators: Comparative studies toward understanding the molecular mechanism of action*, Environ. Health Perspect. 61 (1985), pp. 307–320.

[36] C. Hansch and A. Leo, *Exploring QSAR. Fundamentals and Applications in Chemistry and Biology*, ACS Professional Reference Book Series, Washington, DC, 1995.

[37] S. Bintein, J. Devillers, and W. Karcher, *Nonlinear dependence of fish bioconcentration on n-octanol/water partition coefficient*, SAR QSAR Environ. Res. 1 (1993), pp. 29–39.

[38] S.C. Basak, R. Natarajan, D. Mills, D.M. Hawkins, and J.J. Kraker, *Quantitative structure–activity relationship modeling of insect juvenile hormone activity of 2,4-dienoates using computed molecular descriptors*, SAR QSAR Environ. Res. 16 (2005), pp. 581–606.

[39] S.S. Young and D.M. Hawkins, *Using recursive partitioning to analyze a large SAR data set*, SAR QSAR Environ. Res. 8 (1998), pp. 183–193.

[40] J. Devillers, A. Doucet-Panaye, and J.P. Doucet, *Structure–activity relationship (SAR) modelling of mosquito larvicides*, SAR QSAR Environ. Res. 26 (2015), pp. 263–278.

[41] C.A. Henrick, W.E. Willy, B.A. Garcia, and G.B. Staal, *Insect juvenile hormone activity of the stereoisomers of ethyl 3,7,11-trimethyl-2,4-dodecadienoate*, J. Agric. Food Chem. 23 (1975), pp. 396–400.

[42] C.A. Henrick, R.J. Anderson, G.B. Staal, and G.F. Ludvik, *Insect juvenile hormone activity of optically active alkyl (2E,4E)-3,7,11-trimethyl-2,4-dodecadienoates and of arylterpenoid analogs*, J. Agric. Food Chem. 26 (1978), pp. 542–550.

[43] M. Londershausen, B. Alig, R. Pospischil, and A. Turberg, *Activity of novel juvenoids on arthropods of veterinary importance*, Arch. Insect Biochem. Physiol. 32 (1996), pp. 651–658.

[44] J. Devillers, *Autocorrelation descriptors for modeling (eco)toxicological endpoints*, in *Topological Indices and Related Descriptors in QSAR and QSPR*, J. Devillers and A.T. Balaban, eds., Gordon and Breach, the Netherlands, 1999, pp. 595–612.

[45] R.F. Rekker and R. Mannhold, *Calculation of Drug Lipophilicity. The Hydrophobic Fragmental Constant Approach*, John Wiley & Sons Ltd., Weinheim, Germany, 1992.

[46] C. Hansch and A. Leo, *Substituent Constants for Correlation Analysis in Chemistry and Biology*, John Wiley & Sons Ltd., New York, 1979.

[47] L. Breiman, J. Freidman, R. Olshen, and C. Stone, *Classification and Regression Trees*, Wadsworth, Belmont, CA, 1984, pp. 358.

[48] L. Breiman, *Random forests*, Mach. Learn. 45 (2001), pp. 5–32.

[49] J.H. Friedman, *Stochastic Gradient Boosting*, Stanford University, 1999, p. 10 (http://www-stat.stanford.edu/~jhf/ftp/stobst.pdf).

[50] T. Kohonen, *Self-Organizing Maps*, Springer Series in Information Sciences, Springer, Berlin, 1995.

[51] L. Bruzzone and S.B. Serpico, *Classification of imbalanced remote-sensing data by neural networks*, Patt. Rec. Lett. 18 (1997), pp. 1323–1328.

[52] Y.L. Murphey, H. Guo, and L.A. Feldkamp, *Neural learning from unbalanced data*, J. Appl. Intel. 21 (2004), pp. 117–128.

[53] G. Daqi, L. Chunxia, and Y. Yunfan, *Task decomposition and modular single-hidden-layer perceptron classifiers for multi-class learning problems*, Patt. Rec. 40 (2007), pp. 2226–2236.

[54] J. Devillers, ed. *Endocrine Disruption Modeling*, CRC Press, Boca Raton, FL, 2009.

[55] S. Marcombe, R. Poupardin, F. Darriet, S. Reynaud, J. Bonnet, C. Strode, C. Brengues, A. Yébakima, H. Ranson, V. Corbel, and J.P. David, *Exploring the molecular basis of insecticide resistance in the dengue vector* Aedes aegypti*: A case study in Martinique Island (French West Indies)*, BMC Genomics 10(494) (2009), pp. 1–14.

[56] S. Marcombe, R.B. Mathieu, N. Pocquet, M.A. Riaz, R. Poupardin, S. Sélior, F. Darriet, S. Reynaud, A. Yébakima, V. Corbel, J.P. David, and F. Chandre, *Insecticide resistance in the dengue vector* Aedes aegypti *from Martinique: Distribution, mechanisms and relations with environmental factors*, PLoS ONE 7 (2012), pp. e30989.

[57] G.J. Devine, E.Z. Perea, G.F. Killeen, J.D. Stancil, S.J. Clark, and A.C. Morrison, *Using adult mosquitoes to transfer insecticides to* Aedes aegypti *larval habitats*, Proc. Nat. Acad. Sci. 106 (2009), pp. 11530–11534.

[58] J. Devillers, J.P. Doucet, A. Panaye, N. Marchand-Geneste, and J.M. Porcher, *Structure–activity of a diverse set of androgen receptor ligands*, in *Endocrine Disruption Modeling*, J. Devillers, ed., CRC Press, Boca Raton, FL, 2009, pp. 335–355.

[59] J. Devillers, J.P. Doucet, A. Doucet-Panaye, A. Decourtye, and P. Aupinel, *Linear and non-linear QSAR modelling of juvenile hormone esterase inhibitors*, SAR QSAR Environ. Res. 23 (2012), pp. 357–369.

[60] J.P. Doucet and A. Doucet-Panaye, *Structure–activity relationship study of trifluoromethylketone inhibitors of insect juvenile hormone esterase: Comparison of several classification methods*, SAR QSAR Environ. Res. 25 (2014), pp. 589–616.

10 SAR Predictions of Benzoylphenylurea Chitin Synthesis Inhibitors Active on Larvae of *Aedes aegypti*

James Devillers, Annick Doucet-Panaye, Jean-Pierre Doucet, Armand Lattes, Hubert Matondo, Christophe Lagneau, Sébastien Estaran, Marie-Michelle Clémente, and André Yébakima

CONTENTS

ABSTRACT

An attempt was made to derive structure–activity relationship (SAR) models allowing the prediction of the mosquito larvicidal activity of benzoylphenylureas (BPUs) known to potentially inhibit the chitin synthesis of insects during their development. Activity values (active/inactive) on *Aedes aegypti* larvae for about 200 BPUs were obtained under the same experimental conditions.

Chemicals were described by means of autocorrelation vectors encoding lipophilicity, molar refractivity, H-bonding acceptor ability, and H-bonding donor ability. The data set was randomly split into learning sets and external test sets of 80%/20%, respectively. Feature selection procedures were used to optimally reduce the number of descriptors. A three-layer perceptron was used as statistical tool. The performances of the models were evaluated through the analysis of the prediction results obtained on the different training sets and external test sets. Experimental results on larvae of *Ae. aegypti* obtained on the few commercialized BPUs and, for two published series of 32 and 73 structurally diverse BPUs, were also used to evaluate and select the most interesting configurations. Two models, presenting 91% of good predictions on the whole data set of BPUs, were selected. They included autocorrelation descriptors but also a descriptor encoding the presence of fluorine atoms in the *ortho*-position of the benzene ring linked to the carbonyl group of the BPUs. Interestingly, most of the autocorrelation descriptors were of higher order and half of them encoded lipophilicity. Both models are of interest to rationalize the discovery of new BPUs active on larvae of *Ae. aegypti*.

KEYWORDS

- Benzoylphenylureas (BPUs)
- *Aedes aegypti*
- Larvicide
- SAR model
- Autocorrelation method
- Three-layer perceptron

10.1 INTRODUCTION

The body of an insect is covered by a cuticle, which offers an efficient protection barrier against desiccation, pathogens, and predation. This exoskeleton also allows the attachment of the muscles. The cuticle is divided into three main regions—namely, the procuticle that lies just above the epidermal cells, the epicuticle, and the envelope, which is the outermost layer of the cuticle. The procuticle in turn consists of the endocuticle and exocuticle. The cuticle appears as hard and stiff regions called the sclerites, which can be separated by more flexible regions, termed the arthrodial (or intersegmental) membranes, where the exocuticle is absent. This morphology implies a limited capacity to keep pace with body growth, and the insects are therefore periodically forced to replace their old cuticle with a new one during molting [1].

The insect cuticle is composed largely of proteins, lipids, and chitin, a name coined in 1823 by Odier [2], based on the Greek word "chiton," due to its resemblance to an envelope or tunic. The proteins and chitin interact to give the necessary strength and hardness to the cuticle. The proteins are primarily located in the procuticle, but the epicuticle also includes minor proteins. It is worth noting that sclerotization, also known as tanning, stabilizes the protein matrix to make it harder, more insoluble, and more resistant to degradation [1,3]. Chitin is found in the endocuticle and exocuticle but not in the epicuticle. Lipids are located mostly on the outermost layer of the cuticle, where they prevent desiccation and provide chemical cues for species recognition [1].

Although the larvae of the holometabolous insects can grow between molts, growth mainly occurs during and immediately after the molting process, which starts by apolysis, the separation of the epidermal cells from the old cuticle. This creates an area between the cuticle and epidermis, named the exuvial space and filled with the molting gel that contains inactive chitinase and protease enzymes that will be able to digest the old cuticle once they are activated. Then, the epidermal cells secrete a new envelope whose lipoproteins become tanned and invulnerable to the enzymes when they are activated by factors produced by the epidermal cells. Once activated, the enzymes become the molting fluid that starts the digestion of the old unsclerotized endocuticle; the sclerotized exocuticle is spared. About 90% of the material in the old endocuticle is recycled [1]. As the old cuticle is digested, the epidermal cells start the secretion of the new procuticle using the recycled raw material. The synthesis of the new epicuticle continues with the formation of the inner and then outer epicuticle. Just before ecdysis, the components of the wax layer are released on the surface of the outer epicuticle to secure the waterproofing of the new cuticle before the elimination of the old one. The molting fluid is then totally resorbed. A complex process follows initiated by hormones that will ultimately lead to the release of the old cuticle, now called exuvia. Following ecdysis, the soft and unsclerotized new cuticle can expand to allow the change in the size of the insect [1].

Chitin is a water-insoluble linear amino polysaccharide polymer made up of β-(1-4)-linked *N*-acetyl-D-glucosamine units. It is the most widespread nitrogen-bearing organic compound found in nature. At least 10 gigatons of chitin are synthesized and degraded each year in the biosphere [4]. Chitin exists in three different crystalline forms—α-, β-, and γ-chitin, depending on the orientation of their constitutive chains. In the α-chitin, found in the hydrozoa, nematodes, rotifers, mollusks, arthropods, and fungi, the chains are arranged in an antiparallel fashion. The chains are parallel in β-chitin, which is found mainly in diatoms and annelids. The γ-chitin, which is a combination of the α and β forms, is rare and is found in the stomach of squid and the cocoons of two genera of beetles [4–6]. The antiparallel arrangement of the chains in α-chitin contributes significantly to the mechanical strength and stability of the cuticle in insects.

Chitin can represent up to 60% of the dry weight in some insect species [5]. This shows the importance of chitin for the survival of the insects as well as the rapid mobilization of the metabolic machinery necessary to guarantee its production during the development of the insect and, subsequently, the vulnerable points offered for disrupting growth of pest insects from chemicals applied during their development.

In the early 1970s, an unexpected discovery led to the design of a new class of insecticides targeting the chitin. Dutch scientists at the Philips–Duphar laboratory were initially looking for potent weed control agents by combining the two herbicides dichlobenil and diuron. The resulting acylurea compound, named DU-19111, showed poor herbicidal activity; when tested on insect larvae, there was no immediate knockdown effect, but treated individuals showed various degrees of molt deformities. The normal formation of the cuticle was disrupted through the alteration of chitin synthesis. This discovery sparked research on molt-inhibiting benzoylphenylureas (BPUs) as a new class of insecticides targeting chitin synthesis in insects [5,7]. One of the earliest compounds synthesized presented a difluorobenzoyl group on one end of the urea bridge and a chlorophenyl group on the other side and came to be known as PH 60-40 or diflubenzuron (Figure 10.1) and it was commercialized under the name Dimilin® [5]. Nakagawa et al. [8–15] conducted in-depth quantitative structure–activity relationships (QSARs) from Hansch analysis on structurally

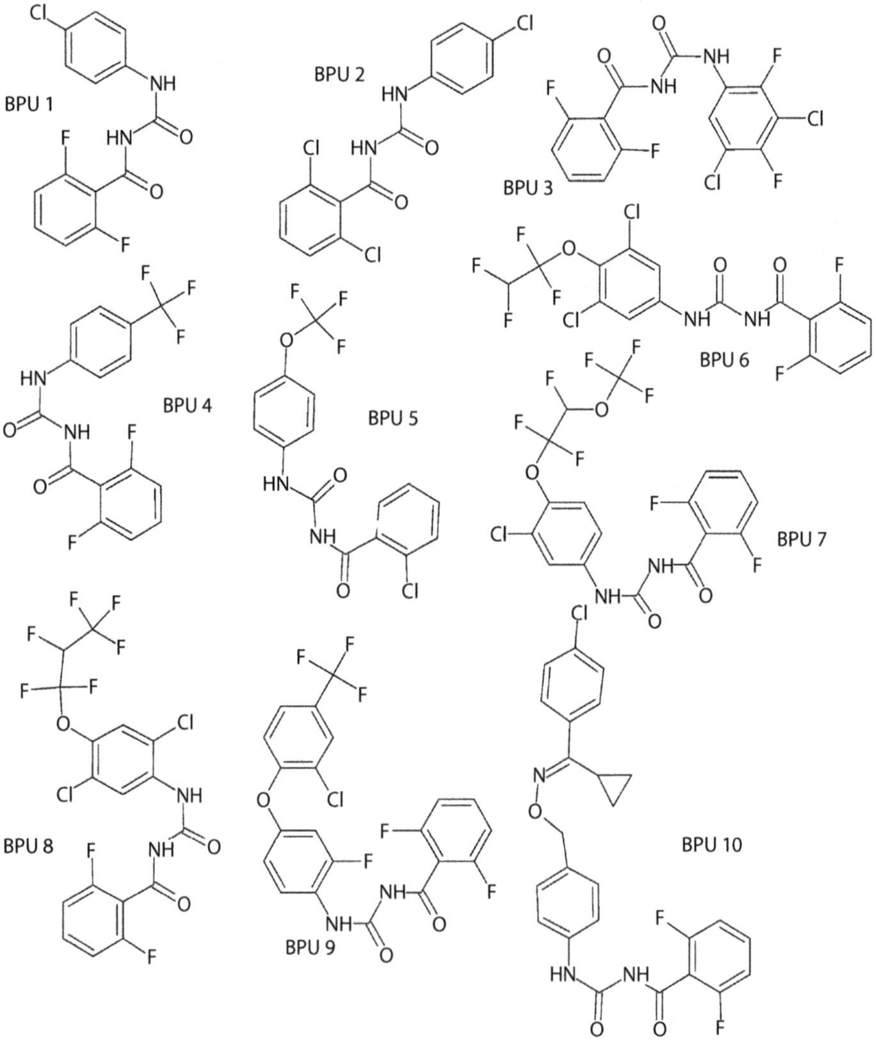

FIGURE 10.1 Structure of commercial benzoylphenylureas tested on larvae of mosquitoes. BPU 1: diflubenzuron; BPU 2: dichlorbenzuron; BPU 3: teflubenzuron; BPU 4: penfluron; BPU 5: triflumuron; BPU 6: hexaflumuron; BPU 7: novaluron; BPU 8: lufenuron; BPU 9: flufenoxuron; and BPU 10: flucycloxuron.

diverse BPUs tested for their larvicidal activity on the rice stem borer, *Chilo suppressalis*, and sometimes on the larvae of silkworm (*Bombyx mori*). Thus, for example, against rice stem borers, electron-withdrawing, hydrophobic, and small substituents are favorable to the activity for BPU derivatives having substituents at the *ortho*-positions in the benzoyl moiety and the *para*-position in the anilide moiety [14]. It is worth noting that an important difference of sensibility can exist between species for the same BPU [16]. Moreover, the mechanism of action of the BPUs is not totally understood [5,16]. This might explain the rather limited number of BPUs commercialized to date. Indeed, to date, more than 10,000 BPUs have been synthesized, while only a limited number of them have been commercialized after more than four decades of research and development [16]. In this context, an attempt was made to study the SAR rules governing the toxicity of BPUs against mosquito larvae.

Aedes aegypti (Linnaeus 1762) is the main vector responsible for dengue, yellow fever, chikungunya, and Zika. Although yellow fever has been reasonably brought under control, no vaccine is available against the other arboviruses transmitted by this mosquito. So far, the only method of controlling or preventing viral transmission is to fight the mosquitoes by reducing their populations by the use of chemicals. However, because of increasing resistance to the different families of insecticides, reduction of *Aedes* populations is becoming increasingly difficult [17,18]. As a result, there is a need to find new chemicals active on the larvae and adult mosquito populations. Wellinga et al. [19,20] have tested a series of BPUs against larvae of *Ae. aegypti*. From their laboratory test results, obtained under the same experimental conditions, an attempt was made to derive various SARs to better understand the mechanism of action of these chemicals against *Ae. aegypti* and to predict the potential activity of untested chemicals on this mosquito species.

10.2 MATERIALS AND METHODS

10.2.1 Biological Test

A series of 203 BPUs sharing the common substructure Phi1-C(=O)-N1-C(=O)-N2-Phi2, where Phi1 and Phi2 are benzene rings, was tested by Wellinga et al. [19,20] on larvae of *Ae. aegypti*. Aliquots of 100 mL of tap water containing respectively 1, 0.3, 0.1, 0.03, 0.01, 0.003, 0.001, 0.0003, or 0 ppm of chemical were supplied with twenty 1-day-old mosquito larvae and kept at 25°C. The larvae were fed with malt yeast powder. After 13 days, when the pupae of the untreated insects had hatched, the mortality percentages were calculated with a correction for the natural mortality according to Abbott [21]. All the tests were performed in triplicate. The result obtained for each tested concentration was recorded according to three levels of toxicity corresponding to 90% to 100% of mortality (L1), 50% to 89% of mortality (L2), and 0% to 49% of mortality (L3). Results obtained for the 203 BPUs are listed in the last three columns of Table 10.1. In practice, molecules that show 90% to 100% of mortality at 1 ppm become of interest for *in silico* optimization; however, it is obvious that the more the molecules are active at much lower concentrations, the more attractive they are for use in vector control. In Table 10.1, each molecule is characterized by the concentration at which the three levels of toxicity were recorded.

TABLE 10.1

Structure of the 203 Benzoylphenyl Ureas (BPUs) and the Concentrations at Which They Lead to 90%–100% (L1), 50%–89% (L2), and/or 0%–49% (L3) Mortality against Larvae of *Ae. aegypti*

Num.	Set[a]	Phi1[b]	N1	N2	Phi2	L1	L2	L3
W001	C	2,6-Cl_2	H	H	H	0	0	0
W002	B	2,6-Cl_2	H	CH_3	H	0	0	0
W003	C	2,6-Cl_2	H	H	4-CH_3	2	3	4
W004	C	2,6-Cl_2	H	H	4-C_2H_5	7	N	8
W005	C	2,6-Cl_2	H	H	4-*n*-C_3H_7	5	N	6
W006	B	2,6-Cl_2	H	CH_3	4-*n*-C_3H_7	0	0	0
W007	C	2,6-Cl_2	H	H	4-*i*-C_3H_7	0	0	0
W008	B	2,6-Cl_2	H	CH_3	4-*i*-C_3H_7	0	0	0
W009	C	2,6-Cl_2	H	H	3-Cyclopropyl	2	N	3
W010	C	2,6-Cl_2	H	H	4-Cyclopropyl	1	2	4
W011	C	2,6-Cl_2	H	H	4-*n*-C_4H_9	0	1	3
W012	B	2,6-Cl_2	H	CH_3	4-*n*-C_4H_9	0	0	0
W013	C	2,6-Cl_2	H	H	4-*i*-C_4H_9	4	5	6
W014	B	2,6-Cl_2	H	CH_3	4-*i*-C_4H_9	0	0	0
W015	C	2,6-Cl_2	H	H	4-*sec*-C_4H_9	3	N	4
W016	B	2,6-Cl_2	H	CH_3	4-*sec*-C_4H_9	0	0	0
W017	C	2,6-Cl_2	H	H	4-*t*-C_4H_9	0	1	2
W018	B	2,6-Cl_2	H	CH_3	4-*t*-C_4H_9	0	0	0
W019	C	2,6-Cl_2	H	H	4-Cyclo-C_6H_{11}	0	1	3
W020	C	2,6-Cl_2	H	H	4-*n*-C_8H_{17}	2	3	4
W021	B	2,6-Cl_2	H	CH_3	4-*n*-C_8H_{17}	0	0	0
W022	C	2,6-Cl_2	H	H	4-*n*-$C_{12}H_{25}$	3	4	5

(Continued)

TABLE 10.1 (CONTINUED)
Structure of the 203 Benzoylphenyl Ureas (BPUs) and the Concentrations at Which They Lead to 90%–100% (L1), 50%–89% (L2), and/or 0%–49% (L3) Mortality against Larvae of *Ae. aegypti*

Num.	Set[a]	Phi1[b]	N1	N2	Phi2	L1	L2	L3
W023	B	2,6-Cl_2	H	CH_3	4-*n*-$C_{12}H_{25}$	0	0	0
W024	C	2,6-Cl_2	H	H	2,3-$(CH_3)_2$	0	0	0
W025	C	2,6-Cl_2	H	H	2,6-$(CH_3)_2$	0	0	0
W026	C	2,6-Cl_2	H	H	3,4-$(CH_3)_2$	0	1	2
W027	C	2,6-Cl_2	H	H	3,4-[$(CH_2)_4$]	0	0	0
W028	C	2,6-Cl_2	H	H	2,3-[$(CH{=}CH)_2$]	0	0	0
W029	C	2,6-Cl_2	H	H	3,4-[$(CH{=}CH)_2$]	0	0	0
W030	C	2,6-Cl_2	H	H	2-CH_3,4-Clr	2	N	3
W031	C	2,6-Cl_2	H	H	3-CH_3,4-Cl	2	N	3
W032	C	2,6-Cl_2	H	H	3-Cl,4-CH_3	0	0	0
W033	C	2,6-Cl_2	H	H	2-CH_3,4,5-Cl_2	0	0	0
W034	C	2,6-Cl_2	H	H	4-CH_2CN	0	0	0
W035	C	2,6-Cl_2	H	H	4-$CH_2N(CH_3)_2$	0	0	0
W036	C	2,6-Cl_2	H	H	4-CH_2SCH_3	0	1	3
W037	C	2,6-Cl_2	H	H	2-CF_3	0	0	0
W038	C	2,6-Cl_2	H	H	3-CF_3	3	4	6
W039	C	2,6-Cl_2	H	H	4-CF_3	7	N	8
W040	C	2,6-Cl_2	H	H	3-(2,2-Dichlorocyclopropyl)	0	0	0
W041	C	2,6-Cl_2	H	H	4-(2,2-Dichlorocyclopropyl)	0	0	0
W042	C	2,6-Cl_2	H	H	4-$C(CH_3)_2CH_2OCH_3$	0	0	0
W043	C	2,6-Cl_2	H	H	4-C_6H_5	5	N	6
W044	C	2,6-Cl_2	H	H	4-(4-ClC_6H_4)	5	N	6

(Continued)

TABLE 10.1 (CONTINUED)
Structure of the 203 Benzoylphenyl Ureas (BPUs) and the Concentrations at Which They Lead to 90%–100% (L1), 50%–89% (L2), and/or 0%–49% (L3) Mortality against Larvae of *Ae. aegypti*

Num.	Set[a]	Phi1[b]	N1	N2	Phi2	L1	L2	L3
W045	C	2,6-Cl_2	H	H	4-(4-BrC_6H_4)	5	6	7
W046	C	2,6-Cl_2	H	H	4-(4-$NO_2C_6H_4$)	0	0	0
W047	C	2,6-Cl_2	H	H	4-CH_2-C_6H_5	3	4	5
W048	C	2,6-Cl_2	H	H	4-CH_2-(4-ClC_6H_4)	5	N	6
W049	C	2,6-Cl_2	H	H	4-$CH_2CH_2C_6H_5$	3	4	5
W050	C	2,6-Cl_2	H	H	2-F	2	N	3
W051	C	2,6-Cl_2	H	H	3-F	0	1	3
W052	C	2,6-Cl_2	H	H	4-F	2	3	4
W053	C	2,6-Cl_2	H	H	3-Cl	0	0	0
W054	C	2,6-Cl_2	H	H	4-Cl	5	N	6
W055	B	2,6-Cl_2	H	CH_3	4-Cl	2	N	3
W056	B	2,6-Cl_2	H	C_2H_5	4-Cl	2	N	3
W057	B	2,6-Cl_2	H	i-C_3H_7	4-Cl	0	1	2
W058	B	2,6-Cl_2	H	n-C_4H_9	4-Cl	1	N	2
W059	B	2,6-Cl_2	H	n-C_5H_{11}	4-Cl	1	N	2
W060	B	2,6-Cl_2	H	CH_2CN	4-Cl	3	4	5
W061	B	2,6-Cl_2	H	CH_2CH_2OH	4-Cl	0	0	0
W062	B	2,6-Cl_2	H	CH_2CF_3	4-Cl	0	0	0
W063	B	2,6-Cl_2	H	CH_2CH_2CN	4-Cl	0	1	2
W064	B	2,6-Cl_2	H	$CH_2CH_2C(=O)NH_2$	4-Cl	0	1	2
W065	B	2,6-Cl_2	H	CH_2CH_2COOH	4-Cl	0	0	0

(Continued)

TABLE 10.1 (CONTINUED)
Structure of the 203 Benzoylphenyl Ureas (BPUs) and the Concentrations at Which They Lead to 90%–100% (L1), 50%–89% (L2), and/or 0%–49% (L3) Mortality against Larvae of *Ae. aegypti*

Num.	Set[a]	Phi1[b]	N1	N2	Phi2	L1	L2	L3
W066	B	2,6-Cl_2	H	$CH_2CH{=}CH_2$	4-Cl	0	0	0
W067	B	2,6-Cl_2	H	$CH_2CBr{=}CH_2$	4-Cl	0	0	0
W068	B	2,6-Cl_2	H	C_6H_5	4-Cl	0	0	0
W069	B	2,6-Cl_2	H	4-ClC_6H_4	4-Cl	0	0	0
W070	B	2,6-Cl_2	H	$CH_2C_6H_5$	4-Cl	2	N	3
W071	B	2,6-Cl_2	H	1-Cyclohexenyl	4-Cl	0	0	0
W072	B	2,6-Cl_2	H	1-Cyano-1-cyclohexyl	4-Cl	0	0	0
W073	B	2,6-Cl_2	⟨CH_2CH_2⟩		4-Cl	0	0	0
W074	B	2,6-Cl_2	⟨$COCH_2$⟩		4-Cl	0	1	2
W075	B	2,6-Cl_2	⟨CO-1-cyclohexyl⟩		4-Cl	0	0	0
W076	B	2,6-Cl_2	⟨COCO⟩		4-Cl	4	5	7
W077	B	2,6-Cl_2	⟨CH_2OCH_2⟩		4-Cl	4	N	5
W078	B	2,6-Cl_2	⟨$COCH_2CH_2$⟩		4-Cl	0	0	0
W079	B	2,6-Cl_2	CH_3	H	4-Cl	4	5	6
W080	B	2,6-Cl_2	OCH_3	H	4-Cl	0	0	0
W081	B	2,6-Cl_2	CH_2OCH_3	H	4-Cl	4	N	5
W082	B	2,6-Cl_2	$CH_2C_6H_5$	H	4-Cl	2	3	4
W083	C	2,6-Cl_2	H	H	4-Br	0	0	0
W084	C	2,6-Cl_2	H	H	4-I	0	0	0
W085	C	2,6-Cl_2	H	H	2,4-F_2	0	1	2

(Continued)

TABLE 10.1 (CONTINUED)
Structure of the 203 Benzoylphenyl Ureas (BPUs) and the Concentrations at Which They Lead to 90%–100% (L1), 50%–89% (L2), and/or 0%–49% (L3) Mortality against Larvae of *Ae. aegypti*

Num.	Set[a]	Phi1[b]	N1	N2	Phi2	L1	L2	L3
W086	C	2,6-Cl_2	H	H	2,5-F_2	0	0	0
W087	C	2,6-Cl_2	H	H	2,6-F_2	0	0	0
W088	C	2,6-Cl_2	H	H	3,4-F_2	0	1	2
W089	C	2,6-Cl_2	H	H	3-F,4-Cl	5	N	6
W090	C	2,6-Cl_2	H	H	3-F,4-Br	0	0	0
W091	C	2,6-Cl_2	H	H	2-F,4-I	5	6	7
W092	C	2,6-Cl_2	H	H	3-F,4-I	5	N	6
W093	C	2,6-Cl_2	H	H	2,4-Cl_2	1	2	4
W094	C	2,6-Cl_2	H	H	2,6-Cl_2	0	0	0
W095	C	2,6-Cl_2	H	H	3,4-Cl_2	5	6	8
W096	B	2,6-Cl_2	H	CH_3	3,4-Cl_2	2	3	4
W097	B	2,6-Cl_2	H	C_2H_5	3,4-Cl_2	2	3	4
W098	B	2,6-Cl_2	H	*n*-C_5H_{11}	3,4-Cl_2	0	0	0
W099	B	2,6-Cl_2	CH_3	H	3,4-Cl_2	4	5	6
W100	B	2,6-Cl_2	CH_3	CH_3	3,4-Cl_2	2	N	3
W101	B	2,6-Cl_2	⟨CH_2OCH_2⟩		3,4-Cl_2	4	5	6
W102	B	2,6-Cl_2	CH_2OCH_3	H	3,4-Cl_2	4	N	5
W103	B	2,6-Cl_2	$CH_2C_6H_5$	H	3,4-Cl_2	3	N	4
W104	B	2,6-Cl_2	OH	H	3,4-Cl_2	0	0	0
W105	B	2,6-Cl_2	H	OH	3,4-Cl_2	5	6	7
W106	B	2,6-Cl_2	H	$COOC_2H_5$	3,4-Cl_2	0	1	2
W107	B	2,6-Cl_2	H	$COCH_3$	3,4-Cl_2	0	0	0

(Continued)

TABLE 10.1 (CONTINUED)
Structure of the 203 Benzoylphenyl Ureas (BPUs) and the Concentrations at Which They Lead to 90%–100% (L1), 50%–89% (L2), and/or 0%–49% (L3) Mortality against Larvae of *Ae. aegypti*

Num.	Set[a]	Phi1[b]	N1	N2	Phi2	L1	L2	L3
W108	C	2,6-Cl_2	H	H	3-Cl,4-Br	1	2	3
W109	C	2,6-Cl_2	H	H	3-Br,4-Cl	0	0	0
W110	C	2,6-Cl_2	H	H	3-Cl,4-I	0	0	0
W111	C	2,6-Cl_2	H	H	3,4-Br_2	0	0	0
W112	C	2,6-Cl_2	H	H	2,5-F_2,4-Br	5	N	6
W113	C	2,6-Cl_2	H	H	2,4,5-Cl_3	0	0	0
W114	C	2,6-Cl_2	H	H	3,4,5-Cl_3	0	0	0
W115	C	2,6-Cl_2	H	H	4-$N(CH_3)_2$	0	0	0
W116	C	2,6-Cl_2	H	H	3-Cl,4-$N(CH_3)_2$	0	0	0
W117	C	2,6-Cl_2	H	H	4-$NHCOCH_3$	0	0	0
W118	C	2,6-Cl_2	H	H	2-OH,3,5-Br_2	0	0	0
W119	C	2,6-Cl_2	H	H	3-OH	0	0	0
W120	C	2,6-Cl_2	H	H	4-OH	0	0	0
W121	C	2,6-Cl_2	H	H	3-OCH_3	0	0	0
W122	C	2,6-Cl_2	H	H	3-OCH_3,4-Cl	0	0	0
W123	C	2,6-Cl_2	H	H	4-OCH_3	0	0	0
W124	C	2,6-Cl_2	H	H	3-Cl,4-OCH_3	0	0	0
W125	C	2,6-Cl_2	H	H	3,4-OCH_2O	0	0	0
W126	C	2,6-Cl_2	H	H	3,4-$CH_2CH_2CH_2O$ 0.5 C_6H_6	0	0	0
W127	C	2,6-Cl_2	H	H	3-$OCOCH_3$	0	0	0
W128	C	2,6-Cl_2	H	H	4-O-(4-ClC_6H_4)	0	0	0
W129	C	2,6-Cl_2	H	H	4-O-(2,6-$Cl_2C_6H_3$)	0	0	0

(Continued)

TABLE 10.1 (CONTINUED)

Structure of the 203 Benzoylphenyl Ureas (BPUs) and the Concentrations at Which They Lead to 90%–100% (L1), 50%–89% (L2), and/or 0%–49% (L3) Mortality against Larvae of *Ae. aegypti*

Num.	Set[a]	Phi1[b]	N1	N2	Phi2	L1	L2	L3
W130	C	2,6-Cl_2	H	H	4-SCH_3	1	2	3
W131	C	2,6-Cl_2	H	H	3-Cl,4-SCH_3	0	1	2
W132	C	2,6-Cl_2	H	H	4-S-*n*-C_5H_{11}	3	N	4
W133	C	2,6-Cl_2	H	H	4-SC_6H_5	0	0	0
W134	C	2,6-Cl_2	H	H	4-S-(2,6-$Cl_2C_6H_3$)	0	0	0
W135	C	2,6-Cl_2	H	H	4-S-(4-ClC_6H_4)	4	N	5
W136	C	2,6-Cl_2	H	H	4-S-(2,4,5-$Cl_3C_6H_2$)	0	0	0
W137	C	2,6-Cl_2	H	H	4-S-(4-$NO_2C_6H_4$)	0	0	0
W138	C	2,6-Cl_2	H	H	4-SCN	0	0	0
W139	C	2,6-Cl_2	H	H	4-$COCH_3$	0	1	2
W140	C	2,6-Cl_2	H	H	4-COC_6H_5	1	N	2
W141	C	2,6-Cl_2	H	H	4-$COOC_2H_5$	0	0	0
W142	C	2,6-Cl_2	H	H	4-CN	3	N	4
W143	C	2,6-Cl_2	H	H	3,5-$(CN)_2$	0	1	2
W144	C	2,6-Cl_2	H	H	4-NO_2	0	1	2
W145	C	2,6-Cl_2	H	H	3-Cl,4-NO_2	0	1	2
W146	C	2,6-Cl_2	H	H	3-CF_3,4-NO_2	0	0	0
W147	C	2,6-Cl_2	H	H	3-NO_2, 4-CH_3	0	0	0
W148	C	2,6-Cl_2	H	H	2,4-$(NO_2)_2$	0	0	0
W149	C	2,6-Cl_2	H	H	4-$N^+(CH_3)3I^-$,1aq	0	0	0
W150	C	2,6-Cl_2	H	H	4-SO_2CH_3	0	0	0
W151	C	2,6-Cl_2	H	H	4-SO_2-*n*-C_5H_{11}	0	0	0

(Continued)

TABLE 10.1 (CONTINUED)
Structure of the 203 Benzoylphenyl Ureas (BPUs) and the Concentrations at Which They Lead to 90%–100% (L1), 50%–89% (L2), and/or 0%–49% (L3) Mortality against Larvae of *Ae. aegypti*

Num.	Set[a]	Phi1[b]	N1	N2	Phi2	L1	L2	L3
W152	C	2,6-Cl_2	H	H	4-$SO_2C_6H_5$	0	0	0
W153	C	2,6-Cl_2	H	H	4-SO_2-(4-ClC_6H_4)	3	4	5
W154	C	2,6-Cl_2	H	H	4-SO_2-(2,4,5-$Cl_3C_6H_2$)	0	0	0
W155	C	2,6-Cl_2	H	H	4-SO_2NH_2	0	0	0
W156	C	2,6-Cl_2	H	H	4-$SO_2N(CH_3)_2$	0	0	0
W157	C	2,6-Cl_2	H	H	4-$SO_2N(n\text{-}C_3H_7)_2$	0	0	0
W158	A	H	H	H	3,4-Cl_2	0	0	0
W159	A	2-Cl	H	H	3,4-Cl_2	0	1	4
W160	A	2-Br	H	H	3,4-Cl_2	0	1	4
W161	A	2-OH	H	H	3,4-Cl_2	0	0	0
W162	A	2-OCH_3	H	H	3,4-Cl_2	0	0	0
W163	A	2-$OCH_2C_6H_5$	H	H	4-Cl	0	0	0
W164	A	4-Cl	H	H	3,4-Cl_2	0	0	0
W165	A	4-OCH_3	H	H	3,4-Cl_2	0	0	0
W166	A	4-NO_2	H	H	3,4-Cl_2	0	0	0
W167	A	2,4-Cl_2	H	H	3,4-Cl_2	0	0	0
W168	A	2-Cl, 4-NO_2	H	H	3,4-Cl_2	0	0	0
W169	A	3,4-Cl_2	H	H	3,4-Cl_2	0	0	0
W170	A	2,4,5-Cl_3	H	H	4-Cl	0	0	0
W171	A	2,3,6-Cl_3	H	H	3,4-Cl_2	0	0	0
W172	A	2,6-Cl_2, 4-OH	H	H	3,4-Cl_2	0	0	0
W173	A	2,6-Cl_2, 4-OCH_3	H	H	3,4-Cl_2	0	0	0

(Continued)

TABLE 10.1 (CONTINUED)

Structure of the 203 Benzoylphenyl Ureas (BPUs) and the Concentrations at Which They Lead to 90%–100% (L1), 50%–89% (L2), and/or 0%–49% (L3) Mortality against Larvae of *Ae. aegypti*

Num.	Set[a]	Phi1[b]	N1	N2	Phi2	L1	L2	L3
W174	A	2,6-Cl_2, 3-NO_2	H	H	3,4-Cl_2	0	0	0
W175	A	2,4,6-$(CH_3)_3$	H	H	3,4-Cl_2	0	0	0
W176	A	2,6-$(CH_3)_2$	H	H	4-Cl	1	N	2
W177	A	2,6-$(CH_3)_2$	H	H	3,4-Cl_2	0	1	2
W178	A	2,6-$(CH_3)_2$	H	H	4-CF_3	2	3	4
W179	A	2,6-$(C_2H_5)_2$	H	H	4-Cl	0	0	0
W180	A	2,6-$(C_2H_5)_2$	H	H	4-CF_3	0	0	0
W181	A	2,6-$(OH)_2$	H	H	3,4-Cl_2	0	0	0
W182	A	2,6-$(OCH_3)_2$	H	H	3,4-Cl_2	0	1	2
W183	A	2,6-$(OCH_2C_6H_5)_2$	H	H	3,4-Cl_2	0	0	0
W184	A	2-Cl, 6-OH	H	H	4-Cl	0	0	0
W185	A	2-Cl, 6-$OCH_2C_6H_5$	H	H	4-Cl	0	0	0
W186	A	2-F, 6-$N(CH_3)_2$	H	H	4-Cl	0	0	0
W187	A	2-Cl, 6-NO_2	H	H	3,4-Cl_2	2	N	3
W188	A	2,6-F_2	H	H	4-Cl	5	6	7
W189	A	2,6-F_2	H	H	4-Br	5	6	7
W190	A	2,6-F_2	H	H	3,4-Cl_2	7	N	N

(Continued)

TABLE 10.1 (CONTINUED)
Structure of the 203 Benzoylphenyl Ureas (BPUs) and the Concentrations at Which They Lead to 90%–100% (L1), 50%–89% (L2), and/or 0%–49% (L3) Mortality against Larvae of *Ae. aegypti*

Num.	Set[a]	Phi1[b]	N1	N2	Phi2	L1	L2	L3
W191	A	2,6-F_2	H	H	4-CF_3	7	N	N
W192	A	2,6-F_2	H	H	4-i-C_3H_7	3	N	4
W193	A	2,6-F_2	H	H	4-n-C_4H_9	5	N	6
W194	A	2,6-F_2	H	H	4-t-C_4H_9	3	4	5
W195	A	2-F, 6-Cl	H	H	4-Cl	5	6	7
W196	A	2-F, 6-Cl	H	H	4-CF_3	5	6	7
W197	A	2-F, 6-Br	H	H	4-Cl	5	N	6
W198	A	2-F, 6-Br	H	H	3,4-Cl_2	5	N	6
W199	A	2-F, 6-Br	H	H	4-CF_3	6	N	7
W200	A	2-Cl, 6-Br	H	H	4-Cl	4	N	5
W201	A	2-Cl, 6-Br	H	H	3,4-Cl_2	5	N	6
W202	A	2-Cl, 6-Br	H	H	4-CF_3	5	6	7
W203	A	2,6-Br_2	H	H	3,4-Cl_2	0	1	5

Source: K. Wellinga, R. Mulder, and J.J. van Daalen, J. Agric. Food Chem. 21 (1973), pp. 348–354 and pp. 993–998.

[a] See text for explanation.

[b] Phi1-C(=O)-N1-C(=O)-N2-Phi2, where Phi1 and Phi2 are benzene rings.

The toxicity was ranked from 1 to 8, corresponding to 1 to 0.0003 ppm, respectively. Thus, for example, molecule W003 is characterized by L1 = 2, L2 = 3, and L3 = 4. This means that 90% to 100% of mortality was observed at 0.3 ppm, 50% to 89% of mortality at 0.1 ppm, and 0% to 49% of mortality at 0.03 ppm. For molecule W001 in Table 10.1, L1, L2, and L3 = 0. This means that until 1 ppm, at best, 0% to 49% of mortality was observed. For molecule W004, 90% to 100% of mortality was observed at 0.001 ppm (L1 = 7) and 0% to 49% at 0.0003 ppm (L3 = 8). Because the level of 50% to 89% of mortality was not observed, the symbol "N" was used for L2. Finally, molecule W010 is characterized by L1 = 1, L2 = 2, and L3 = 4. This means that 90% to 100% of mortality was observed at 1 ppm, 50% to 89% of mortality was recorded at 0.3 and 0.1 ppm, and 0% to 49% of mortality was observed at 0.03 ppm. This strategy allows a better characterization of the larvicidal profile of the studied chemicals. Chemicals that sharply change of level of toxicity from a concentration to another one are rapidly identified. It is also easy to detect chemicals having the same level of toxicity for different tested concentrations.

Although the analysis of the substitution pattern on the BPU skeleton provides invaluable structure–activity rules, the rational prediction of potential active BPUs needs the design of SAR models encoding these rules. As a result, the BPUs were split into two groups: those showing no significant activity even at 1 ppm and the others. The choice of this threshold value is not fortuitous. In practice, this concentration represents the upper limit below which the molecules can be of interest for optimization of their larvicidal activity against *Ae. aegypti*. Moreover, because the test results are rather roughly expressed, it is preferable to select a simple and nonambiguous partitioning. Consider, as inactive molecules, those belonging to the lowest class of activity for the tested concentrations that satisfy these requirements. Last, it is worth noting that expression on a millimolar basis does not change the selected portioning. Thus, the SAR models were computed on a data set including 129 inactive (0) and 72 active (1) BPUs after exclusion of two chemicals.

10.2.2 Molecular Descriptors

BPUs were encoded from the autocorrelation method—except molecules W126 and W149 (Table 10.1), which were impossible to encode. With the autocorrelation method, an organic molecule is represented by means of a graph where the atoms are displayed by nodes (or vertices) and the bonds are depicted by edges. In this graph, the distance between two nodes is defined as the smallest number of edges between them [22]. Considering that there are physicochemical properties that can be calculated rather accurately from atomic contributions (ACs), the autocorrelation method allows us to generate topological descriptors encoding specific physicochemical properties [23]. This leads to descriptors that are much more informative than topological indices solely based on the counting of the number of neighbors of each node in the molecules or on any kind of purely topological algorithm. For a property, calculable from ACs, the classical algorithm of the autocorrelation method allows the computation of all the products ($AC_i \times AC_i$, $AC_j \times AC_j$, etc.) corresponding to the different smallest internodal distances (i.e., 0 to n) in the molecular graph of each molecule. The sum of these products for the same distance in the graph gives an autocorrelation component (AC) of the autocorrelation vector (AV) for the selected property. The number of ACs different from zero in an AV depends on the size of the molecule. This is the reason why, very often, prior to the design of a model, the AVs have to be truncated to avoid having ACs with too many zero values [23]. While the autocorrelation method presents numerous advantages, the classical algorithm described before also shows some shortcomings. The product of two values attributed to atoms constitutes the basis of the calculation of an AC with the classical method. If the sign of the two contribution values is negative, the product becomes positive and, as a result, the physicochemical meaning of the AC is lost. To overcome this problem and optimize the descriptive power and the weak redundancy of the AVs, we have proposed a new algorithm in which the physicochemical information carried out by each atomic contribution is kept [24].

Thus, from the fragmental constants of Rekker and Mannhold [25], an autocorrelation vector, L, representing lipophilicity was derived. Second, an autocorrelation vector, MR, encoding molar refractivity, was designed from the fragmental constants of Hansch and Leo [26] or directly from the Lorentz–Lorenz equation. Last, autocorrelation vectors encoding the H-bonding acceptor (HBA) ability and H-bonding donor (HBD) ability of the molecules were also calculated by means of Boolean contributions (i.e., 0/1). The autocorrelation vectors were calculated by means of AUTOCOR™ 2.4 (CTIS, France) using SMILES (Simplified Molecular Input Line Entry System) notation as inputs. For the autocorrelation vectors L and MR, a truncation of the vectors was performed in order to obtain 18 autocorrelation components (distances 0–17 in the graphs of the molecules). For the HBA autocorrelation vector, the three first components were calculated (distances 0–2) while, for the HBD autocorrelation vector, only the first component was selected.

Feature selections by forward and backward procedures and also with the genetic algorithm/neural network procedure of the Statistica™ software (StatSoft, France) were used to retrieve the most discriminating autocorrelation descriptors.

10.2.3 Statistical Analysis

A three-layer perceptron (TLP) was used for deriving SAR models with the BPU activity values (Table 10.1). The TLP is perhaps the most popular supervised artificial neural network (ANN) used in QSAR and QSPR (quantitative structure–property relationship) modeling (see, for example, references 27–33). Because its functioning has been widely described in the literature (see reference 34), only the basic principles are recalled here. A TLP includes one input layer with a number of neurons corresponding to the number of selected molecular descriptors, one output layer of generally one neuron corresponding to the modeled activity, and one hidden layer, between the two preceding layers, with an adjustable number of neurons for distributing the information within the network. Too many hidden neurons often lead to over-fitting and, hence, to avoid problems, their number has to be limited. This is done from a trial-and-error procedure and also by the use of pruning algorithms. The neurons of each layer are connected in the forward direction (i.e., input to output) and are activated by means of activation functions. Each connection is associated with a weight. The weights are adjusted during the learning process that aims to minimize an error computed from the target and calculated outputs. Numerous learning algorithms are available and, among them, the backpropagation, the Levenberg–Marquardt, and the Broyden–Fletcher–Goldfarb–Shanno (BFGS) algorithms were tested alone or in combination [34].

All the calculations were performed with the data mining module of the Statistica™ software (version 10).

10.2.4 Estimation Performances and Selection of the Most Interesting TLP Model

The data set (Table 10.1) was randomly split into different training and external test sets of 161 and 40 BPUs, respectively. In addition, an extensive bibliographical investigation was performed in order to collect all the available toxicity results on the commercialized BPUs (Figure 10.1) tested against larvae of *Ae. aegypti*, but also against other species of mosquitoes [35–72]. The retrieved results (Table 10.2) were used as a kind of out-sample test set [73] in order to fully evaluate the prediction performances of the models. Indeed, it is always interesting to confront the predictions of a model to laboratory results obtained according to different experimental conditions and/or for chemicals that do not exactly belong to the applicability domain of the selected model. In this context, the simulation results obtained with the models were also compared with the experimental results obtained by Carmellino et al. [74] (Table 10.3) and by Meazza et al. [75] (Table 10.4) from groups of twenty-five 4-day-old larvae of *Ae. aegypti*. Ultimately, the goal of these comparison exercises was the selection of the SAR model showing the best performances on the largest set of structurally diverse BPUs.

10.3 RESULTS AND DISCUSSION

10.3.1 Structural Features Governing Activity of BPUs

In order to facilitate the search for structural features that govern the activity of BPUs against larvae of *Ae. aegypti*, the common substructure (Phi1-C(=O)-N1-C(=O)-N2-Phi2) was subdivided into three parts named P1, P2, and P3. They correspond to the phenyl ring bound to the carbonyl group, the di-urea central fragment, and the phenyl ring bound to the nitrogen atom, respectively. This allowed us to consider three different sets of molecules.

In set A, the substituents on P1 are variable but the cycle never bears only two chlorine atoms in positions 2 and 6. The nitrogen atoms of P2 are not substituted, and P3 includes different substituents. In set B, P1 is a 2,6-dichlorophenyl ring with no other substituent and P2 and P3 include various substituents. In set C, P1 is a 2,6-dichlorophenyl ring with no other substituents, P2 does not include substituents, and P3 includes various substituents. There are 46, 49, and 108 chemicals in sets A, B, and C, respectively (Table 10.1).

The 46 molecules constituting set A show 31 different substitution patterns on P1 and 7 on P3. P1 can have zero to three substituents, which can be in the *ortho*-, *meta*-, or *para*-position. A di-*ortho*-substitution is found in about 72% of the molecules and

TABLE 10.2
Toxicity[a] of Commercial Benzoylphenylureas against Mosquito Larvae

Chemical[b]	CAS RN	Species	Stage[c]	Toxicity Result[d]	Formulation	Ref.
Diflubenzuron	35367-38-5	*Ae. aegypti*	L3	EI_{50} = 390 (290–450)	TM	35
				EI_{90} = 940 (870–1030)		
			LL3/EL4	EI_{50} = 1.59 (1.11–2.11)	TM 96.1%	36
				EI_{50} = 0.02 (0.001–0.04)	10% EC	
			EL4	EI_{30} = 0.26 (0.19–0.33)	25% WP	37
				EI_{50} = 0.50 (0.42–0.61)		
				EI_{90} = 3.5 (2.6–5.0)		
			EL4	LC_{50} = 3		38
			NHL	LC_{50} = 0.3	WP	39
			EL4	LC_{50} = 1 (0.9–1.2)	25% WP	40
				LC_{90} = 1.6 (1.4–1.9)		
				EI_{100} = 6.25		
				LC_{50} = 0.9 (0.9–1.1)	22% SL	40
				LC_{90} = 1.6 (1.5–1.7)		
				EI_{100} = 6.25		
			LL3/EL4	EI_{50} = 0.048 (0.012–0.101)	10% EC	41
			EL4	EI_{50} = 0.36	4% G	42
			EL4	LC_{50} = 5.19 (4.59–5.79)	TM 99.4%	43
				LC_{95} = 12.24 (10.77–14.30)		
			L3	LC_{90} = 0.706	5% EC	44
		Ae. albopictus	LL3/EL4	LC_{50} = 0.45 (0.39–0.49)	TM 90%	45
				LC_{90} = 0.84 (0.76–0.97)		
		Ae. melanimon	L4	LC_{90} = 4–10	25% WP	44

(Continued)

TABLE 10.2 (CONTINUED)
Toxicity[a] of Commercial Benzoylphenylureas against Mosquito Larvae

Chemical[b]	CAS RN	Species	Stage[c]	Toxicity Result[d]	Formulation	Ref.
		Ae. nigromaculis	L4	LC_{90} = 1–4	25% WP	44
		Ae. sollicitans	L3	LC_{90} = 0.036	5% EC	44
		Ae. taeniorhynchus	L3	LC_{90} = 0.69	25% WP	44
			L3	LC_{90} = 0.045	5% EC	44
			L4	LC_{90} = 2.05	25% WP	44
		Ae. triseriatus	L3	LC_{90} = 0.718	5% EC	44
		An. albimanus	L4	EI_{50} = 0.70 (0.59–0.83)	25% WP	46
			L4	LC_{90} = 1		44
		An. culicifacies	EL4	LC_{50} = 0.8 (0.7–0.9)	25% WP	40
				LC_{90} = 1.5 (1.3–1.7)		
				LC_{50} = 1.1 (1–1.2)	22% SL	40
				LC_{90} = 2.3 (2–2.7)		
		An. gambiae	L4	EI_{50} = 3.7 (2.6–5.2)	25% WP	46
			Eggs	LC_{50} = 15		38
			L1	LC_{50} = 5		38
			EL4	LC_{50} = 3		38
			EP	LC_{50} = 4		38
		An. quadrimaculatus	L3	LC_{90} = 0.086	5% EC	44
			EL4	LC_{50} = 1.1		38
		An. stephensi	EL4	LC_{50} = 0.7 (0.6–0.8)	25% WP	40
				LC_{90} = 1.4 (1.2–2)		
				EI_{100} = 6.25		
				LC_{50} = 0.9 (0.9–1.2)	22% SL	40

(Continued)

TABLE 10.2 (CONTINUED)
Toxicity[a] of Commercial Benzoylphenylureas against Mosquito Larvae

Chemical[b]	CAS RN	Species	Stage[c]	Toxicity Result[d]	Formulation	Ref.
				$LC_{90} = 1.5$ (1.3–1.6)		
				$EI_{100} = 6.25$		
			L4	$EI_{50} = 0.84$ (0.71–0.99)	25% WP	46
		Cx. nigripalpus	L3	$LC_{90} = 0.16$	25% WP	44
			L4	$LC_{90} = 0.80$		
		Cx. pipiens	L4	$LC_{50} = 0.016$		47
		Cx. pipiens pallens	L3	$LC_{50} = 0.035$	25% WP	48
			L4	$LC_{50} = 0.062$		
			L4	$EI_{100} = 0.5$		
		Cx. quinquefasciatus	L4	$EI_{90} = 0.20$		49
			EL4	$LC_{50} = 1.1$ (0.9–1.2)	25% WP	40
				$LC_{90} = 1.7$ (1.6–1.9)		
				$LC_{50} = 0.8$ (0.7–0.9)	22% SL	40
				$LC_{90} = 1.6$ (1.3–1.8)		
			LL3/EL4	$LC_{50} = 1.4$ (1.2–1.7)	TM 90%	50
				$LC_{90} = 3.4$ (2.7–4.8)		
			LL3/EL4	$EI_{50} = 0.5$ (0.4–0.6)	25% WP	51
				$EI_{90} = 2.7$ (2–3.5)		
			Eggs	$LC_{50} = 25$		38
			L1	$LC_{50} = 2$		38
			L2	$LC_{50} = 2.5$		38
			EL4	$LC_{50} = 1.3$		38
			LL4	$LC_{50} = 2$		38

(Continued)

TABLE 10.2 (CONTINUED)
Toxicity[a] of Commercial Benzoylphenylureas against Mosquito Larvae

Chemical[b]	CAS RN	Species	Stage[c]	Toxicity Result[d]	Formulation	Ref.
			EP	$LC_{50} = 10$		38
		Cx. salinarius	L3	$LC_{90} = 0.121$	5% EC	44
		Cx. tarsalis	L3	$LC_{90} = 1.049$	5% EC	44
		Cx. tritaeniorhynchus	L3	$LC_{50} = 0.042$	25% WP	48
			L4	$LC_{50} = 0.060$		
		Culiseta inornata	L3	$LC_{90} = 1.639$	5%EC	44
Dichlorbenzuron	35409-97-3	*Ae. aegypti*	L2/L3	$LC_{90\text{–}100} = 10$		52
			EL4	$LC_{50} = 4$		38
		An. gambiae	L1	$LC_{50} = 7$		38
			EL4	$LC_{50} = 5$		38
		An. quadrimaculatus	EL4	$LC_{50} = 2.5$		38
		Cx. quinquefasciatus	L1	$LC_{50} = 6$		38
			L2	$LC_{50} = 10$		38
			EL4	$LC_{50} = 5$		38
			LL4	$LC_{50} = 10$		38
Teflubenzuron	83121-18-0	*Ae. aegypti*	L4	$LC_{50} = 0.60$ (0.55–0.65) $LC_{95} = 3.7$ (3.04–4.75)	TM	53
			L4	$LC_{50} = 0.60$ (0.53–0.66) $LC_{95} = 4.06$ (3.5–4.85)	TM	54
			L4	$LC_{50}\text{–}LC_{90} = 0.30\text{–}0.60$	10% EC	55
				$LC_{50}\text{–}LC_{90} = 0.40\text{–}0.70$	TM	55
		Cx. pipiens	L4	$LC_{50} = 0.0059$		47
		Cx. quinquefasciatus	L2	$LC_{50}\text{–}LC_{90} = 0.30\text{–}0.60$	10% EC	55

(Continued)

TABLE 10.2 (CONTINUED)
Toxicity[a] of Commercial Benzoylphenylureas against Mosquito Larvae

Chemical[b]	CAS RN	Species	Stage[c]	Toxicity Result[d]	Formulation	Ref.
				LC_{50}–LC_{90} = 0.20–0.40	TM	55
			L4	LC_{50}–LC_{90} = 0.40–0.70	10% EC	55
				LC_{50}–LC_{90} = 0.30–0.50	TM	55
Triflumuron	64628-44-0	*Ae. aegypti*	LL3	EI_{50} = 0.2 (0.13–0.34)	48% SC	56
				EI_{90} = 2.6		
			L3	EI_{50} = 0.86 (0.80–0.93)	48% SC	57–59
				EI_{90} = 1.80 (1.51–2.17)		
				EI_{99} = 3.95 (2.46–4.49)		
		Ae. albopictus	L3	EI_{50} = 1.59 (1.54–1.63)	48% SC	57
				EI_{90} = 2.63 (2.42–2.64)		
				EI_{99} = 3.95 (3.35–4.68)		
		Cx. pipiens	L4	LC_{50} = 0.036		47
		Cx. pipiens pipiens	L3	EI_{50} = 0.0165 (0.0117–0.0233)	48% EC	60,61
			L4	EI_{50} = 0.0355 (0.0269–0.0468)		
		Cx. quinquefasciatus	LL3	EI_{50} = 0.3 (0.23–0.59)	48% SC	56
				EI_{90} = 10.2		
			L3	EI_{50} = 5.28 (4.15–6.72)	48% SC	57
				EI_{90} = 12.47 (8.85–17.60)		
				EI_{99} = 25.11 (12.9–48.86)		
			L4	EI_{90} = 0.70		49
			LL3/EL4	EI_{50} = 0.2 (0.1–0.21)	48% EC	51
				EI_{90} = 0.9 (0.7–1.2)		
		An. stephensi	LL3	EI_{50} = 0.1 (0.11–0.25)	48% SC	56

(Continued)

TABLE 10.2 (CONTINUED)
Toxicity[a] of Commercial Benzoylphenylureas against Mosquito Larvae

Chemical[b]	CAS RN	Species	Stage[c]	Toxicity Result[d]	Formulation	Ref.
				$EI_{90} = 2.4$		
Hexaflumuron	86479-06-3	*Ae. aegypti*	L3	$EI_{50} = 0.11$	5% EC	62
				$EI_{90} = 3.06$		
			L4	$EI_{50} = 6.273$	5% EC	63
				$EI_{90} = 3.242$		
			L4	LC_{50}–LC_{90} = 0.47–0.84	5% EC	55
				LC_{50}–LC_{90} = 0.50–1.40	TM	55
		Ae. albopictus	L3	$EI_{50} = 0.191$	5% EC	64
				$EI_{90} = 0.615$		
		An. stephensi	L3	$EI_{50} = 0.22$	5% EC	62
				$EI_{90} = 1.52$		
			L4	$EI_{50} = 3.43$	5% EC	63
				$EI_{90} = 15.1$		
		Cx. quinquefasciatus	L3	$EI_{50} = 0.09$	5% EC	62
				$EI_{90} = 2.69$		
			L4	$EI_{50} = 0.4$	5% EC	63
				$EI_{90} = 3.317$		
			L4	LC_{50}–LC_{90} = 0.45–0.92	10% EC	55
				LC_{50}–LC_{90} = 0.60–1.45	TM	55
Novaluron	116714-46-6	*Ae. aegypti*	L3	EI_{50} = 0.10–0.14	10% EC	65
				$EI_{99} = 0.30$		
			LL3/EL4	EI_{50} = 0.038 (0.030–0.046)	10% EC	41
			L2	$EI_{50} = 0.026$	TM 99.4%	66

(Continued)

TABLE 10.2 (CONTINUED)
Toxicity[a] of Commercial Benzoylphenylureas against Mosquito Larvae

Chemical[b]	CAS RN	Species	Stage[c]	Toxicity Result[d]	Formulation	Ref.
				EI_{90} = 0.144		
			L4	EI_{50} = 0.045 EI_{90} = 0.160	TM 99.4%	66
			L1	LC_{50} = 15.36 (1.69–54.65) LC_{99} = 69.51 (36.65–89.32)	10% EC	67
			L3	LC_{50} = 25.35 (13.32–57.82) LC_{99} = 70.86 (45.55–95.58)	10% EC	67
			Pupae	LC_{50} = 8.91 (5.68–11.23) LC_{99} = 67.23 (48.67–93.78)	10% EC	67
		Ae. albopictus	L1	LC_{50} = 19.17 (1.85–39.34) LC_{99} = 68.39 (37.87–97.67)	10% EC	67
			L3	LC_{50} = 34.99 (29–61.21) LC_{99} = 94.39 (54.57–127.43)	10% EC	67
			Pupae	LC_{50} = 2.05 (1.35–3.06) LC_{99} = 50.06 (26.04–86.75)	10% EC	67
		An. albimanus	L1	LC_{50} = 18.99 (1.88–27.81) LC_{99} = 68.62 (38.22–86.53)	10% EC	67
			L3	LC_{50} = 31.03 (9.67–78.17) LC_{99} = 88.83 (55.29–118.83)	10% EC	67
			Pupae	LC_{50} = 2.92 (2.08–4.25) LC_{99} = 8.68 (6.57–13.18)	10% EC	67
		An. pseudopunctipennis	L1	LC_{50} = 16.56 (1.4–63.67) LC_{99} = 73.70 (42.54–107.5)	10% EC	67

(*Continued*)

TABLE 10.2 (CONTINUED)
Toxicity[a] of Commercial Benzoylphenylureas against Mosquito Larvae

Chemical[b]	CAS RN	Species	Stage[c]	Toxicity Result[d]	Formulation	Ref.
			L3	LC_{50} = 32.61 (18.44–59.01) LC_{99} = 99.65 (68.83–129.3)	10% EC	67
			Pupae	LC_{50} = 6.64 (4.73–10.19) LC_{99} = 15.18 (11.19–26.19)	10% EC	67
		Cs. longiareolata	L3	LC_{50} = 0.51 (0.41–0.61) LC_{90} = 2.32	10% EC	68
			L4	LC_{50} = 0.91 (0.69–1.19) LC_{90} = 4.30 (3.28–5.63)	10% EC	68
		Cx. pipiens	L3	LC_{50} = 0.32 (0.27–0.37) LC_{90} = 1.16 (0.98–1.36)	10% EC	69
			L4	LC_{50} = 0.58 (0.47–0.71) LC_{90} = 2.22 (1.80–2.73)	10% EC	69
		Cx. quinquefasciatus	L4	EI_{50} = 0.47 (0.38–0.59) EI_{90} = 2.36 (1.7–3.68)	0.12% CRD	70
			L1	LC_{50} = 9.45 (4.93–17.13) LC_{99} = 145.3 (77.6–223.12)	10% EC	67
			L3	LC_{50} = 1.31 (0.9–1.71) LC_{99} = 161.11 (88–254.41)	10% EC	67
			Pupae	LC_{50} = 9.56 (5.43–13.24)	10% EC	67

(Continued)

TABLE 10.2 (CONTINUED)
Toxicity[a] of Commercial Benzoylphenylureas against Mosquito Larvae

Chemical[b]	CAS RN	Species	Stage[c]	Toxicity Result[d]	Formulation	Ref.
				LC_{99} = 159.61 (92.14–228.78)		
Lufenuron	103055-07-8	*Ae. aegypti*	L2	LC_{50} = 5		71
				LC_{40} = 2		
				LC_{20} = 0.2		
			L4	LC_{50} = 6		71
				LC_{40} = 20		
				LC_{20} = 1		
		Cx. quinquefasciatus	LL3/EL4	EI_{50} = 0.4 (0.3–0.41)	5% EC	51
				EI_{90} = 1.4 (1.1–1.8)		
Flufenoxuron	101463-69-8	*Cx. pipiens*	L3	LC_{50} = 54 (44–63)	5% EC	72
Flucycloxuron	113036-88-7	*Ae. aegypti*	NHL	LC_{50} = 3.8	DC	39
		Cx. pipiens	L4	LC_{50} = 0.036		47

[a] Micrograms per liter.
[b] See structures in Figure 10.1. The same order has been used.
[c] L1 to L4 are the larval stages. E and L before the larval stage mean early and late, respectively. NHL = newly hatched larvae.
[d] LC = lethal concentration and EI = emergence inhibition.

TABLE 10.3
Observed and Calculated Toxicity of BPUs against Larvae of *Ae. aegypti*

Num.	Phi1[a]	N1	N2	Phi2	% Mortality (20 ppb)	% Mortality (2 ppb)	TLP cal.[b]	
							No. 429	No. 169
C03	2,6-F_2	H	H	4-C(=O)-C_6H_5	100	15	1	1
C04	2,6-Cl_2	H	H	4-C(=O)-C_6H_5	100	71	1	1
C05	2,6-F_2	H	H	4-C(=O)-(4-CH_3-C_6H_4)	100	22	1	1
C06	2,6-Cl_2	H	H	4-C(=O)-(4-CH_3-C_6H_4)	68	47	0	1
C07	2,6-F_2	H	H	4-C(=O)-(4-*i*-C_3H_7-C_6H_4)	100	40	1	1
C08	2,6-Cl_2	H	H	4-C(=O)-(4-*i*-C_3H_7-C_6H_4)	78	51	0	0
C09	2,6-F_2	H	H	4-C(=O)-(4-*n*-C_4H_9-C_6H_4)	100	85	1	1
C10	2,6-Cl_2	H	H	4-C(=O)-(4-*n*-C_4H_9-C_6H_4)	58	41	0	0
C11	2,6-F_2	H	H	4-C(=O)-(4-*sec*-C_4H_9-C_6H_4)	100	69	1	1
C12	2,6-Cl_2	H	H	4-C(=O)-(4-*sec*-C_4H_9-C_6H_4)	73	31	0	0
C13	2,6-F_2	H	H	4-C(=O)-(4-*t*-C_4H_9-C_6H_4)	90	40	1	1
C14	2,6-Cl_2	H	H	4-C(=O)-(4-*t*-C_4H_9-C_6H_4)	27	8	0	0
C15	2,6-F_2	H	H	4-C(=O)-(4-F-C_6H_4)	100	90	1	1
C16	2,6-Cl_2	H	H	4-C(=O)-(4-F-C_6H_4)	100	8	1	1
C17	2,6-F_2	H	H	4-C(=O)-(4-Cl-C_6H_4)	100	100	1	1
C18	2,6-Cl_2	H	H	4-C(=O)-(4-Cl-C_6H_4)	100	100	0	1
C19	2,6-F_2	H	H	4-C(=O)-(4-Br-C_6H_4)	100	100	1	1

(Continued)

TABLE 10.3 (CONTINUED)
Observed and Calculated Toxicity of BPUs against Larvae of *Ae. aegypti*

Num.	Phi1[a]	N1	N2	Phi2	% Mortality (20 ppb)	% Mortality (2 ppb)	TLP cal.[b] No. 429	TLP cal.[b] No. 169
C20	2,6-Cl_2	H	H	4-C(=O)-(4-Br-C_6H_4)	100	68	0	1
C21	2,6-F_2	H	H	4-C(=O)-(3,4-Cl_2-C_6H_3)	100	100	1	1
C22	2,6-Cl_2	H	H	4-C(=O)-(3,4-Cl_2-C_6H_3)	100	39	0	1
C23	2,6-F_2	H	H	4-C(=O)-(3-Cl,4-F-C_6H_3)	100	100	1	1
C24	2,6-Cl_2	H	H	4-C(=O)-(3-Cl,4-F-C_6H_3)	100	100	0	1
C25	2,6-F_2	H	H	4-C(=O)-(3-Br,4-F-C_6H_3)	100	100	1	1
C26	2,6-Cl_2	H	H	4-C(=O)-(3-Br,4-F-C_6H_3)	100	100	0	1
C27	2,6-F_2	H	H	4-C(=O)-(2,4-F_2-C_6H_3)	100	100	1	1
C28	2,6-Cl_2	H	H	4-C(=O)-(2,4-F_2-C_6H_3)	100	100	0	1
C29	2,6-F_2	H	H	4-C(=O)-(4-pyridyl)	28	8	1	1
C30	2,6-Cl_2	H	H	4-C(=O)-(4-pyridyl)	20	0	1	0
C31	2,6-F_2	H	H	4-CH_2-(4-pyridyl)	16	0	1	1
C32	2,6-Cl_2	H	H	4-CH_2-(4-pyridyl)	28	0	1	0
C33	2,6-F_2	H	H	8-CH_3-quinolyl	0	0	1	1
C34	2,6-Cl_2	H	H	8-CH_3-quinolyl	0	0	0	0

Source: M.L. Carmellino et al., Pestic. Sci. 45 (1995), pp. 227–236.

[a] Phi1-C(=O)-N1-C(=O)-N2-Phi2, where Phi1 and Phi2 are benzene rings.

[b] BPU predicted by the TLP as active (1) or inactive (0).

TABLE 10.4
Observed and Calculated Toxicity of BPUs against Larvae of *Ae. aegypti*

					Rating Toxicity[b]		TLP cal.[c]	
Num.	Phi1[a]	N1	N2	Phi2	10 ppb	1 ppb	No. 429	No. 169
IIa	2-Cl	H	H	4-CH=C(CF_3)(Cl)	4	2	0	1
IIb	2-Cl	H	H	4-CH=C(CF_3)(Br)	2	0	0	1
IIc	2-Cl	H	H	4-CH=C(CF_3)(F)	4	1	0	0
IId	2-Cl	H	H	4-CH=C(CF_3)(CF_3)	3	0	0	1
IIe	2,6-F_2	H	H	4-CH=C(CF_3)(Cl)	3	1	1	1
IIf	2-Cl	H	H	3-CH_3,4-CH=C(CF_3)(Cl)	0	nt[d]	1	0
IIg	2-Cl	H	H	3-Cl,4-CH=C(CF_3)(Cl)	2	1	1	0
IIh	2,6-F_2	H	H	3-Cl,4-CH=C(CF_3)(Cl)	3	2	1	1
IIi	2-Cl	H	H	3-OCH_3,4-CH=C(CF_3)(Cl)	0	nt	0	0
IIj	2-Cl	H	H	2-Cl,4-CH=C(CF_3)(Cl)	0	nt	0	0
IIk	2,6-F_2	H	H	3,5-Cl_2,4-CH=C(CF_3)(Cl)	4	4	1	1
IIl	2-Cl	H	H	3,5-Cl_2,4-CH=C(CF_3)(Cl)	4	3	0	0
IIm	2-Cl	H	H	2-Br,5-OH,4-CH=C(CF_3)(Cl)	0	nt	0	0
IIIa	2-Cl	H	H	3-Cl,4-C(F)=C(CF_3)(F)	0	nt	1	0
IIIb	2-Cl	H	H	3-CH_3,4-C(F)=C(CF_3)(F)	0	nt	1	0
IIIc	2-Cl	H	H	3-OCH_3,4-C(F)=C(CF_3)(F)	0	nt	0	0
IIId	2-Cl	H	H	4-CH=CH-CF_3	4	1	0	0
IIIe	2-Cl	H	H	3-Cl,4-CH=CH-CF_3	4	3	1	0
IIIf	2-Cl	H	H	3,5-Cl_2,4-CH=CH-CF_3	4	3	0	0
IIIg	2,6-F_2	H	H	3,5-Cl_2,4-CH=CH-CF_3	4	4	1	1
IIIh[e]	2,6-F_2	H	H	4-CH=CH-CF_3	1	0	1	1
IVa	2,6-H_2	H	H	4-$CH_2CH_2CF_3$	0	nt	0	0

(Continued)

TABLE 10.4 (CONTINUED)
Observed and Calculated Toxicity of BPUs against Larvae of *Ae. aegypti*

					Rating Toxicity[b]		TLP cal.[c]	
Num.	**Phi1[a]**	**N1**	**N2**	**Phi2**	**10 ppb**	**1 ppb**	**No. 429**	**No. 169**
IVb	2-Cl	H	H	4-$CH_2CH_2CF_3$	2	0	0	1
IVc	2-Br	H	H	4-$CH_2CH_2CF_3$	1	0	0	1
IVd	2-CH_3	H	H	4-$CH_2CH_2CF_3$	0	nt	0	1
IVe	2-CF_3	H	H	4-$CH_2CH_2CF_3$	0	nt	0	1
IVf	2,6-OCH_3	H	H	4-$CH_2CH_2CF_3$	0	nt	0	0
IVg	2,6-F_2	H	H	4-$CH_2CH_2CF_3$	4	3	1	1
IVh	2,6-Cl_2	H	H	4-$CH_2CH_2CF_3$	3	0	1	1
IVi	2-Cl	H	H	3-CH_3,4-$CH_2CH_2CF_3$	3	0	1	0
IVj	2-Cl	H	H	3-Cl,4-$CH_2CH_2CF_3$	4	3	1	0
IVk	2-Cl	H	H	3,5-Cl_2,4-$CH_2CH_2CF_3$	4	4	0	0
IVl	2-Cl	H	H	2,3,5-Cl_3,4-$CH_2CH_2CF_3$	4	3	0	0
Va	2,6-Cl_2	H	H	3-CH=C(CF_3)(Cl)	2	0	0	0
Vb	2-Cl	H	H	3-CH=C(CF_3)(Cl)	1	0	0	0
Vc	2,6-Cl_2	H	H	3-CH=C(CF_3)(Br)	0	nt	0	0
Vd	2-Cl	H	H	3-CH=C(CF_3)(Br)	0	nt	0	0
Ve	2-Cl	H	H	2-CH_3,3-C(F)=C(CF_3)(F)	0	nt	0	0
Vf	2-Cl	H	H	2-OCH_3,3-CH=C(CF_3)(Cl)	0	nt	0	0
Vg	2-Cl	H	H	2-OCH_3,3-C(F)=C(CF_3)(F)	0	nt	0	0
Vh	2-Cl	H	H	3-CH=C(CF_3)(Cl),4-OCH_3	0	nt	0	0
Vi	2-Cl	H	H	2-F,3-CH=C(CF_3)(Cl)	2	1	0	0
Vj	2,6-Cl_2	H	H	2-F,3-CH=C(CF_3)(Cl)	0	nt	0	0
Vk	2-Cl	H	H	3-CH=C(CF_3)(Cl),4-F	0	nt	0	0

(Continued)

TABLE 10.4 (CONTINUED)
Observed and Calculated Toxicity of BPUs against Larvae of *Ae. aegypti*

					Rating Toxicity[b]		TLP cal.[c]	
Num.	**Phi1[a]**	**N1**	**N2**	**Phi2**	**10 ppb**	**1 ppb**	**No. 429**	**No. 169**
Vl	2,6-Cl_2	H	H	3-CH=C(CF_3)(Cl),4-F	0	nt	0	0
Vm	2-Cl	H	H	3-CH=C(CF_3)(Cl),4-Cl	0	nt	0	0
Vn	2-Cl	H	H	3-CH=C(CF_3)(F),4-Cl	0	nt	0	0
Vo	2-Cl	H	H	3-CH=C(CF_3)(Br),4-Cl	0	nt	0	0
Vp	2,6-F_2	H	H	3-CH=C(CF_3)(F),4-Cl	3	1	1	1
Vq	2-Cl	H	H	2,4-Cl_2,3-CH=C(CF_3)(Cl)	0	nt	0	0
VIa	2-Cl	H	H	2-CH=C(CF_3)(Cl)	0	nt	0	0
VIb	2-Cl	H	H	2-CH=C(CF_3)(Cl),4-Cl	3	0	0	0
VIIa	2-Cl	H	H	3-$CH_2CH_2CF_3$,4-F	0	nt	0	0
VIIb	2,6-Cl_2	H	H	3-$CH_2CH_2CF_3$,4-F	0	nt	1	0
VIIc	2-Cl	H	H	3-$CH_2CH_2CF_3$,4-Cl	0	nt	0	0
VIId	2,6-F_2	H	H	3-$CH_2CH_2CF_3$,4-Cl	3	0	1	1
VIIe	2-Cl	H	H	3-$CH_2CH_2CF_3$,6-F	0	nt	0	0
VIIf	2,6-Cl_2	H	H	3-$CH_2CH_2CF_3$,6-F	0	nt	1	0
VIIg	2-Cl	H	H	2,4-Cl_2,3-$CH_2CH_2CF_3$	0	nt	0	0
VIIIa	2-Cl	H	H	4-$CH_2CH_2CH_2CF_3$	4	2	0	0
VIIIb	2,6-F_2	H	H	4-$CH_2CH_2CH_2CF_3$	4	3	1	1
VIIIc	2-Cl	H	H	4-$CH_2CH_2CH(Br)CF_3$	2	0	0	0
VIIId	2,6-F_2	H	H	4-$CH_2CH_2CH(Br)CF_3$	4	2	1	1

(*Continued*)

TABLE 10.4 (CONTINUED)
Observed and Calculated Toxicity of BPUs against Larvae of *Ae. aegypti*

					Rating Toxicity[b]		TLP cal.[c]	
Num.	Phi1[a]	N1	N2	Phi2	10 ppb	1 ppb	No. 429	No. 169
VIIIe	2-Cl	H	H	4-$CH_2CH_2CH(Cl)CF_3$	nt	nt	0	0
VIIIf	2,6-F_2	H	H	4-$CH_2CH_2CH(Cl)CF_3$	nt	nt	1	1
VIIIg	2-Cl	H	H	4-$CH_2CH_2C(Br)(Br)CF_3$	1	0	0	0
VIIIh	2-Cl	H	H	2-Cl,4-$CH_2CH_2CH(Br)CF_3$	0	nt	0	0
VIIIi	2-Cl	H	H	3-Cl,4-$CH_2CH_2C(Br)(Br)CF_3$	4	2	0	1
VIIIj	2-Cl	H	H	3-Cl,4-$CH_2CH_2CH(Br)CF_3$	4	1	0	1
VIIIk	2,6-F_2	H	H	3-Cl,4-$CH_2CH_2CH(Br)CF_3$	4	1	1	1
VIIIl	2,6-F_2	H	H	3-Cl,4-$CH_2CH_2C(Br)(Br)CF_3$	4	1	1	1
VIIIm	2-Cl	H	H	2,5-Cl_2,4-$CH_2CH_2CH(Br)CF_3$	3	0	0	0
VIIIn	2,6-F_2	H	H	2,5-Cl_2,4-$CH_2CH_2C(Br)(Br)CF_3$	4	1	1	1
VIIIo	2-Cl	H	H	3,5-Cl_2,4-$CH_2CH_2C(Br)(Br)CF_3$	3	1	0	0
VIIIp	2-Cl	H	H	2,3,5-Cl_3,4-$CH_2CH_2C(Br)(Br)CF_3$	4	3	0	0

Source: G. Meazza et al., Pestic. Sci. 35 (1992), pp. 137–144.

[a] Phi1-C(=O)-N1-C(=O)-N2-Phi2, where Phi1 and Phi2 are benzene rings.

[b] Mortality: 0 = 0%–25%; 1 = 25.1%–50%; 2 = 50.1%–75%; 3 = 75.1%–95%; 4 = >95%.

[c] BPU predicted by the TLP as active (1) or inactive (0).

[d] nt = not tested.

[e] Ph2 is a pyridine with its nitrogen atom near N2.

a *para*-substitution in about 22% of them. Only three molecules have their P1 trisubstituted in the 2,4,6-position. In P3, the substitutions only occur in positions 3 and/or 4. Inspection of Table 10.1 shows that the presence of a fluorine or a chlorine in one *ortho*-position of P1 and a fluorine, bromine, or chlorine in the other *ortho*-position of this cycle leads to significant toxicity. The presence of two fluorine atoms in position 2,6 of P1 and a –3,4-Cl_2 on P3 or a –CF_3 in the *para*-position of this second benzene ring lead to molecules with the highest toxicity. It is the case of molecules W190 and W191 for which 90% to 100% of mortality was observed at 0.001 ppm. Methyl groups in the 2,6-position of P1 (chemicals W176 and W178) or a chlorine atom and a nitro-group (chemical W187) are of lower interest (Table 10.1). In the same way, bromine atoms in the two *ortho*-positions of P1 (chemical W203) do not favor an activity of interest even if this one seems to be difficult to precisely define. Indeed, inspection of Table 10.1 shows that with chemical W203, 50% to 89% of mortality was recorded at 1, 0.3, 0.1, and 0.03 ppm.

Set B, which includes 49 compounds, shows eight substitutions only on N1 by means of five different structural features, 33 only on N2 by means of 21 different atoms or functional groups, one on both nitrogen atoms, and seven substitutions are cyclic with six different patterns (Table 10.1). Substitutions on N1 or N2 very often decrease toxicity in comparison with their unsubstituted analogs. Thus, for example, chemical W006 only differs from W005 by the presence of a methyl group on its N2. Nonetheless, Table 10.1 shows that at 1 ppm, W006 leads to 0% to 49% of mortality at 1 ppm, while W005 induces 90% to 100% of mortality at 0.01 ppm. Notice that the difference in toxicity is not always so important; however, this example clearly illustrates how a limited change in the structure of a BPU can drastically change its larvicidal activity against *Ae. aegypti*. It is difficult to determine whether a substitution on N1 is less penalizing than on N2 or the converse. Sometimes, a substitution on N1 seems to have a lower impact on the activity than on N2. Thus, for example, chemical W054, which presents two chlorine atoms in position 2,6 of P1, no substitution on N1 and N2, and a chlorine in *para*-position of P3, induces 90% to 100% of mortality at 0.01 ppm. Chemical W079, which only differs from W054 by the presence of a CH_3 on N1, leads to 90% to 100% of mortality at 0.03 ppm. On the other side, chemical W055, which differs from W054 and W079 by the presence of a CH_3 on N2, induces 90% to 100% of mortality at 0.3 ppm. The situation is the same with chemical W095 (unsubstituted on N1 and N2), chemical W099 (with a methyl group on N1), and chemical W096 (with a methyl group on N2) because they lead to 90% to 100% of mortality at 0.01, 0.03, and 0.3 ppm, respectively. However, notice that 90% to 100% of mortality is also observed at 0.3 ppm when a methyl group is added both on N1 and N2, as shown in chemical W100 (Table 10.1). The substitution of an –OH group on N1 or N2 does not have the same effect on the activity. Thus, W105, which only differs from W095 by the presence of a hydroxyl group on N2, shows the same toxicity as W095. For both chemicals, a mortality of 90% to 100% is obtained at 0.01 ppm. On the other hand, addition of an –OH group on N1 decreases the toxicity. Indeed, Table 10.1 shows that only 0% to 49% of mortality is reached with 1 ppm of chemical W104.

The 108 molecules constituting set C are all substituted only by a chlorine atom in positions 2 and 6 of P1 and do not bear substituents on N1 and N2. They vary only by the number and position of the atoms and/or functional groups on the phenyl ring bound to the nitrogen atom (P3). About 56% of the substitutions on P3 lead to 0% to 49% mortality at 1 ppm. Conversely, only two compounds, W004 and W039, lead to 90% to 100% mortality at 0.001 ppm. The former presents an ethyl group in *para*-position of Phi2 and the latter a $-CF_3$ also in *para*-position (Table 10.1). Interestingly, drastic changes in the toxicity of the molecules of this set are noted, depending on the length of the *n*-alkyl chain in *para*-position of P3. Thus, molecules W003, W004, W005, and W011 have a methyl, *n*-ethyl, *n*-propyl, and *n*-butyl group in *para*-position on Phi2, respectively. W003, W004, and W005 respectively lead to 90% to 100% mortality at 0.3, 0.001, and 0.01 ppm, while W011 induces only 50% to 89% mortality at 1 ppm as well as at 0.3 ppm. Surprisingly, chemical W020, bearing a 4-*n*-C_8H_{17} on Phi2, leads to 90% to 100% mortality at 0.3 ppm, and chemical W022, having a 4-*n*-$C_{12}H_{25}$, induces 90% to 100% mortality at 0.1 ppm. Inspection of Table 10.1 shows that branched alkyls also often present a different toxicity behavior. Thus, if 90% to 100% mortality is obtained at 0.01 ppm with chemical W005 (4-*n*-C_3H_7), chemical W007, which bears a 4-*i*-C_3H_7, only leads to 0% to 49% mortality at 1 ppm. Substitution by 4-*i*-C_4H_9 (chemical W013) is much more interesting than by 4-*s*-C_4H_9 (chemical W015), which is better than 4-*n*-C_4H_9 (chemical W011), which seems a little bit more toxic than 4-*t*-C_4H_9 (chemical W017).

Chemical W039, with a 4-CF_3 on Phi2, leads to 90% to 100% mortality at 0.001 ppm. When the functional group is in position 2 or 3 of the phenyl group, this decreases the toxicity of the corresponding chemical. Chemical W038 (3-CF_3) leads to 90% to 100% mortality at 0.1 ppm while chemical W037 (2-CF_3) induces 0% to 49% mortality at 1 ppm. The presence of only two fluorine atoms in positions 2,4-, 2,5-, 2,6-, and 3,4- is highly detrimental to the toxicity. The situation can vary drastically with other halogens. Thus, for example, chemical W095, which is substituted by two chlorine atoms in positions 3 and 4 of Phi2, leads to 90% to 100% mortality at 0.01 ppm. Substitutions by different halogens also induce various toxicity behaviors. Thus, chemical W089 (3-F,4-Cl) and chemical W092 (3-F,4-I) also lead to 90% to 100% mortality at 0.01 ppm, while chemical W090 (3-F,4-Br) only induces 0% to 49% mortality at 1 ppm. Inspection of Table 10.1 shows that the presence of an additional phenyl group in *para*-position is beneficial for the larvicidal activity. On the other hand, the presence of a NO_2 or SO_2 group is highly detrimental to the toxicity. Thus, for example, chemical W038 (3-CF_3) leads to 90% to 100% mortality at 0.1 ppm, while chemical W146 with an additional NO_2 in *para*-position of Phi2 leads to 0% to 49% mortality at 1 ppm.

Even if it is more convenient to split the structures of BPUs into three parts in order to better detect the influential structural features on the toxicity of these chemicals against *Ae. aegypti*, it is obvious that the observed activity is the result of the action of the whole structure of the BPUs on the larvae. The spatial structure of the molecules also intervenes in their activity [76]. Thus, the phenyl ring P1 and the carbonyl of part B form an angle varying from about 50° to 90° according to the atoms and functional groups found in the two *ortho*-positions. Thus, for example, the angle with 2F = 49° and between 2Cl = 89° when N1 and N2 are unsubstituted and the phenyl ring C includes a $-CF_3$ in *para*-position.

10.3.2 Neural Network Models

The feature selection process of the data mining module of the Statistica™ software (version 10) was used to reduce the number of descriptors to use as input neurons in the TLP models. This led to the selection of 15 descriptors, among the 40 autocorrelation components characterizing each molecule. They are ranked in Figure 10.2 by decreasing order of importance for their ability to discriminate active from inactive BPUs. Inspection of Figure 10.2 shows that the autocorrelation components computed with the smallest interatomic distances within the molecular graphs are not the first selected. Thus, L0 and L1 occupy rank numbers 14 and 8, respectively. MR0 and MR1 were not selected. It is worthy to note that with the modified autocorrelation method, the components of zero order for a property of interest (e.g., L0, MR0) simply correspond to the summation of the contribution values attributed to the different atoms constituting each molecule of the studied data set. Because the BPUs share a common structural skeleton, it is not surprising that autocorrelation components of order 0 and 1 are not selected in priority. The situation with HBA0, which is ranked in second position in Figure 10.2, is different because the corresponding autocorrelation vector is computed from Boolean contributions; this is not the case for the autocorrelation vectors L and MR, which are computed from contribution values. In fact, Figure 10.2 clearly shows that descriptors that allow encoding the location and characteristics of substituents on P1, P2, and P3 are selected in priority. Thus, for example, L12, which occupies the first rank in Figure 10.2, allows the characterization of molecules having substituents on the two phenyl rings. For example, it

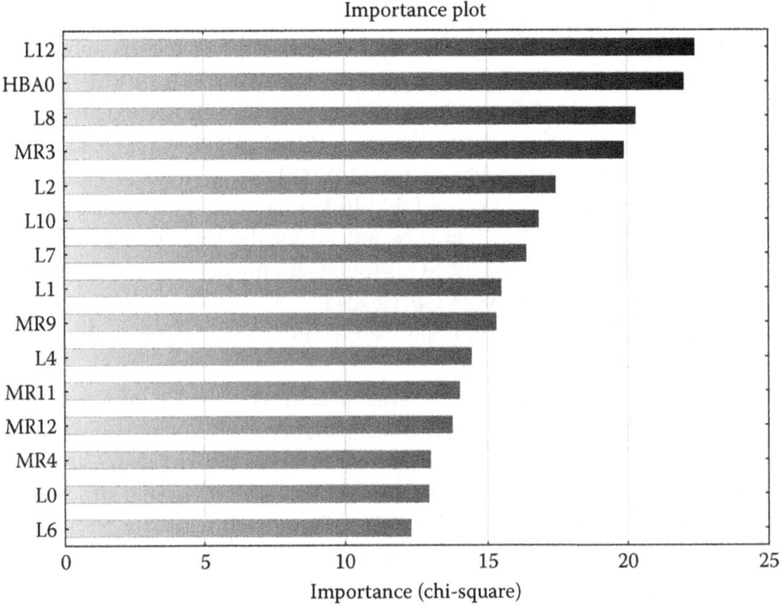

FIGURE 10.2 List of the 15 most discriminating autocorrelation descriptors.

is suitable to encode both the *ortho*-position on Phl and specific substituents in the *para*-position on Phi2. The autocorrelation method outperforms most of the other topological approaches because, in addition to accounting for shape and branching information, the autocorrelation vectors have a physicochemical meaning. Thus, inspection of Figure 10.2 shows that among the 15 selected descriptors, 9 encode lipophilicity. This means that hydrophobicity seems to be a key property for modeling the toxicity of BPUs against larvae of *Ae. aegypti*.

The set of 15 descriptors listed in Figure 10.2 was used to derive TLP models. Each time, 8 to 11 of them were selected as input neurons with 4 to 6 hidden neurons and 2 output neurons. Indeed, in classification problems, the neural network module of the Statistica™ software allocates one neuron per class of activity. In the same way, an input bias and a hidden bias are always included in the architecture of the TLP. Attempts were also made to replace one or two selected descriptors in the input layer by nonselected autocorrelation descriptors (e.g., HBD0). More than 100,000 different runs were performed by changing the number and composition of the input neurons, the number of hidden neurons, the learning algorithm, the error function, and the type of activation functions. The composition of the training and test sets (80%/20%) was also randomly changed. The TLP model with the fewest number of connections was an 8/4/2 (TLP no. 24) that led to 90% of good predictions on both the training and test sets. The input neurons were L2, L6, L10, L12, MR3, MR11, HBA0, and HBD0. The BFGS algorithm was used during the learning phase and the convergence was obtained after only 105 cycles. The sum of squares was used as error function for the training phase. The hidden and output activation functions were both tanh (hyperbolic tangent). An increase in the number of input and/or output neurons generally led to a slight increase in the percentage of good predictions on the training and test sets. Because the presence of two fluorine atoms in the *ortho*-position of the benzene ring linked to the carbonyl group tends to increase the larvicidal activity of BPUs, a descriptor encoding the number of fluorine atoms in the *ortho*-position of Phil was added to the set of autocorrelation descriptors. It is worth noting that this descriptor was independent on the presence of other substituents on P1, while in practice the 2,6-F_2 substitution increases toxicity when there is no other substituent on P1. Thus, a value of 0, 1, or 2 was allocated to each BPU depending on the presence of 0, 1, or 2 fluorine atoms in the *ortho*-position of P1. Selection of the best configuration was first based on the prediction results obtained on the training and test sets but also on the commercial BPUs for which toxicity results are available on *Ae. aegypti* and/or on other mosquito species (Table 10.2). Although the experimental conditions were different from those used by Wellinga et al. [19,20], the larvicidal activity against *Ae. aegypti* of the BPUs listed in Table 10.2 is high enough to allow this kind of comparison. Moreover, the first two BPUs in Table 10.2 also belong to the data set used to derive the SAR models and show comparable levels of toxicity. These chemicals are named W054 (dichlorbenzuron) and W188 (diflubenzuron) in Table 10.1. Last, it is worth noting that Table 10.1 includes a third commercial BPU (W191), named penfluron (Figure 10.1), for which no other toxicity results have been found in the literature.

Thus, more than 700,000 runs by batches of 2,000 were performed by changing the characteristics of the TLP and/or the set of autocorrelation descriptors as described before, while the 2,6-F_2 descriptor was always used as input neuron. The number of input neurons ranged from 10 to 12 and the number of hidden neurons was set to 4 or 5. Different learning algorithms, error functions, and activation functions were tested. In each batch of 2,000 runs, only the 50 configurations with the best overall predictive performances were kept; among them, only those with at least 90% of good predictions on both the training and test sets were saved and analyzed. These conditions were satisfied by 76 configurations. They were analyzed and, among them, TLP no. 169 and TLP no. 429 were considered the most interesting models.

The TLP no. 429 model includes 11 input neurons (2,6-F_2, L2, L4, L6, L8, L10, L12, MR4, MR11, MR15, and HBD0), 5 hidden neurons, and 2 output neurons. The model was trained by the BFGS algorithm. Cross entropy (CE) was used as an error function and the hidden and the output activation functions were tanh and softmax, respectively. It is noteworthy that the CE error function assumes that the data are drawn from the multinomial family of distributions and support a direct probabilistic interpretation of the outputs. While using this error function, the output activation functions are automatically set to softmax. This restriction ensures that the outputs are true class membership probabilities, which is known to enhance the performance of classification artificial neural networks. With the TLP no. 429 model, convergence was obtained in only 50 cycles. The percentages of good predictions on the training and test sets were equal to 90.10% and 95%, respectively. The external test set included 20 active and 20 inactive BPUs and only one of them was badly predicted by the model in each category. Indeed, the active BPU W178 was predicted as inactive and the inactive BPU W075 was predicted as active (Table 10.1). The training set included 52 active BPUs and 6 of them were predicted inactive by the model: W020, W043, W108, W130, W176, and W201. Inspection of Table 10.1 shows that, except for BPU W043 and BPU W201, the four other chemicals are only slightly active. Among the 109 inactive BPUs, 10 were predicted as active by the model: W011, W041, W062, W085, W090, W098, W144, W147, W148, and W171 (Table 10.1). The toxicity of dichlorbenzuron (W054) belonging to the external test set as well as the toxicity of diflubenzuron (W188) and penfluron (W191) belonging to the training set was correctly predicted by the TLP no. 429 model. This is also the case for the teflubenzuron, hexaflumuron, novaluron, lufenuron, flufenoxuron, and flucycloxuron listed in Table 10.2. Conversely, triflumuron was badly predicted. The performances of the TLP no. 429 model were also estimated from the experimental data produced by Carmellino et al. [74] (Table 10.3) and by Meazza et al. [75] (Table 10.4), even though the comparison of the data is not straightforward because the authors did not test their chemicals at 1 ppm. Indeed, Carmellino et al. [74] tested 32 BPUs from 2×10^{-2} ppm to 2×10^{-9} ppm for the most toxic of them, but only the results obtained at 20 ppb and 2 ppb are reported in Table 10.3. Twenty BPUs show 100% mortality at 20 ppb; among them, 14 were correctly predicted by the model. With regard to inactive compounds, only chemicals C33 and C34 (Table 10.3) are obligatorily inactive at 1 ppm and only the latter is correctly predicted by the model. Meazza et al. [75] tested 73 BPUs at 10 and 1 ppb, and 32 showed 0% to 25% mortality at 10 ppb and were not tested at 1 ppb due to this lack of activity. Among them, 26 were predicted as inactive by the TLP no. 429 model. On the other hand, 21 BPUs showed more than

95% mortality at 10 ppb; however, among them, only 9 were predicted as active by the TLP no. 429 model.

The TLP no. 169 model also provided interesting results. It included 12 input neurons (2,6-F_2, L2, L4, L6, L8, L10, L12, L15, MR4, MR11, MR15, and HBD0), 5 hidden neurons, and 2 output neurons. This model was also trained by the BFGS algorithm. The sum of squares was used as the error function and the hidden and output activation functions were exponential and logistic, respectively. The convergence was obtained with 62 cycles and the percentages of good predictions on the training and test sets were equal to 91.30% and 90%, respectively. Thus, the TLP no. 169 model included more connections than the TLP no. 429 model. Moreover, even if its prediction performances were slightly better on the training set, they were significantly lower on the external test. Nevertheless, we assume that the simulation performances of the TLP no. 169 model are also of interest.

The external test set included the same 20 active and 20 inactive BPUs because this configuration and the TLP no. 429 model belonged to the same series of runs. Only the toxicity of two BPUs was badly predicted in each category. Indeed, the active BPUs W010 and W142 were predicted as inactive and the inactive BPUs W021 and W063 were predicted as active (Table 10.1). The activity of these BPUs was correctly predicted by the TLP no. 429 model. Among the 52 active BPUs of the training set, the toxicity of the 8 following BPUs was badly predicted by the model: W009, W015, W038, W043, W050, W105, W153, and W187. Among the 109 inactive BPUs of the training set, only the following six were badly predicted by the model: W002, W083, W084, W090, W098, and W186. Only BPUs W043, W090, and W098 were badly predicted by both models, but there is no obvious reason for these results. The toxicity of dichlorbenzuron (W054), diflubenzuron (W188), and penfluron (W191) was correctly predicted by the TLP no. 169 model. It was also the case for all the other commercial BPUs listed in Table 10.2, expect for triflumuron, which was again badly predicted by the model.

Inspection of Table 10.3 shows that the TLP no. 169 model outperforms the TLP no. 429 model in the prediction of the toxicity of the 32 BPUs tested by Carmellino et al. [74]. This is particularly true with regard to the 20 BPUs showing 100% mortality at 20 ppb. They are all correctly predicted by the TLP no. 169 model (Table 10.3). On the other hand, again, only the absence of toxicity of the BPU C34 is correctly simulated by the models.

Among the 32 BPUs tested by Meazza et al. [75] that induced 0% to 25% mortality at 10 ppb to larvae of *Ae. aegypti*, 30 were predicted as inactive by the TLP no. 169 model. On the other hand, among the 21 BPUs showing more than 95% mortality at 10 ppb, 11 were predicted as active by the TLP no. 169 model. Thus, inspection of Table 10.4 shows that the simulation performances of the TLP no. 169 model outperform those of the TLP no. 429 model.

More generally, analysis of the prediction results obtained on the training and test sets (Table 10.1) as well as those on the BPUs listed in Tables 10.3 and 10.4 reveals that both models are flexible enough to correctly predict the toxicity of structurally diverse BPUs against larvae of *Ae. aegypti*. The highest difficulty remains the prediction of the larvicidal activity of BPUs only showing one chlorine atom in the *ortho*-position of Phi1. If the corresponding BPUs are inactive, such as chemicals IIm, IIIa to IIIc, Vc to Vh, Vk, and Vm to Vo in Table 10.4, very often they are correctly predicted, especially by the TLP no. 169 model. Conversely, if these 2-Cl-BPUs are active, both models very often fail in the prediction of their larvicidal activity. The wrong prediction obtained with triflumuron with both models relies on the presence of a unique chlorine atom in the *ortho*-position of Phi1 (Figure 10.1). It is worth noting that among the TLP configurations analyzed, it was exceptional to obtain a good prediction for the triflumuron. In that case, the toxicity of other commercial BPUs listed in Table 10.2 was badly predicted (e.g., hexaflumuron, Figure 10.1). The poor performances of the TLP models for predicting active 2-Cl-BPUs rely on the fact that the data set used for deriving them [19,20] includes only one 2-Cl-BPU that is inactive (W159 in Table 10.1). This chemical was in the training set of the models and was predicted as inactive by the TLP no. 169 and no. 429 models.

Attempts were made to hybridize interesting configurations, but the improvement in the prediction performances was not significant enough to justify a complexification of the structure of the simulation tools.

10.4 CONCLUSIONS

In this study, an attempt was made to develop a simple but efficient classification system allowing the detection of benzoylphenylureas (BPUs) potentially active on larvae of *Ae. aegypti*. A large data set of 201 BPUs, presenting various structures and tested on larvae of *Ae. aegypti* under the same experimental conditions, was used to derive the models. BPUs were described by means of the modified autocorrelation method. A supervised artificial neural network—namely, a three-layer perceptron (TLP)—was employed as a statistical tool for computing the SAR models due to its capacity to find powerful, even complex relationships between activity and descriptor data. The performances of the models were evaluated through the analysis of the prediction results obtained on different training sets and external test sets, which were randomly selected (80%/20%). Available experimental results on larvae of *Ae. aegypti* for commercialized BPUs and for two series of structurally diverse BPUs were also used to evaluate and select the most interesting configurations. Two models (i.e., no. 169, no. 429) were selected. They both presented 91% of good predictions on the whole data set of 201 BPUs. They included autocorrelation descriptors but also a descriptor encoding the presence of fluorine atoms in the *ortho*-positions of the benzene ring linked to the carbonyl group of the BPUs. Interestingly, most of the autocorrelation descriptors were of higher order and half of them encoded lipophilicity. Although TLP no. 429 should be selected in priority because it includes a lower number of connections than TLP no. 169 model and also gives better predictions on the external test set, the TLP no. 169 model must also be considered due to its capacity to better extrapolate than the TLP no. 429 models. As a result, both models are of interest to rationalize the discovery of new BPUs active on larvae of *Ae. aegypti*. To our knowledge, they are the very first SAR models derived on mosquitoes for this important family of chemicals acting on the synthesis of chitin.

ACKNOWLEDGMENT

The financial support from the French Agency for Food, Environmental and Occupational Health & Safety (Anses) is gratefully acknowledged (contract #EST-2012/2/64).

REFERENCES

[1] M.J. Klowden, *Physiology Systems in Insects*, 2nd ed., Elsevier, Amsterdam, 2007.

[2] A. Odier, *Mémoire sur la composition chimique des parties cornées des insectes*, Mem. Soc. Hist. Nat. Paris 1 (1823), pp. 29–42.

[3] S.O. Andersen, *Cuticular sclerotisation and tanning*, in *Insect Molecular Biology and Biochemistry*, L.I. Gilbert, ed., Academic Press, San Diego, 2012.

[4] R.A.A. Muzzarelli, *Native, industrial and fossil chitins*, in *Chitin and Chitinases*, P. Jollès and R.A.A. Muzzarelli, eds., Birkhäuser Verlag, Basel, 1999, pp. 1–6.

[5] D. Doucet and A. Retnakaran, *Insect chitin: Metabolism, genomics and pest management*, Adv. Insect Physiol. 43 (2012), pp. 437–511.

[6] H. Merzendorfer and L. Zimoch, *Chitin metabolism in insects: Structure, function and regulation of chitin synthases and chitinases*, J. Exp. Biol. 206 (2003), pp. 4393–4412.

[7] E. Cohen, *Chitin biochemistry: Synthesis, hydrolysis and inhibition*, Adv. Insect Physiol. 38 (2010), pp. 5–73.

[8] Y. Nakagawa, K. Kitahara, T. Nishioka, H. Iwamura, and T. Fujita, *Quantitative structure–activity studies of benzoylphenylurea larvicides. I. Effects of substituents at aniline moiety against* Chilo suppressalis *Walker*, Pest. Biochem. Physiol. 21 (1984), pp. 309–325.

[9] Y. Nakagawa, H. Iwamura, and T. Fujita, *Quantitative structure–activity studies of benzoylphenylurea larvicides. II. Effects of benzyloxy substituents at aniline moiety against* Chilo suppressalis *Walker*, Pest. Biochem. Physiol. 23 (1985), pp. 7–12.

[10] Y. Nakagawa, T. Sotomatsu, K. Irie, K. Kitahara, H. Iwamura, and T. Fujita, *Quantitative structure–activity studies of benzoylphenylurea larvicides. III. Effects of substituents on benzyloxy moiety*, Pest. Biochem. Physiol. 27 (1987), pp. 143–155.

[11] T. Sotomatsu, Y. Nakagawa, and T. Fujita, *Quantitative structure–activity studies of benzoylphenylurea larvicides. IV. Benzoyl ortho substituent effects and molecular conformation*, Pest. Biochem. Physiol. 27 (1987), pp. 156–164.

[12] Y. Nakagawa, T. Akagi, H. Iwamura, and T. Fujita, *Quantitative structure–activity studies of benzoylphenylurea larvicides. V. Substituted pyridyloxyphenyl and related derivatives*, Pest. Biochem. Physiol. 30 (1988), pp. 67–78.

[13] Y. Nakagawa, T. Akagi, H. Iwamura, and T. Fujita, *Quantitative structure–activity studies of benzoylphenylurea larvicides. VI. Comparison of substituent effects among activities against different insect species*, Pestic. Biochem. Physiol. 33 (1989), pp. 144–157.

[14] Y. Nakagawa, K. Izumi, N. Oikawa, A. Kurozumi, H. Iwamura, and T. Fujita, *Quantitative structure–activity studies of benzoylphenylurea larvicides. VII. Separation of effects of substituents in the multisubstituted anilide moiety on the larvicidal activity against* Chilo suppressalis, Pestic. Biochem. Physiol. 40 (1991), pp. 12–26.

[15] Y. Nakagawa, M. Matsutani, N. Kurihara, K. Nishimura, and T. Fujita, *Quantitative structure–activity studies of benzoylphenylurea larvicides. VIII. Inhibition of N-acetlyglucosamine incorporation into the cultured integument of* Chilo suppressalis *Walker*, Pestic. Biochem. Physiol. 43 (1992), pp. 141–151.

[16] R. Sun, C. Liu, H. Zhang, and Q. Wang, *Benzoylurea chitin synthesis inhibitors*, J. Agric. Food Chem. 63 (2015), pp. 6847–6865.
[17] J. Devillers, C. Lagneau, A. Lattes, J.C. Garrigues, M.M. Clémenté, and A. Yébakima, *In silico models for predicting vector control chemicals targeting* Aedes aegypti, SAR QSAR Environ. Res. 25 (2014), pp. 805–835.
[18] J. Devillers, L. Lagadic, O. Yamada, F. Darriet, R. Delorme, X. Deparis, J.P. Jaeg, C. Lagneau, B. Lapied, F. Quiniou, and A. Yébakima, *Use of multicriteria analysis for selecting candidate insecticides for vector control,* in *Juvenile Hormones and Juvenoids. Modeling Biological Effects and Environmental Fate*, J. Devillers, ed., CRC Press, Boca Raton, FL, 2013, pp. 341–381.
[19] K. Wellinga, R. Mulder, and J.J. van Daalen, *Synthesis and laboratory evaluation of 1-(2,6-disubstituted benzoyl)-3-phenylureas, a new class of insecticides. I. 1-(2,6-dichlorobenzoyl)-3-phenylureas*, J. Agric. Food Chem. 21 (1973), pp. 348–354.
[20] K. Wellinga, R. Mulder, and J.J. van Daalen, *Synthesis and laboratory evaluation of 1-(2,6-disubstituted benzoyl)-3-phenylureas, a new class of insecticides. II. Influence of the acyl moiety on insecticidal activity*, J. Agric. Food Chem. 21 (1973), pp. 993–998.
[21] W.S. Abbott, *A method of computing the effectiveness of an insecticide*, J. Econ. Entomol. 18 (1925), pp. 265–267.
[22] J. Devillers and A.T. Balaban, *Topological Indices and Related Descriptors in QSAR and QSPR*, Gordon and Breach Science Publishers, the Netherlands, 1999.
[23] P. Broto and J. Devillers, *Autocorrelation of properties distributed on molecular graphs*, in *Practical Applications of Quantitative Structure–Activity Relationships (QSAR) in Environmental Chemistry and Toxicology*, W. Karcher and J. Devillers, eds., Kluwer Academic Publishers, Dordrecht, the Netherlands, 1990, pp. 105–127.
[24] J. Devillers, *Autocorrelation descriptors for modeling (eco)toxicological endpoints*, in *Topological Indices and Related Descriptors in QSAR and QSPR*, J. Devillers and A.T. Balaban, eds., Gordon and Breach, The Netherlands, 1999, pp. 595–612.
[25] R.F. Rekker and R. Mannhold, *Calculation of Drug Lipophilicity. The Hydrophobic Fragmental Constant Approach*, John Wiley & Sons Ltd., Weinheim, Germany, 1992.
[26] C. Hansch and A. Leo, *Substituent Constants for Correlation Analysis in Chemistry and Biology*, John Wiley & Sons Ltd., New York, 1979.
[27] T. Aoyama and H. Ichikawa, *Neural networks as nonlinear structure–activity relationship analyzers. Useful functions of the partial derivative method in multilayer neural networks*, J. Chem. Inf. Comput. Sci. 32 (1992), pp. 492–500.
[28] J. Devillers, D. Domine, C. Guillon, and W. Karcher, *Simulating lipophilicity of organic molecules with a back-propagation neural network*, J. Pharm. Sci. 87 (1998), pp. 1086–1090.
[29] D.V. Eldred, C.L. Weikel, P.C. Jurs, and K.L.E. Kaiser, *Prediction of fathead minnow acute toxicity of organic compounds from molecular structure*, Chem. Res. Toxicol. 12 (1999), pp. 670–678.
[30] J. Devillers, M.H. Pham-Delègue, A. Decourtye, H. Budzinski, S. Cluzeau, and G. Maurin, *Structure–toxicity modeling of pesticides to honey bees*, SAR QSAR Environ. Res. 13 (2002), pp. 641–648.
[31] J. Devillers, *A new strategy for using supervised artificial neural networks in QSAR*, SAR QSAR Environ. Res. 16 (2005), pp. 433–442.
[32] J. Devillers, J.P. Doucet, A. Doucet-Panaye, A. Decourtye, and P. Aupinel, *Linear and non-linear QSAR modelling of juvenile hormone esterase inhibitors*, SAR QSAR Environ. Res. 23 (2012), pp. 357–369.
[33] J. Devillers, A. Doucet-Panaye, and J.P. Doucet, *Structure–activity relationship (SAR) modelling of mosquito larvicides*, SAR QSAR Environ. Res. 26 (2015), pp. 263–278.

[34] J. Devillers, *Strengths and weaknesses of the backpropagation neural network in QSAR and QSPR studies*, in *Neural Networks in QSAR and Drug Design*, J. Devillers, ed., Academic Press, London, 1996, pp. 1–46.
[35] H.D.R. Rocha, M.H.S. Paiva, N.M. Silva, A.P. de Araújo, D.d.R.d.R.d.A. Camacho, A.J.F.d. Moura, L.F. Gómez, C.F.J. Ayres, and M.A.V.d.M. Santos, *Susceptibility profile of* Aedes aegypti *from Santiago Island, Cabo Verde, to insecticides*, Acta Trop. 152 (2015), pp. 66–73.
[36] E. Seccacini, A. Lucia, L. Harburguer, E. Zerba, S. Licastro, and H. Masuh, *Effectiveness of pyriproxyfen and diflubenzuron formulations as larvicides against* Aedes aegypti, J. Am. Mosq. Cont. Assoc. 24 (2008), pp. 398–403.
[37] F. Fournet, C. Sannier, and N. Monteny, *Effects of the insect growth regulators OMS 2017 and diflubenzuron on the reproductive potential of* Aedes aegypti, J. Am. Mosq. Cont. Assoc. 9 (1993), pp. 426–430.
[38] J.R. Busvine, Y. Rongsriyam, and D. Bruno, *Effects of some insect development inhibitors on mosquito larvae*, Pestic. Sci. 7 (1976), pp. 153–160.
[39] A.C. Grosscurt and R.A.J. Wixley, *Effects of temperature on acaricidal and insecticidal activities of the benzoylureas flucycloxuron and diflubenzuron*, Entomol. Exp. Appl. 59 (1991), pp. 239–248.
[40] M.A. Ansari, R.K. Razdan, and U. Sreehari, *Laboratory and field evaluation of hilmilin against mosquitoes*, J. Am. Mosq. Cont. Assoc. 21 (2005), pp. 432–436.
[41] V. Sfara, S.A. de Licastro, H.M. Masuh, E.A. Seccacini, R.A. Alzogaray, and E.N. Zerba, *Synergism between cis-permethrin and benzoyl phenyl urea insect growth regulators against* Aedes aegypti *larvae*, J. Am. Mosq. Cont. Assoc. 23 (2007), pp. 24–28.
[42] H.A. Kamal and E.I.M. Khater, *The biological effects of the insect growth regulators; pyriproxyfen and diflubenzuron on the mosquitoes* Aedes aegypti, J. Egypt. Soc. Parasitol. 40 (2010), pp. 565–574.
[43] J.J. da Silva and J. Mendes, *Susceptibility of* Aedes aegypti (L) *to the insect growth regulators diflubenzuron and methoprene in Uberlândia, State of Minas Gerais*, Rev. Soc. Brasil. Med. Trop. 40 (2007), pp. 612–616.
[44] A.C. Grosscurt, *Diflubenzuron: Some aspects of its ovicidal and larvicidal mode of action and an evaluation of its practical possibilities*, Pestic. Sci. 9 (1978), pp. 373–386.
[45] A. Ali, J.K. Nayar, and R. Xue, *Comparative toxicity of selected larvicides and insect growth regulators to a Florida laboratory population of* Aedes albopictus, J. Am. Mosq. Cont. Assoc. 11 (1995), pp. 72–76.
[46] H. Kawada, Y. Shono, T. Ito, and Y. Abe, *Laboratory evaluation of insect growth regulators against several species of anopheline mosquitoes*, Jpn. J. Sanit. Zool. 44 (1993), pp. 349–353.
[47] N. Soltani, *Les moustiques: Risques sanitaires, bioessais et strategies de controle*, 1er Séminaire National sur l'Entomologie Médicale et la Lutte Biologique, Tébessa, 19–20 Octobre 2015.
[48] M. Takahashi and T. Ohtaki, *A laboratory evaluation of the IGR, PH 60-40, against* Culex pipiens pallens *and* Culex tritaeniorhynchus, Eisei-Dobutsu 27 (1976), pp. 361–365.
[49] M.S. Mulla, *Insect growth regulators for the control of mosquito pests and disease vectors*, Chinese J. Entomol. 6 (1991), pp. 81–91.
[50] A. Ali, M.A. Chowdhury, M.I. Hossain, M.U. Ameen, D.B. Habiba, and A.F.M. Aslam, *Laboratory evaluation of selected larvicides and insect growth regulators against field-collected* Culex quinquefasciatus *larvae from urban Dhaka, Bangladesh*, J. Am. Mosq. Cont. Assoc. 15 (1999), pp. 43–47.
[51] D.S. Suman, B.D. Parashar, and S. Prakash, *Efficacy of various insect growth regulators on organophosphate resistant immatures of* Culex quinquefasciatus *(Diptera: Culicidae) from different geographical areas of India*, J. Entomol. 7 (2010), pp. 33–43.

[52] R.M. Mulder and M.J. Gijswijt, *The laboratory evaluation of two promising new insecticides which interfere with cuticle deposition*, Pestic. Sci. 4 (1973), pp. 737–745.
[53] V.W.D. Chui, C.W. Koo, W.M. Lo, and X.J. Qiu, *Laboratory evaluation of Vectobac® 12AS and teflubenzuron against* Culex *and* Aedes *mosquito larvae under different physical conditions*, Environ. Int. 19 (1993), pp. 193–202.
[54] V.W.D. Chui, K.W. Wong, and K.W. Tsoi, *Control of mosquito larvae (Diptera: Culicidae) using Bti and teflubenzuron: Laboratory evaluation and semi-field test*, Environ. Int. 21 (1995), pp. 433–440.
[55] M.S. Mulla, H.A. Darwazeh, and E.T. Schreiber, *Impact of new insect growth regulators and their formulations on mosquito larval development in impundment and floodwater habitats*, J. Am. Mosq. Cont. Assoc. 5 (1989), pp. 15–20.
[56] C.P. Batra, P.K. Mittal, T. Adak, and M.A. Ansari, *Efficacy of IGR compound Starycide 480 SC (triflumuron) against mosquito larvae in clear and polluted water*, J. Vect.-Borne Dis. 42 (2005), pp. 109–116.
[57] T.A. Belinato, A.J. Martins, J.B.P. Lima, and D. Valle, *Effect of triflumuron, a chitin synthesis inhibitor, on* Aedes aegypti, Aedes albopictus and Culex quinquefasciatus *under laboratory conditions*, Paras. Vect. 6 (2013), pp. 83.
[58] A.J. Martins, T.A. Belinato, J.B.P. Lima, and D. Valle, *Chitin synthesis inhibitor effect on* Aedes aegypti *populations susceptible and resistant to the organophosphate temephos*, Pest. Manag. Sci. 64 (2008), pp. 676–680.
[59] T.A. Belinato, A.J. Martins, J.B.P. Lima, T.N. de Lima-Camara, A.A. Peixoto, and D. Valle, *Effect of the chitin synthesis inhibitor triflumuron on the development, viability and reproduction of* Aedes aegypti, Mem. Inst. Oswaldo Cruz, Rio de Janeiro 104 (2009), pp. 43–47.
[60] N. Rehimi and N. Soltani, *Laboratory evaluation of Alsystin, a chitin synthesis inhibitor, against* Culex pipiens pipiens *L. (Dip., Culicidae): Effects on development and cuticle secretion*, J. Appl. Ent. 123 (1999), pp. 437–441.
[61] N. Soltani, N. Rehimi, H. Beldi, and F. Bendali, *Activité du triflumuron sur* Culex pipiens pipiens *(Diptera: Culicidae) et impacts sur deux espèces larvivores non visées*, Ann. Soc. Entomol. 35 (1999), pp. 502–508.
[62] D. Amalraj and R. Velayudhan, *Insect growth regulator XRD-473 (OMS 3031), a prospective compound for control of mosquito vectors*, Proc. Indian Acad. Sci. Anim. Sci. 98 (1989), pp. 325–329.
[63] V. Vasuki, *Influence of IGR treatment on oviposition of three species of vector mosquitos at sublethal concentrations*, Southeast Asian J. Trop. Med. Public Health 30 (1999), pp. 200–203.
[64] D. Montada, A.R. Rajavel, and V. Vasuki, *Use of hexaflumuron, an insect growth regulator in the control of* Aedes albopictus *(Skuse)*, Southeast Asian J. Trop. Med. Public Health 25 (1994), pp. 374–377.
[65] L.C. Farnesi, J.M. Brito, J.G. Linss, M. Pelajo-Machado, D. Valle, and G.L. Rezende, *Physiological and morphological aspects of* Aedes aegypti *developing larvae: Effects of the chitin synthesis inhibitor novaluron*, PLoS ONE 7 (2012), pp. e30363.
[66] M.S. Mulla, U. Thavara, A. Tawatsin, J. Chompoosri, M. Zaim, and T. Su, *Laboratory and field evaluation of novaluron a new acylurea insect growth regulator against* Aedes aegypti *(Diptera: Culicidae)*, J. Vect. Ecol. 28 (2003), pp. 241–254.
[67] J.I. Arredondo-Jiménez and K.M. Valdez-Delgado, *Effect of Novaluron (Rimon ® 10 EC) on the mosquitoes* Anopheles albimanus, Anopheles pseudopunctipennis, Aedes aegypti, Aedes albopictus and Culex quinquefasciatus *from Chiapas, Mexico*, Med. Vet. Entomol. 20 (2006), pp. 377–387.
[68] A. Bouaziz, H. Boudjelida, and N. Soltani, *Toxicity and perturbation of the metabolite contents by a chitin synthesis inhibitor in the mosquito larvae of* Culiseta longiareolata, Ann. Biol. Res. 2 (2011), pp. 134–142.

[69] N. Djeghader, H. Boudjelida, A. Bouaziz, and N. Soltani, *Biological effects of a benzoylphenylurea derivative (novaluron) on larvae of* Culex pipiens *(Diptera: Culicidae)*, Adv. Appl. Sci. Res. 4 (2013), pp. 449–456.

[70] T. Su, M.L. Cheng, A. Melgoza, and J. Thieme, *Laboratory and field evaluations of Mosquiron® 0.12CRD, a new formulation of novaluron, against* Culex *mosquitoes*, J. Am. Mosq. Cont. Assoc. 30 (2014), pp. 284–290.

[71] S.G. Salokhe, S.N. Mukherjee, S.G. Deshpande, V.P. Ghule, and J.R. Mathad, *Effect of sub-lethal concentrations of insect growth regulator, lufenuron on larval growth and development of* Aedes aegypti, Curr. Sci. 99 (2010), pp. 1256–1259.

[72] A. El-Samie Emtithal and A. El-Baset Thanaa, *Efficacy of some insecticides on field populations of* Culex pipiens *(Linnaeus) from Egypt*, J. Basic Appl. Zool. 65 (2012), pp. 62–73.

[73] J. Devillers, S. Bintein, D. Domine, and W. Karcher, *A general QSAR model for predicting the toxicity of organic chemicals to luminescent bacteria (Microtox test)*, SAR QSAR Environ. Res. 4 (1995), pp. 29–38.

[74] M.L. Carmellino, G. Pagani, M. Pregnolato, M. Terreni, V. Caprioli, and F. Zani, *Studies on the insecticidal activities of some new N-benzoyl-N′-arylureas*, Pestic. Sci. 45 (1995), pp. 227–236.

[75] G. Meazza, F. Rama, F. Bettarini, P. Piccardi, P. Massardo, and V. Caprioli, *Synthesis and bioactivity of some fluorine-containing benzoyl arylureas. Part 1. Insecticidal–acaricidal products in which the aryl group bears a trifluoromethyl-substituted alkyl or alkenyl side chain*, Pestic. Sci. 35 (1992), pp. 137–144.

[76] J.M. Luteijn and J. Tipker, *A note on the effects of fluorine substituents on the biological activity and environmental chemistry of benzoylphenylureas*, Pestic. Sci. 17 (1986), pp. 456–458.

11 Predicting Toxicity of Piperidines against Female Adults of *Aedes aegypti*

James Devillers, Annick Doucet-Panaye, Jean-Pierre Doucet, Christophe Lagneau, Sébastien Estaran, and André Yébakima

CONTENTS

ABSTRACT

Nonlinear quantitative structure–activity relationship (QSAR) models were derived for predicting the adulticidal activity of piperidines against mosquitoes. Thirty-three chemicals with their 24-hour LD_{50} values, obtained under the same experimental conditions on *Aedes aegypti* females, were used. Energy optimization allowed us to consider two conformational series of structures that were described by means of CODESSA descriptors. Only those encoding physicochemical information were considered. The 1-octanol/water partition coefficient (log *P*) of the molecules was also computed from the AUTOLOGP™ software. Each series was subject to modeling from a three-layer perceptron (TLP). The models were derived from training sets including 70% of the molecules, which were randomly selected. The remaining molecules were used as external test sets. Only 3/2/1 TLP models were considered. A consensus model was computed with an interesting configuration coming from each series. The selected hybrid TLP model showed correlation coefficients of 0.93 and 0.97 on the training set and external test set, respectively. From the simulation results, various molecules were identified and searched for their availability in existing free databases. A list of molecules of potential interest was elaborated that will be further refined in order to find new molecules having a synergistic activity with pyrethroids, especially deltamethrin.

KEYWORDS

- *Aedes aegypti*
- Adulticide
- Synergist
- QSAR
- Piperidines
- Three-layer perceptron
- Deltamethrin

11.1 INTRODUCTION

Chemicals extracted from various botanical sources have provided numerous beneficial applications ranging from pharmaceuticals to insecticides used in agriculture and vector control [1–3]. Although synthetic organic insecticides have shown their efficacy against a huge number of target species, they are very often detrimental for the biota in the ecosystems and they can induce various adverse effects in humans. In addition to adverse environmental and health effects from conventional insecticides, most major mosquito disease vector and pest species have become resistant to many of these xenobiotics. These factors have created the need for environmentally safe, degradable, and more target-specific insecticides. Thus, the search for such compounds active on mosquitoes has been directed extensively to the plant kingdom. Sukumar et al. [4] reviewed the ovicidal, larvicidal, adulticidal, and repellency activity of a huge number of plant extracts to the different mosquito species. About 340 plant species belonging to 100 families were investigated. They showed that the differential responses induced by phytochemicals on various species of mosquitoes were influenced by extrinsic and intrinsic factors such as the plant species, the parts of the plant (e.g., stem, leaf, root, seed), the solvents used for extractions, the geographical location where the plants were grown, and the methods employed for evaluation. Extracts from different parts of the same plant species can show various levels of toxicity to mosquitoes [4].

Thus, with about 2,000 species distributed pantropically, the genus *Piper*, belonging to the family *Piperaceae*, is one of the largest genera of basal angiosperms [5]. Patterns of distribution of *Piper* species vary from being locally endemic to widespread [6]. Members of this family have been used for the control of mosquitoes. A methanol extract of *Piper longum* fruits was found to be active against larvae of *Culex pipiens pallens*. A piperidine alkaloid, pipernonaline, was responsible for the toxicity, with 24-hour LD_{50} and LD_{95} values of 0.21 mg/L (0.17–0.28) and 0.52 mg/L (0.48–0.57) in third instar larvae, respectively [7]. The results were equivalent to those found for pirimiphos-methyl under the same experimental conditions.

The 24-hour LD_{50} and LD_{95} values for this organophosphorus insecticide were equal to 0.11 mg/L (0.09–0.14) and 0.17 mg/L (0.15–0.18), respectively [7]. Results were the same with fourth-instar larvae of *Aedes aegypti* [8]. Indeed, the 24-hour LD_{50} values for pipernonaline and pirimiphos-methyl were equal to 0.25 mg/L and 0.13 mg/L, respectively [8]. Pohlit et al. [9] showed that methanol extracts of leaves and roots of *P. aduncum* and leaves, fruits, branches, and stems of *P. tuberculatum* were significantly more active against third-instar larvae of *Ae. aegypti* than the methanol extracts of leaves of *P. amapense* and *P. capitarianum*, branches and stems of *P. dilatatum*, and leaves, branches, and stems of *P. hostmanianum*. When tested on the same species, the water extracts were always less toxic than the methanol extracts after 24 hours of exposure [9]. Water extracts of *P. retrofractum* were tested on late third- and early fourth-instar larvae of *Cx. quinquefasciatus* and *Ae. aegypti*. The LC_{50} and LC_{90} values after 48 hours of exposure were respectively equal to 135 mg/L and 1,079 mg/L for the former mosquito species and 79 mg/L and 229 mg/L for the latter [10]. Ethanolic extracts derived from *P. longum* (fruit), *P. ribesoides* (wood), and *P. sarmentosum* (whole plant) were evaluated for efficacy against early fourth-instar larvae of *Ae. aegypti* [11]. LD_{50} values of 2.23 mg/L (2.11–2.37), 8.13 mg/L (7.84–8.42), and 4.06 mg/L (3.68–4.43) were obtained with *P. longum*, *P. ribesoides*, and *P. sarmentosum*, respectively. The LD_{95} values were equal to 4.80 mg/L (4.10–6.18), 14.01 mg/L (12.87–15.78), and 12.06 mg/L (10.05–15.69), respectively [11]. These extracts were also tested on adults of *Stegomyia aegypti* [12]. LD_{50} values of 0.26 µg/mg female (0.23–0.28), 0.15 µg/mg female (0.13–0.17), and 0.14 µg/mg female (0.12–0.15) were obtained with *P. longum*, *P. ribesoides*, and *P. sarmentosum*, respectively. The LD_{95} values were equal to 0.71 µg/mg female (0.60–0.95), 0.83 µg/mg female (0.60–1.38), and 0.42 µg/mg female (0.34–0.58), respectively [12]. LC_{50} values of essential oils derived from leaves, stems, and inflorescences of *P. marginatum* against fourth-instar larvae of *Ae. aegypti* after 48 hours of exposure were equal to 23.8 ± 0.4 ppm, 19.9 ± 0.5 ppm, and 19.9 ± 0.5 ppm, respectively [13]. The essential oil from leaves and stems exhibited an oviposition deterrent effect at 50 and 100 ppm, in that significantly lower numbers of eggs (<50%) were laid in glass vessels containing the test solutions compared with the control solution [13]. LC_{50} (LC_{95}) values for third-instar larvae of *Ae. aegypti* exposed to methanol, ethyl acetate, hexane, and aqueous extracts of *P. retrofractum* for 24 hours were equal to 2.21 ppm (4.45 ppm), 2.96 ppm (5.61 ppm), 2.77 ppm (5.79 ppm), and 297.39 ppm (629.65 ppm), respectively. In the same way, LC_{50} (LC_{95}) values for third-instar larvae of *Cx. quinquefasciatus* were equal to 2.11 ppm (5.70 ppm), 2.47 ppm (6.30 ppm), 2.90 ppm (4.18 ppm), and 431.28 ppm (1,621 ppm), respectively [14].

If *Piper* extracts offer a valuable source of potential insecticides, they are also of interest for reducing the likelihood of resistance development when applied as synergists with pyrethroid insecticides such as pyrethrum [15]. The synergistic action of piperamides has a rather long history. Miyakado et al. [16] showed that binary mixtures of pipercide (I), dihydropipercide (II), or guineensine (III) were more toxic than one of the single chemicals tested alone against male adults of the Adzuki bean weevil (*Callosobruchus chinensis* L.). A mixture of the three chemicals in a 1: 1: 1 ratio demonstrated the highest synergistic action. The authors were also able to simulate the toxicity of a crude natural pipercide with a 75 (I): 5 (II): 20 (III) (w/w) mixture of these chemicals [16]. Piperine, 4,5-dihydropiperlonguminine, 4,5-dihydropiperine, and piperlonguminine isolated from *P. tuberculatum* were tested alone or in combination against the rock-hole breeding mosquito *Ae. atropalpus* (L) [17]. Binary mixtures including 4,5-dihydropiperlonguminine with one of the three other chemicals produced less than the expected 50% mortality. Binary mixtures with the other chemicals were equal in toxicity to the single piperamides tested alone. However, tertiary mixtures including 4,5-dihydropiperlonguminine were more toxic than any of the binary mixtures that did not contain this chemical. All equimolar tertiary mixtures and the quaternary mixture led to mortality greater than the 50% mortality produced by the piperamides tested alone. The 24-hour EC_{50} values of 4,5-dihydropiperlonguminine, 4,5-dihydropiperine, piperine, and piperlonguminine on fourth-instar larvae of *Ae. atropalpus* were equal to 6 μg/mL (3.7–12.7), 13.5 μg/mL (9–18.9), 11.1 μg/mL (7.7–16.9), and >45 μg/mL, respectively [17]. *Drosophila melanogaster* adults were treated with pyrethrum alone or pyrethrum supplemented with 0.1 mg/mL *P. nigrum* ethyl acetate extract [18]. A synergist ratio of 11.6 was obtained with the addition of 0.1 mg/mL *P. nigrum* to pyrethrum, leading to the conclusion that the *P. nigrum* extract might be as effective a synergist as piperonyl butoxide (PBO) [18].

The search for piperidine derivatives having a potential synergistic activity with pyrethroids can be done by means of QSAR modeling [19,20]. However, the sine qua non condition for deriving a QSAR model is to have a structurally diverse set of molecules of interest that have been tested under the same experimental conditions. To our knowledge, only the experiments made by Pridgeon et al. [21] on *Ae. aegypti* fulfil these conditions. Indeed, the acute toxicity of 33 piperidines was evaluated on adult mosquito females under the same laboratory conditions. As a result, the LD_{50} values obtained after 24 hours of exposure were used in conjunction with molecular descriptors to derive meaningful QSAR models.

11.2 MATERIALS AND METHODS

11.2.1 Biological Data

The 33 molecules tested on *Ae. aegypti* are listed in Table 11.1 and represented in Figure 11.1. Each piperidine derivative was serially diluted in acetone and topically applied to individual female mosquitoes. Before application of the compound, the 5- to 7-day-old females were anesthetized for 30 seconds with carbon dioxide and placed on a 4°C chill table [21]. A droplet of 0.5 μL of chemical solution was applied to the dorsal thorax of the mosquito. Six concentrations leading to 0%–100% mortality were used on 25–30 females per concentration. Tests were replicated three times. Controls with 0.5 μL acetone were validated only if they led to <10% mortality. After the treatments, mosquitoes were kept in plastic cups and supplied with 10% sucrose solution for 24 hours before mortality was recorded [21]. LD_{50} values were expressed in micrograms per mosquito (Table 11.1) and then transformed into log ($1/LD_{50}$, μmol), noted as pLD_{50}, for modeling purposes.

TABLE 11.1
Acute Toxicity in 24 Hours[a] of Piperidine Derivatives to *Ae. aegypti* Females after Topical Application

No.	Chemical	LD_{50} (95% C.I.[b])
1	1-(Cyclohexylacetyl)-2-methyl-piperidine	1.77 (1.46–2.27)
2	(2*R*)-1-Decanoyl-2-methyl-piperidine	2.74 (1.97–4.34)
3	1-Dodecanoyl-2-methyl-piperidine	8.76 (6.62–12.68)
4	(2*R*)-1-Heptanoyl-2-methyl-piperidine	1.20 (0.94–1.78)
5	1-(3-Cyclohexylpropanoyl)-2-methyl-piperidine	1.09 (0.83–1.68)
6	1-[(4-Methylcyclohexyl)carbonyl]-2-methyl-piperidine	1.13 (0.86–1.44)
7	(3*S*)-1-(1-Methylcyclohexyl)carbonyl-3-methyl-piperidine	4.14 (3.32–5.88)
8	(3*S*)-1-(3-Cyclohexylpropanoyl)-3-methyl-piperidine	1.92 (1.58–2.47)
9	(3*S*)-1-Heptanoyl-3-methyl-piperidine	2.07 (1.85–2.38)
10	(3*S*)-1-(Cyclohexylcarbonyl)-3-methyl-piperidine	1.80 (1.37–2.75)
11	1-Decanoyl-4-methyl-piperidine	4.90 (3.96–5.78)
12	1-(4-Cyclohexylbutanoyl)-4-methyl-piperidine	4.25 (3.11–6.00)
13	1-(Cyclohexylcarbonyl)-4-methyl-piperidine	2.63 (2.21–3.09)
14	1-(3-Cyclohexylpropanoyl)-4-methyl-piperidine	1.22 (0.94–1.93)
15	1-Dodecanoyl-4-methyl-piperidine	6.71 (5.16–9.14)
16	1-(Cyclohexylcarbonyl)-2-ethyl-piperidine	1.67 (1.44–2.08)
17	1-(3-cyclohexylpropanoyl)-2-ethyl-piperidine	0.94 (0.70–1.40)
18	1-Propionyl-2-ethyl-piperidine	1.56 (1.33–1.78)
19	1-(3-Cyclopentylpropanoyl)-2-ethyl-piperidine	1.83 (1.14–2.63)
20	1-Nonanoyl-2-ethyl-piperidine	0.84 (0.60–1.10)
21	1-Octanoyl-3-benzyl-piperidine	29.20 (19.82–49.09)
22	1-Undec-10-enoyl-4-benzyl-piperidine	14.72 (10.59–25.29)
23	1-Cyclohexylacetyl-4-benzyl-piperidine	19.22 (12.68–42.67)
24	1-(3-Cyclohexylpropanoyl)-4-benzyl-piperidine	12.89 (10.11–17.45)
25	2-Methyl-1-undec-10-enoyl-piperidine	1.38 (1.12–1.98)
26	2-Ethyl-1-undec-10-enoyl-piperidine	0.80 (0.65–0.97)
27	2-Benzyl-1-undec-10-enoyl-piperidine	3.59 (2.68–7.03)
28	3-Methyl-1-undec-10-enoyl-piperidine	2.07 (1.84–2.33)
29	3-Ethyl-1-undec-10-enoyl-piperidine	1.32 (0.97–1.69)
30	3-Benzyl-1-undec-10-enoyl-piperidine	7.43 (6.02–9.68)
31	4-Methyl-1-undec-10-enoyl-piperidine	2.72 (2.07–3.85)
32	4-Ethyl-1-undec-10-enoyl-piperidine	1.54 (1.18–2.71)
33	Piperine, (*E*,*E*)-1-piperoyl-piperidine	8.13 (6.10–12.99)

Source: J.W. Pridgeon et al., J. Med. Entomol. 44 (2007), pp. 263–269.

[a] Micrograms per mosquito.

[b] Confidence interval.

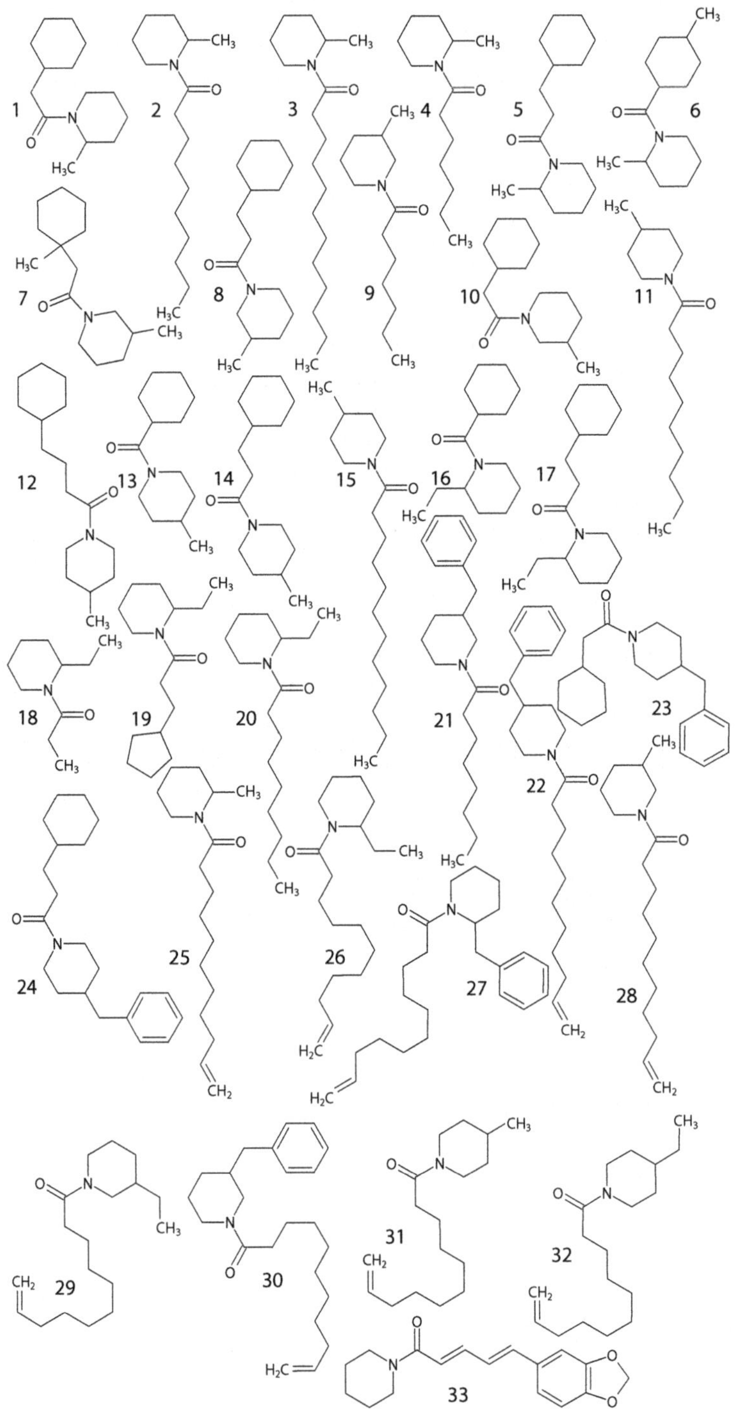

FIGURE 11.1 Structure of the 33 piperidines (see Table 11.1 for their chemical names).

11.2.2 Molecular Descriptors

All the 33 chemical structures listed in Figure 11.1 are *N*-acetyl piperidine derivatives with aliphatic and/or cyclic substituents on the carbonyl group as well as aliphatic and/or cyclic substituents in positions 2, 3, or 4 of the piperidine nucleus. The chemicals were geometry optimized with the AM1 semiempirical method in HyperChem (Hypercube, Gainesville, Florida) and exported to the CODESSA (comprehensive descriptors for structural and statistical analysis) Pro software (http://www.codessa-pro.com).

Substituents in position two or three lead to the presence of an asymmetric carbon atom (Figure 11.2). In Pridgeon et al. [21], when the configuration was given, the *R* enantiomer was tested for the 2-substituted compounds and the *S* enantiomer for the 3-substituted chemicals. Consequently, for the 2- or 3-substituted molecules that were tested as racemic (Table 11.1), the same choice—namely, 2*R* and 3*S*, was applied for modeling. Regarding the 4 position, because the nitrogen of the cycle is conjugated, it is almost planar and there is no geometric isomerism. Each substituent on the piperidine cycle can occupy an axial or equatorial position (Figure 11.3). The position of the piperidine cycle respective to the carbonyl group also has to be considered. This leads to two conformational series P1 P2, according to the cyclic bond eclipsing or opposite to the carbonyl. Thus, after the selection of the bond O=**C**–**N**–C in transoid position, the sign of the angle (O)C–**N**–**C**–C was examined. It was positive for the series P1 and negative for the series P2. So, for 2*R*-, 3*S*-, and 4-substituted compounds, four conformations were studied.

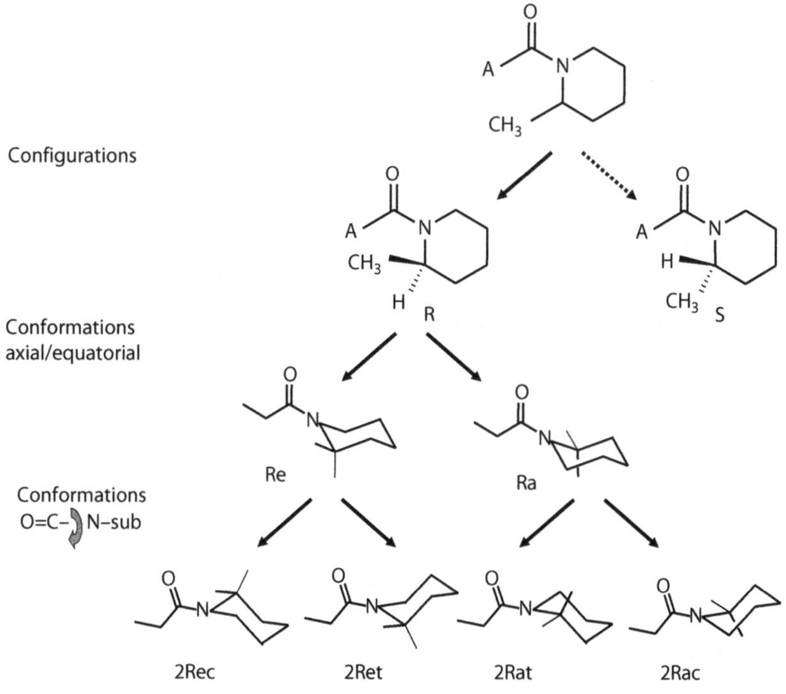

FIGURE 11.2 Configurations and conformations of the studied molecules.

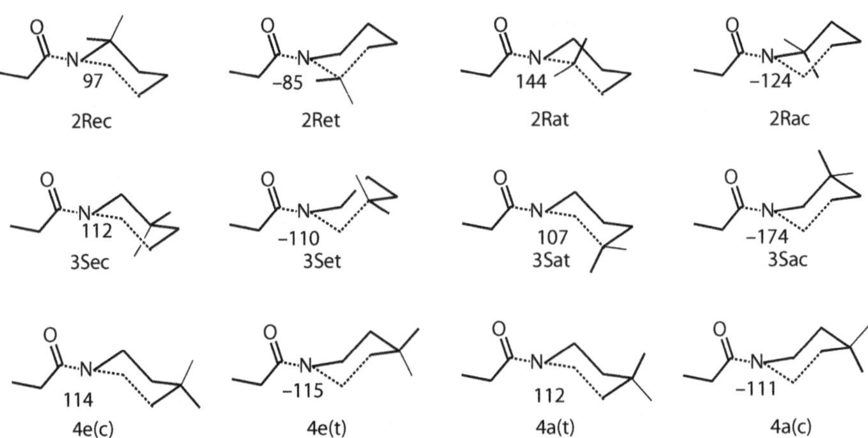

FIGURE 11.3 Different conformers for the substitutions in positions 2, 3, and 4 and extension to the cisoid and transoid positions based on the value of the angles.

It is easy to label the different conformers of a given configuration 2*R* and 3*S* according to the axial/equatorial and cisoid/transoid positions for the 2- or 3-substituents with respect to the carbonyl (e, a, c, and t mean equatorial, axial, cisoid, and transoid, respectively). But the cisoid/transoid terminology is meaningless with regard to the 4-substituents. Nevertheless, it was artifactually represented in Figure 11.3. It is possible to extend the cisoid and transoid terms since, in the series P1, 2*R* and 3*S* in axial position are transoid and the equatorial position is cisoid. For the series P2 it is the reverse.

The conformer with the highest stability was then kept. We showed that the substituent in position two of the piperidine cycle had to be in an axial position while those in three and four positions had to be in equatorial position. But the rotation of the piperidinic cycle around the (O)C–N bond leads to conformers of very close energy. This led to selecting the series 2Rat, 3Sec, 4e(c) for P1 and 2Rac, 3Set, 4e(t) for P2. Both final selected conformational series were used to calculate topological and physicochemical descriptors from the CODESSA Pro software. In the current study, only the physicochemical descriptors were considered in order to describe differently the molecules corresponding to the two series. Descriptors with poor variance were discarded. The 1-octanol/water partition coefficient (log *P*), calculated from AUTOLOGP [22], was also considered even if it uses canonical SMILES (simplified molecular input line entry specification) as input. Based on the hybridization of TLP [23] models, it is particularly suited for predicting with accuracy the log *P* values of structurally complex molecules as well as those with lengthy chains of carbon atoms, with polysubstitutions, and so on [24,25]. At the end, nine physicochemical descriptors were considered in the modeling process.

11.2.3 Statistical Analyses

The QSAR models were derived from a TLP that is a supervised artificial neuron network [23] including an input layer (descriptors), a hidden layer with an adjustable number of neurons for distributing the information, and an output layer with one neuron corresponding to the pLD_{50} data. Different algorithms can be used to train a TLP. In this study, the Broyden–Fletcher–Goldfarb–Shanno algorithm was selected because this second-order training algorithm guarantees a very fast convergence [23]. Different error functions and activation functions were tested. The data set was randomly split into training sets and test sets of 60% and 40%, 70% and 30%, and 80% and 20%, respectively. Due to the limited sizes of the different training sets, the number of hidden neurons was fixed to two.

All the calculations were performed with the data mining module of the Statistica™ software (version 10, StatSoft, France).

11.3 RESULTS AND DISCUSSION

The automatic feature selection process of Statistica™ software identified log *P*, MinRI (minimum 1-electron reactivity index for a C atom), and RNCS (relative negative charged SA (SAMNEG*RNCG) [Zefirov's PC]) as an interesting combination of descriptors for modeling the P1 series. MinRI belongs to the quantum chemical descriptors and encodes reactivity at the site of C atoms. RNCS belongs to the category of the electrostatic descriptors and it encodes the nucleophilic portion of the molecule's surface [26,27]. Obviously, log *P* encodes the hydrophobicity of the whole molecule.

Ten runs of 10,000 TLP models were computed with a training set including 60% of the molecules; the remaining compounds were used as external test set. Molecules of the P1 series were described by their log *P*, MinRI, and RNCS. Among the 10,000 TLPs of a run, the 100 models showing the best compromise between the prediction performances of the two sets were automatically saved. Only the models showing at least a correlation coefficient (*r*) equal to 0.90 on both the training and test sets were analyzed. Moreover, the *r* value of the test set had to be equal or superior to the *r* value of the training set. This second condition was applied to avoid problems of over-fitting with the TLP models. Among the 10 runs, only six configurations were found satisfying the preceding two conditions. It is worth noting that different error and activation functions were automatically tested during the modeling process. Ten runs of 10,000 TLPs performed from the molecules of the P2 series led to broadly the same results. Attempts were also made to find good configurations with log *P*, ZX shadow/ZX rectangle, and PNSA-1 partial negative surface area [Zefirov's PC], which were the descriptors selected by the automatic feature selection process of Statistica™ software for the P2 series. No improvement was obtained. The number of configurations with at least an *r* value equal to 0.90 on both the training and test sets remained very low. Conversely, when the trials of 10,000 TLPs were run with a training set and a test set including respectively 80% and 20% of the molecules, it was easier to have *r* values ≥ 0.90% on both sets for the molecules of the P1 series as well as those of P2 (with the two different triplets of descriptors). Use of the 70% versus 30% proportion had made it possible to transform rare events into low occurrences but with a significant frequency during the tens of runs of 10,000 TLPs, thus leading to enough interest for selecting this splitting for deriving the final models. Indeed, it is necessary to have a test sample with the largest possible size, especially with a TLP, to optimize the predictive performances of the models. Thus, the 33 molecules (Table 11.1, Figure 11.1) with their pLD_{50} values were randomly split into training sets of 24 molecules and external test sets of 9 molecules. Molecules of the two conformational series P1 and P2 were described by their log *P*, MinRI, and RNCS because these descriptors led to the best overall predictive performances. It is worth noting that the log *P* values were the same for both series, while those of MinRI and RNCS were very often different.

Tens of runs of 5,000 TLP models were computed on the two selected conformational series of molecules P1 and P2 with randomly selected training and test sets of 70% and 30%, respectively. All the models were 3/2/1 TLPs with an input bias and a hidden bias. Each run was automatically filtered in order to select the best 200 models in terms of performances on both sets. In a second step, only the models showing at least *r* values equal to 0.90 on both sets were kept. Moreover, the prediction performances on the test sets had to be equal or superior to those of the training sets. This led to keeping 95 models computed on the conformational series P1 and 99 models on P2. They were respectively termed models P1 and P2 for the sake of simplicity. Convergence of the TLPs was always obtained with less than 100 cycles. Analyses of the residuals, experimental pLD_{50} values minus those calculated from the models, are shown in Tables 11.2 and 11.3, respectively. In each table, the minimum and maximum absolute residual values obtained for each chemical used as training and test sets are given. Inspection of these tables shows that, very often, the chemicals are correctly predicted irrespective of their presence in the training set or test set. However, the presence of a lengthy alkane or alkene chain linked to the carbonyl group and a benzyl moiety in position two, three, or four of the piperidine nucleus can lead to large residuals, especially if the corresponding chemical belongs to the test sets. This is the case, for example, of chemical no. 22 (Figure 11.1), for which the highest absolute residual value equals 0.47 in P1 (Table 11.2) and 0.51 in P2 (Table 11.3). In both categories of models, this chemical presents the highest residual value when it is included in the test set. It is also the case for chemical no. 21 (Figure 11.1), which shows the lowest toxicity (Table 11.1). Absolute residual values of 0.44 and 0.42 were obtained at least once with P1 models (Table 11.2) and P2 models (Table 11.3), respectively. This chemical also presents a large residual value when it belongs to the training sets of the P2 models (Table 11.3). High absolute residual values can also be found when chemical no. 33 belongs to the external test sets of both categories of models (Tables 11.2 and 11.3). The largest minimum residual value was also found for this chemical in the P1 models (Table 11.2). This is not surprising because its structure is different from those of the other piperidines (Figure 11.1). Piperine was generally well predicted when the molecule was included in the training set. This is particularly true with the P2 models (Table 11.3). Despite this difference of structure, we claim that the presence of piperine in the models is allowed to increase their domain of prediction.

Attempts were made to choose a model in each selected conformational series and to hybridize them in order to create a unique model for optimizing the toxicity prediction of new molecules as the conformational requirements for rendering the piperidines active on female mosquitoes are unknown.

TABLE 11.2
Lowest and Highest Absolute Residual Values for 33 Chemicals Used as Training and Test Sets in 95 Selected P1 TLP Models

	Test Sets		Training Sets	
No.	Minimum	Maximum	Minimum	Maximum
1	0.05	0.16	0.02	0.17
2	0.16	0.24	0	0.29
3	0.02	0.32	0	0.26
4	0.03	0.20	0.01	0.20
5	0.12	0.26	0.08	0.20
6	0.02	0.17	0.03	0.17
7	0	0.31	0	0.16
8	0.07	0.12	0.01	0.18
9	0.07	0.25	0.03	0.23
10	0.04	0.16	0	0.12
11	0	0.38	0.03	0.18
12	0.01	0.31	0	0.22
13	0.11	0.17	0.05	0.34
14	0.14	0.16	0.02	0.15
15	0	0.28	0	0.18
16	0	0.13	0	0.10
17	0.16	0.19	0.12	0.32
18	0.02	0.21	0	0.18
19	0.03	0.15	0.02	0.16
20	0.20	0.25	0.17	0.36
21	0.09	0.44	0	0.36
22	0.01	0.47	0.09	0.33
23	0	0.26	0	0.17
24	0	0.24	0	0.24
25	0	0.14	0	0.12
26	0.03	0.40	0	0.33
27	0.21	0.41	0.15	0.51
28	0.03	0.20	0	0.18
29	0	0.21	0.01	0.12
30	0	0.25	0	0.33
31	0.14	0.31	0.14	0.36
32	0	0.05	0	0.12
33	0.26	0.45	0	0.37

TABLE 11.3
Lowest and Highest Absolute Residual Values for 33 Chemicals Used as Training and Test Sets in 99 Selected P2 TLP Models

	Test Sets		Training Sets	
No.	Minimum	Maximum	Minimum	Maximum
1	0	0.11	0	0.10
2	0.03	0.24	0	0.35
3	0.02	0.36	0.01	0.14
4	0.15	0.27	0.09	0.27
5	0.24	0.25	0	0.27
6	0.19	0.25	0.01	0.25
7	0.15	0.50	0.01	0.49
8	0.01	0.17	0	0.42
9	0.01	0.12	0	0.15
10	0.01	0.08	0	0.10
11	0.17	0.44	0.22	0.30
12	0.26	0.42	0.17	0.37
13	0.14	0.22	0.12	0.29
14	0.22	0.27	0.02	0.27
15	0	0.36	0	0.17
16	0	0.15	0	0.12
17	0.10	0.24	0.04	0.28
18	0.04	0.10	0	0.26
19	0	0.2	0	0.13
20	0.13	0.27	0.06	0.29
21	0	0.42	0.01	0.47
22	0.11	0.51	0.20	0.38
23	0.01	0.11	0	0.30
24	0.01	0.16	0	0.32
25	0.03	0.17	0	0.19
26	0.03	0.24	0	0.19
27	0.09	0.39	0	0.43
28	0	0.11	0	0.16
29	0.02	0.21	0.01	0.31
30	0	0.20	0.01	0.28
31	0.06	0.22	0.05	0.24
32	0	0.16	0	0.16
33	0.01	0.45	0	0.27

Among the P1 models, the selected 3/2/1 TLP showed *r* values equal to 0.91 and 0.96 on the training set and external test set, respectively. This test set included chemical nos. 1, 9, 11, 16, 19, 22, 25, 28, and 31. This random selection was preferred because it allowed us to account for the structural diversity of the molecules and their difference of toxicity (Table 11.1, Figure 11.1). With this model, the convergence was obtained after only 42 cycles. The sum of squares was used as error function. The hidden activation function was tanh, while the output activation function was identity. With the identity activation function used during the training phase of the neural network model, the activation level is passed on directly as the output. Residual values obtained with this model are displayed in Figure 11.4. The highest residual value was equal to −0.27 (no. 22) for the test set chemicals and 0.35 (no. 27) for the training set chemicals.

Among the P2 models, the selected 3/2/1 TLP showed *r* values equal to 0.91 and 0.94 on the training set and external test set, respectively. Obviously, the composition of both sets was the same as that for the selected P1 model. The convergence was obtained after only 59 cycles. The sum of squares was used as error function. The

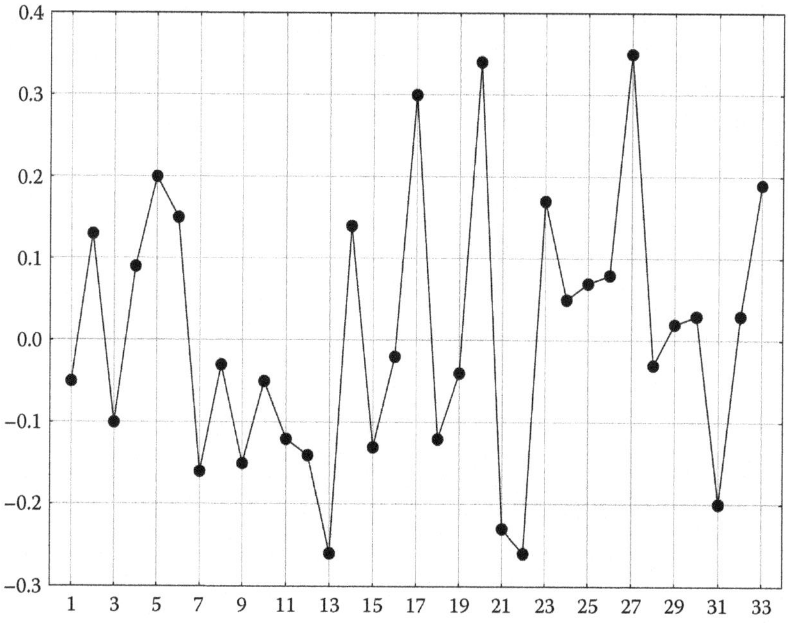

FIGURE 11.4 Residuals obtained with the selected P1 TLP model.

hidden and activation functions were both tanh. Residual values obtained with this model are displayed in Figure 11.5. The highest residual value was equal to –0.27 (no. 11) for the test set chemicals and –0.40 (no. 7) for the training set chemicals.

The P1 and P2 models were hybridized to form a general model with a correlation coefficient of 0.93 for the training set and 0.97 for the external test set. The calculated pLD_{50} values obtained with this P1P2 hybrid model are listed in Table 11.4 with the corresponding residual values. The scatter plot of the observed versus calculated pLD_{50} values is displayed in Figure 11.6. Inspection of Table 11.4 and Figure 11.6 shows that all the piperidines are well predicted except chemical no. 27. This is not surprising because this chemical was also poorly predicted by the individual models and, in both cases, the toxicity was underestimated.

From the analysis of our different modeling results as well as the structures and activity values of the molecules listed in Table 11.1, different chemical structures were selected. After their description by AUTOLOGP and the two selected CODESSA descriptors, their calculated pLD_{50} values were estimated from the selected P1P2 hybrid model. Molecules showing different structures and potential toxicity were searched for in the Sigma Aldrich database (https://www.sigmaaldrich.com/france.html), the Super Natural II database (http://bioinformatics.charite.de/supernatural) [28], the ZINC database (http://zinc.docking.org/) [29], and MolPort (https://www.molport.com/shop/index) to see whether they were available and, if so, whether it was possible to obtain them under acceptable conditions of purity and cost. Indeed, the goal of our study was not to design a molecule with the highest

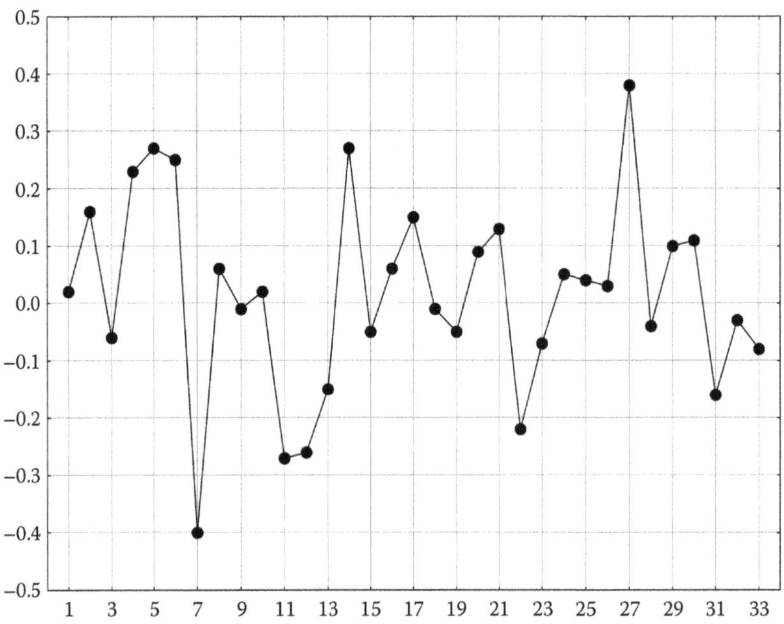

FIGURE 11.5 Residuals obtained with the selected P2 TLP model.

TABLE 11.4
Observed and Calculated pLD_{50} Values Obtained with Selected P1P2 Hybrid Model

No.	Obs.	Cal.	Res.
1[a]	2.10	2.12	−0.02
2	1.97	1.83	0.14
3	1.51	1.59	−0.08
4	2.25	2.09	0.16
5	2.34	2.11	0.23
6	2.30	2.10	0.20
7	1.76	2.04	−0.28
8	2.09	2.08	0.01
9[a]	2.01	2.09	−0.08
10	2.09	2.11	−0.02
11[a]	1.71	1.91	−0.20
12	1.77	1.97	−0.20
13	1.90	2.10	−0.20
14	2.29	2.09	0.20
15	1.62	1.71	−0.09
16[a]	2.13	2.11	0.02
17	2.43	2.20	0.23
18	2.04	2.10	−0.06
19[a]	2.11	2.16	−0.05
20	2.48	2.26	0.22
21	1.01	1.06	−0.05
22[a]	1.37	1.61	−0.24
23	1.19	1.14	0.05
24	1.39	1.34	0.05
25[a]	2.28	2.22	0.06
26	2.54	2.48	0.06
27	1.98	1.61	0.37
28[a]	2.11	2.15	−0.04
29	2.33	2.27	0.06
30	1.66	1.59	0.07
31[a]	1.99	2.17	−0.18
32	2.26	2.26	0
33	1.55	1.49	0.06

[a] Test set chemicals.

possible acute toxicity against *Ae. aegypti* females, but rather to identify a molecule to use in combination with pyrethroids, especially deltamethrin, against strains of *Ae. aegypti* resistant to this family of insecticides, which is widely used worldwide in vector control. A list of molecules of potential interest has been elaborated and will be further refined from other *in silico* approaches.

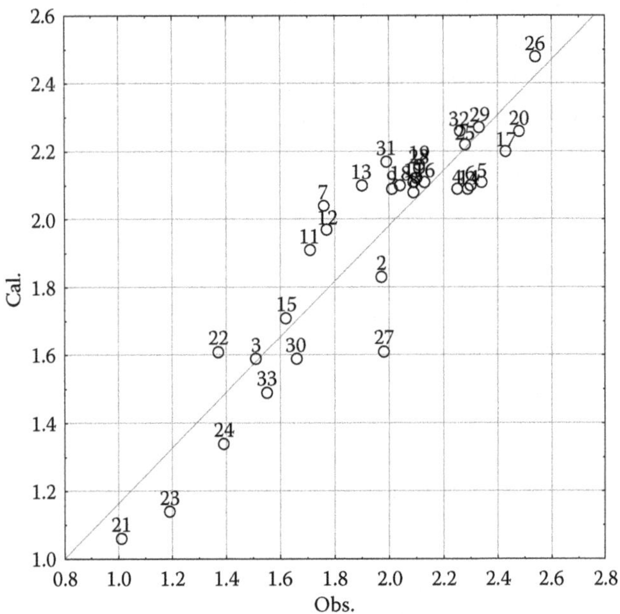

FIGURE 11.6 Observed versus calculated pLD_{50} values with the P1P2 TLP hybrid model.

11.4 CONCLUSION

Mosquitoes are not only nuisances because of their bites but these dipterans can also be the origin of numerous diseases considerably impacting human and animal populations worldwide. The repeated use of insecticides for mosquito control has fostered contamination of ecosystems, undesirable effects on aquatic and terrestrial nontarget organisms, and increasing resistance of mosquitoes to insecticides.

Discovering a way to reduce mosquito resistance to insecticides is as important as designing new molecules with new mechanisms of action—and maybe less time consuming and thus less costly—especially if *in silico* methods are integrated into the process.

In this context, an attempt was made to derive QSAR models allowing us to predict the adulticidal activity of piperidines against mosquitoes. A set of 33 chemicals with their acute toxicity on *Aedes aegypti* females was used. Energy optimization allowed us to consider two conformational series of structures that were described by CODESSA descriptors. The 1-octanol/water partition coefficient (log *P*) of the molecules was also computed. Each series was subject to modeling from a TLP. The models were derived from training sets including 70% of the molecules, which were randomly selected. The remaining molecules were used as external test sets. The TLP models included three physicochemical descriptors as input neurons and two hidden neurons. A consensus model was computed with an interesting configuration coming from each series. From the simulation results, various molecules were identified and searched for their availability in existing databases. Interesting candidates were retrieved for the potential evaluation of their interest as synergists.

ACKNOWLEDGMENT

This research was supported by the French Ministry of Ecology, Sustainable Development and Energy (DeltaSyn, #2101587412).

REFERENCES

[1] A.D. Kinghorn and M.F. Balandrin, *Human Medicinal Agents from Plants*, ACS Symposium Series 534, American Chemical Society, Washington, DC, 1993.

[2] C.W. Wright, *Plant derived antimalarial agents: New leads and challenges*, Phytochem. Rev. 4 (2005), pp. 55–61.

[3] S.O. Duke, C.L. Cantrell, K.M. Meepagala, D.E. Wedge, N. Tabanca, and K.K. Schrader, *Natural toxins for use in pest management*, Toxins 2 (2010), pp. 1943–1962.

[4] K. Sukumar, M.J. Perich, and L.R. Boobar, *Botanical derivatives in mosquito control: A review*, J. Am. Mosq. Cont. Assoc. 7 (1991), pp. 210–237.

[5] M.A. Quijano-Abril, R. Callejas-Posada, and D.R. Miranda-Esquivel, *Areas of endemism and distribution patterns for Neotropical* Piper *species (Piperaceae)*, J. Biogeogr. 33 (2006), pp. 1266–1278.

[6] M.A. Jaramillo and P.S. Manos, *Phylogeny and patterns of floral diversity in the genus* Piper *(Piperaceae)*, Am. J. Bot. 88 (2001), pp. 706–716.

[7] S.E. Lee, *Mosquito larvicidal activity of pipernonaline, a piperidine alkaloid derived from long pepper*, Piper longum, J. Am. Mosq. Cont. Assoc. 16 (2000), pp. 245–247.

[8] Y.C. Yang, S.G. Lee, H.K. Lee, M.K. Kim, S.H. Lee, and H.S. Lee, *A piperidine amide extracted from* Piper longum L. *fruit shows activity against* Aedes aegypti *mosquito larvae*, J. Agric. Food Chem. 50 (2002), pp. 3765–3767.

[9] A.M. Pohlit, E.L.J. Quignard, S.M. Nunomura, W.P. Tadei, A. de Freitas Hidalgo, A.C. da Silva Pinto, E.V.M. dos Santos, S.K.R. de Morais, R. De Cássia Guedes Saraiva, L.C. Ming, A.M. Alecrim, A. de Barros Ferraz, A.C. da Silva Pedroso, E. la Vieira Diniz, E.K. Finney, E. de Oliveira Gomes, H.B. Dias, K. dos Santos de Souza, L.C. Pereira de Oliveira, L. de Castro Don, M.M.A. Queiroz, M.C. Henrique, M. dos Santos, O. da Silva Lacerda Júnior, P. de Souza Pinto, S.G. Silva, and Y.R. Graça, *Screening of plants found in the State of Amazonas, Brazil for larvicidal activity against* Aedes aegypti *larvae*, Acta Amazon. 34 (2004), pp. 97–105.

[10] U. Chansang, N.S. Zahiri, J. Bansiddhi, T. Boonruad, P. Thongsrirak, J. Mingmuang, N. Benjapong, and M.S. Mulla, *Mosquito larvicidal activity of aqueous extracts of long pepper* (Piper retrofractum *Vahl) from Thailand*, J. Vect. Ecol. 30 (2005), pp. 195–200.

[11] U. Chaithong, W. Choochote, K. Kamsuk, A. Jitpakdi, P. Tippawangkosol, D. Chaiyasit, D. Champakaew, B. Tuetun, and B. Pitasawat, *Larvicidal effect of pepper plants on* Aedes aegypti *(L.) (Diptera: Culicidae)*, J. Vect. Ecol. 31 (2006), pp. 138–144.

[12] W. Choochote, U. Chaithong, K. Kamsuk, E. Rattanachanpichai, A. Jitpakdi, P. Tippawangkosol, D. Chaiyasit, D. Champakaew, B. Tuetun, and B. Pitasawat, *Adulticidal activity against* Stegomyia aegypti *(Diptera: Culicidae) of three* Piper *spp.*, Rev. Inst. Med. trop. S. Paulo 48 (2006), pp. 33–37.

[13] E.S. Autran, I.A. Neves, C.S.B. da Silva, G.K.N. Santos, C.A.G. da Câmara, and D.M.A.F. Navarro, *Chemical composition, oviposition deterrent and larvicidal activities against* Aedes aegypti *of essential oils from* Piper marginatum *Jacq. (Piperaceae)*, Biores. Technol. 100 (2009), pp. 2284–2288.

[14] T. Subsuebwong, S. Attrapadung, R. Potiwat, R. Srisawat, and N. Komalamisra, *Insecticidal Activities of* Piper retrofractum *Extracts against* Aedes aegypti *and* Culex quinquefasciatus *(Diptera: Culicidae)*, The National and International Graduate Research Conference 2016, Universitas Muhammadiyah Yogyakarta, Indonesia, NIGRC 2016.

[15] I.M. Scott, H.R. Jensen, B.J.R. Philogène, and J.T. Arnason, *A review of* Piper *spp. (Piperaceae) phytochemistry, insecticidal activity and mode of action*, Phytochem. Rev. 7 (2008), pp. 65–75.

[16] M. Miyakado, I. Nakayama, and H. Yoshioka, *Insecticidal joint action of pipercide and co-occurring compounds isolated from* Piper nigrum *L*, Agric. Biol. Chem. 44 (1980), pp. 1701–1703.

[17] I.M. Scott, E. Puniani, T. Durst, D. Phelps, S. Merali, R.A. Assabgui, P. Sanchez-Vindas, L. Poveda, B.J.R. Philogène, and J.T. Arnason, *Insecticidal activity of* Piper tuberculatum *Jacq. extracts: Synergistic interaction of piperamides*, Agri. For. Entomol. 4 (2002), pp. 137–144.

[18] H.R. Jensen, I.M. Scott, S.R. Sims, V.L. Trudeau, and J.T. Arnason, *The effect of a synergistic concentration of a* Piper nigrum *extract used in conjunction with pyrethrum upon gene expression in* Drosophila melanogaster, Insect Molec. Biol. 15 (2006), pp. 329–339.

[19] W. Karcher and J. Devillers, *Practical Applications of Quantitative Structure–Activity Relationships (QSAR) in Environmental Chemistry and Toxicology*, Kluwer Academic Publishers, Dordrecht, the Netherlands, 1990.

[20] J. Devillers and W. Karcher, *Applied Multivariate Analysis in SAR and Environmental Studies*, Kluwer Academic Publishers, Dordrecht, the Netherlands, 1991.

[21] J.W. Pridgeon, K.M. Meepagala, J.J. Becnel, G.G. Clark, R.M. Pereira, and K.J. Linthicum, *Structure–activity relationships of 33 piperidines as toxicants against female adults of* Aedes aegypti *(Diptera: Culicidae)*, J. Med. Entomol. 44 (2007), pp. 263–269.

[22] J. Devillers, *AUTOLOGP™: A computer tool for simulating n-octanol-water partition coefficients*, Analusis 27 (1999), pp. 23–29.

[23] J. Devillers, *Neural Networks in QSAR and Drug Design*, Academic Press, London, 1996.

[24] J. Devillers, D. Domine, C. Guillon, and W. Karcher, *Simulating lipophilicity of organic molecules with a back-propagation neural network*, J. Pharm. Sci. 87 (1998), pp. 1086–1090.

[25] J. Devillers, *Calculation of octanol/water partition coefficients for pesticides. A comparative study*, SAR QSAR Environ. Res. 10 (1999), pp. 249–262.

[26] A.R. Katritzky, *CODESSA PRO User's Manual*, University of Florida, Gainesville, 2005.

[27] K.M. Yerramsetty, B.J. Neely, S.V. Madihally, and K.A.M. Gasem, *A skin permeability model of insulin in the presence of chemical penetration enhancer*, Int. J. Pharm. 388 (2010), pp. 13–23.

[28] M. Dunkel, M. Fullbeck, S. Neumann, and R. Preissner, *SuperNatural: A searchable database of available natural compounds*, Nucleic Acids Res. 34 (2006), pp. D678–D683.

[29] J.J. Irwin and B.K. Shoichet, *ZINC—A free database of commercially available compounds for virtual screening*, J. Chem. Inf. Model. 45 (2005), pp. 177–182.

12 Molecular Modeling Studies of Inhibition by Natural Compounds of a Mosquito Detoxification System

Implication in Mosquito Vector Control

Pornpimol Rongnoparut, Aruna Prasopthum, and Phisit Pouyfung

CONTENTS

ABSTRACT

Detoxification enzymes play a role in defense systems against insecticides in insects including mosquitoes. The heme-containing cytochrome P450 monooxygenases (P450s), one of the major classes of insecticide detoxification enzymes, have been recognized as playing a key role in pyrethroid resistance in mosquitoes through their overexpression. Thus, inhibition of P450-mediated defense mechanisms may overcome or restrict insecticide resistance. Inhibitory compounds of P450 enzymes, in particular from plant sources, could serve as insecticide synergists to increase efficacy of insecticides and/or preserve the existent effective synthetic insecticides. Two mosquito P450s, CYP6AA3 and CYP6P7, were isolated from the deltamethrin-resistant *Anopheles minimus* malaria vector and have shown capability in pyrethroid metabolisms. Homology models of CYP6AA3 and CYP6P7 have been built and inhibition of both P450s by compounds of different chemical structures and compounds isolated from selected plant extracts have been reported. This chapter describes molecular docking to CYP6AA3 and CYP6P7 models and explores binding modes of enzyme models with pyrethroid and fluorescent substrates, as well as different inhibitory compound groups isolated from plant sources. We demonstrate that the amino acids within both enzyme pockets interact with various substrates and inhibitors. Structure–activity relationships of different chemical compound groups with inhibition effects and modes can be delineated and explained by molecular docking analysis. Insights into molecular mechanisms of interactions of natural compounds with both mosquito P450s can be valuable for the rational design of pyrethroid insecticide synergists and for insecticide resistance management control of malaria vectors.

KEYWORDS

- Molecular modeling
- Cytochrome P450 monooxygenases
- Detoxification
- Phytochemicals
- Inhibitors
- Mosquitoes

12.1 INTRODUCTION

Mosquitoes are the vectors of important human pathogens causing mosquito-borne diseases, including malaria, dengue, encephalitis, and yellow fever. Malaria transmitted by anopheline mosquitoes remains widespread in the tropical and subtropical regions of the world. In 2012, an estimated 3.4 billion people were at risk of malaria; of these, 1.2 billion were at high risk (greater than one case per 10,000 population) and nearly 627,000 people died [1]. Among people at high risk of malaria, approximately 37% were living in Southeast Asia and 47% in Africa [1]. In the absence of malarial vaccine or effective drugs for malaria treatment, as well as problems of drug affordability, vector control using insecticides, in particular pyrethroids, has been the primary means of reducing malaria [2]. Pyrethroids are the insecticide class recommended by the World Health Organization (WHO) to be used for the impregnation of insecticidal mosquito nets and indoor residual spraying, because they are safe, cheap, effective, and long lasting, and their use successfully has resulted in significant reduction of malaria mortality and morbidity [1]. However, due to long-term and large-scale pyrethroid use, resistance of malaria mosquito populations to pyrethroids is on the rise and has hindered control efforts of malaria vectors [3–5]. In addition, widespread applications of the synthetic pyrethroid insecticides can lead to environmental contamination and adverse effects on nontarget species, including humans. Thus, effective resistance management strategies are critical and required to delay insecticide resistance in mosquito vectors, as well as to preserve effectiveness of the existing insecticides, especially pyrethroids, or to reverse insecticide resistance.

The target site of pyrethroid insecticides involves the voltage-gated sodium channel on insect neurons [6]. Through insecticide binding, pyrethroids delay the closing of the sodium channel and prolong the action potential, leading to insect paralysis and death [6]. Three major mechanisms have been recorded responsible for insecticide resistance in insects. These are (1) target-site insensitivity, caused by mutations of the proteins targeted by the insecticide and consequently reduced binding of insecticide; (2) reduced cuticular penetration, leading to reduced uptake of insecticides; and (3) metabolic resistance, caused by increased insecticide degradation activities by detoxification enzymes and thus lowering the amounts of insecticides that reach the target site [7–9]. In mosquitoes, target-site insensitivity and metabolic resistance are mainly detected in insecticide-resistance populations [7,10]. Target-site insensitivity in malaria mosquitoes conferring knockdown resistance (kdr) is caused by mutations of amino acids in the voltage-gated sodium channel, resulting in the ability of mosquitoes to withstand prolonged exposure to insecticides without being knocked down [10]. Metabolic resistance, on the other hand, involves alteration in enzymatic activities of detoxification enzymes including esterases, glutathione *S*-transferases, and cytochrome P450 monooxygenases (P450s), resulting in the increased metabolisms of insecticides [8,11]. Of the three sets of enzymes, P450s are primarily associated with resistance of insects to various insecticides, especially pyrethroids [11,12]. Increased P450 activity, as a result of overproduction of P450s, was noted in insecticide-resistant insects including malaria mosquito vectors; in susceptible insects, these P450s were generally expressed at basal level [7,8,11]. Thus, P450s have been a target for development of insecticide synergists, via inhibition of insecticide detoxifications in target-resistant insects in order to overcome or restrict insecticide resistance. In general insecticide synergists are nontoxic compounds to be combined with insecticides to increase the efficacy of insecticides [13]. Insecticide synergistic compounds inhibiting P450 detoxification systems and resulting in enhanced insecticide metabolisms in resistant insects have been successfully employed to prevent development of resistant races of insects, including mosquitoes. Piperonyl butoxide (PBO), for example, which is a P450 inhibitor, has been applied together with pyrethroids to increase pyrethroid efficiency in pyrethroid-resistant insects such as *Anopheles arabiensis* and *Culex tritaeniorhynchus* [14,15]. An alternative strategy could be the use of naturally occurring phytochemicals as synergists, as their applications might reduce chemical insecticide uses and they might be environmentally safer than their synthetic counterparts and readily degraded.

We previously isolated two cytochrome P450s, CYP6AA3 and CYP6P7, from the deltamethrin-resistant *Anopheles minimus* mosquito and demonstrated their capability in pyrethroid metabolisms [16,17]. Molecular models of both enzymes were built and docking with pyrethroids of both enzyme models supported their roles in pyrethroid metabolisms [18]. Inhibition of both P450s was examined with several compounds of different chemical structures and compounds isolated from selected plant extracts [17,19–21]. In this chapter, we utilized molecular docking to CYP6AA3 and CYP6P7 models to investigate and delineate inhibition effects and modes of different inhibitory compound groups isolated from plant sources that expressed competitive inhibition and mixed type between competitive and noncompetitive inhibitions. Moreover, to gain insight into inhibitory natures of these inhibitory compounds, the amino acids within both enzyme pockets that interacted with inhibitors were determined. Knowledge of interactions of amino acids with inhibitors within mosquito P450 active sites might serve as a driver for the design of pyrethroid insecticide synergists and be fruitful for a resistance management program of malaria mosquito vectors.

12.2 MOSQUITO CYTOCHROME P450 MONOOXYGENASES

The P450s are heme-containing monooxygenases that catalyze the metabolisms of various endogenous compounds, such as pheromones, fatty acids, hormones, and exogenous compounds such as insecticides and plant toxins [12]. P450s constitute a superfamily of enzymes, each of which possesses a different substrate preference, allowing the P450 system to metabolize many types of substrate compounds. P450s have been recognized as playing a key role in pyrethroid resistance in various mosquitoes, as they were overexpressed in various pyrethroid-resistant mosquito species [22–27]. These P450s have accordingly shown ability to metabolize pyrethroid insecticides. For instance, in malaria vectors in Africa, the *An. gambiae* CYP6P3 and CYP6M2 and the *An. funestus* CYP6P9 can metabolize permethrin and deltamethrin pyrethroid insecticides [28–30]. In *Cx. quinquefasciatus*, the CYP9M10 overexpression and its ability to metabolize permethrin and deltamethrin have been elucidated in the pyrethroid-resistant mosquitoes [31]. Investigation into pyrethroid-resistant *Aedes aegypti* has also found increased expression and the ability to metabolize pyrethroids of CYP9J32 and CYP9M6 [32,33]. In *An. minimus*, a major malaria vector in Thailand, CYP6AA3 and CYP6P7 have been found up-regulated in laboratory-selected pyrethroid-resistant mosquitoes and possess activities toward pyrethroid degradation [16,17,27].

12.3 INHIBITION OF INSECT CYTOCHROME P450S

Studies of inhibition of P450-catalyzed insecticide metabolisms by various compound groups have been conducted in insects, including the house fly *Musca domestica,* and *An. gambiae* and *An. minimus* mosquitoes, but mechanisms of their inhibitions have been investigated infrequently [17,34–36]. The mechanisms of P450 inhibitions can be divided in general into two categories: reversible and irreversible inhibitions. Reversible inhibition involves reversible association and dissociation of inhibitory compounds with enzymes, and it is divided into competitive, noncompetitive, uncompetitive, and mixed-type between competitive and noncompetitive inhibitions [37]. Reversible inhibitions of P450 enzymes comply with the standard characteristics of enzyme inhibitions, of which competitive inhibitors compete with substrate for the same target site of the enzyme. In noncompetitive inhibition, the inhibitor binds the enzyme at a different site from the substrate and can bind to both free enzyme and the enzyme–substrate complex to prevent substrate binding or complex formation. In uncompetitive inhibition, the inhibitor binds to the enzyme–substrate complex to prevent product formation.

Irreversible inhibition of P450s is referred to as mechanism-based inhibition (MBI) [38]. MBI is a process that requires a P450 catalysis cycle to bioactivate the compound into a reactive intermediate inhibitor; the resulting inhibitor further inactivates the enzyme. The process requires time and NADPH to generate the inhibitory molecule; inhibition becomes more prominent with time of exposure and concentration of inhibitor [38]. MBI thus exhibits NADPH-, concentration-, and time-dependence characteristics [38]. For development of insecticide synergists that targets detoxifying P450s, the irreversible inhibitory mechanism is essential because the inactivated enzyme needs to be replaced by newly synthesized enzymes *in vivo* in order to recover enzymatic activity [38]. Several functional groups—for instance, the methylenedioxyphenyl group, epoxides, and alkenes—have been shown susceptible to MBI mediated by mammalian P450s [39]. Flavonoids are another functional group that displays high inhibition potencies against various P450s, in both humans and insects. In a study of function and structure relationship of inhibition of human P450s by flavonoid derivatives, most flavonoids displayed effective inhibition effects on a few human P450s, with the IC_{50} values within micromolar range [40]. Synthetic flavonoids and methylenedioxyphenyls have shown inhibition with high efficiencies on insect P450s, including *M. domestica* and *An. gambiae* [34–36]; however, their kinetics and mechanisms of inhibition have not been explored.

12.4 *AN. MINIMUS* CYP6AA3 AND CYP6P7 ENZYMES AND PYRETHROID METABOLISMS

An. minimus is one of the malaria vectors in Southeast Asia including Laos, Cambodia, Vietnam, and Thailand. Two P450s, CYP6AA3 and CYP6P7, were found overexpressed and were isolated from a laboratory-selected deltamethrin-resistant *An. minimus* mosquito [26,27]. The increase of expression of both P450s coincided with mosquitoes' increased resistance to deltamethrin, implicating their role in deltamethrin resistance in the *An. minimus* mosquito [27]. Through heterologous expression of CYP6AA3 and CYP6P7 in *Spodoptera frugiperda* (Sf9) insect cells via a baculovirus-mediated expression system, both microsomal enzymes were observed to be able to metabolize pyrethroids, but with different metabolic profiles against different pyrethroids [17]. Figures 12.1 and 12.2 show chemical structures of substrates and inhibitors investigated in this chapter, with structures of pyrethroids

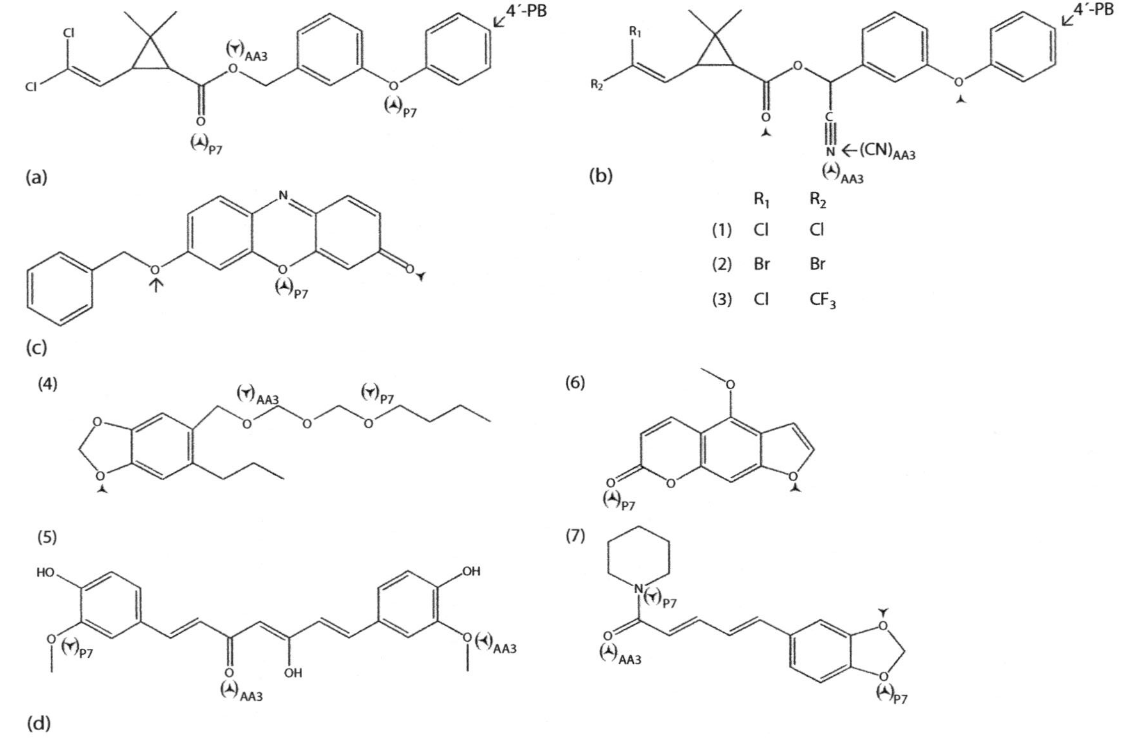

FIGURE 12.1 Chemical structures of permethrin type I pyrethroid (a), type II pyrethroids (b), benzyloxyresorufin (c) and test inhibitors (d). Type II pyrethroids include cypermethrin (1), deltamethrin (2), and λ-cyhalothrin (3). Inhibitors are PBO (4), curcumin (5), bergapten (6), and piperine (7). Sites of substrates attacked by CYP6AA3 and CYP6P7 enzymes are indicated by arrows. Sites of compounds interacted with amino acids of both enzymes are indicated by triangles, and those bound differently by each enzyme are shown as triangles specified with the name of the enzyme.

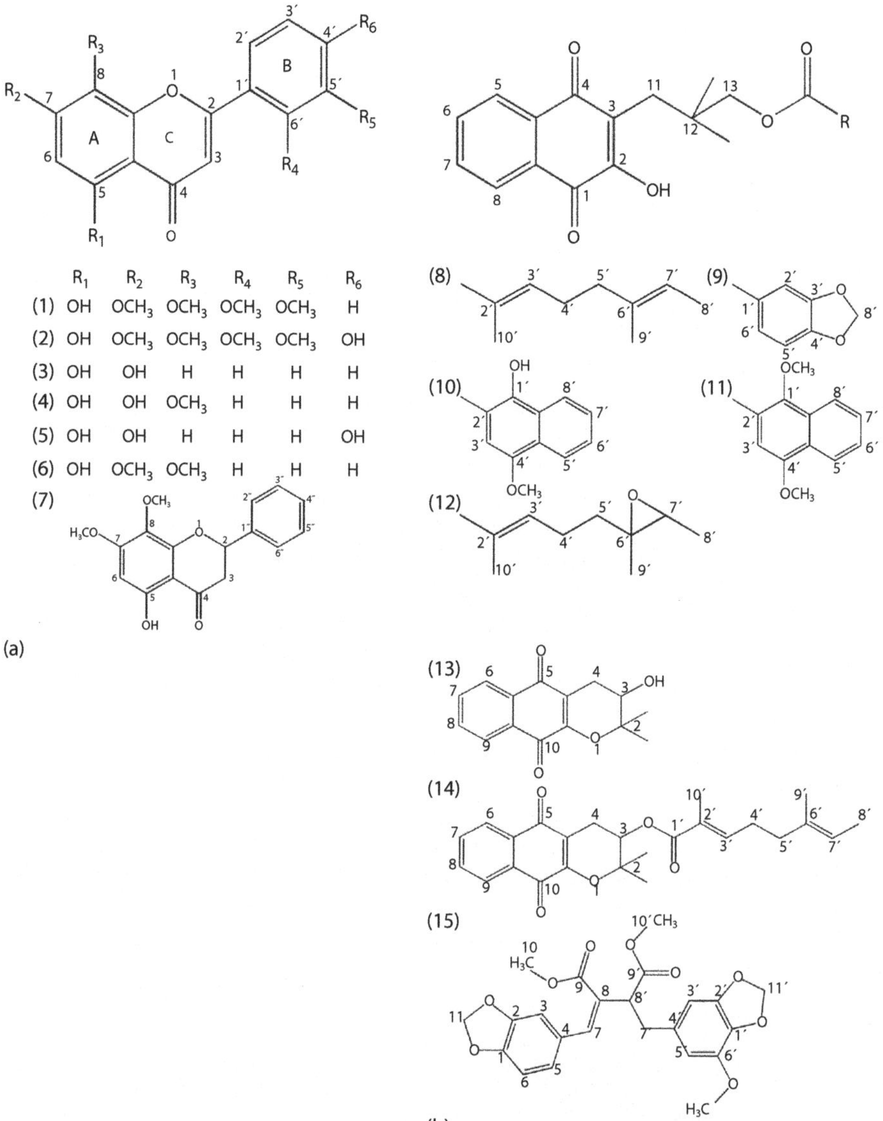

	R_1	R_2	R_3	R_4	R_5	R_6
(1)	OH	OCH_3	OCH_3	OCH_3	OCH_3	H
(2)	OH	OCH_3	OCH_3	OCH_3	OCH_3	OH
(3)	OH	OH	H	H	H	H
(4)	OH	OH	OCH_3	H	H	H
(5)	OH	OH	H	H	H	OH
(6)	OH	OCH_3	OCH_3	H	H	H

(7)

(a)

(8) (9) (10) (11) (12) (13) (14) (15)

(b)

FIGURE 12.2 Chemical structures of flavonoids and rhinacanthins. Flavonoids with flavone core (a) employed in molecular docking are 5-hydroxy-7,8,2′,3′-tetramethoxy flavone (1), 5,4′-dihydroxy-7,8,2′,3′-tetramethoxyflavone (2), 5,7-dihydroxyflavone (chrysin, 3), 5,7-dihydroxy-8-methoxyflavone (4), 5,7,4′-trihydroxyflavone (apigenin, 5), and 5-hydroxy-7,8-dimethoxyflavone (6). A flavanone 5-hydroxy-7,8-dimethoxyflavanone (7) is shown. (b) Rhinacanthins comprising naphthoquinone core structure are rhinacanthin-C (8), -D (9), -N (10), -Q (11), -G (12). Other rhinacanthins include rhinacanthin-A (13), -B (14), -E (15).

and inhibitory compounds primarily investigated shown in Figure 12.1, and various inhibitory compounds isolated from plant extracts displayed in Figure 12.2. The common feature of pyrethroids comprises a cyclopropane acid moiety, a central ester bond, and a phenoxybenzyl moiety (Figure 12.1). There are generally two types of pyrethroids: type I pyrethroids (i.e, permethrin) that do not contain a cyano group and type II pyrethoids (i.e., cypermethrin, deltamethrin, and λ-cyhalothrin) that comprise the cyano group. Both CYP6AA3 and CYP6P7 were reported to be able to metabolize permethrin, cypermethrin, and deltamethrin. However there was a difference in the ability of CYP6AA3 and CYP6P7 in the metabolism of λ-cyhalothrin, as CYP6AA3 could metabolize λ-cyhalothrin, but CYP6P7 had limited ability to metabolize it [17].

The *in silico* homology modeling of both CYP6AA3 and CYP6P7 has been successfully built in the absence of insect P450 crystal structures, and the models were used to gain insight into the molecular basis of their pyrethroid metabolisms [18]. Both models were generated based on the known human P450 crystal structures, including CYP2C8, CYP2C9, and CYP3A4, capable of pyrethroid metabolisms [41], as templates using the multiple amino acid sequence alignment method [18]. CYP6AA3 and CYP6P7 homology models comprised conserved P450 folds and cysteine-pocket attaching heme; however, they were different attributing to their substrate recognition sites (SRSs) that spanned access channels and active sites of both enzymes (Figure 12.3). As a result, both models are different in the geometry of their active-site cavities, with CYP6AA3 having a large active site and CYP6P7 comprising a smaller size and a narrow channel opening to its active site (Figure 12.3 [18]). Moreover there were six depicted SRSs for both enzyme models as previously described [42], with SRSs 1, 4, 5, and 6 predicted to participate in formation of the catalytic site and SRSs 2 and 3 in formation of a substrate access channel (Figure 12.3). Most residues of different SRSs in both CYP6AA3 and CYP6P7 cavities that were predicted to interact with pyrethroid substrates were nonpolar, consistent with the enzymes' ability to metabolize both types of I and II pyrethroids better than those more polar nonsubstrate insecticides such as chlorpyrifos and propoxur

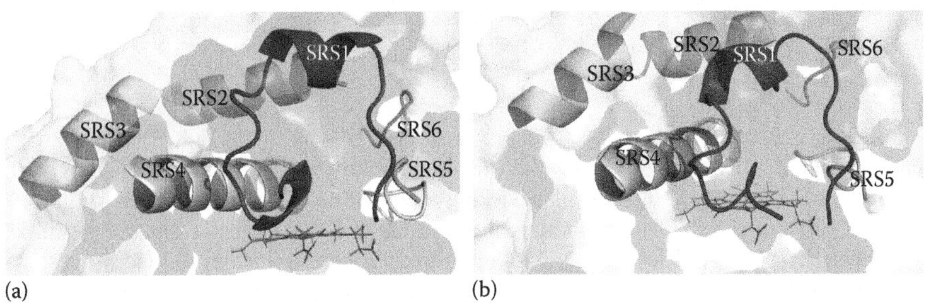

FIGURE 12.3 **(See color insert.)** Active sites and predicted substrate recognition sites (SRSs) designated 1-6 of CYP6AA3 (a) and CYP6P7 (b) homology models (P. Lertkiatmongkol, E. Jenwitheesuk, and P. Rongnoparut, BMC Res. Notes 4 (2011), pp. 321). SRSs are depicted in cartoon. The heme group is shown at the lower part of the structure.

[17–18]. Docking of pyrethroids including permethrin, cypermethrin, and deltamethrin to the active sites of CYP6AA3 and CYP6P7 models revealed two major sites of pyrethroids, which were the alpha carbon at the cyano group of type II pyrethroids and the 4′-phenoxybenzyl carbon of both type I and II pyrethroids, which could be attacked by CYP6AA3 to generate metabolites (Figure 12.1, reference 18). In contrast, only the 4′-phenoxybenzyl carbon of pyrethroids was found as the major attacking site by CYP6P7 (Figure 12.1, reference 18). These were consistent with the metabolites formed from CYP6AA3- and CYP6P7-mediated metabolisms of pyrethroids [16]. Moreover, it appeared that the ability of CYP6AA3 to metabolize λ-cyhalothrin (type II pyrethroid) in contrast to the inability of CYP6P7 to metabolize λ-cyhalothrin accounted for the topology differences of both enzymes [18]. Thus, the *in silico* homology models have the potential to enhance the understanding of pyrethroid metabolisms mediated by the mosquito P450s. To better understand the nature of interaction of pyrethroid substrates with both CYP6AA3 and CYP6P7 enzymes, we performed molecular docking to further determine amino acid residues that were predicted to interact with pyrethroids. In addition, amino acid residues of both enzyme models that were important in interaction and might help orient the inhibitory compounds in the enzyme active sites were determined utilizing molecular docking.

12.4.1 Molecular Docking Methods

The CYP6AA3 and CYP6P7 homology models were generated using the multiple template alignment method based on the human ligand-free CYP3A4 (PDB: 1TQN), CYP2C8 (1PQ2), and CYP2C9 (1OG2) [18]. Substrate-binding state of the P450s was generated by assigning a high-spin state of five-coordinated ferrous heme complex using a restrained electrostatic potential atomic charge method [43]. Three-dimensional structures of all of the substrates and inhibitors were generated using Chem3D Ultra 8.0 (CambridgeSoft Corporation, MA). Partial charges and polar hydrogen atoms of protein models and ligands were generated and random rotational torsions of ligands were set using the Autodock tool. A docking box of 40Å X 40Å X 40Å covering SRSs of each P450 enzyme was defined and centered in point (x, y, z) = (–15.89, –23.06, –11.44) for CYP6AA3 and (–15.63, –22.94, –11.39) for CYP6P7. All prepared ligand compounds were docked within the active sites of both enzyme models using Autodock Vina and the Lamarckian genetic algorithm [44]. In general, docking parameters were kept to their default values except that the exhaustiveness value was set as 100. The docking runs were carried out to generate a maximum of twenty possible conformations for each ligand with a maximum energy difference of 10 kcal/mol. Twenty conformations per ligand were obtained and ranked according to the sum of the ligand's internal energy [$\Delta G_{binding}$, kcal/mol; equation follows]. The best binding modes for each ligand were selected based on low $\Delta G_{binding}$ values with closest interactions to heme iron, together with those displaying the maximum number of contact residues within 1.0 scaling factor of van der Waals radii visualized on PyMoL 1.20 (Delano Scientific, LLC, CA). The critical amino acid residues of the models possibly interacted with docked ligands as hydrogen bonds (maximum distance = 2.9 Å) and π–π interactions (maximum distance = 4.0 Å) were determined as previously described [45]. The free binding energy ΔG of the enzyme–ligand complex is given by

$$\Delta G_{binding} = \Delta G_{vdW} + \Delta G_{elec} + \Delta G_{hbond} + \Delta G_{desolv} + \Delta G_{tors}$$

where

ΔG_{vdW} = Lennard–Jones potential
ΔG_{elec} = coulombic with Solmajer dielectric
ΔG_{hbond} = 12–10 H-bonding potential with Goodford directionality
ΔG_{desolv} = Stouten pairwise atomic solvation parameters
ΔG_{desolv} = number of rotational bonds

12.4.2 Amino Acid Residues within Enzyme Active Sites Involved in Binding to Pyrethroids

The predicted amino acid residues of CYP6AA3 and CYP6P7 surrounding the cypermethrin substrate as an example of pyrethroid substrate are displayed in Figures 12.4(a) and 12.4(b). Residues identified for CYP6AA3 were (Figure 12.4a and Table 12.1) the following:

Asn111, Glu112, Asp115, Pro116, His120, and Phe122 in SRS1
Pro217, Arg220, and Asn221 in SRS2
Phe305, Val306, Phe309, Ala310, Glu313, Thr314, and Thr317 in SRS4
Pro375, Val376, Ile380, and Arg381 in SRS5
Met491 and Leu492 in SRS6

Residues in CYP6P7 included (Figure 12.4b and Table 12.1):

Phe110, Pro113, Asp116, Pro117, and Phe123 in SRS1
Gln216, Lys217, Thr220, Met221, and Phe226 in SRS2
Leu310, Leu313, Ala314, Glu317, Thr318, and Thr321 in SRS4
Leu380, Glu381, Ser382, Ile383, and Arg385 in SRS5
Phe494, Ile495, Leu496, and Ser497 in SRS6

Relatively similar residues of both enzyme models were also found surrounding other pyrethroid substrates (Table 12.1). These hydrophobic residues within SRSs of both P450s, including Phe, Ala, Val, Ile, Met, Pro, and Leu, were predicted to play significant roles in formation of a hydrophobic network, together with polar and charged residues including His, Asn, Asp, Ser, Arg, Lys, Thr, Gln, and Glu, which were anticipated to be involved in hydrogen bonding and polar interactions with pyrethroids (Figure 12.4). Further examination of binding contact residues with substrates and inhibitors investigated in this study indicated that the hydrogen bonds and hydrophobic contacts, especially π–π interactions, might be critical in binding, recognition, and orientation of substrates/inhibitors at the active sites of both CYP6AA3 and CYP6P7. Thus, such key residues involved in hydrogen bond formation and π–π interactions,

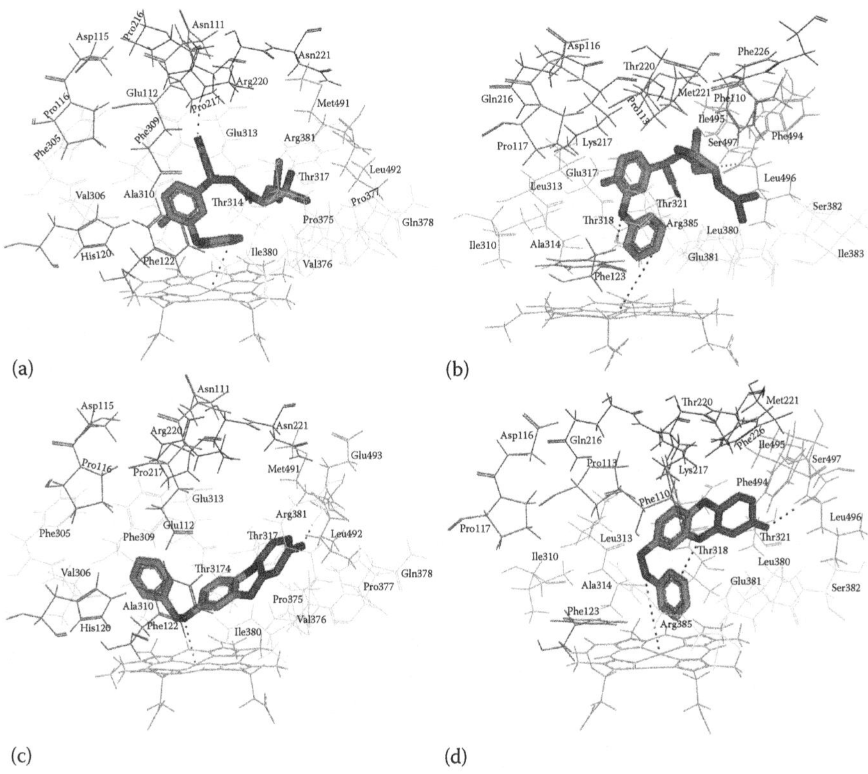

FIGURE 12.4 **(See color insert.)** Docking simulation of cypermethrin (a, b) and benzyloxyresorufin (c, d) into CYP6AA3 (a, c) and CYP6P7 (b, d). The heme group of P450s is shown at the lower part of each figure. Amino acid residues interacted with ligands are presented.

together with contact residues within the active sites of both enzymes, are presented for bindings of substrates and inhibitors (Tables 12.1 through 12.4).

Table 12.1 shows binding modes of one conformer of permethrin type I pyrethroid and two different conformers of each of the type II pyrethroids that were attacked at either the cyano group or the 4′-phenoxybenzyl carbon of deltamethrin, cypermethrin, and λ-cyhalothrin by the heme center of CYP6AA3. In the CYP6AA3 model, Phe122 and His120 in SRS1, Arg220 in SRS2, and Phe309 in SRS4 of CYP6AA3 were predicted to participate in holding pyrethroid substrates in the optimal conformations (Table 12.1). The interacting sites of pyrethroids with amino acids of both enzymes are indicated in Figure 12.1. In this respect, the side chain of Phe122 and/or Phe309 was predicted to form π–π interactions with the benzene ring of cypermethrin, λ-cyhalothrin, deltamethrin, and permethrin pyrethroids, while His120 and Arg220 were conserved in binding pyrethroids via hydrogen bonds, except deltamethrin whose 4′-phenoxybenzyl carbon that was oriented toward CYP6AA3 heme site was bound to Arg381 (SRS5) via hydrogen bond without His120 and Arg220 contacts (Table 12.1). Moreover, the distance from heme and binding energies were

TABLE 12.1
Kinetic Values and Binding Modes of Pyrethroids toward CYP6AA3 and CYP6P7

Compounds (apparent K_m[a] and predicted binding sites[b])	$\Delta G_{binding}$ (kcal/mol)	Contact Residues	Residues		Binding Conformation
			H-Bonds	π–π	
CYP6AA3					
Deltamethrin (K_m = 80.2 μM, CN)	–10.8	Glu112, Pro116, His120, Phe122, Pro217, Arg220, Asn221, Phe305, Val306, Phe309, Ala310, Glu313, Thr314, Thr317, Pro375, Val376, Gln378, Ile380, Arg381, Met491, Leu492	His120 (SRS1) – O (ether)	Phe309 (SRS4)	
Deltamethrin (K_m = 80.2 μM, C4′-PB)	–9.9	Asn111, Glu112, Asn115, Pro116, His120, Phe122, Pro217, Arg220, Asn221, Phe305, Val306, Phe309, Ala310, Glu313, Thr314, Thr317, Pro375, Val376, Pro377, Gln378, Ile380, Arg381, Met491, Leu492	Arg381 (SRS5) – O (ester)	Phe122 (SRS1)	

(Continued)

TABLE 12.1 (CONTINUED)
Kinetic Values and Binding Modes of Pyrethroids toward CYP6AA3 and CYP6P7

Compounds (apparent K_m[a] and predicted binding sites[b])	$\Delta G_{binding}$ (kcal/mol)	Contact Residues	Residues: H-Bonds	Residues: π–π	Binding Conformation
Cypermethrin (K_m = 70.0 μM, CN)	–10.7	Asn111, Glu112, Asn115, Pro116, His120, Phe122, Pro217, Arg220, Asn221, Phe305, Val306, Phe309, Ala310, Glu313, Thr314, Thr317, Pro375, Val376, Ile380, Arg381, Met491, Leu492	His120 (SRS1) – O (ether)	Phe122 (SRS1) Phe309 (SRS4)	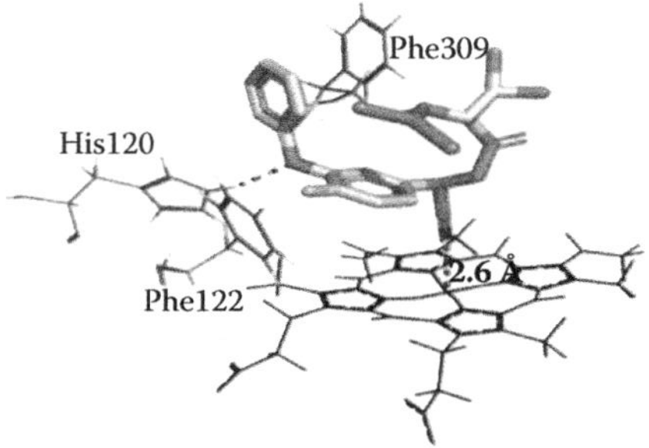
Cypermethrin (K_m = 70.0 μM, C4′-PB)	–10.2	Asn111, Glu112, Asn115, Pro116, His120, Phe122, Pro216, Pro217, Arg220, Asn221, Phe305, Val306, Phe309, Ala310, Glu313, Thr314, Thr317, Pro375, Val376, Pro377, Gln378, Ile380, Arg381, Met491, Leu492	Arg220 (SRS2) – CN	Phe122 (SRS1)	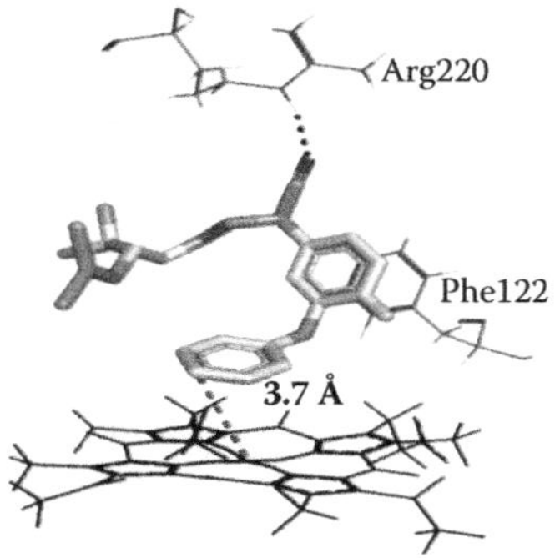

(Continued)

TABLE 12.1 (CONTINUED)
Kinetic Values and Binding Modes of Pyrethroids toward CYP6AA3 and CYP6P7

Compounds (apparent K_m[a] and predicted binding sites[b])	$\Delta G_{binding}$ (kcal/mol)	Contact Residues	Residues: H-Bonds	Residues: π–π	Binding Conformation
λ-Cyhalothrin (K_m = 78.3 μM, CN)	−11.5	Glu112, Pro116, His120, Phe122, Pro217, Arg220, Asn221, Phe305, Val306, Phe309, Ala310, Glu313, Thr314, Thr317, Pro375, Val376, Pro377, Gln378, Ile380, Arg381, Val382, Met491, Leu492	His120 (SRS1) – O (ether)	Phe122 (SRS1) Phe309 (SRS4)	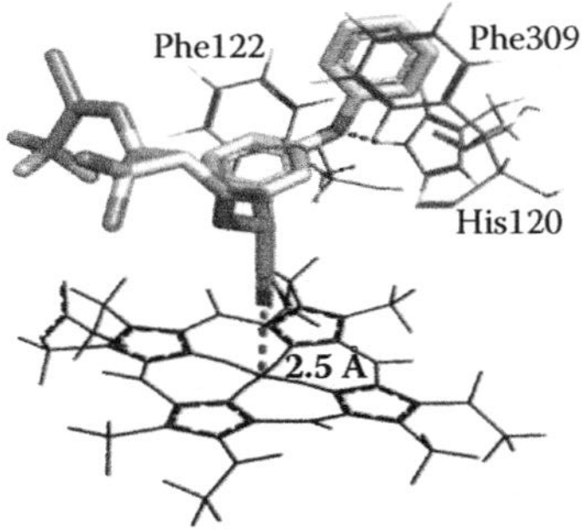
λ-Cyhalothrin (K_m = 78.3 μM, C4′-PB)	−10.6	Glu112, Pro116, His120, Phe122, Pro217, Arg220, Asn221, Phe305, Val306, Phe309, Ala310, Glu313, Thr314, Thr317, Pro375, Val376, Pro377, Gln378, Ile380, Arg381, Val382, Met491, Leu492	Arg220 (SRS2) – CN	Phe122 (SRS1) Phe309 (SRS4)	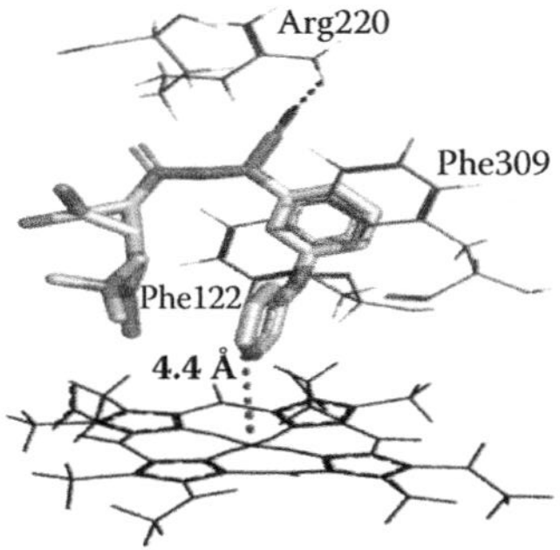

(Continued)

TABLE 12.1 (CONTINUED)
Kinetic Values and Binding Modes of Pyrethroids toward CYP6AA3 and CYP6P7

Compounds (apparent K_m[a] and predicted binding sites[b])	$\Delta G_{binding}$ (kcal/mol)	Contact Residues	Residues		Binding Conformation
			H-Bonds	π–π	
Permethrin (K_m = 41.1 μM, C4′-PB)	−10.5	Glu112, Asn115, Pro116, His120, Phe122, Pro217, Arg220, Phe305, Val306, Phe309, Ala310, Glu313, Thr314, Thr317, Pro375, Val376, Pro377, Gln378, Ile380, Arg381, Val382, Met491, Leu492	His120 (SRS1) – O (ester)	Phe122 (SRS1)	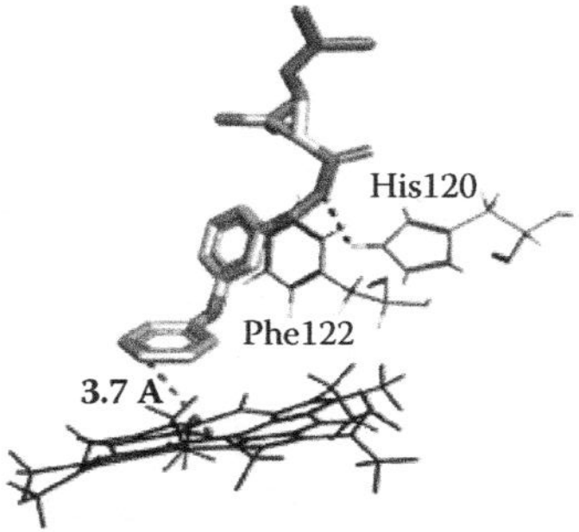
CYP6P7					
Deltamethrin (K_m = 73.3 μM, C4′-PB)	−9.9	Phe110, Pro113, Asp116, Pro117, Phe123, Gln216, Lys217, Thr220, Met221, Phe226, Ile310, Leu313, Ala314, Glu317, Thr318, Leu380, Glu381, Ser382, Ile383, Arg385, Phe494, Ile495, Leu496, Ser497	Thr318 (SRS4) – O (ether) Leu496 (SRS6) – O (ester)	Phe123 (SRS1)	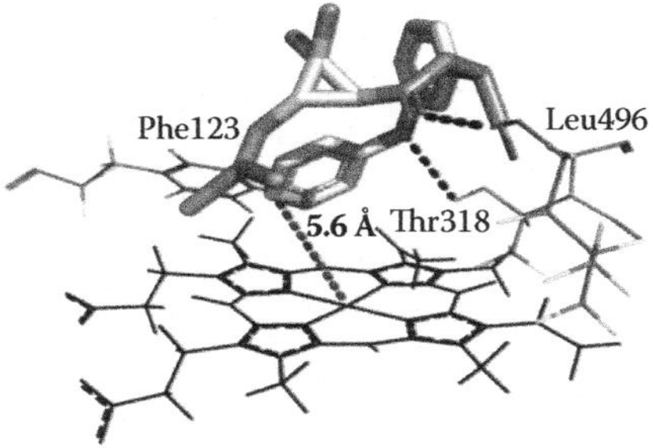

(Continued)

TABLE 12.1 (CONTINUED)
Kinetic Values and Binding Modes of Pyrethroids toward CYP6AA3 and CYP6P7

Compounds (apparent K_m[a] and predicted binding sites[b])	$\Delta G_{binding}$ (kcal/mol)	Contact Residues	Residues: H-Bonds	Residues: π–π	Binding Conformation
Cypermethrin (K_m = 97.3 μM, C4′-PB)	–9.3	Phe110, Pro113, Asp116, Pro117, Phe123, Gln216, Lys217, Thr220, Met221, Phe226, Ile310, Leu313, Ala314, Glu317, Thr318, Thr321, Leu380, Glu381, Ser382, Ile383, Arg385, Phe494, Ile495, Leu496, Ser497	Thr318 (SRS4) – O (ether) Leu496 (SRS6) – O (ester)	Phe123 (SRS1)	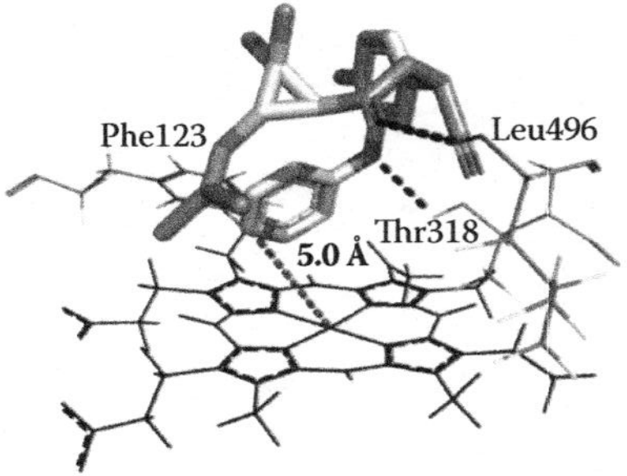
Permethrin (K_m = 69.7 μM, C4′-PB)	–9.2	Phe110, Pro113, Asp116, Pro117, Phe123, Val213, Gln216, Lys217, Thr220, Met221, Phe226, Ile310, Leu313, Ala314, Leu317, Thr318, Leu380, Glu381, Ser382, Ile383, Arg385, Phe494, Ile495, Leu496, Ser497	Thr318 (SRS4) – O (ether) Leu496 (SRS6) – O (ester)	Phe123 (SRS1)	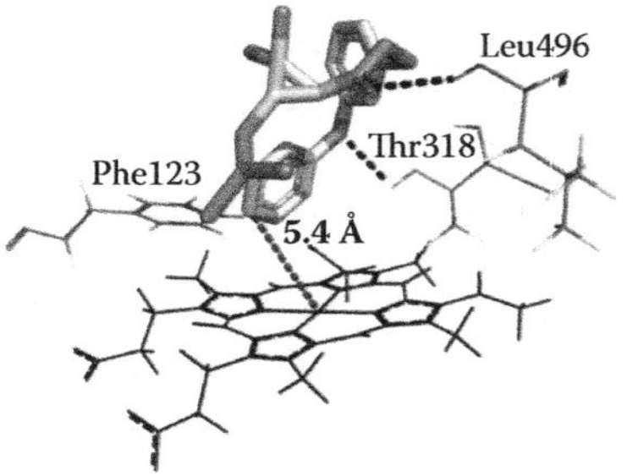

[a] Values obtained from P. Duangkaew et al., Arch. Insect Biochem. Physiol. 76 (2011), pp. 236–248.

[b] Predicted binding sites of pyrethroids toward heme moeity of P450s according to P. Lertkiatmongkol et al., BMC Res. Notes 4 (2011), pp. 321. CN = cyano group; C4′-PB = C4′ of phenoxybenzyl group.

TABLE 12.2
Inhibition Kinetics and Binding Modes of Inhibitors toward CYP6AA3 and CYP6P7

Compounds (apparent K_i[a] and mode of inhibition[b])	$\Delta G_{binding}$ (kcal/mol)	Contact Residues	Residues: H-Bonds	Residues: π–π	Binding Conformation
CYP6AA3					
PBO (K_i = 4.95 μM, MBI)	–7.8	Asn111, Glu112, Asp115, Pro116, His120, Phe122, Pro217, Arg220, Asn221, Phe305, Val306, Phe309, Ala310, Glu313, Thr314, Pro375, Val376, Ile380, Arg381, Val382, Met491, Leu492	His120 (SRS1) – O (ether) Thr314 (SRS4) – O (ether)	–	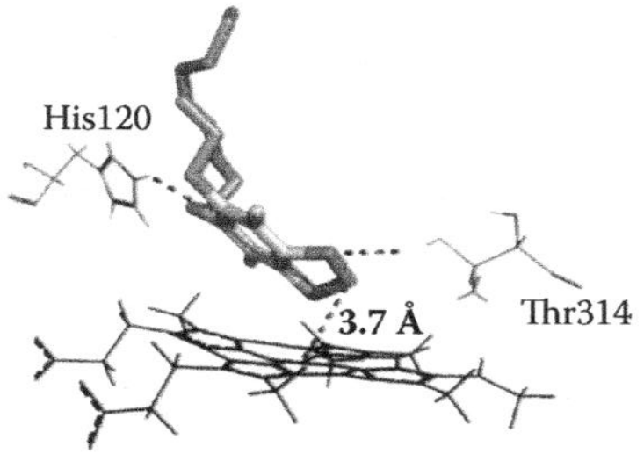 His120 Thr314 3.7 Å
Curcumin (K_i = 0.40 μM, mixed type)	–9.9	Glu112, Asp115, Pro116, Leu117, Ser118, Gly119, His120, Phe122, Pro216, Pro217, Arg220, Val246, Gly247, Met251, Ala302, Phe305, Val306, Phe309, Ala310, Glu313, Thr314, Pro375, Val376, Ile380, Arg381, Val382, Met491, Leu492	Arg220 (SRS2) – O (ketone) Thr314 (SRS4) – OCH_3 (benzene)	Phe305 (SRS4)	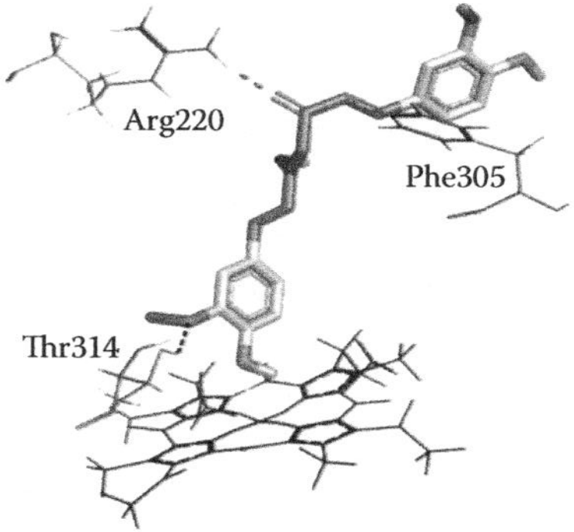 Arg220 Phe305 Thr314

(*Continued*)

TABLE 12.2 (CONTINUED)
Inhibition Kinetics and Binding Modes of Inhibitors toward CYP6AA3 and CYP6P7

Compounds (apparent K_i[a] and mode of inhibition[b])	$\Delta G_{binding}$ (kcal/mol)	Contact Residues	Residues: H-Bonds	Residues: π–π	Binding Conformation
Bergapten (K_i = 46.9 μM, mixed type)	–8.3	Glu112, His120, Phe122, Arg220, Phe305, Val306, Phe309, Ala310, Thr314, Pro375, Val376, Pro377, Gln378, Ile380, Arg381, Val382, Met491, Leu492	His120 (SRS1) – O (ether)	Phe122 (SRS1)	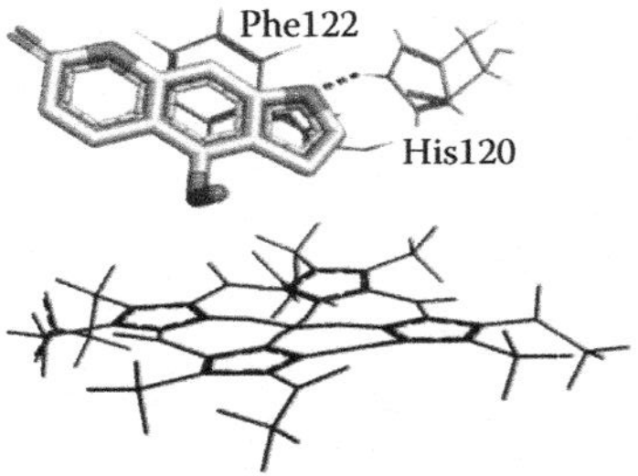
Piperine (K_i = 9.13 μM, MBI)	–9.8	Glu112, Asp115, Pro116, Leu117, Ser118, Gly119, His120, Phe122, Ala123, Pro216, Pro217, Arg220, Val246, Gly247, Met251, Ala302, Phe305, Val306, Phe309, Ala310, Glu313, Thr314, Pro375, Val376, Ile380, Arg381, Met491, Leu492	Arg220 (SRS2) – O (amide) Thr314 (SRS4) – O (ether)	Phe122 (SRS1)	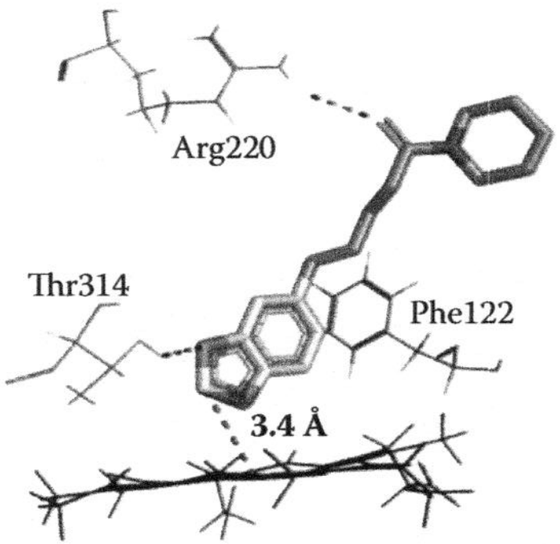

(Continued)

TABLE 12.2 (CONTINUED)
Inhibition Kinetics and Binding Modes of Inhibitors toward CYP6AA3 and CYP6P7

Compounds (apparent K_i[a] and mode of inhibition[b])	$\Delta G_{binding}$ (kcal/mol)	Contact Residues	Residues: H-Bonds	Residues: π–π	Binding Conformation
CYP6P7					
PBO (K_i = 15.9 μM, MBI)	–6.7	Phe110, Pro113, Asp116, Pro117, Phe123, Gln216, Lys217, Thr220, Met221, Phe226, Leu313, Ala314, Glu317, Thr318, Thr321, Leu380, Glu381, Ser382, Ile383, Arg385, Phe494, Ile495, Leu496	Thr318 (SRS4) – O (ether) Leu496 (SRS6) – O (ether)	–	Leu496 Thr318 5.0 Å
Curcumin (K_i = 2.05 μM, mixed type)	–8.8	Phe110, Asp112, Pro113, Asp116, Pro117, Phe123, Gln216, Lys217, Ser219, Thr220, Ile225, Phe226, Val229, Asp245, Val246, Phe250, Leu313, Ala314, Glu317, Thr318, Leu380, Glu381, Ile383, Arg385, Phe494, Ile495, Leu496	Thr318 (SRS4) – OCH_3 (benzene)	Phe110 (SRS1) Phe123 (SRS1)	Phe110 Thr318 Phe123

(*Continued*)

TABLE 12.2 (CONTINUED)
Inhibition Kinetics and Binding Modes of Inhibitors toward CYP6AA3 and CYP6P7

Compounds (apparent K_i[a] and mode of inhibition[b])	$\Delta G_{binding}$ (kcal/mol)	Contact Residues	Residues: H-Bonds	Residues: π–π	Binding Conformation
Bergapten (K_i = 65.6 μM, mixed type)	–7.0	Phe123, Gln216, Lys217, Thr220, Met221, Leu313, Ala314, Glu317, Thr318, Thr321 Leu380, Glu381, Ser382, Ile383, Arg385, Phe494, Ile495, Leu496, Ser497	Thr220 (SRS2) – O (ether) Leu496 (SRS6) – O (ester)	–	Thr220 Leu496
Piperine (K_i = 26.4 μM, MBI)	–8.6	Phe110, Asp112, Pro113, Asp116, Pro117, Phe123, Gln216, Lys217, Ser219, Thr220, Met221, Ile225, Phe226, Asp245, Val246, Phe250, Leu313, Ala314, Glu317, Thr318, Leu380, Glu381, Arg385, Phe494, Ile495, Leu496	Thr220 (SRS2) – N (amide) Thr318 (SRS4) – O (ether) Arg385 (SRS5) – O (ether)	Phe123 (SRS1)	Thr220 Phe123 Arg385 Thr318 4.1 Å

[a] Values obtained from P. Duangkaew et al., Arch. Insect Biochem. Physiol. 76 (2011), pp. 236–248.
[b] Mixed type = mixed-type inhibition; MBI = mechanism-based inactivation.

TABLE 12.3
Inhibition Kinetics and Binding Modes of Flavonoids toward CYP6AA3 and CYP6P7

Compounds[a] (apparent K_i and mode of inhibition[b])	$\Delta G_{binding}$ (kcal/mol)	Contact Residues	Residues: H-Bonds	Residues: π–π	Binding Conformation
CYP6AA3					
5H782′3′TMF (K_i = 4.73 μM, MBI)	−9.4	Glu112, Asn115, Pro116, His120, Phe122, Pro217, Arg220, Asn221, Phe305, Val306, Phe309, Ala310, Glu313, Thr314, Pro375, Val376, Pro377, Gln378, Ile380, Arg381, Val382, Met491, Leu492	His120 (SRS1) – OCH_3 (C7) Thr314 (SRS4) – O (C4)	–	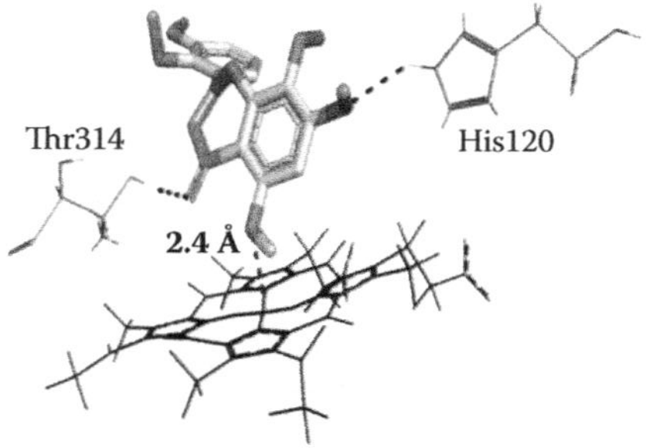
54′DH782′3′TMF (K_i = 2.32 μM, MBI)	−9.1	Glu112, Pro116, His120, Phe122, Pro217, Arg220, Asn221, Phe305, Val306, Phe309, Ala310, Glu313, Thr314, Pro375, Val376, Pro377, Gln378, Ile380, Arg381, Val382, Met491, Leu492	His120 (SRS1) – OCH_3 (C7) Thr314 (SRS4) – O (C4) Arg381 (SRS5) – OCH_3 (C3′)	–	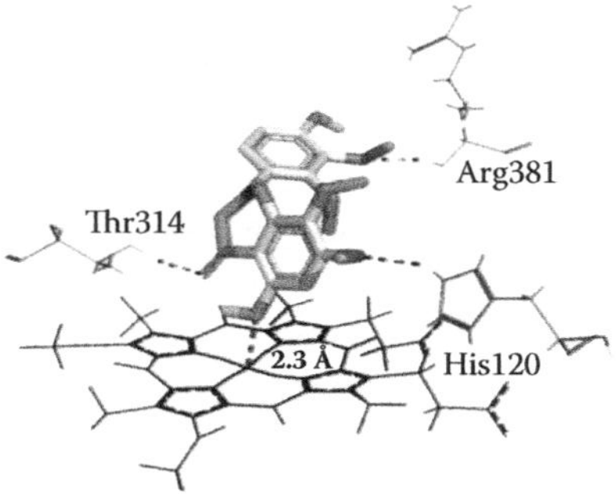

(*Continued*)

TABLE 12.3 (CONTINUED)
Inhibition Kinetics and Binding Modes of Flavonoids toward CYP6AA3 and CYP6P7

Compounds[a] (apparent K_i and mode of inhibition[b])	$\Delta G_{binding}$ (kcal/mol)	Contact Residues	Residues		Binding Conformation
			H-Bonds	π–π	
5H78DMF (K_i = 6.50 μM, MBI)	–9.6	Glu112, Pro116, His120, Phe122, Pro217, Arg220, Asn221, Val306, Phe309, Ala310, Glu313, Thr314, Pro375, Val376, Pro377, Gln378, Ile380, Arg381, Val382, Met491, Leu492	His120 (SRS1) – OCH_3 (C7) Thr314 (SRS4) – O (C4)	–	
57DH8MF (K_i = 9.70 μM, mixed type)	–9.2	Glu112, Asp115, Pro116, His120, Phe122, Pro217, Arg220, Asn221, Phe305, Val306, Phe309, Ala310, Glu313, Thr314, Pro375, Val376, Pro377, Ile380, Arg381, Val382, Met491, Leu492	Thr314 (SRS4) – OH (C7)	Phe305 (SRS4) Phe309 (SRS4)	

(*Continued*)

TABLE 12.3 (CONTINUED)
Inhibition Kinetics and Binding Modes of Flavonoids toward CYP6AA3 and CYP6P7

Compounds[a] (apparent K_i and mode of inhibition[b])	$\Delta G_{binding}$ (kcal/mol)	Contact Residues	Residues: H-Bonds	Residues: π–π	Binding Conformation
5H78DMFN (K_i = 15.3 μM, mixed type)	–9.2	Tyr109, Asn111, Glu112, Asp115, Pro116, His120, Phe122, Pro217, Arg220, Phe305, Val306, Phe309, Ala310, Thr314, Pro375, Val376, Pro377, Gln378, Ile380, Arg381, Val382, Met491	Arg220 (SRS2) – OH (C5)	Phe122 (SRS1)	
CYP6P7					
5H782′3′TMF (K_i = 5.59 μM, noncompetitive)[c]	–7.9, –8.6				

(*Continued*)

TABLE 12.3 (CONTINUED)
Inhibition Kinetics and Binding Modes of Flavonoids toward CYP6AA3 and CYP6P7

Compounds[a] (apparent K_i and mode of inhibition[b])	$\Delta G_{binding}$ (kcal/mol)	Contact Residues	Residues: H-Bonds	Residues: π–π	Binding Conformation
54′DH782′3′TMF (K_i = 11.0 μM, noncompetitive)[c]	−7.8, −8.1				
574′THF (Apigenin) (K_i = 1.35 μM, mixed type)	−8.9	Phe110, Pro113, Phe123, Lys217, Thr220, Met221, Phe226, Leu313, Ala314, Glu317, Thr318, Leu380, Glu381, Ser382, Ile383, Arg385, Phe494, Ile495, Leu496, Ser497	Leu496 (SRS6) – OH (C7)	Phe123 (SRS1)	

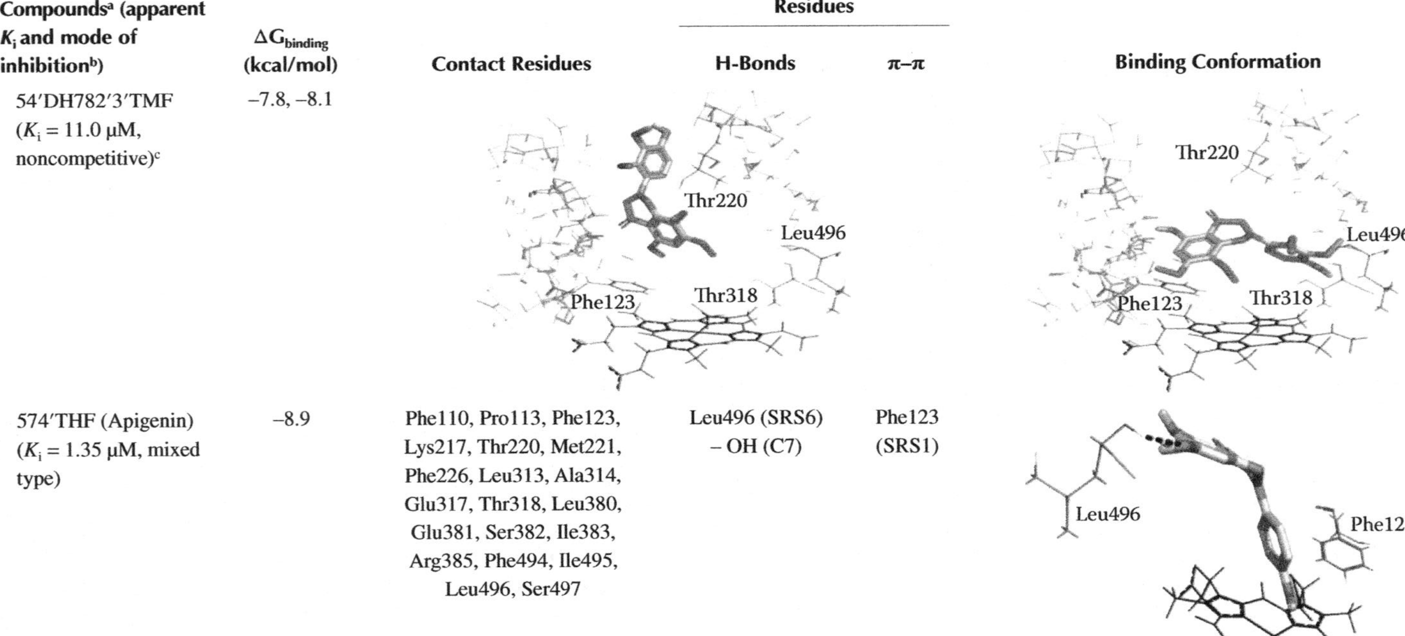

(Continued)

TABLE 12.3 (CONTINUED)
Inhibition Kinetics and Binding Modes of Flavonoids toward CYP6AA3 and CYP6P7

Compounds[a] (apparent K_i and mode of inhibition[b])	$\Delta G_{binding}$ (kcal/mol)	Contact Residues	Residues		Binding Conformation
			H-Bonds	π–π	
57DHF (Chrysin) (K_i = 1.35 μM, mixed type)	–8.5	Phe110, Pro113, Asp116, Pro117, Phe123, Val213, Gln216, Lys217, Thr220, Met221, Leu313, Ala314, Glu317, Thr318, Thr321, Leu380, Glu381, Arg385, Phe494, Ile495, Leu496, Ser497	Leu496 (SRS6) – OH (C7)	Phe123 (SRS1)	

[a] 5H782′3′TMF = 5-hydroxy-7,8,2′,3′-tetramethoxyflavone; 54′DH782′3′TMF = 5,4′-dihydroxy-7,8,2′,3′-tetramethoxyflavone; 5H78DMF = 5-hydroxy-7,8-dimethoxyflavone; 57DH8MF = 5,7-dihydroxy-8-methoxyflavone; 5H78DMFN = 5-hydroxy-7,8-dimethoxyflavanone; 574′THF = 5,7,4′-trihydroxyflavone; 57DHF = 5,7-dihydroxyflavone.

[b] Inhibition kinetics and modes obtained from R. Kotewong et al., Parasitol. Res. 113 (2014), pp. 3381–3392.

[c] The two best binding modes are displayed.

TABLE 12.4
Inhibition Kinetics and Binding Modes of Rhinacanthins toward CYP6AA3 and CYP6P7

Compounds (apparent K_i and mode of inhibition[a])	$\Delta G_{binding}$ (kcal/mol)	Contact Residues	Residues: H-Bonds	Residues: π–π	Binding Conformation
CYP6AA3					
Rhinacanthin–A (K_i = 5.1 μM, mixed type)	–10.5	Glu112, Asp115, Pro116, His120, Phe122, Pro217, Arg220, Asn221, Phe305, Val306, Phe309, Ala310, Glu313, Thr314, Pro375, Val376, Ile380, Arg381, Val382, Met491, Leu492	His120 (SRS1) – O (C5) Arg220 (SRS2) – OH (C3)	Phe122 (SRS1) Phe309 (SRS4)	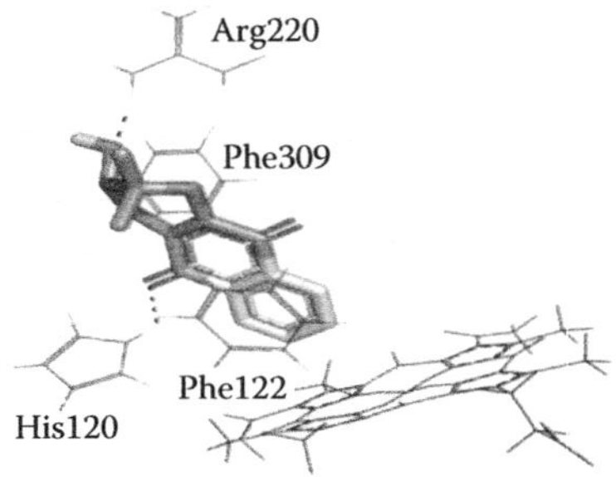 Arg220 Phe309 Phe122 His120
Rhinacanthin–B (K_i = 1.7 μM, MBI)	–10.7	Glu112, Asp115, Pro116, His120, Phe122, Pro217, Arg220, Asn221, Phe305, Val306, Phe309, Ala310, Glu313, Thr314, Thr317, Pro375, Val376, Pro377, Ile380, Arg381, Val382, Met491, Leu492, Glu493	His120 (SRS1) – O (C10) Arg220 (SRS2) – O (C5)	Phe122 (SRS1) Phe305 (SRS4) Phe309 (SRS4)	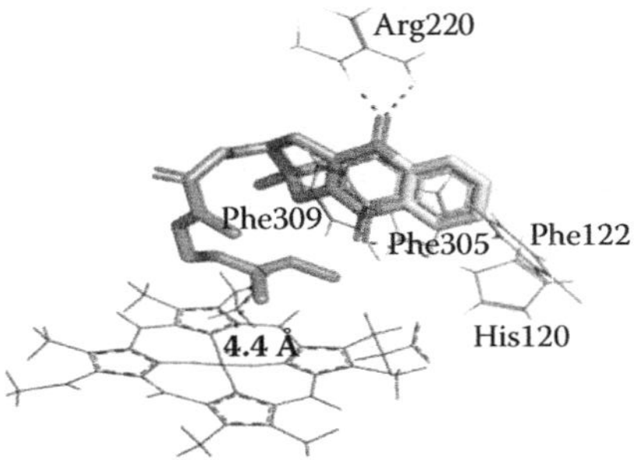 Arg220 Phe309 Phe305 Phe122 4.4 Å His120

(Continued)

TABLE 12.4 (CONTINUED)
Inhibition Kinetics and Binding Modes of Rhinacanthins toward CYP6AA3 and CYP6P7

Compounds (apparent K_i and mode of inhibition[a])	$\Delta G_{binding}$ (kcal/mol)	Contact Residues	Residues: H-Bonds	Residues: π–π	Binding Conformation
Rhinacanthin–D (K_i = 2.1 μM, MBI)	−11.2	Glu112, Asp115, Pro116, His120, Phe122, Pro217, Arg220, Asn221, Phe305, Val306, Phe309, Ala310, Glu313, Thr314, Thr317, Pro375, Val376, Pro377, Gln378, Ile380, Arg381, Val382, Met491, Leu492	His120 (SRS1) – O (C1) Arg220 (SRS2) – O (C4) Thr314 (SRS4) – O (MDP)	Phe122 (SRS1) Phe309 (SRS4)	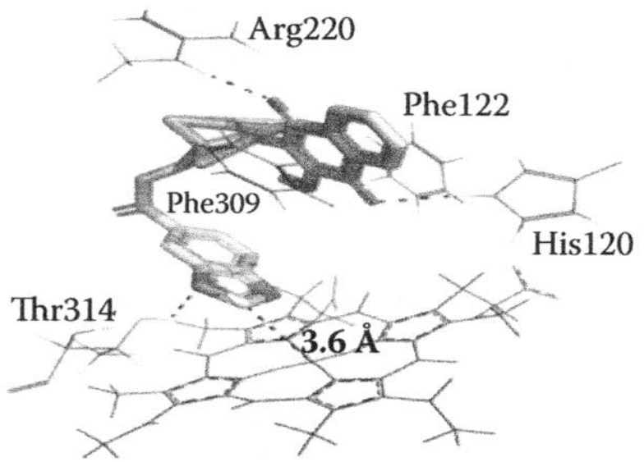
Rhinacanthin–E (K_i = 13.4 μM, MBI)	−10.8	Asn111, Glu112, Asp115, Pro116, Leu117, His120, Phe122, Ala123, Pro216, Pro217, Arg220, Asn221, Val246, Gly247, Met251, Phe305, Val306, Phe309, Ala310, Glu313, Thr314, Thr317, Pro375, Val376, Pro377, Ile380, Arg381, Val382, Met491, Leu492	Arg220 (SRS2) – O (ester, C9′)	Phe122 (SRS1) Phe309 (SRS4)	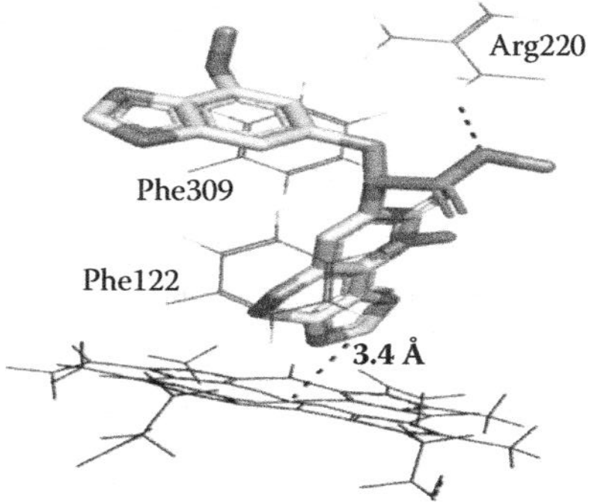

(Continued)

TABLE 12.4 (CONTINUED)
Inhibition Kinetics and Binding Modes of Rhinacanthins toward CYP6AA3 and CYP6P7

Compounds (apparent K_i and mode of inhibition[a])	$\Delta G_{binding}$ (kcal/mol)	Contact Residues	Residues		Binding Conformation
			H-Bonds	π–π	
Rhinacanthin–G (K_i = 14.9 μM, MBI)	−10.3	Glu112, Asp115, Pro116, Leu117, Gly119, His120, Phe122, Pro216, Pro217, Trp219, Arg220, Asn221, Val246, Gly247, Met251, Phe305, Val306, Phe309, Ala310, Glu313, Thr314, Thr317, Pro375, Val376, Pro377, Ile380, Arg381, Val382, Met491, Leu492	His120 (SRS1) – O (ester) Arg220 (SRS2) – O (epoxide) Thr314 (SRS5) – O (C1)	–	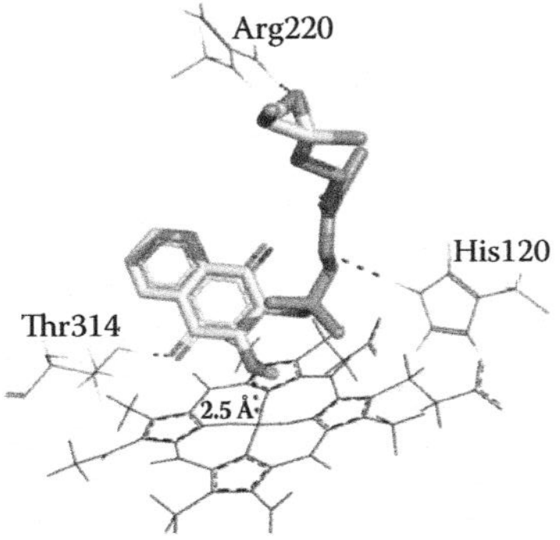 Arg220 His120 Thr314 2.5 Å
Rhinacanthin–N (K_i = 12.6 μM, MBI)	−11.8	Asn111, Glu112, Pro116, His120, Phe122, Pro216, Pro217, Arg220, Asn221, Phe305, Val306, Phe309, Ala310, Glu313, Thr314, Pro375, Val376, Pro377, Ile380, Arg381, Val382, Met491, Leu492	His120 (SRS1) – OCH_3 (C4′)	Phe122 (SRS1) Phe309 (SRS4)	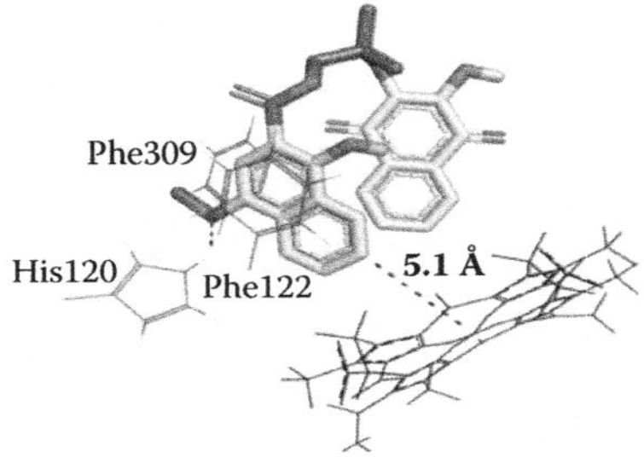 Phe309 His120 Phe122 5.1 Å

(Continued)

TABLE 12.4 (CONTINUED)
Inhibition Kinetics and Binding Modes of Rhinacanthins toward CYP6AA3 and CYP6P7

Compounds (apparent K_i and mode of inhibition[a])	$\Delta G_{binding}$ (kcal/mol)	Contact Residues	Residues: H-Bonds	Residues: π–π	Binding Conformation
Rhinacanthin–Q (K_i = 8.8 μM, mixed type)	−11.4	Glu112, His120, Phe122, Pro216, Pro217, Arg220, Asn221, Phe305, Val306, Phe309, Ala310, Glu313, Thr314, Thr317, Pro375, Val376, Pro377, Ile380, Arg381, Val382, Met491, Leu492	His120 (SRS1) – OCH_3 (C1′) Arg220 (SRS2) – OCH_3 (C4′)	Phe122 (SRS1) Phe309 (SRS4)	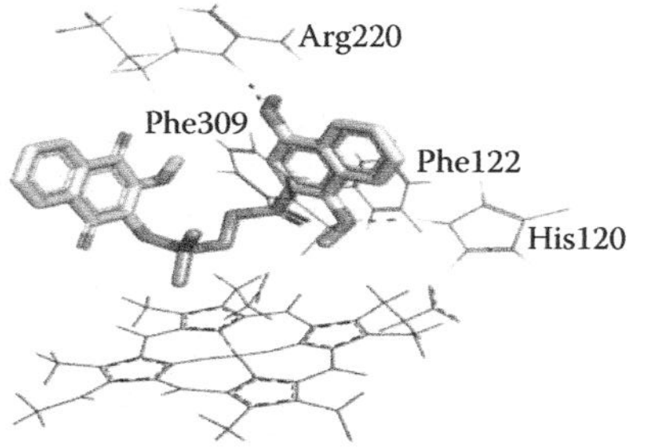 Arg220 Phe309 Phe122 His120
CYP6P7					
Rhinacanthin–A (K_i = 17.6 μM, MBI)	−8.2	Leu109, Phe110, Pro113, Phe123, Gln216, Lys217, Thr220, Met221, Phe226, Leu313, Ala314, Glu317, Thr318, Thr321, Leu380, Glu381, Ser382, Ile383, Arg385, Phe494, Ile495, Leu496, Ser497	Thr220 (SRS2) – OH (C3)	Phe123 (SRS1)	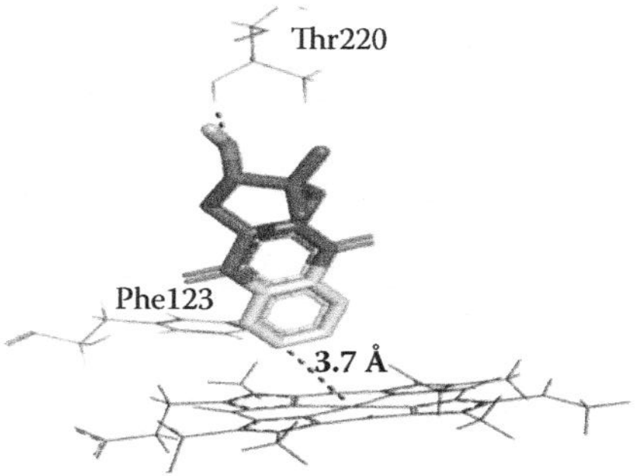 Thr220 Phe123 3.7 Å

(Continued)

TABLE 12.4 (CONTINUED)
Inhibition Kinetics and Binding Modes of Rhinacanthins toward CYP6AA3 and CYP6P7

Compounds (apparent K_i and mode of inhibition[a])	$\Delta G_{binding}$ (kcal/mol)	Contact Residues	Residues: H-Bonds	Residues: π–π	Binding Conformation
Rhinacanthin–B (K_i = 2.7 μM, MBI)	−10.3	Phe110, Asp112, Pro113, Asp116, Pro117, Phe123, Gln216, Lys217, Thr220, Met221, Phe226, Phe250, Leu313, Ala314, Glu317, Thr318, Leu380, Glu381, Ile383, Arg385, Phe494, Ile495, Leu496, Ser497	Thr220 (SRS2) – O (ester) Arg385 (SRS5) – O (C5)	Phe123 (SRS1)	Thr220 Arg385 Phe123 3.8 Å
Rhinacanthin–C (K_i = 7.3 μM, competitive)	−8.9	Leu109, Phe110, Pro113, Asp116, Pro117, Phe123, Gln216, Lys217, Thr220, Met221, Phe226, Leu313, Ala314, Glu317, Thr318, Thr321, Leu380, Glu381, Ser382, Ile383, Arg385, Phe494, Ile495, Leu496, Ser497	Thr220 (SRS2) – O (C4) Ser382 (SRS5) – O (C1) Lue496 (SRS6) – OH (C2)	Phe110 (SRS1) Phe494 (SRS6)	Thr220 Phe494 Lue496 Phe110 Ser382

(Continued)

TABLE 12.4 (CONTINUED)
Inhibition Kinetics and Binding Modes of Rhinacanthins toward CYP6AA3 and CYP6P7

Compounds (apparent K_i and mode of inhibition[a])	$\Delta G_{binding}$ (kcal/mol)	Contact Residues	Residues: H-Bonds	Residues: π–π	Binding Conformation
Rhinacanthin–D (K_i = 3.6 μM, MBI)	–9.8	Gly108, Leu109, Phe110, Pro113, Asp116, Pro117, Phe123, Gln216, Lys217, Thr220, Met221, Phe226, Ile310, Leu313, Ala314, Glu317, Thr318, Thr321, Leu380, Glu381, Ser382, Ile383, Arg385, Phe494, Ile495, Leu496,	Thr220 (SRS2) – O (C4) Ser382 (SRS5) – OH (C2) Arg385 (SRS5) – O (MDP)	Phe110 (SRS1) Phe123 (SRS1) Phe494 (SRS6)	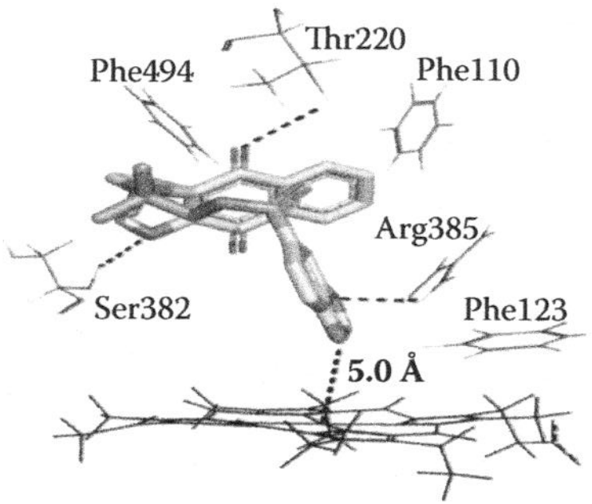
Rhinacanthin–G (K_i = 15.3 μM, MBI)	–9.0	Phe110, Asp112, Pro113, Asp116, Pro117, Phe123, Gln216, Lys217, Ser219, Thr220, Met221, Ile225, Phe226, Asp245, Val246, Phe250, Leu313, Ala314, Glu317, Thr318, Leu380, Glu381, Arg385, Phe494, Ile495, Leu496	Thr220 (SRS2) – O (ester)	Phe123 (SRS1)	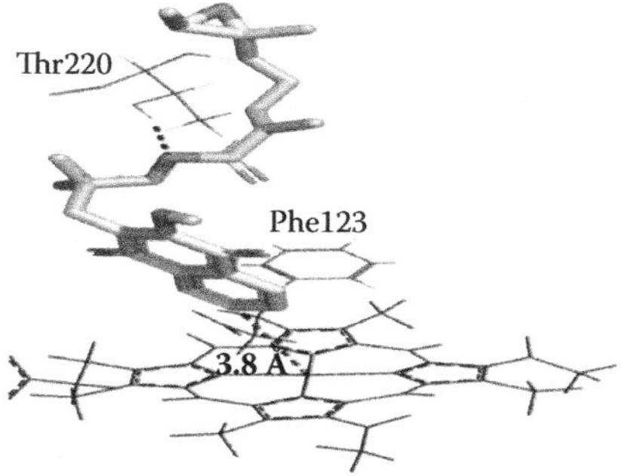

(*Continued*)

TABLE 12.4 (CONTINUED)
Inhibition Kinetics and Binding Modes of Rhinacanthins toward CYP6AA3 and CYP6P7

Compounds (apparent K_i and mode of inhibition[a])	$\Delta G_{binding}$ (kcal/mol)	Contact Residues	Residues		Binding Conformation
			H-Bonds	π–π	
Rhinacanthin–N (K_i = 5.8 μM, MBI)	–9.3	Phe110, Pro113, Asp116, Pro117, Phe123, Gln216, Lys217, Thr220, Met221, Phe226, Leu313, Ala314, Glu317, Thr318, Thr321, Leu380, Glu381, Ser382, Ile383, Arg385, Phe494, Ile495, Leu496	Thr220 (SRS2) – O (ester)	Phe110 (SRS1) Phe123 (SRS1) Phe494 (SRS6)	

[a] Inhibition kinetics and modes of inhibition obtained from S. Pethuan et al., J. Med. Entomol. 49 (2012), pp. 993–1000 and R. Kotewong et al., Parasitol. Res. 114 (2015), pp. 2567–2579.

comparable among the same attacking sites (either the cyano group or the 4′-phenoxybenzyl carbon) of type II pyrethroids, consistent with their approximately similar K_m values to the CYP6AA3 enzyme (Table 12.1). In type I permethrin pyrethroid, its binding to CYP6AA3 depicted a hydrogen bond formed by side chain of His120 with permethrin and a π–π interaction with Phe122 to allow the 4′-phenoxybenzyl carbon to be positioned toward the CYP6AA3 heme iron (Table 12.1).

For CYP6P7, the pyrethroids that are substrates, including permethrin, cypermethrin, and deltamethrin, arranged their 4′-phenoxybenzyl carbons positioned toward the heme center of the enzyme with approximately the same distance, in correlation to their approximately similar K_i values (Table 12.1). The amino acids Thr318 and Leu496 in SRS4 and SRS6, respectively, were predicted to bind to pyrethroids via hydrogen bonds and Phe123 in SRS1 via π–π interaction (Table 12.1).

12.5 INHIBITION OF *AN. MINIMUS* CYP6AA3 AND CYP6P7 ENZYMES BY TEST COMPOUNDS

Compounds with different chemical structures, including flavonoids, furanocoumarins, and the methylenedioxyphenyl group, have been investigated in inhibition against P450s of insects such as *M. domestica*, *Papilio polyxenes*, and *An. gambiae* [35,36,46]. However, their modes of inhibition were rarely explored. We have previously investigated the inhibition of these different compound structures, including α- and β-naphthoquinone flavonoids, bergapten and xanthotoxin furanocoumarins, piperine and PBO methylenedioxyphenyl compounds, and curcumin, on CYP6AA3- and CYP6P7-mediated benzyloxyresorufin-*O*-debenzylation using benzyloxyresorufin as the probe substrate of both enzymes [17]. The naphthoquinone flavonoids were noncompetitive inhibitors of both enzymes, while curcumin, bergapten, piperine, and PBO (both piperine and PBO also displayed MBI) exhibited mixed-type inhibition against CYP6AA3 and CYP6P7 [17 and unreported data for curcumin].

To examine contact residues of CYP6AA3 and CYP6P7 with benzyloxyresorufin substrate, docking was performed. Benzyloxyresorufin was surrounded by the following residues (Figure 12.4c):

Asn111, Glu112, Asp115, Pro116, His120, and Phe122 in SRS1
Pro217, Arg220, and Asn221 in SRS2
Phe305, Val306, Phe309, Ala310, Glu313, Thr314, and Thr317 in SRS4
Val376, Pro377, Gln378, Ile380, and Arg381 in SRS5
Met491, Leu492, and Glu493 in SRS6 for CYP6AA3

For CYP6P7, the following residues were found surrounding benzyloxyresorufin (Figure 12.4d):

Phe110, Pro113, Asp116, Pro117 and Phe123 in SRS1
Gln216, Lys217, Thr220, Met 221, and Phe226 in SRS2
Ile310, Leu313, Ala314, Thr318, and Thr321 in SRS4
Leu380, Glu381, Ser382, and Arg385 in SRS5
Phe494, Ile495, Leu496, and Ser497 in SRS6

Figure 12.4(c) also reveals that benzyloxyresorufin formed π–π interaction with Phe122 (SRS1) of CYP6AA3 and a hydrogen bond with Arg381 (SRS5). In contrast, in CYP6P7, whose pocket site was smaller, two hydrogen bonds formed by Thr318 (SRS4) and Leu496 (SRS6) and a π–π interaction formed by Phe123 (SRS1) were found in contact with the benzyloxyresorufin substrate (Figure 12.4d).

In docking with the aforementioned inhibitors (shown in Table 12.2), it appeared for CYP6AA3 that Thr314 (SRS4), together with Arg220 (SRS2) and His120 (SRS1), formed hydrogen bonds with curcumin, bergapten, piperine, and PBO, and π–π interactions formed by either Phe122 (SRS1) or Phe305 (SRS4) (Table 12.2). The distance from heme of these compounds was correlated with their K_i values, with longer distance correlated with higher K_i values. Moreover, it could be noted in PBO and piperine that the methylenic carbon of each compound was situated toward the heme center of CYP6AA3, supporting its MBI pattern against the enzyme. The methylenic carbon might be oxidized to carbene, bound to the heme of the CYP6AA3 enzyme as complex, and resulted in irreversible inhibition as described [38,39,47].

For CYP6P7, conformations of these inhibitors (curcumin, bergapten, PBO, and piperine) similar to those for CYP6AA3 were placed toward the CYP6P7 heme center. The distance of these compounds' positioning from CYP6P7 heme center was ranking in concomitance with their K_i values, with larger distance correlated with higher K_i values (Table 12.2). Hydrogen bond formation with these compounds was found for Thr318 (SRS4), Thr220 (SRS2), Arg385 (SRS5), and Leu496 (SRS6), and π–π interactions were formed by Phe123 and Phe110 in SRS1 (Table 12.2).

12.6 INHIBITION OF *AN. MINIMUS* CYP6AA3 AND CYP6P7 ENZYMES BY PLANT-DERIVED COMPOUNDS

We recently reported the investigation of the structures and inhibition potency of the compound series isolated from *Andrographis paniculata* Nees (Acanthaceae) and *Rhinacanthus nasutus* (Linn.) Kurz (Acanthaceae) plant extracts that could be delineated with inhibition effects and be explained by the groups attached to the core structures of each compound type (Table 12.3) [19–21]. In particular, inhibition of those compounds with MBI patterns could be promising candidate inhibitory leads against both mosquito P450s. In *A. paniculata*, the flavone derivatives were contributors of good inhibitory potency against both P450s, while in *R. nasutus* the pyranonaphthoquinones (rhinacanthin-A and -B), lignan (rhinacanthin-E), and naphthoquinones (rhinacanthin-C, -D, -E, -G, -N, and -Q) were responsible for inhibitory activities observed against both mosquito P450s [19–21]. Moreover, rhinacanthins possessed different functional groups attached to the naphthoquinone core structure that included methylenedioxyphenyl, epoxide, alkene chain, and naphthalene groups (Figure 12.2). Further molecular docking with homology models of both mosquito P450s to explain the inhibition nature of flavone derivatives and rhinacanthins isolated from *A. paniculata* and *R. nasutus*, respectively, could well explain the binding properties of these inhibitors to both enzymes.

12.6.1 Inhibition of *An. minimus* CYP6AA3 and CYP6P7 Enzymes by Compounds Isolated from *A. paniculata*

As displayed in Table 12.3, five flavonoids consisting of four 5-hydroxyflavones (5-hydroxy-7,8,2′,3′-tetramethoxyflavone; 5,4′-dihydroxy-7,8,2′,3′-tetramethoxyflavone; 5-hydroxy-7,8-dimethoxyflavone; and 5,7-dihydroxy-8-methoxyflavone) and one 5-hydroxyflavanone (5-hydroxy-7,8-dimethoxyflavanone) efficiently inhibited both mosquito P450s in different modes and different degrees of inhibition effects (see chemical structures in Figure 12.2). Among these, three methoxylated flavones—including 5-hydroxy-7,8,2′,3′-tetramethoxyflavone; 5,4′-dihydroxy-7,8,2′,3′-tetramethoxyflavone; and 5-hydroxy-7,8-dimethoxyflavone—inhibited CYP6AA3 in the MBI mode. Docking analysis revealed that these methoxylated flavones with MBI mode gave similar orientations by turning their 5-hydroxy group of A ring interacted with the iron center of CYP6AA3 heme (Table 12.3). It has also been suggested that the free hydroxyl group at the position C5 of flavonoids prefers to interact with heme iron of human P450s due to its stearic availability [48]. Docking analysis also revealed important residues of the CYP6AA3 enzyme that might help direct and accommodate MBI flavonoids in the pocket. Important residues of CYP6AA3 surrounding 5,4′-dihydroxy-7,8,2′3′-tetramethoxyflavone are shown in Figure 12.5(a) as an example of flavonoids.

The three methoxylated flavones with MBI mode were contacted with His120 and Thr314 of CYP6AA3 via hydrogen bonds at A and C rings and with slightly different surrounding contact residues (Table 12.3). It could be noted for CYP6AA3 that an additional binding by hydrogen bond of Arg381 (SRS5) to 5,4′-dihydroxy-7,8,2′,3′-tetramethoxyflavone might be associated with the lower K_i value compared to the 5-hydroxy-7,8-dimethoxyflavone and 5-hydroxy-7,8,2′,3′-tetramethoxyflavone (Table 12.3), possibly to ensure the accurate conformation was placed within CYP6AA3 pocket. All together, the flavonoids that exhibited MBI had their A and C rings contacted via hydrogen bonds with His120 and Thr314 residues, respectively, possibly to ensure formation of inhibitory metabolites (Table 12.3). In contrast, the weaker reversible mixed-type 5,7-dihydroxy-8-methoxyflavone inhibitor formed one hydrogen bond at the A ring with Thr314 and two π–π interactions with Phe305 and Phe309 of CYP6AA3 and the compound located at a further distance than the flavones with MBI mode, coherent with its higher K_i value (Table 12.3). Finally, the 5-hydroxy-7,8-dimethoxyflavanone, of which the C2–C3 double bond was absent, was interacted with Phe122 by a π–π interaction and with Arg220 via hydrogen bond (Table 12.3). The flavanone compound was docked in a different orientation than seen in the four former flavones as it had its B ring placed toward the heme center of CYP6AA3 in a distance, in correlation to its weak mixed-type inhibition (Table 12.3). However, the ligand–P450 interaction energies of these flavonoids were not significantly correlated with their K_i values for the inhibition of the CYP6AA3 enzyme (Table 12.3). It could thus be summarized for small inhibitory flavonoids that two hydrogen bonds mediated by His120 and Thr314 were important in orienting the potent MBI flavones toward the CYP6AA3 heme without π–π interaction, while weak reversible inhibitors were occupied within the CYP6AA3 pocket by one hydrogen bond and π–π interaction.

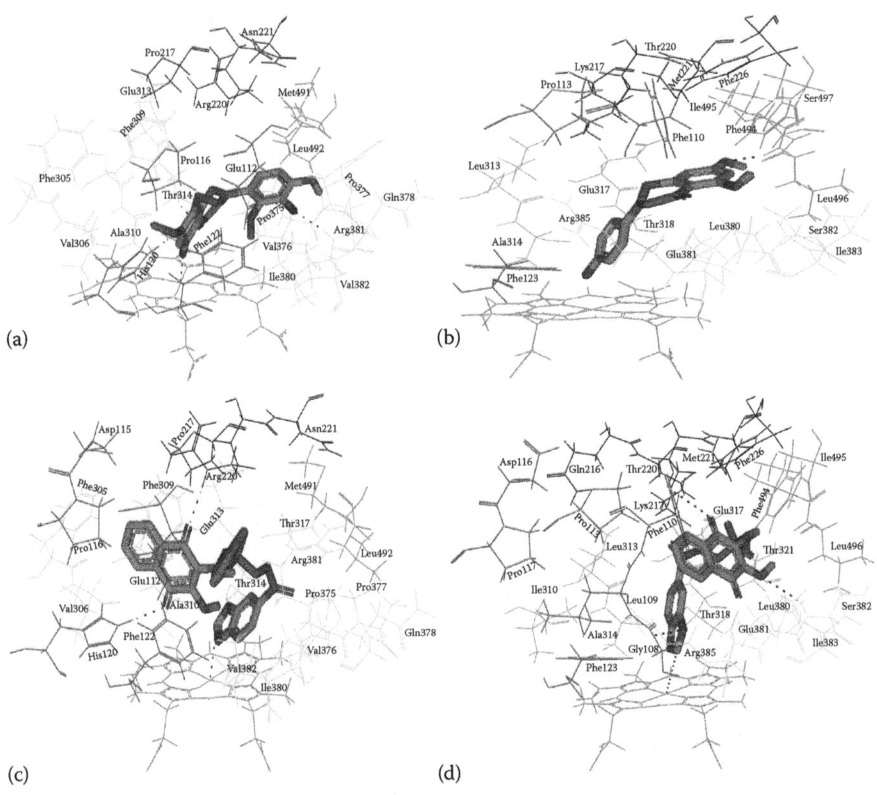

FIGURE 12.5 **(See color insert.)** Docking simulation of 5,4′-dihydroxy-7,8,2′3′-tetramethoxyflavone (a), 5,7,4′-trihydroxyflavone (apigenin, b), and rhinacanthin-D (c, d) into CYP6AA3 (a, c) and CYP6P7 (b, d). Other details are the same as in the caption of Figure 12.4.

In contrast in CYP6P7, the 5-hydroxyflavone derivatives with methoxy groups (5-hydroxy-7,8,2′,3′-tetramethoxyflavone and 5,4′-dihydroxy-7,8,2′,3′-tetramethoxyflavone) were inhibitors of noncompetitive mode, and 5,7,4′-trihydroxyflavone (apigenin) and 5,7-dihydroxyflavone (chrysin) were mixed-type inhibitors. Relevant important residues of CYP6P7 surrounding the apigenin flavone are shown in Figure 12.5(b). The presence of four methoxylated group of 5-hydroxyflavones hindered these tetramethoxylated flavones from being accommodated within the CYP6P7 narrow pocket (see the example of 5-hydroxy-7,8,2′,3′-tetramethoxy flavone and 5,4′-dihydroxy-7,8,2′,3′-tetramethoxy flavone in Table 12.3), while the absence of the methoxy groups in flavones (apigenin and chrysin) could fit in the pocket (Table 12.3). Blocking of flavonoids (3′,4′-dimethoxy-5,7-dihydroxyflavone, for example) from entering the active site cavity of a human CYP1A2 by amino acids in SRSs has also been reported [40]. As shown in Table 12.3, apigenin and chrysin with their B rings turning toward the heme center were interacted with Leu496 and Phe123 of CYP6P7 by hydrogen bonds and π–π interaction, respectively (Table 12.3). These results indicated that the two small-size compounds could reside within the narrow pocket of CYP6P7 via a couple of interactions with amino acids of the enzyme.

12.6.2 Inhibition of *An. minimus* CYP6AA3 and CYP6P7 Enzymes by Compounds Isolated from *R. nasutus*

Molecular docking of CYP6AA3 with rhinacanthins series (rhinacanthin-A, -B, -D, -E, -G, -N, and -Q), of which all MBI-inhibited CYP6AA3, except rhinacanthin-A and -Q that were reversibly mixed-type inhibitors, is displayed in Table 12.4 [20,21]. These rhinacanthins were surrounded by amino acids of CYP6AA3 and CYP6P7 slightly similar to those found for flavonoids, with mostly additional amino acids bound to rhinacanthins compared to flavonoids (Figures 12.5c and 12.5d). Figures 12.5(c) and 12.5(d) show amino acids of CYP6AA3 and CYP6P7, respectively, surrounding rhinacanthin-D as an example. The pyranonaphthoquinone rhinacanthin-A, having its naphthoquinone ring posing toward the CYP6AA3 heme center, was interacted with Phe122 and Phe309 by π–π interactions together with surrounding amino acids and with two hydrogen bonds formed with His120 and Arg220 (Table 12.4). In contrast, in rhinacanthin-B pyranonaphthoquinone compound, which was MBI of CYP6AA3, the C7′ at the alkene chain was positioned toward the catalytic heme center at a closer distance than rhinacanthin-A for catalysis and might form a 6′,7′-epoxide product. The epoxide might covalently bind amino acids of CYP6AA3 and result in its inactivation. Inactivation of the human CYP2E1 or binding of epoxide on proteins (e.g., microsomal proteins and hemoglobin) have been known [49,50]. Rhinacanthin-B was bound by two hydrogen bonds formed with His120 and Arg220, together with Phe122- and Phe309-mediated π–π interactions similar to with rhinacanthin-A, but with an additional π–π interaction with Phe305 (SRS4). Rhinacanthin-B was placed closer to the heme moiety of CYP6AA3 than the rhinacanthin-A pyranonaphthoquinone counterpart, in congruence with its approximately threefold lower K_i value than rhinacanthin-A (Table 12.4). Docking to CYP6AA3 of rhinacanthin-D and -E, of which structures comprised a methylenedioxyphenyl group attached with naphthoquinone ring via ester bond (for rhinacanthin-D) and two methylenedioxyphenyl groups forming a lignan (rhinacanthin-E), respectively, revealed that the methylenic carbon of both compounds was predicted posing toward the heme center of the enzyme (Table 12.4). The methylenic carbon was predicted to form carbene carbon upon CYP6AA3-mediated catalysis and further formed heme-inhibitory complex in compliance with MBI inhibitory modes (Table 12.4, reference 21). Both were bound by two π–π interactions with Phe122 and Phe309, and one hydrogen bond with Arg220, but with two additional hydrogen bonds formed by His120 and Thr314 for rhinacanthin-D to result in a distinct lower K_i value for rhinacanthin-D than rhinacanthin-E (Table 12.4, reference 21). This might imply that more interactions were required to assure an accurate conformation and perhaps more tight binding of inhibitors that displayed lower K_i value than their counterparts, as seen with rhinacanthin-B versus rhinacanthin-A and rhinacanthin-D versus rhinacanthin-E. For rhinacanthin-G, whose side chain was an alkene containing an epoxide group, it inhibited CYP6AA3 in MBI mode and also inhibited the enzyme in the absence of NADPH [21]. Prior results suggested that the epoxide group on the alkene side chain of rhinacanthin-G directly inhibited CYP6AA3, in addition to forming an MBI reactive inhibitor [21]. Docking analysis of rhinacanthin-G to the CYP6AA3 active site gave a conformation of its 2-hydroxy group positioning toward heme iron, thus possibly generating an additional epoxide product—a CYP6AA3 inactivator, in addition to the existing epoxide group that displayed direct inhibition on CYP6AA3. Rhinacanthin-G formed three hydrogen bonds with His120, Arg220, and Thr314 residues of CYP6AA3 without π–π interaction (Table 12.4). Finally, another pair of rhinacanthins, rhinacanthin-N that comprised a 1′-hydroxy,4′-methoxynaphthalene ring and rhinacanthin-Q that possessed a 1′,4′-dimethoxynaphthalene ring (Figure 12.2), were docked into the active site of CYP6AA3. With the nature of MBI by rhinacanthin-N, in contrast to rhinacanthin-Q, it was predicted to place its naphthalene ring toward the heme center to generate an epoxide product, as previously reported for CYP1A2 and CYP2A13 [51], and consequently inactivated CYP6AA3 enzyme [21]. The binding of rhinacanthin-N onto the enzyme cavity was depicted by two π–π interactions with Phe122 and Phe309 and a hydrogen bond with His120, while there was an additional hydrogen bond formation by Arg220 interacted with rhinacanthin-Q, whose K_i value was lower than that of rhinacanthin-N (Table 12.3). Moreover, it could be observed that rhinacanthin-Q had its naphthalene ring moving away from the heme center in order to fit within the enzyme cavity, supporting its inability to be catalyzed by CYP6AA3 to form a mechanism-based inhibitor (Table 12.4).

In conclusion, it could be observed that the seven rhinacanthins could form several hydrogen bonds (up to three bonds) with CYP6AA3 amino acid residues; His120 (SRS1) and Arg220 (SRS2) mainly bound six rhinacanthins and Thr314 (SRS4) bound two rhinacanthins (Table 12.4). Moreover, His120 of CYP6AA3 was a common amino acid that formed hydrogen bonds with both flavonoids and rhinacanthins. In addition, Phe122 (SRS1) and Phe309 (SRS4) were residues that primarily formed π–π interactions with pyrethroid substrates and rhinacanthins, but less frequently with flavonoids. These differences implied that the compound nature and molecular size might affect their bindings to the specific residues in the active site cavity of CYP6AA3.

For CYP6P7, rhinacanthin-A, -B, -D, -G, and -N were MBI inhibitors and rhinacanthin-C was a competitive inhibitor [20,21]. Similarly to CYP6AA3, rhinacanthin-G possessed both direct and MBI inhibitions on CYP6P7 [21]. Molecular docking of these compounds in the CYP6P7 pocket revealed that the orientations of rhinacanthin-A and -D in interacting with the heme in CYP6P7 were similar to CYP6AA3, with rhinacanthin-A posing naphthoquinone and rhinacanthin-D having its methylenic carbon of the methylenedioxyphenyl group positioned toward the CYP6P7 heme (Table 12.4). Thus, rhinacanthin-A and -D could form a CYP6P7-mediated epoxide product of the naphthoquinone ring and methylenic carbene, respectively, that could inactivate the CYP6P7 enzyme, similarly to inactivation of CYP6AA3. The remaining rhinacanthin-B, -G, and -N arranged their naphthoquinone rings toward the CYP6P7 heme center and were able to fit within the narrow CYP6P7 active site (Table 12.4). Such orientations of rhinacanthin-B, -G, and -N implied that their intermediate MBI inhibitors mediated by CYP6P7 catalysis might form epoxide products of the naphthoquinone ring and consequently inactivated the enzyme. The K_i values of the five MBI rhinacanthins (rhinacanthin-A, -B, -D, -G, and -N) were correlated with their binding energies, but there was absence of correlation with distance from heme (Table 12.4). In contrast, in rhinacanthin-C (a competitive inhibitor), unlike the MBI rhinacanthins, its alkyl chain was placed pointing toward the active site of CYP6P7 (Table 12.4). These different conformations of inhibitors, observed in CYP6P7 compared with CYP6AA3, implied that different mechanisms in interaction with rhinacanthins were employed by CYP6AA3 and CYP6P7. Similar different effects of amino acids in binding and directing ligands in particular conformations have been found for various P450s in insects [52].

Contact residues of CYP6P7 that played a predominant role in accommodating rhinacanthins to fit in the CYP6P7 active site were Thr220 (SRS2), which formed hydrogen bonds with all six rhinacanthin inhibitors, and Phe123 (SRS1), which formed π–π interaction. Other residues within the CYP6P7 active site, including Arg385 (SRS5), Ser382 (SRS5), Leu496 (SRS6), Phe110 (SRS1), and Phe494 (SRS6), were observed interacting with less frequency with rhinacanthins (Table 12.4). Moreover, we found that Thr318 (SRS4) and Leu496 (SRS6) of CYP6P7 predominantly formed hydrogen bonds, and mainly Phe123 (SRS1) formed π–π interaction, with pyrethroids, benzyloxyrufin, test inhibitors, and two flavonoids, but not rhinacanthins. It might also be possible that inhibitors with different molecular sizes and/or low K_i values formed multiple hydrogen bonds with amino acid residues of both CYP6AA3 and CYP6P7, resulting in good interaction between inhibitors and enzymes.

In this study, docking to the active site cavities of CYP6AA3 and CYP6P7 of pyrethroid and benzyloxyresorufin substrates, and inhibitory compounds—mainly flavonoids and naphthoquinone derivatives—isolated from *A. paniculata* and *R. nasutus*, respectively, was performed. All of the compounds were contacted with amino acids, of which some residues were shared in binding with substrates and inhibitors. Essentially, these compounds formed hydrogen bonds with residues of the active sites of both CYP6AA3 and CYP6P7—mostly with His120 (SRS1), Arg220 (SRS2), and Thr314 (SRS4) for CYP6AA3, and Thr220 (SRS2), Thr318 (SRS4), and Leu496 (SRS6) for CYP6P7. In addition, most of these compounds formed π–π interactions with Phe122 (SRS1) and, less frequently, with Phe309 (SRS4) of CYP6AA3 (Table 12.3). The residues His120 and Arg220 bound both substrates and inhibitors, while Thr314 preferred to bind inhibitory compounds. For CYP6P7, the residue Phe123 in SRS1 was the main contact residue forming π–π interaction with most compounds, while Thr220 preferentially formed hydrogen bonds with most rhinacanthin inhibitors and Thr318 and Leu496 preferred to bind most substrates and some inhibitors (Tables 12.1 through 12.4). Nevertheless, contact residues of both mosquito enzymes were shared in interaction with substrates and inhibitors, implying that the inhibitors exert their roles partly by blocking or competing with substrates in entering the enzyme active sites. The results also indicated that substrates and inhibitors were located near the SRSs 1, 2, and 4 of CYP6AA3, as its active site was large and not all SRSs had contacts with their ligands. In contrast, CYP6P7 substrates and inhibitors were oriented near SRSs 1, 2, 4, and 6. All compounds were located at a distance from SRS3 of both enzymes, which was a part of the substrate access channel site. Moreover, these compounds were hydrophobic and that might increase favorable interactions with the hydrophobic networking of the active sites of both CYP6AA3 and CYP6P7 enzymes. The nature of residues and differences in topology of both enzyme binding sites might thus exert effects on their bindings by the substrates and inhibitors studied.

12.7 CONCLUSION

In summary, there were a total of seven and eight residues in the active site of CYP6AA3 and CYP6P7, respectively, involved in hydrogen bond formations and π–π interactions with the substrates and inhibitory compounds. These included Phe122 (SRS1), Phe305 (SRS4), Phe309 (SRS4), His120 (SRS1), Arg220 (SRS2), Thr314 (SRS4), and Arg381 (SRS5) in CYP6AA3; and residues Phe110 (SRS1), Phe123 (SRS1), Phe494 (SRS6), Thr220 (SRS2), Thr318 (SRS4), Ser382 (SRS5), Arg385 (SRS5), and Leu496 (SRS6) in CYP6P7. These amino acids displayed distinct roles in bonding formations for different compounds. Among CYP6AA3 residues, His120 was predicted predominantly to form hydrogen bonds with four pyrethroids and 11 inhibitors out of 16 inhibitors, Thr314 formed hydrogen bonds with 9 inhibitors, and Arg220 formed hydrogen bonds with two pyrethroids and 9 inhibitors. In addition, Phe122 and Phe309 of CYP6AA3 were key interactions for substrates and especially rhinacanthins inhibitors via π–π interactions. For CYP6P7, the residues Thr318 and Leu496 preferentially formed hydrogen bonds with pyrethroids, benzyloxyresorufin, and test inhibitors, while Thr220 preferred to bind rhinacanthins. Moreover, Phe123 of CYP6P7 was the key π–π interaction for all substrates and most inhibitors. These results thus reflect the critical role of His120, Phe122, Arg220, Phe309, and Thr314 of CYP6AA3 and Phe123, Thr318, Leu496, and Thr220 of CYP6P7 in binding, orientation, and recognition of substrates and inhibitors. These interactions shared between substrates and inhibitors might support inhibitory effects of inhibitors by blocking the binding site of substrates within the enzyme active site. Thus, molecular modeling has identified important amino acid residues within the active sites of CYP6AA3 and CYP6P7 enzymes that were predicted interacting with different compound groups as substrates and inhibitors. The relationships between the chemical characteristics of these compounds and their inhibition modes could be explained by molecular docking analysis. This study thus could provide valuable information for the rational design of mosquito P450s inhibitors and the preparation of natural compounds that might be potentially developed as pyrethroid synergists for insecticide resistance management control strategies in malaria mosquito vectors.

REFERENCES

[1] World Health Organization, *WHO malaria report: 2013*, in *WHO Global Malaria Programme*, WHO, Geneva, 2013, p. 199. Available at www.who.int/entity/malaria/publications/world_malaria_report_2013/en/-6k

[2] F.O. Okumu and S.J. Moore, *Combining indoor residual spraying and insecticide-treated nets for malaria control in Africa: A review of possible outcomes and an outline of suggestions for the future*, Malar. J. 10 (2011), pp. 208.

[3] K.H. Toe, C.M. Jones, S. N'Fale, H.M. Ismail, R.K. Dabire, and H. Ranson, *Increased pyrethroid resistance in malaria vectors and decreased bed net effectiveness, Burkina Faso*, Emerg. Infect. Dis. 20 (2014), pp. 1691–1696.

[4] K.S. Choi, R. Christian, L. Nardini, O.R. Wood, E. Agubuzo, M. Muleba, S. Munyati, A. Makuwaza, L.L. Koekemoer, B.D. Brooke, R.H. Hunt, and M. Coetzee, *Insecticide resistance and role in malaria transmission of* Anopheles funestus *populations from Zambia and Zimbabwe*, Parasit. Vectors 7 (2014), pp. 464.

[5] E. Ochomo, N.M. Bayoh, L. Kamau, F. Atieli, J. Vulule, C. Ouma, M. Ombok, K. Njagi, D. Soti, E. Mathenge, L. Muthami, T. Kinyari, K. Subramaniam, I. Kleinschmidt, M.J. Donnelly, and C. Mbogo, *Pyrethroid susceptibility of malaria vectors in four districts of western Kenya*, Parasit. Vectors 7 (2014), pp. 310.

[6] T.G. Davies, L.M. Field, P.N. Usherwood, and M.S. Williamson, *DDT, pyrethrins, pyrethroids and insect sodium channels*, IUBMB Life 59 (2007), pp. 151–162.

[7] J. Hemingway and H. Ranson, *Insecticide resistance in insect vectors of human disease*, Annu. Rev. Entomol. 45 (2000), pp. 371–391.

[8] J. Hemingway, N.J. Hawkes, L. McCarroll, and H. Ranson, *The molecular basis of insecticide resistance in mosquitoes*, Insect Biochem. Mol. Biol. 34 (2004), pp. 653–665.

[9] N.R. Price, *Insect resistance to insecticides: Mechanisms and diagnosis*, Comp. Biochem. Physiol. C 100 (1991), pp. 319–326.

[10] M.J. Donnelly, V. Corbel, D. Weetman, C.S. Wilding, M.S. Williamson, and W.C. Black, *Does* kdr *genotype predict insecticide-resistance phenotype in mosquitoes?* Trends Parasitol. 25 (2009), pp. 213–219.

[11] J.P. David, H.M. Ismail, A. Chandor-Proust, and M.J. Paine, *Role of cytochrome P450s in insecticide resistance: Impact on the control of mosquito-borne diseases and use of insecticides on Earth*, Philos. Trans. R. Soc. Lond. B Biol. Sci. 368 (2013), p. 20120429.

[12] R. Feyereisen, *Insect P450 enzymes*, Annu. Rev. Entomol. 44 (1999), pp. 507–533.

[13] R.L. Metcalf, Mode of action of insecticide synergists, Annu. Rev. Entomol. 12 (1967), pp. 229–256.

[14] M.R. Fakoorziba, F. Eghbal, and V.A. Vijayan, *Synergist efficacy of piperonyl butoxide with deltamethrin as pyrethroid insecticide on* Culex tritaeniorhynchus *(Diptera: Culicidae) and other mosquitoe species*, Environ. Toxicol. 24 (2009), pp. 19–24.

[15] J.C. Mouatcho, G. Munhenga, K. Hargreaves, B.D. Brooke, M. Coetzee, and L.L. Koekemoer, *Pyrethroid resistance in a major African malaria vector* Anopheles arabiensis *from Mamfene, northern KwaZulu—Natal, South Africa*, S. Afr. J. Sci. 105 (2009), pp. 127–131.

[16] S. Boonsuepsakul, E. Luepromchai, and P. Rongnoparut, *Characterization of* Anopheles minimus *CYP6AA3 expressed in a recombinant baculovirus system*, Arch. Insect Biochem. Physiol. 69 (2008), pp. 13–21.

[17] P. Duangkaew, S. Pethuan, D. Kaewpa, S. Boonsuepsakul, S. Sarapusit, and P. Rongnoparut, *Characterization of mosquito CYP6P7 and CYP6AA3: Differences in substrate preference and kinetic properties*, Arch. Insect Biochem. Physiol. 76 (2011), pp. 236–248.

[18] P. Lertkiatmongkol, E. Jenwitheesuk, and P. Rongnoparut, *Homology modeling of mosquito cytochrome P450 enzymes involved in pyrethroid metabolism: Insights into differences in substrate selectivity*, BMC Res. Notes 4 (2011), pp. 321.

[19] R. Kotewong, P. Duangkaew, E. Srisook, S. Sarapusit, and P. Rongnoparut, *Structure–function relationships of inhibition of mosquito cytochrome P450 enzymes by flavonoids of* Andrographis paniculata, Parasitol. Res. 113 (2014), pp. 3381–3392.

[20] S. Pethuan, P. Duangkaew, S. Sarapusit, E. Srisook, and P. Rongnoparut, *Inhibition against mosquito cytochrome P450 enzymes by rhinacanthin-A, -B, and -C elicits synergism on cypermethrin cytotoxicity in* Spodoptera frugiperda *cells*, J. Med. Entomol. 49 (2012), pp. 993–1000.

[21] R. Kotewong, P. Pouyfung, P. Duangkaew, A. Prasopthum, and P. Rongnoparut, *Synergy between rhinacanthins from* Rhinacanthus nasutus *in inhibition against mosquito cytochrome P450 enzymes*, Parasitol. Res. 114 (2015), pp. 2567–2579.

[22] G. Bingham, C. Strode, L. Tran, P.T. Khoa, and H.P. Jamet, *Can piperonyl butoxide enhance the efficacy of pyrethroids against pyrethroid-resistant* Aedes aegypti? Trop. Med. Int. Health 16 (2011), pp. 492–500.

[23] R.F. Djouaka, A.A. Bakare, O.N. Coulibaly, M.C. Akogbeto, H. Ranson, J. Hemingway, and C. Strode, *Expression of the cytochrome P450s, CYP6P3 and CYP6M2 is significantly elevated in multiple pyrethroid resistant populations of* Anopheles gambiae *s.s. from Southern Benin and Nigeria*, BMC Genomics 9 (2008), p. 538.

[24] P. Muller, M. Chouaibou, P. Pignatelli, J. Etang, E.D. Walker, M.J. Donnelly, F. Simard, and H. Ranson, *Pyrethroid tolerance is associated with elevated expression of antioxidants and agricultural practice in* Anopheles arabiensis *sampled from an area of cotton fields in Northern Cameroon*, Mol. Ecol. 17 (2008), pp. 1145–1155.

[25] D.A. Amenya, R. Naguran, T.C. Lo, H. Ranson, B.L. Spillings, O.R. Wood, B.D. Brooke, M. Coetzee, and L.L. Koekemoer, *Over expression of a cytochrome P450 (CYP6P9) in a major African malaria vector,* Anopheles funestus, *resistant to pyrethroids*, Insect Mol. Biol. 17 (2008), pp. 19–25.

[26] P. Rongnoparut, S. Boonsuepsakul, T. Chareonviriyaphap, and N. Thanomsing, *Cloning of cytochrome P450, CYP6P5, and CYP6AA2 from* Anopheles minimus *resistant to deltamethrin*, J. Vector Ecol. 28 (2003), pp. 150–158.

[27] P. Rodpradit, S. Boonsuepsakul, T. Chareonviriyaphap, M.J. Bangs, and P. Rongnoparut, *Cytochrome P450 genes: Molecular cloning and overexpression in a pyrethroid-resistant strain of* Anopheles minimus *mosquito*, J. Am. Mosq. Control Assoc. 21 (2005), pp. 71–79.

[28] P. Muller, E. Warr, B.J. Stevenson, P.M. Pignatelli, J.C. Morgan, A. Steven, A.E. Yawson, S.N. Mitchell, H. Ranson, J. Hemingway, M.J. Paine, and M.J. Donnelly, *Field-caught permethrin-resistant* Anopheles gambiae *overexpress CYP6P3, a P450 that metabolises pyrethroids*, PLoS Genet. 4 (2008), pp. e1000286.

[29] B.J. Stevenson, J. Bibby, P. Pignatelli, S. Muangnoicharoen, P.M. O'Neill, L.Y. Lian, P. Muller, D. Nikou, A. Steven, J. Hemingway, M.J. Sutcliffe, and M.J. Paine, *Cytochrome P450 6M2 from the malaria vector* Anopheles gambiae *metabolizes pyrethroids: Sequential metabolism of deltamethrin revealed*, Insect Biochem. Mol. Biol. 41 (2011), pp. 492–502.

[30] J.M. Riveron, H. Irving, M. Ndula, K.G. Barnes, S.S. Ibrahim, M.J. Paine, and C.S. Wondji, *Directionally selected cytochrome P450 alleles are driving the spread of pyrethroid resistance in the major malaria vector* Anopheles funestus, Proc. Natl. Acad. Sci. U S A 110 (2013), pp. 252–257.

[31] C.S. Wilding, I. Smith, A. Lynd, A.E. Yawson, D. Weetman, M.J. Paine, and M.J. Donnelly, *A cis-regulatory sequence driving metabolic insecticide resistance in mosquitoes: Functional characterisation and signatures of selection*, Insect Biochem. Mol. Biol. 42 (2012), pp. 699–707.

[32] B.J. Stevenson, P. Pignatelli, D. Nikou, and M.J. Paine, *Pinpointing P450s associated with pyrethroid metabolism in the dengue vector,* Aedes aegypti*: Developing new tools to combat insecticide resistance*, PLoS Negl. Trop. Dis. 6 (2012), pp. e1595.

[33] S. Kasai, O. Komagata, K. Itokawa, T. Shono, L.C. Ng, M. Kobayashi, and T. Tomita, *Mechanisms of pyrethroid resistance in the dengue mosquito vector,* Aedes aegypti*: Target site insensitivity, penetration, and metabolism*, PLoS Negl. Trop. Dis. 8 (2014), pp. e2948.

[34] J.G. Scott, *Inhibitors of CYP6D1 in house fly microsomes*, Insect Biochem. Mol. Biol. 26 (1996), pp. 645–649.

[35] J.G. Scott, M. Foroozesh, N.E. Hopkins, T.G. Alefantis, and W.L. Alworth, *Inhibition of cytochrome P450 6D1 by alkynylarenes, methylenedioxyarenes, and other substituted aromatics*, Pestic. Biochem. Physiol. 67 (2000), pp. 63–71.

[36] L.A. McLaughlin, U. Niazi, J. Bibby, J.P. David, J. Vontas, J. Hemingway, H. Ranson, M.J. Sutcliffe, and M.J. Paine, *Characterization of inhibitors and substrates of* Anopheles gambiae *CYP6Z2*, Insect Mol. Biol. 17 (2008), pp. 125–135.

[37] A. Comish-Bowden, *Reversible inhibition and inactivation,* in *Fundamentals of Enzyme Kinetics*, A. Comish-Bowden, ed., 3 rd ed., Portland Press, London, 2004, pp. 144–200.

[38] M.A. Correia and P.R. Ortiz de Montellano, *Inhibition of cytochrome P450 enzymes*, in *Cytochrome P450: Structure, Mechanism, and Biochemistry*, P.R. Ortiz de Montellano, ed., 3rd ed., Kluwer Academic/Plenum Publisher, New York, 2005, pp. 247–322.

[39] S.T. Orr, S.L. Ripp, T.E. Ballard, J.L. Henderson, D.O. Scott, R.S. Obach, H. Sun, and A.S. Kalgutkar, *Mechanism-based inactivation (MBI) of cytochrome P450 enzymes: Structure–activity relationships and discovery strategies to mitigate drug–drug interaction risks*, J. Med. Chem. 55 (2012), pp. 4896–4933.

[40] T. Shimada, K. Tanaka, S. Takenaka, N. Murayama, M.V. Martin, M.K. Foroozesh, H. Yamazaki, F.P. Guengerich, and M. Komori, *Structure–function relationships of inhibition of human cytochromes P450 1A1, 1A2, 1B1, 2C9, and 3A4 by 33 flavonoid derivatives*, Chem. Res. Toxicol. 23 (2010), pp. 1921–1935.

[41] S.J. Godin, J.A. Crow, E.J. Scollon, M.F. Hughes, M.J. DeVito, and M.K. Ross, *Identification of rat and human cytochrome p450 isoforms and a rat serum esterase that metabolize the pyrethroid insecticides deltamethrin and esfenvalerate*, Drug Metab. Dispos. 35 (2007), pp. 1664–1671.

[42] O. Gotoh, *Substrate recognition sites in cytochrome P450 family 2 (CYP2) proteins inferred from comparative analyses of amino acid and coding nucleotide sequences*, J. Biol. Chem. 267 (1992), pp. 83–90.

[43] A. Oda, N. Yamaotsu, and S. Hirono, *New AMBER force field parameters of heme iron for cytochrome P450s determined by quantum chemical calculations of simplified models*, J. Comput. Chem. 26 (2005), pp. 818–826.

[44] O. Trott, and A.J. Olson, *AutoDock Vina: Improving the speed and accuracy of docking with a new scoring function, efficient optimization, and multithreading*, J. Comput. Chem. 31 (2010), pp. 455–461.

[45] C. Bissantz, B. Kuhn, and M. Stahl, *A medicinal chemist's guide to molecular interactions*, J. Med. Chem. 53 (2010), pp. 5061–5084.

[46] Z. Wen, M.R. Berenbaum, and M.A. Schuler, *Inhibition of CYP6B1-mediated detoxification of xanthotoxin by plant allelochemicals in the black swallowtail (*Papilio polyxenes*)*, J. Chem. Ecol. 32 (2006), pp. 507–522.

[47] T. Usia, T. Watabe, S. Kadota, and Y. Tezuka, *Metabolite-cytochrome P450 complex formation by methylenedioxyphenyl lignans of* Piper cubeba: *Mechanism-based inhibition*, Life Sci. 76 (2005), pp. 2381–2391.

[48] Y. Li, E. Wang, C.J. Patten, L. Chen, and C.S. Yang, *Effects of flavonoids on cytochrome P450-dependent acetaminophen metabolism in rats and human liver microsomes*, Drug Metab. Dispos. 22 (1994), pp. 566–571.

[49] P.D. Premdas, R.J. Bowers, and P.G. Forkert, *Inactivation of hepatic CYP2E1 by an epoxide of diallyl sulfone*, J. Pharmacol. Exp. Ther. 293 (2000), pp. 1112–1120.

[50] S.S. Lau, and V.G. Zannoni, *Bromobenzene epoxidation leading to binding on macromolecular protein sites*, J. Pharmacol. Exp. Ther. 219 (1981), pp. 563–572.

[51] T. Fukami, M. Katoh, H. Yamazaki, T. Yokoi, and M. Nakajima, *Human cytochrome P450 2A13 efficiently metabolizes chemicals in air pollutants: Naphthalene, styrene, and toluene*, Chem. Res. Toxicol. 21 (2008), pp. 720–725.

[52] M.A. Schuler and M.R. Berenbaum, *Structure and function of cytochrome P450s in insect adaptation to natural and synthetic toxins: Insights gained from molecular modeling*, J. Chem. Ecol. 39 (2013), pp. 1232–1245.

13 Critical Review of Models for Assessment of Resident and Bystander Exposure to Adulticides Used in Mosquito Control

James Devillers

CONTENTS

ABSTRACT

The goal of this study was to review the different models available for assessing the exposure of residents and bystanders to adulticides used in vector control. Because the exposure to a chemical needs first to know its concentration and fate in the environment, the first part of the chapter is focused on the different models that can be used for evaluating the concentrations of adulticides found in the environmental compartments after a treatment. Then, the exposure models and equations are described mainly through their practical use in diverse scenarios and with insecticides presenting different physicochemical properties. The chapter ends by an inventory of the main gaps in the domain and the necessary work to do to fill them.

KEYWORDS

- Resident
- Bystander
- Exposure models
- Ultralow-volume (ULV) applications
- Age categories

13.1 INTRODUCTION

Human health risk assessment for chemicals used in vector control is a classical four-step process leading to the characterization of the adverse effects that may be caused by these chemicals during the treatments. After hazard identification in the first step of this process, exposure and dose–response assessment have to be considered. The former can be simply viewed as the contact with the chemical at the outer boundary of human, while the latter quantifies the health effects caused by a range of dose levels. The crucial role for the exposure assessment is to provide results that can be meaningfully coupled to the dose–response assessment, such as risk characterization, so that the fourth step in the risk assessment paradigm can be done.

Once the chemical used in vector control is released in the environment, it is transported through space and time. Various transformation reactions may occur, such as degradation, volatilization, deposition, and so forth. Moreover, the topography of the medium and the abiotic factors intervene in the fate of the chemical. These processes will result in various concentrations of the chemical within the environment as functions of time. Exposure then occurs when humans contact the chemical-bearing media in the course of their activities. If the chemical subsequently crosses the outer boundary and enters the human body, a dose occurs that may then lead to a health effect [1,2].

Different approaches exist for quantifying human exposure. Direct methods involve measurements of exposure taken at the point of contact or uptake. Indirect methods encompass the exposure estimates from measurements and data by means of models of different levels of complexity.

An exposure model is an empirical or statistical device which allows estimation of individual sub-population or population exposure from a set of input data. Exposure models represent important tools for indirect exposure assessments. They are typically used where direct measurements of exposure or biological monitoring data are not available or where these techniques are not appropriate for the exposure assessment situation [3]. Additionally, there are a number of benefits associated with the use of models for quantifying human exposures:

- They allow us to better understand the exposure process. The modeler forms a conceptual work flow of the exposure process. It is algorithmically transformed into a set of equations and a functional scheme combining these equations with the data needed to solve them. The algorithm must describe the process with sufficient accuracy to fulfill the purposes of the assessment, but its complexity has to be optimized in order to respect the principle of parsimony. It is then converted into a functional computer program that executes the computations to provide results that can be effectively used together with dose–response information to complete the risk characterization. This process of reducing the exposure process to a model requires a clear understanding of the exposure process and of the influential factors.

- They are suited to perform "what-if" analyses and can predict potential exposures for future or hypothetical releases following different scenarios, which are combinations of facts, assumptions, and inferences that define a discrete situation where potential exposures may occur.
- They reduce the need for costly and time-consuming monitoring programs.
- They allow us to maximize the use of existing measurement data.
- They allow the evaluation of the importance of each input datum on the quality and significance of the output (sensitivity analysis).
- They allow us to consider exposures via multiple routes and pathways.
- They are particularly suited to compare the effects of different chemicals from a same set of input data [1,3,4].

The goal of this study was to review the available models for assessing the exposure of residents and bystanders to adulticides used in mosquito control. This analysis should answer the following general questions: What approaches can be adopted to assess exposures in a given context? What strategies and parameters should be favored or adjusted to describe satisfactorily the exposures to adulticide treatments in normal situations and during outbreaks?

Because studying the exposure to a chemical needs first to know its concentration and fate in the environment, the first part of this review is focused on the different models that have been used for evaluating the concentrations of adulticides found in the environment after a treatment. Then, the exposure models and equations are described mainly through their practical use for estimating the exposure of residents and bystanders to various adulticides.

13.2 ESTIMATION OF ENVIRONMENTAL CONCENTRATIONS AND FATE OF ADULTICIDES

13.2.1 Ultralow-Volume Applications

One of the most effective and used techniques of managing high densities of flying adult mosquitoes is ultralow-volume spray application of insecticides (Figure 13.1). The very essence of the ULV method is the use of a minimum application volume [5]. The space spray, or aerosol, is a liquid insecticide dispersed into the air in the form of hundreds of millions of tiny droplets of less than 50 µm in diameter. It is only effective while the droplets remain airborne. Space sprays are applied as either thermal or cold fogs [6]. The insecticide used in thermal fogs is diluted in a carrier liquid, which is

FIGURE 13.1 Ground vehicle-mounted ULV space spraying application of adulticide.

usually oil based. With cold fogs, the droplets are formed by the mechanical breaking up of the spray mixture, either by passing it through high-pressure nozzles or by passing a slow stream of the mixture through a high-velocity vortex of air [7]. The speed at which droplets fall is determined by their mass. Thus, for example, a droplet of 20 μm diameter will fall at 0.012 min/s, taking 14 min to fall 10 m in still air, whereas a 100 μm droplet falls at 0.279 min/s and will take only 36 s to fall the same distance [7].

Droplet size is the main factor affecting the efficiency of space sprays for the control of adult mosquitoes. There are different parameters that can be used to characterize the size of the droplets:

- The volume median diameter (VMD, in micrometers), also known as diameter volume 50% ($Dv_{0.5}$) [8], is the number that divides the spray into two equal parts by volume: one-half containing droplets smaller than this diameter, the other half containing larger droplets [7]. The optimum droplet size (VMD) for adult mosquito control is 5–25 μm [9]. Droplets smaller than 5 μm do not impinge well on flying mosquitoes and often drift far beyond the intended target sites, while droplets larger than 25 μm rapidly fall out of the aerosol due to their heavier weights.
- The number median diameter (NMD) is the value that divides the spray into two equal parts by number of droplets, so that half the droplets are smaller and half larger.

- Span is determined from the diameter of the 90% value (V_{90}) by volume minus the 10% value by volume (V_{10}), divided by the VMD value. Span gives an indication of the range of droplet sizes and is ideally <2 [7].

For ULV ground aerosols of adulticides against mosquitoes, the critical factors are wind speed and direction, temperature, and atmospheric stability and turbulence. Wind speed has a strong effect on droplet distribution and impingement on mosquitoes. In most situations a wind speed of 1–4 m/s (about 3.6–15 km/h) is needed to drift the droplets downwind from the line of travel. Spraying should not take place when wind speed exceeds 15 km/h [7]. It is noteworthy that the topography of the terrain and the presence of vegetation impact air movement and hence the distribution of the droplets. Generally, people avoid treatments when wind direction varies >45° from perpendicular to swath direction [5].

Atmospheric stability and turbulence can be evaluated by the calculation of the stability ratio (Eq. 13.1).

$$\text{Stability ratio} = (\text{T2} - \text{T1}/\mu^2) \times 10^5 \tag{13.1}$$

where T2 is the temperature in °C at 10 m, T1 is the temperature in °C at 3 m, and μ is the average wind velocity (cm/s) [5].

A limited number of experimental studies has been performed to examine the air concentration and/or the terrestrial deposition of adulticides after truck-mounted ULV applications for the control of mosquitoes. Perich et al. [10] performed ground ULV applications of malathion (91%) at 438.8 mL/ha (500 g active ingredient [a.i.]/ha) in urban residential areas, each being about 11.5 ha with 24 blocks. Homes had a small courtyard area with various plants (0.5–2 m) and one or two tropical trees (10–15 m). The truck operated at 8 km/h. Doors and windows of homes were open before the beginning of the treatments. The mean amounts of malathion deposited on the filter papers beneath the bed of a bedroom and in a backroom kitchen, which were collected 30 min after the treatment outside, were equal to 0.69, 0.38, and 0.27 μg/cm^2, respectively, in a first experiment, and 0.47, 0.21, and 0.07 μg/cm^2 in a second one. The VMDs of malathion aerosol collected from outside and inside homes were respectively equal to 19.26 μm (range of 14.39 to 24.23) and 17.69 μm (7 to 42.12) in the first treatment and 12.95 μm (9.82 to 17.67) and 21.96 μm (17.59 to 61.65) in the second [10]. In another study [11], malathion (95%, 104 mL/min) and permethrin (4% with 12% piperonyl butoxide [PBO]), 148 mL/min) were sprayed using a truck-mounted ULV applicator (16.1 km/h) in a suburban area. Sod-grass blocks (sampling stations) were placed in the front and back yards of homes in the neighborhood at 7.6, 15.2, 30.5, and 91.4 m from the road where the treatments were made. Sampling was made before application and at 15 min, 12 h, 24 h, and 36 h after application. During the insecticide applications, temperature was 18°C, relative humidity ranged from 52% to 62% and wind speed was 1.6 km/h. Ranges of detection were 0 to 16.6 mg/0.18 m^2 and 0 to 25.9 mg/0.18 m^2 for malathion and permethrin, respectively. The highest concentrations were measured 15 min after the spray operation. Concentrations of permethrin for blocks located in front yards at 7.6, 15.2, 30.5, and

91.4 m from the road were equal to 25.9, 10.2, 7.2, and 1.3 mg/0.18 m^2, respectively. For the blocks located in back yards, the concentrations were equal to 23.7, 12.1, 3.8, and 0.9 mg/0.18 m^2, respectively. In the same way, 15 min after application, concentrations of malathion for blocks located in front yards at 7.6, 15.2, 30.5, and 91.4 m from the road were equal to 16.6, 4.3, 1.4, and 0.3 mg/0.18 m^2, respectively. For the blocks located in back yards, the concentrations were equal to 15.7, 5.1, 1.6, and 0.4 mg/0.18 m^2, respectively. At 7.6 and 15.2 m, 24 h after the spraying, permethrin and malathion were found in sod blocks located in front and back yards. However, it is worth noting that the detection limits for malathion and permethrin were only 0.05 ppm and 0.1 ppm, respectively [11]. Tietze et al. [12,13] compared deposition of ULV malathion in open versus residential areas. In open areas [12], applications of malathion (95%) were made with a ULV applicator (blower pressure = 0.421 kg/cm^2) mounted on a pickup truck driven at 16.1 km/h. The application rate was 127 mL/min. Sampling sites were aligned to prevailing wind direction, while the path of the spray vehicle varied from 90° to about 45° of that of the wind. Filter papers were placed 3 to 4 m apart and parallel to the course of the sprayer for sampling. Average wind speed, relative humidity, and ground and air temperature were 4.8 km/h, 79.6%, 25°C, and 26.5°C, respectively. The mass deposited, droplet abundance, and droplet size (mass median diameter) were determined at distances of 5, 25, 100, and 500 m from the spray head. It was shown that deposited mass of malathion decreased with distance. Deposit concentrations were equal to 33.36, 16.75, 15.72, and 2.10 ng/cm^2 at 5, 25, 100, and 500 m from the spray head, respectively. The greatest variation in droplet abundance was detected closest to the spray head [12]. Assuming 100% deposition within the area (4,047 m^2) and a homogeneous distribution, an estimated mass of 577 ng/cm^2 malathion was applied. Actually, the maximum deposition rate was 33.4 ng/cm^2 (5.8%), at 5 m from the spray head. Tietze et al. [12] have proposed a simple model (Eq. 13.2, $r^2 = 0.93$) to estimate the mean mass of malathion (MASS in nanograms per square centimeter) deposited in function of the distance (DIST in meters) from the spray head.

$$\text{MASS} = -14.47 \log(\text{DIST}) + 41.57 \tag{13.2}$$

Tietze et al. [13] also investigated the deposition behavior of malathion around a residential area, which consisted in four houses moderately vegetated within rectangular city blocks. Sampling sites for the front yard and beside and behind the houses were put on the ground at the middle of each facade and about 2.4 m away from the structures; the back yard sampling site was positioned about 2.4 m from the fence delineating the property lines [13]. The application rate of malathion (95%) was 127 mL/min. The spray head on the pickup truck extended to 2.13 m above the ground and angled at 45° above horizontal. Five spray tests were performed. Wind speed ranged from 0 to 4 km/h [13]. Mean air temperature was 25.5°C and relative humidity was 70.8%. Distance from the street to sampling sites, in front of, beside, and behind house and back yard averaged 11, 19.9, 28.6, and 42.8 m, respectively. Average malathion deposition for the front yard, side of house, behind house, and back yard was 88.8 (±24.9), 56.8 (±11.7), 62.5 (±23.4), and 29.9 (±7.8) ng/cm^2. (Eq. 13.3) ($r^2 = 0.864$) shows the relationship between the mean mass of malathion

(MASS in nanograms per square centimeter) deposited in function of the distance (DIST in meters) [13].

$$\text{MASS} = -1.66\,\text{DIST} + 101.88 \tag{13.3}$$

Permethrin (Permanone® 10% EC, 7.85 g/ha, 205 mL/min) and naled (Trumpet® EC, 22.42 g/ha, 44.36 mL/min) were sprayed using a truck-mounted ULV applicator (16.1 km/h) [14]. The applications took place in 2007 and 2008 in open fields with no vegetation taller than 20 cm. Sample collectors were placed at 25, 50, and 75 m from the spray source as well as at 1.25 m above the ground. Air concentration samplers were placed 25 m from the spray source. There were three sample replicates with 200 m buffer zones between them. During permethrin applications in 2007 (and 2008), temperature was 22°C (27°C), relative humidity was 35% (27.5%), and wind speed ranged from 8 to 17.7 km/h (7 to 12.9 km/h). During naled applications in 2007 (and 2008), temperature was 21°C (24°C), relative humidity was 33% (23%), and wind speed ranged from 2.4 to 4.8 km/h (8 to 12.9 km/h). The detection limits for *cis*- and *trans*-permethrin and naled were 30 and 1.5 ng, respectively. For permethrin, there was no significant difference between concentrations detected at the ground level and at 1.25 m above it. Mean concentrations of permethrin at 25, 50, and 75 m from the spray source 1 h after application were 2.3 ± 1.2, 3.8 ± 1.1, and 1.1 ± 0.63 ng/cm^2 in 2007 and 4.6 ± 0.67, 2.3 ± 1, and 0.94 ± 0.18 ng/cm^2 in 2008, respectively. Concentrations of permethrin at 25, 50, and 75 m from the spray source 12 h after spraying were 2 ± 1.2, 3.1 ± 0.46, and 0.8 ± 0.5 ng/cm^2 in 2007 and 3.7 ± 0.69, 1.3 ± 0.39, and 0.86 ± 0.2 ng/cm^2 in 2008, respectively. Concentrations of permethrin at 25, 50, and 75 m from the spray source 24 h after spraying were 1.8 ± 1.2, 1.3 ± 0.4, and 0.2 ± 0.1 ng/cm^2 in 2007 and 3.9 ± 0.78, 0.7 ± 0.32, and 0.42 ± 0.36 ng/cm^2 in 2008, respectively. Mean concentrations of naled on collectors placed at ground level and at 25, 50, and 75 m from the spray source 1 h after application were 47 ± 0.1, 66 ± 9.6, and 67 ± 11 ng/cm^2 in 2007 and 15 ± 2.9, 6.1 ± 2.1, and nondetectable ng/cm^2 in 2008, respectively. The mean concentrations of naled at 25, 50, and 75 m 12 h after application were 51 ± 6.7, 74 ± 7, and 71 ± 5.8 ng/cm^2 in 2007 and 20 ± 2.1, 7.7 ± 2.9, and 0.57 ± 0.56 ng/cm^2 in 2008, respectively. In 2007, mean naled depositions at 1.25 m above the ground at 25, 50, and 75 m from the spray source and 1 h after the spraying application were 11 ± 2.2, 6.5 ± 1.5, and 4.8 ± 3.7 ng/cm^2. Depositions 12 h after spraying were equal to 9.7 ± 1.2, 4.9 ± 0.57, and 5.2 ± 3.9 ng/cm^2, respectively. In 2008, 1 h after application, the mean naled depositions at 1.25 m above the ground were equal to 23 ± 5, 13 ± 5.4, and 0.54 ± 0.53 ng/cm^2 at 25, 50, and 75 m. Depositions 12 h after spraying were equal to 14 ± 1.2, 12 ± 3.3, and 1.6 ± 0.95 ng/cm^2 at 25, 50, and 75 m, respectively. The difference found between the 2 years of measurements for naled is certainly due to the difference in wind speed between years [14].

In 2007 and 2008 air concentrations of 375 and 397 ng/m^3 were observed 1 h after application of permethrin, respectively. In 2007, there was one quantifiable air concentration of 3,910 ng/m^3 1 h after application of naled, and in 2008 there were three quantifiable concentrations of 2,300, 2,900, and 4,000 ng/m^3 1 h after the application of naled. There was no quantifiable air concentration measured between 1 and 12 h after application in either 2007 or 2008 for both insecticides [14].

Faraji et al. [15] evaluated the penetration behavior and droplet dynamics of a ULV application of the DUET™ adulticide, which combines sumithrin—also called phenothrin (5%, 43.02 a.i. g/L)—and prallethrin (1%, 8.59 a.i. g/L) with PBO (5%, 43.02 a.i. g/L). Experimental sites consisted of an urban site with about 26 residential blocks, each containing a residential street on all four sides, and divided by a drivable alley between parallel parcels, each being either part of row home or a duplex. A suburban site was also selected. It consisted of roughly 60 residential blocks, many of which did not include a residential street on all four sides, and none of which were divided by a drivable alley. The sprayer was mounted on a flatbed truck at a height of 1.8 m, and the spray boom was angled 45.5° backward. Applications were made between 2:30 and 5:00 a.m. at 46.77 mL/ha (0.40 g a.i./ha prallethrin, 2.02 a.i. g/ha sumithrin and PBO) and 93.53 mL/ha (0.81 g a.i./ha prallethrin, 4.04 g a.i./ha sumithrin and PBO). Rotating impactors versus common droplet size and density determination were tested. Collected droplet sizes showed VMDs ranging between 10.68 ± 0.15 and 18.79 ± 0.57 μm, despite location, rate, and collection method. These differences are not operationally meaningful because they are in the range of the recommended values for adulticide treatments. The results showed that the aerosol plumes penetrated efficiently even into sheltered, cryptic habitats. Droplets were collected from all habitats without detected significance differences between locations within the same application rate or collection method. No differences were found in droplet densities between the two rates within the urban site but more droplets were collected in urban (149.93 ± 11.07 drops per square millimeter) than suburban (114.37 ± 11.32 drops per square millimeter) sites at the maximum rate [15].

Deposition of insecticides on residents and bystanders from ground ULV mosquito sprays has been the subject of very few experimental studies. Deposition of malathion (91% a.i., 128 mL/min) was monitored on body surfaces of three human subjects by Moore et al. [16]. Two standing subjects were exposed downwind to the malathion spray at 7.6 and 15.2 m. The third person jogged in the same direction as the spray vehicle (16.1 km/h) and 1.5 m from the spray path. Wind speed during the five experiments was between 1.5 and 5.5 km/h, temperature was between 21°C and 28°C, and all applications were made between 17:15 and 19:15 h. Average malathion concentrations deposited on gauze surfaces placed on torso, arms, legs, and head of the jogger equaled 1.19, 1.50, 0.39, and 0.20 μg/cm^2, respectively. For the subjects in a stationary position, the concentrations equaled 0.27, 0.33, 0.22, and 0.03 with the spray source at 7.6 m and 0.23, 0.26, 0.14, and 0.04 μg/cm^2 with the spray source at 15.2 m for the gauze surfaces placed on their torso, arms, legs, and head, respectively.

Preftakes et al. [17] used two mannequins as substitutes for human bystanders to measure deposition of two permethrin formulations (Permanone® 30-30 and Aqua-Reslin®, 192 mL/min, 7.85 g/ha of permethrin). The sprayer nozzle was at 135° with respect to the ground and the truck speed was 16.1 km/h. The average wind speed, temperature, and relative humidity for all applications were 213 cm/s, 19°C, and 48%, respectively. Mannequins were placed at 25 and 50 m from the spray source. Average permethrin concentrations deposited on the body from Aqua-Reslin and Permanone 30-30 were equal to 4.2 and 2.1 ng/cm^2, respectively [17].

Due to the limited amount of data dealing with the air concentration and/or the ground deposition of adulticides after truck-mounted ULV applications, *in silico* tools can be used as surrogates for estimating these data.

13.2.2 Air Dispersion Models

Holmes and Morawska [18] have evaluated different air dispersion model packages, from simple box to complex fluid dynamics models, for their capacity to estimate the dispersion of particles in the atmosphere. They showed that it was not possible to rank the models because the order depended on the selected scenario. They stressed that major weaknesses in particle dispersion modeling were the lack of studies that simultaneously measured particle number concentration and gaseous pollutant concentrations and the lack of validation studies that compared the performances of the various models against validation data sets. A limited number of spray dispersion models have been used in the frame of insecticide treatments for the control of mosquitoes. They are described next.

13.2.2.1 AERMOD

In 1991, the American Meteorological Society (AMS) and the US Environmental Protection Agency (EPA) initiated a working group (AMS/EPA Regulatory Model Improvement Committee, AERMIC) that led to the AERMIC model (AERMOD) [19], which is one of the most widely used regulatory dispersion models in the United States. It is a steady-state plume model aiming at short-range (up to 50 km) dispersion from stationary sources. In the stable boundary layer (SBL), which is a layer where negative buoyancy flux dampens turbulence, AERMOD assumes the concentration distribution to be Gaussian (i.e., bell shaped) both vertically and horizontally. In the convective boundary layer (CBL), which is a layer where thermal instability is created and turbulence is generated by positive buoyancy flux, the horizontal distribution is also assumed to be Gaussian, but the vertical distribution is described with a bi-Gaussian probability density function [19,20]. Two preprocessors, AERMET and AERMAP, are required to run AERMOD. AERMET is a meteorological stand-alone program that requires, as input, surface characteristics, cloud cover, a morning upper-air temperature sounding, and one near-surface measurement of wind speed, wind direction, and temperature. With this information, the model computes the friction velocity, Monin–Obukhov length, convective velocity scale, temperature scale, mixing height, and surface heat flux needed by AERMOD [21]. AERMAP uses gridded terrain data for the modeling area to calculate a representative terrain-influence height associated with each receptor location. The gridded data are supplied to AERMAP in the format of the digital elevation model (DEM) data [19]. AERMOD has been successfully used in numerous industrial scenarios [22–25]. It is freely available with an implementation guide, the source code, a complete technical documentation, and evaluation databases (https://www3.epa.gov/scram001/dispersion_prefrec.htm#aermod).

13.2.2.2 ISCST3 Model

Industrial source complex model version 3 (ISC3) is a steady-state Gaussian plume model that can be used to assess pollutant concentrations from a wide variety of sources [26]. It operates in both long-term (ISCLT3) and short-term (ISCST3) modes, the latter being more suited for vector control scenarios. It was the preferred EPA model until it was superseded by AERMOD in 2006. Thereafter, ISC3 has been known as an "alternative model" (https://www3.epa.gov/scram001/dispersion_alt.htm), but it remains widely used, as a result of its robustness, adaptability to different situations, availability of required data, and relative simplicity of use [27]. ISCST3 accepts hourly meteorological data to define the conditions for plume rise, transport, diffusion, and deposition. The meteorological data are cloud cover, temperature, relative humidity, pressure, wind direction and speed, ceiling height, precipitation, global horizontal radiation, and Pasquill–Gifford stability class. The model estimates the concentration or deposition value for each source and receptor combination for each hour of input meteorology, and it calculates user-selected short-term averages. For deposition values, the dry deposition flux, the wet deposition flux, or the total deposition flux can be estimated. The latter is simply the sum of the dry and wet deposition fluxes at a particular receptor location. The user also has the possibility of selecting averages for the entire period of input meteorology [26,28].

13.2.2.3 AGDISP/AgDRIFT

The AGricultural DISPersal model (AGDISP), version 8.26, was developed by the USDA Forest Service. It is a Lagrangian-based model designed to optimize agricultural spraying operations and has detailed algorithms for characterizing the release, dispersion, and deposition over and downwind of the application area. AGDISP predicts the motion of spray material released from aircraft and ground sprayer with booms, including the mean position of the material and the position variance about the mean as a result of turbulent fluctuations, pointing toward a prediction of spray drift. The released spray material is modeled as a discrete set of droplets, collected into categories, and called a drop size distribution. Each drop size category is defined by its average diameter and volume fraction, and is examined sequentially by the model [29]. AGDISP was originally developed for estimating airborne concentrations and off-target deposition of pesticide spray drift arising from aerial applications, and numerous studies have successfully validated the performance of the model in this regard as well as for the control of adult mosquitoes [30,31]. Only recently, a feature for simulating spray drift from ground application of pesticides with boom sprayers was incorporated into the model [32,33]. The AGDISP model forms the computational engine of the AgDRIFT (version 2.1.1), which was developed for use in regulatory assessments of off-site drift associated with agricultural use of pesticides through aerial, ground, or orchard/airblast applications [34]. Like AGDISP, it is freely available (https://www.epa.gov/pesticide-science-and-assessing-pesticide-risks/models-pesticide-risk-assessment). AgDRIFT has three basic levels of operation (tiers I to III), each successive level requiring fewer default assumptions and more inputs from the user [35]. ULV spraying was not specifically addressed in AGDISP and AgDRIFT because the models are physics based. However, Teske et al. [36] claim that there is no reason not to use them for such an application after some modifications. The results obtained by Mickle et al. [37] tend to support this statement.

13.2.2.4 MULV-Disp

Schleier et al. [38] performed field studies in three US states during the summers of 2009 to 2011 to measure environmental concentrations of insecticides after ULV applications. The selected sites of 200 m long showed a poor vegetative and flat topography and included two receptor lines of 25 receptors each, allowing a sampling from 5 to 180 m from the spray source, which was either a Guardian 95 ES (ADAPCO, Sanford, FL) or a London Fogger model 18 (London Fogger Inc., Long Lake, MN). Nozzle orientation of the sprayers was a +45° angle from the horizon, which is the most commonly used angle for mosquito management. The five following oil-based formulations were tested: Permanone® 30–30 (30% permethrin), Scourge® 18+54 (18% resmethrin), Permanone® 31–66 (31% permethrin), Zenivex® E20 (20% etofenprox), and Pyronyl™ Crop Spray (6% pyrethrins). The water-based formulations Aqua-Reslin® (20% permethrin) and Aqua-Kontrol (20% permethrin) were also applied. Chemicals were applied at the maximum application rate of 7.85 g/ha, except for Pyronyl Crop Spray, which was applied at 2.8 g/ha of active ingredient. About 100 spray events were performed—never before 18:00 h and most of the time after 20:00 h. Truck speed was 16.1 km/h.

Spray deposition was quantified from the use of petri dishes disposed on the ground and collected 10 min after each application. The detection limits for water- and oil-based insecticides were 0.00076 and 0.0002 μg/cm^2, respectively. A DC-III portable droplet measurement system (KLD Labs, Inc., Hauppauge, NY) was used to determine the VMD and CMD values for each formulation in all locations. The system was held 2 m from the nozzle in the center of the spray plume and sampling was terminated at 15 s or when 10,000 droplets were sampled [38]. Wind speed, air temperature, wet-bulb depression, and relative humidity were measured with a Hobo Micro Station Data Logger (Onset Computer Corporation, Bourne, MA) attached to temperature and relative humidity sensors with a solar radiation shield and a wind speed and direction smart sensor positioned at 2.5 m from the ground. Temperature and relative humidity readings were also taken 10 m above the ground using the Hobo model H08-032-08. Meteorological measurements were taken upwind of the spray site. Wind speed, wet bulb depression, temperature, and relative humidity at both 2.5 and 10 m above the ground were averaged over an interval of 5 min. Temperature readings taken 2.5 and 10 m and the mean wind speed at 2.5 m above the ground were used to calculate the stability ratio (Eq. 13.1), which was categorized into four classes. Formulation density was determined by averaging the weight of five 1 mL samples of each formulation on a calibrated Mettler AM100 analytical balance (Mettler Toledo AG, Switzerland) [38]. A total of 1,067 data points was collected from the three US states and from 2009 to 2011. They were used for deriving the following regression model (Eq. 13.4) [39]:

$$\begin{aligned} C = {} & -3.714348 + 0.008408058\ D - 1.152668\ AR - 0.005150439\ FR \\ & - 24.57204\ DEN + 3.947896\ CMD + 1.127382\ VMD \\ & - 0.01449041\ WS + 0.3348307\ T + 0.1457546\ RH \\ & + 1.443940\ SC - 0.007430822\ (D \times CMD) - 0.00001502679\ (D \times WS) \\ & + 18.95394\ (DEN \times CMD) - 0.007007278\ (DEN \times WS) \\ & - 1.661\ (DEN \times SC) - 0.1547870\ (CMD \times T) - 0.06394411\ (CMD \times RH) \\ & - 0.3194389\ (CMD \times SC) - 0.7759371\ (CMD \times VMD) - 0.004159109 \\ & (VMD \times T) + 0.03392584\ (VMD \times SC) + 0.0005936624\ (WS \times T) \\ & + 0.0001489897\ (WS \times RH) + 0.0003989388\ (WS \times SC) \\ & - 0.002531922\ (T \times RH) \end{aligned} \tag{13.4}$$

where

C is the log of the concentration (μg/cm^2)
WS is the wind speed (cm/s)
T is the temperature (°C)
RH is the relative humidity (%)
SC is the stability category
D is the distance from spray source (m)
AR is the application rate (g/ha)
FR is the flow rate (mL/min)
DEN is the density (g/mL)
CMD is the count median diameter (μm)
VMD is the volume median diameter (μm)

Schleier et al. [38] justify the use of interaction terms by the fact that they indirectly take into account parameters that cannot be directly measured, citing the example of $T \times RH$, which could account for the evaporation. The adjusted coefficient of determination (r^2) of Eq. (13.4) equaled 0.4 [38]. It is noteworthy that the full data matrix used to compute Eq. (13.4) is available in Schleier [40]. The minimum and maximum values for the 10 variables in Eq. (13.4) are given in Table 13.1 [39,40]. The MULV-Disp model has been recently criticized by Teske et al. [36,41] focusing on the data used for deriving the model, some model assumptions, and so on. They also performed a sensitivity analysis, which was criticized by Schleier and Peterson [42].

It is obvious that the extensive field studies of ground-based ULV spraying operations performed by Schleier [40] have produced invaluable data that could be used for deriving models aimed at predicting spray behaviors during ULV ground applications for adult mosquito control. Schleier et al. [39] used these data to derive the

TABLE 13.1
Minimum and Maximum Values for Variables Used in Computation of Eq. (13.4)

Response	Minimum	Maximum
Distance (m)	5	180
Application rate (g/ha)	2.8	7.85
Flow rate (mL/min)	74	240
Density (g/mL)	0.88	1.02
CMD (μm)	1.19	2.3
VMD (μm)	6.99	31.3
Wind speed (cm/s)	31.76	1,267.24
Temperature (°C)	13.94	32.17
Relative humidity (%)	13.06	77.97
Stability category	1	4

Sources: J.J. Schleier III and R.K.D. Peterson, J. Am. Mosq. Control Assoc. 30 (2014), pp. 223–227; J.J. Schleier III, thesis, Montana State University, Bozeman, Montana, 2012.

MULV-Disp model, which bridges a gap by offering the first ULV-dispersion model for estimating outdoor concentrations of adulticides during mosquito control operations. However, the model presents poor statistics with an r^2 of about 0.4. Rather than using a multiple regression analysis, it would be fruitful to test more powerful statistical methods, such as PLS regression analysis [43] or an artificial neural network [44,45].

13.2.3 Dietary Exposure with DEEM

The dietary exposure evaluation model (DEEM™) provides probabilistic assessments of dietary pesticide exposure [46]. It consists of four modules: a main module, the acute and chronic analysis modules, and the RDFgen™ residue distribution module. The main module is used to create and edit residue files for specific chemical or cumulative applications, and to launch the other modules. The RDFgen module automates single analyte and cumulative residue distribution adjustments and the creation of summary statistics and residue distribution files based upon the USDA Pesticide Data Program (PDP) monitoring data or user-provided residue data. The acute and chronic analysis modules provide dietary exposure assessment models based on USDA consumption data. There are three general exposure assessment models—namely, point estimate, simple distribution, and probabilistic (Monte Carlo). Data used by the modules are of two types: those that cannot be changed by the user, which are the consumption and demographic profiles of the individuals in the USDA's consumption surveys, and the translation factors that translate foods as consumed (e.g., pizza) into the corresponding raw agricultural commodities and food forms (e.g., wheat, tomatoes, etc.). The second type can be modified or provided by the user. Exposure estimates derived by DEEM can be compared to compound specific toxicity measures to derive risk estimates. The user provides the toxicity measures to be used in the comparison. The toxicity measure used by the model depends on the type of assessment being conducted. The source code of the acute and chronic modules is freely available as well as the whole program [46].

13.3 RISK ASSESSMENTS FOR RESIDENT AND BYSTANDER OUTDOOR EXPOSURE TO ADULTICIDES

Human-health risks associated with the use of six insecticides (malathion, naled, permethrin, resmethrin, phenothrin, and pyrethrins) and PBO in mosquito control after truck-mounted ULV sprayings were estimated by Peterson et al. [47]. Acute exposures were defined as single-day exposures immediately after a spray event. Subchronic exposures were defined as exposures per day over a 90-day seasonal multispray event. A total of 10 spray events were assumed to occur on days 1, 4, 14, 17, 27, 30, 40, 43, 53, and 56. AERMOD (version 1.0) [19] was used to predict the 7.6 and 91.4 m air concentrations of each insecticide within time ranges of 1 and 6 h after applications. It was assumed that each insecticide had a 24 h half-life in air, except naled (18 h). The insecticides were applied at their maximum application rate, under the same weather conditions, and at 21:00 h. Each spray release was at 1.5 m. Receptors were set within the model on a Cartesian grid at five intervals of 25 m, with a height of 1.5 m. This allowed the authors to estimate the 1 and 6 h average concentration at 7.6 m, and the peak value at 91.4 m. Two additional assumptions were made: Of the emitted particles, 3% were greater than those allowed as stated on the label, and their density was in accordance with the specific gravity of each insecticide. The ISCST3 model [26] was used to estimate droplet deposition at 7.6 and 91.4 m from the spray area at a 1 h average. The Cartesian grid used for ISCST3 was similar to that used with AERMOD. This allowed to estimate the deposition at 7.6 and 91.4 m as well as the average deposition within 91.4 m of the spray source. For estimating subchronic exposures, an exponential decay model was used to characterize the persistence of insecticides on surfaces accounting for the 10 spray dates. Insecticide concentrations for each spray event were followed through day 90 using Eq. (13.5) [47].

$$D = \sum_{j=1}^{90} P \times e^{(r1 + r2)t} \tag{13.5}$$

where

- D is the sum of the deposition over one spray
- P is the peak deposition after a spray event
- $r1$ is the rate of decay calculated by using the aerobic soil half-life of each insecticide
- $r2$ is the rate of decay calculated by using the soil photolysis half-life of each insecticide
- t is the time in hours
- j is the spray day

The average daily exposure was then obtained by dividing the deposition sum by 90.

The application rates, chemical properties, predicted environmental concentrations, and acute and subchronic toxicity values for each active chemical are listed in Table 13.2. The Kenaga nomogram was used to estimate depositions on garden plants. This is a graphical diagram that allows the prediction of the maximum pesticide residues in parts per million immediately following insecticide application to different categories of plants and the parts of them [48,49].

Regarding acute exposure, it was assumed that multiroute exposures immediately after a single-spray event were limited to 24 h. Routes of insecticide exposure included inhalation, dermal contact with spray, hand-to-mouth ingestion by infants and toddlers from spray deposition on hands, and ingestion of garden produce. It was also considered that residents did nothing to limit their exposure to the spray and also that exposures from potable water and swimming were negligible [47].

Acute inhalation exposure was estimated from Eq. (13.6) [47].

$$PE = (EEC \times RR \times D \times CF)/BW \tag{13.6}$$

where

PE is the potential acute exposure from inhalation (mg/kg of body weight [BW])
EEC is the 6 h average estimated environmental concentration of an active chemical in the air 1.5 m high at 7.6 m from the spray source (μg/m^3)
RR is the respiratory rate under moderate activity (m^3/h)
D is the duration of exposure (h)
CF is the conversion factor to account for the conversion of units from micrograms per cubic meter to milligrams per cubic meter
BW is the body weight (kg)

RRs were assumed to be 1.6 m^3/h for adults and 1.2 m^3/h for children and infants [47]. These rates correspond to moderate activity. This is broadly in agreement with the values generally used in exposure modeling (Table 13.3) [50].

The duration of exposure was 6 h and it was assumed that the person stayed 7.6 m from the emission source during the period.

Acute dermal exposure was estimated from Eq. (13.7) [47].

$$PE = (TDE \times AB)/BW \tag{13.7}$$

where

PE is the potential acute dermal exposure (mg/kg of BW)
TDE is the total dermal exposure (mg)
AB is the dermal absorption rate
BW is the body weight (kg)

Because values for percent dermal absorption after ULV application were missing for the studied chemicals, Peterson et al. [47] selected the following conservative rates: 0.22% for pyrethrins, 2% for PBO, 10% for malathion and resmethrin, 15% for permethrin, 70% for phenothrin, and 100% for naled.

TABLE 13.2
Properties, Predicted Environmental Concentrations, Acute and Subchronic Toxicity Values, and Risk Quotients (RQs × 10^{-3}) for Chemicals Studied

Property/Activity	PBO	Phenothrin	Permethrin	Resmethrin	Malathion	Pyrethrins	Naled
Application rate (kg a.i./ha)	0.0392	0.004	0.0078	0.0078	0.0639	0.009	0.0224
Density (g/mL)	0.898	0.898	0.866	0.87	1.23	0.81	1.67
Surface photolysis half-life (day)	NA	6	23	0.14	6.5	0.5	2.4
Soil aerobic half-life (day)	14	7	37	30	1	1	1
Air concentration (μg/m^3)[a]	7.39	0.81	1.55	1.61	9.76	1.7	1.68
1-Day mean produce concentration (mg/kg dry weight)	0.525	0.054	0.105	0.105	0.855	0.12	0.3
90-Day mean surface concentration (mg/m^2)[b]	15.42	0.43	4.14	0.22	2.18	0.54	0.65
90-Day mean produce concentration (mg/kg dry weight)	2.88	0.055	0.096	0.012	0.73	0.21	0.13
Acute NOAEL[c] or NOEL[d] (mg/kg/day)	630	70[f]	25	10[f]	50	20	1
Acute ingestion RfD[e] (mg/kg/day)	6.3	0.7	0.25	0.1	0.5	0.07	0.01
Subchronic NOAEL or NOEL (mg/kg/day)	89	70[f]	25	10[f]	2.4	4.37	1
Subchronic ingestion RfD (mg/kg/day)	0.89	0.7	0.25	0.1	0.024	0.044	0.01

(Continued)

TABLE 13.2 (CONTINUED)

Properties, Predicted Environmental Concentrations, Acute and Subchronic Toxicity Values, and Risk Quotients (RQs × 10^{-3}) for Chemicals Studied

Property/Activity	PBO	Phenothrin	Permethrin	Resmethrin	Malathion	Pyrethrins	Naled
	Acute/Subchronic RQs × 10^{-3}						
Adult males	0.4/3.2	0.4/0.1	2.0/0.7	5.2/0.4	7.6/36	8.1/5.6	149.6/25.9
Adult females	0.4/3.2	0.4/0.1	2.1/0.7	5.5/0.4	7.9/36.3	8.5/5.6	157.6/26.9
Children (10–12 years)	0.6/4.3	0.6/0.1	2.9/0.8	7.2/0.5	10.5/47	11.3/7.4	212.3/29
Children (5–6 years)	0.9/5.9	1.0/0.2	4.9/1.2	12.3/0.9	17.7/67.6	19/10.4	363.1/44.7
Toddlers (2–3 years)	1.2/26.2	1.3/0.9	6.3/20.4	15.9/3.7	22.5/18.15	24.5/27	472.6/129.4
Infants (0.5–1.5 years)	1.0/32.5	1.2/1.3	5.8/30.1	14.7/5.4	18.8/20.74	21.8/29.2	449.5/166.1

Source: R.K.D. Peterson, P.A. Macedo, and R.S. Davis, Environ. Health Perspect. 114 (2006), pp. 366–372.

Note: The acute and subchronic RfDs were obtained by dividing the NOAEL or NOEL by 100 except for pyrethrins (acute), for which the uncertainty factor was 300.

[a] 6 Hour mean concentration at 7.6 m from the spray source.

[b] 90-Day mean surface concentration within 91.4 m of the spray source.

[c] NOAEL = no observed adverse effect level.

[d] NOEL = no observed effect level.

[e] RfD = reference dose.

[f] NOEL (otherwise NOAEL).

TABLE 13.3
Anthropometric and Physiological Values Commonly Used in Exposure Modeling

Parameter	Adult	Child (6–11 y)	Toddler (2–3 y)	Newborn
Weight (kg)	62	32	14	4.8
Body surface total (m^2)	1.69	1.08	0.61	0.29
Hands (m^2)	0.093	0.054	0.032	0.015
Arms (m^2)	0.264	0.137	0.072	0.040
Forearms (m^2)	0.111	0.058	NA	NA
Legs (m^2)	0.556	0.301	0.142	0.060
Lower legs (m^2)	0.189	0.102	NA	NA
Feet (m^2)	0.123	0.078	0.043	0.019
Head (m^2)	0.135	0.136	0.087	0.053
		Respiration Rate		
Sleeping (m^3/h)	0.40	0.38	0.38	0.28
Sedentary (m^3/h)	0.40	0.38	0.40	0.28
Light activity (m^3/h)	0.89	0.90	1.0	0.66
Moderate activity (m^3/h)	1.9	1.7	1.7	1.3
Food consumption (g/d)	1,100	1,100	1,000	NA
Water consumption (L/d)	2	1	1	NA

Source: Anonymous, *Generic Risk Assessment Model for Indoor and Outdoor Space Spraying of Insecticides*, WHO Pesticide Evaluation Scheme Department of Control of Neglected Tropical Diseases Cluster of HIV/AIDS, Tuberculosis, Malaria and Neglected Tropical Diseases and Chemical Safety Department of Public Health and Environment Cluster of Health Security and Environment, WHO Press, Geneva, Switzerland, 2011.

Acute hand-to-mouth exposure for toddlers and infants was estimated from Eq. (13.8) [47].

$$\mathrm{PE} = [(\mathrm{THD}/\mathrm{HSA}) \times \mathrm{AHS} \times \mathrm{SEF}]/\mathrm{BW} \tag{13.8}$$

where

PE is the potential acute hand-to-mouth exposure (mg/kg of BW)
THD is the total hand dermal exposure (mg)
HSA is the adult hand surface area (m^2)
AHS is the adjusted hand surface area for each subgroup (m^2)
SEF is the saliva extraction factor
BW is the body weight (kg)

The hand surface areas of toddlers (2 to 3 years old) and infants were assumed to be 0.035 and 0.007 m^2, respectively. The latter value, selected by Peterson et al. [47], is underestimated (Table 13.3). The authors assumed that, on the day of application, 50% of the insecticide deposited on the hand was available through saliva extraction.

Acute ingestion of garden produce was focused on tomatoes, which were not washed after picking. The residue concentration was considered as constant during their processing. The amount of insecticide ingested was estimated as the product of the residue concentration and the quantity of food consumed. Tomato consumption patterns were determined from DEEM and the Food Commodity Intake Database (FCID) [47].

Subchronic inhalation, dermal, and hand-to-mouth exposures were calculated with Eq. (13.9) [47].

$$PE = (PE_{\text{acute type}} \times SE)/D \tag{13.9}$$

where

- PE is the potential subchronic inhalation, dermal, and hand-to-mouth exposure (mg/kg BW/day)
- $PE_{\text{acute type}}$ is the acute exposure type from each spray event (mg/kg BW)
- SE is the number of spray events
- D is the duration of exposure (days)

SE and D were equal to 10 and 90, respectively.

Subchronic hand-to-mouth exposure from deposition on surfaces for toddlers and infants was calculated from Eq. (13.10) [47].

$$PE = (EEC \times SEF \times SA \times DR \times FA \times D)/BW \tag{13.10}$$

where

- PE is the potential subchronic hand-to-mouth exposure from deposition on surfaces (mg/kg BW/day)
- EEC is the 90-day average environmental concentration of the insecticide deposited on soil or turf within 91.4 m from the spray source (mg/m^2)
- SEF is the saliva extraction factor
- SA is the surface area for three fingers (m^2)
- DR is the dislodgeable residue
- FA is the frequency of activity (events per hour)
- D is the exposure duration (h)
- BW is the body weight (kg)

SEF and SA were set to 50% and 20 cm^2, respectively. Dislodgeable insecticide residue from soil or turf grass was assumed to be 20%. The frequency of hand-to-mouth activity in children was assumed to be 20.5 events per hour [47].

On the basis of the meta-analysis of 429 subjects and more than 2,000 h of behavioral observations, Xue et al. [51] showed that the average indoor hand-to-mouth behavior ranged from 6.7 to 28.0 contacts per hour, with the lowest value corresponding to the 6- to <11-year-olds and the highest value corresponding to the 3- to <6-month-olds. The average outdoor hand-to-mouth frequency ranged from 2.9 to 14.5 contacts per hour, with the lowest value corresponding to the 6- to <11-year-olds and the highest value corresponding to the 6- to <12-month-olds.

As a result, the value selected by Peterson et al. [47] is a little bit too high. The duration of exposure was assumed to be 4 h/day. In other words, the toddlers or infants were engaged in hand-to-mouth activities outside each day for 4 h over a 90-day period [47].

The assumptions made for subchronic ingestion of garden produce were the same as for the acute scenario, except that Peterson et al. [47] considered that the insecticide was deposited onto both tomatoes and head- and leaf lettuces and all consumption by the residents over the 90 days were from the garden using chronic food consumption patterns (3-day average).

Subchronic dermal contact with soil, turf, and other outdoor surfaces was estimated from Eq. (13.11) [47].

$$PE = (EEC \times SA \times SS \times AB \times DR \times CF)/BW \quad (13.11)$$

where

- PE is the potential exposure from contact with soil, turf, and other outdoor surfaces (mg/kg BW/day)
- EEC is defined as in Eq. (13.10) (mg/m^2)
- SA is the body surfaces in contact with surface (cm^2) and assumed to be the sum of surface areas for face (head/2), hands, arms, legs, and feet
- SS is the weight of soil adhered to skin (mg/cm^2)
- BW is the body weight (kg)

Contact with surfaces was associated with gardening for adults (0.55 mg soil/cm^2 skin) and soccer for children, including infants (0.164 mg soil/cm^2 skin). Activities occurred each day over 90 days. AB is the dermal absorption rate, DR is the dislodgeable residue, and CF is the conversion factor (m^2 vs. cm^2).

Subchronic soil ingestion was calculated from Eq. (13.12) [47].

$$PE = [(EEC/SW) \times SI]/BW \quad (13.12)$$

where

- PE is the potential exposure from incidental ingestion of soil (mg/kg BW/day)
- EEC is defined as in Eq. (13.10) (mg/m^2)
- SW is the soil weight (mg/m^3)
- SI is the soil ingestion (mg/d)
- BW is the body weight (kg)

Peterson et al. [47] assumed that milligrams per square meter were equivalent to milligrams per cubic meter and that soil weight was equal to 3.86 kg/m^3. They estimated that soil ingestion rates were 100 mg/day and 50 mg/day for children and adults, respectively.

Risk assessment for each chemical was made from the calculation of the risk quotient (RQ). The acute RQs were calculated for each population subgroup by dividing the total acute PE by the reference dose (RfD). The subchronic RQs were calculated by the same method with the total subchronic PEs and the subchronic RfDs. The obtained values are listed in Table 13.2. Inspection of this table shows that all the RQ values are <1. As a result, the human health risks from residential exposure to the studied insecticides are low and not likely to exceed levels of concern.

Schleier et al. [52] performed broadly the same analysis for the acute human health risk of the six insecticides and PBO and, by using the same models for estimating air concentration and depositions but as novelty, they used a sensitivity analysis for identifying the most influential input parameters. They showed that the air concentrations and surface depositions contributed the largest amount of variability and uncertainty to the model outputs.

Macedo et al. [53] tried to evaluate health risks to deployed US military personnel from adult mosquito management tactics, focusing on acute, subchronic, and chronic exposures after insecticide application and use of personal protective measures in different scenarios.

Acute exposures corresponded with single-day exposures after a single chemical application. Subchronic exposures were defined as the exposure per day over 180 days with multiple spray events. For chronic exposures, it was assumed that personnel might be deployed for 250 days per year for 10 years. Outdoor truck-mounted ULV applications were made with permethrin, resmethrin, *d*-phenothrin, and PBO. AERMOD [19] and ISCST3 [26] were used to estimate air concentrations and particle deposition with the same assumptions than in Peterson et al. [47].

The formula used to determine acute inhalation exposure was equivalent to Eq. (13.6) except that it was assumed that the person was outside, within 91.44 m of the spray truck, and that he or she remained outside for 1 h, respiring as during moderate physical activity (1.6 m^3/h) for the entire time. The body weight (BW) was estimated to be equal to 71.8 kg for adult males (≥18 years) and 60 kg for females (mean between 13 and 54 years), respectively.

The formula used to determine acute dermal exposure was equivalent to Eq. (13.7) with the same assumptions as in Peterson et al. [47]. It is noteworthy that dietary exposure was not taken into consideration because it was judged negligible for the scenarios modeled [53]. Subchronic inhalation and dermal exposures from spray particles were estimated by assuming 21 spray events and an exposure duration of 180 days.

Subchronic inhalation exposures from resuspended outdoor soil particles were estimated from Eq. (13.13).

$$\mathrm{PE} = [(\mathrm{EEC/SW}) \times \mathrm{CA} \times \mathrm{CF} \times \mathrm{RR}]/\mathrm{BW} \qquad (13.13)$$

where

EEC is the 180-day average concentration of the chemical deposited on soil (mg/mg soil)

SW is the soil weight, which was assumed to be 481 kg/m^3

CA is the concentration of particulate matter in air (μg/m^3)

CF is the conversion factor for micrograms versus milligrams

RR is the respiratory rate (m^3/day)

BW is the body weight (kg)

It was assumed that CA was the PM_{10} standard of 60 μg/m³ and that this concentration was in the breathing zone. It was further assumed that all the suspended PM was from soil, 100% was retained in the lungs, and 100% was absorbed. The respiratory rate used was 15.2 and 11.3 m³/day for adult males and females, respectively [53].

Subchronic dermal exposures from contact with soil were estimated from Eq. (13.14).

$$PE = (CS \times SA \times SS \times AbR \times DR)/BW \tag{13.14}$$

where

PE is the potential exposure (mg/kg BW/day)
CS is the 180-day average concentration of the insecticide deposited on soil (mg/mg soil)
SA is the body surface area in contact with soil (cm^2)
SS is the mass of soil adhered to skin (mg/cm^2)
AbR is the dermal absorption rate
DR is the dislodgeable residue assumed to be 20%
BW is the body weight (kg)

It was assumed that the insecticide was deposited within the first-centimeter layer of soil (1 m × 1 m × 1 cm = 0.01 m³), and that soil weight was equal to 481 kg/m³. The body surface area in contact with soil was assumed to be the sum of surface areas for hands, arms, trunk, legs, and face (head/2). Therefore, it was assumed that people were only in shorts and shoes. Contact with soil was estimated to be 1.089 mg soil/cm² skin [54].

Subchronic dermal exposures from contact with outdoor surfaces were estimated from Eq. (13.15), the equation parameters being those defined in Eqs. (13.10) and (13.11).

$$PE = (EEC \times SA \times AbR \times DR)/BW \tag{13.15}$$

Ground ULV spraying of malathion (95%), resmethrin without (40%) or with (18%, 54%) PBO was investigated from the ISCST3 model [55]. Applications were made at 16 km/h and with a spraying distance of 500 m. Fence spraying of permethrin (50%) by means of individual sprayers was also simulated with the ISCST3 model. Applications were made at 1 km/h and with a spraying distance of 100 m. Wind speed ranged from 1 to 4.5 m/s and the atmospheric stability was equal to four or five. Wind direction was perpendicular (90°), quasi-parallel (175°), or intermediary (135°) to the transect. Air concentrations were those modeled at 8 m from the emission source, downwind, and at a height of 1.5 m. This corresponds to an exposure where houses are located near the road used by the spreader truck. The height of 1.5 m was considered the average inhalation route, taking into account children and people sleeping on the ground floors. Droplet sizes were equal to 20 and 55 μm for the ground ULV spraying and fence spraying, respectively.

It is worth noting that AgDrift [34,35] was also tested, but it was rapidly discarded because the concentrations on the ground were underestimated while those found in air were overestimated [55].

The results given by ISCST3 correspond to concentrations resulting from a single spraying event (Table 13.4). However, at worst, a total of 10 sprays would be possible during a season, over a total period of 56 days with 5 × 2 sprays with an interval of 3 days between two sprays, and a 10-day period between each two-spray series. This means that the spraying operations were made on days 1, 4, 14, 17, 27, 30, 40, 43, 53, and 56. If necessary, sprays would be repeated year after year, resulting in an exposure that may be considered chronic.

Exposure doses were estimated from the maximum concentration values except for permethrin, for which mean values were used.

In the acute exposure assessment, in order to account for the maximal concentrations of insecticide found on the soil throughout the season, the accumulation of the chemical coming from previous spraying events and the degradation process were taken into account to estimate the possibility of reaching a toxicity threshold or not.

Each day when a spraying event occurs, an amount equivalent to the modeled one is added to the residual amount in the environment that results from the previous sprays. The residual amount is calculated from Eq. (13.16) [55].

$$C_J = C_0 \times e^{-KJ} \tag{13.16}$$

where

C_J is the residual concentration at day J

C_0 is the initial modeled concentration

K is the elimination rate constant in days of the substance in the soil or on its surface (0.693/half-life)

J is the number of days

The day of the treatment, it is assumed that significant concentrations of insecticides are found in air for 4 h after the spraying, and for the whole day inside and

TABLE 13.4
Maximum and Mean Concentrations in Air[a] and on Soil[b] Estimated from the ISCST3 Model

	Maximum Concentration		Mean Concentration	
Adulticide	**In Air**	**On Soil**	**In Air**	**On Soil**
Malathion, 95%	63.3	23.0	43.2	7.3
Resmethrin, 18%	10.1	3.5	6.9	1.1
PBO, 54%	30.3	10.5	20.7	3.3
Resmethrin, 40%	22.8	7.9	15.6	2.5
Permethrin, 50%	11.7	10	6.3[c]	1.3[c]

[a] Micrograms per cubic meter.
[b] Milligrams per cubic meter.
[c] After exclusion of the receptors with a zero value.

on the surface of soil and vegetable products. This represents a potential source for exposure by inhalation, ingestion, and dermal contact. Because the simulation exercise was done in urban areas, the concentration in groundwater was not considered. Use of J in hours instead of days in Eq. (13.16) allows the estimation of mean concentrations during the 4 h after the spraying events. Eq. (13.16) is useful to estimate the concentrations found on and within the soil after one season of treatments [55].

Evaluation of the exposure to the adulticides accounts for the so-called exposure due to the background noise, which is the exposure in relation to the pollution of the environment by these molecules. The report [55] only included information for malathion. Thus, it was estimated that the actual daily intake of malathion by Canadian adults through food was 0.012 μg/kg/day. The mean concentration of malathion in food can be estimated by Eq. (13.17) [55].

$$C_{Food} = DIC \times BW/CTF \tag{13.17}$$

where

C_{Food} is the mean concentration of insecticide in food (mg/kg)
DIC is the mean daily intake concentration of insecticide by an adult (mg/kg/day)
BW is the body weight (Canadian adult = 70.7 kg)
CTF is the average daily consumption of total food by an adult (Canadian adult = 2.214 kg/day)

Use of Eq. (13.17) leads to an average concentration of 0.000383 mg malathion per kilogram of food. The same reasoning can be made for residential air in which an estimate of 220 ng/m^3 of malathion were found, as well as for potable water in which 100 ng/L of this organophosphorus insecticide were found due to the contamination of the environment.

The background noise by age category was calculated from Eq. (13.18).

$$BN_{age} = (C_{Food} \times IF_{age}/BW_{age}) + (C_{PW} \times IPW_{age}/BW_{age}) \tag{13.18}$$

where

BN_{age} is the background noise by age category (mg/kg/day)
C_{Food} is the mean concentration of insecticide found in food (mg/kg)
IF_{age} is the average daily intake of total food by age category (kg/day)
BW_{age} is the body weight (kg) by age category
C_{PW} is the mean concentration of insecticide found in drinking water (mg/L)
IPW_{age} is the average daily intake of drinking water by age category (L/day)

The background noise exposures found for malathion for the different categories of ages in the general population are given in Table 13.5 [55].

Acute exposure included inhalation of droplets found in air; dermal absorption of droplets found in air, on soil, and on turf; and ingestion of residues deposited on garden products. In addition, because the treatments are made during the summertime,

TABLE 13.5
Background Noise Exposures (BNEs) for Malathion

	BNE (mg/kg/day)		
	Ingestion	Inhalation	Total
0 to 6 months	4.64E-05[a]	1.21E-04	1.67E-04
7 months to 4 years	3.72E-05	1.24E-04	1.61E-04
5 to 11 years	2.41E-05	9.70E-05	1.21E-04
12 to 19 years	1.54E-05	1.94E-05	3.48E-05
20 years and more	1.42E-05	4.92E-05	6.34E-05

[a] Does not account for potential contamination by maternal milk.

it was assumed that contamination was also possible via swimming pools. The time spent in a swimming pool was estimated to be 2.5, 1.5, and 0.5 h for children, adolescents, and adults, respectively [55].

Acute exposure by inhalation was estimated by using an equation similar to Eq. (13.6), except that the time of exposure was estimated to be 4 h.

Inhalation of vapors from the swimming pool was calculated from Eq. (13.19) [55].

$$ID_{sw} = C_{vap} \times I_{age} \times D/BW_{age} \tag{13.19}$$

where

ID_{sw} is the inhaled dose of swimming pool vapors by age category (mg/kg/day)
C_{vap} is the concentration due to the evaporation (mg/m^3)
I_{age} is the inhalation rate by age category (m^3/h)
D is the swimming activity duration (h/day)
BW_{age} is the body weight (kg) by age category

C_{vap} was calculated from Eq. (13.20).

$$C_{vap} = (C_{water} \times VP \times 273\text{ K} \times MW_{water} \times 1{,}000\text{ L/m}^3 \times \text{L}/1{,}000\text{ g})/(760\text{ mmHg} \times T \times 22.4\text{ L/mol}) \tag{13.20}$$

where

C_{water} is the concentration of insecticide in water (μg/L)
VP is the vapor pressure of the adulticide at the temperature of the swimming pool (mmHg)
T is the swimming pool temperature (K)
MW_{water} is the water molecular weight (18 g/mol)

C_{water} was calculated from Eq. (13.21).

$$C_{water} = C_{surf} \times S/Vol \tag{13.21}$$

where C_{surf} is the concentration of adulticide deposited by surface unit (μg/cm^2), S is the mean swimming pool surface (488,581 cm^2), and Vol is the mean volume of the swimming pool (528,000 L) [55].

Total dermal exposure is the sum of exposures by dermal contact with the pool water, with suspended droplets, soil particles, those deposited on external objects, and the droplets deposited on the surface of grass.

Exposure by dermal contact with the pool water was estimated from Eq. (13.22) [55].

$$D_{water} = C_{water} \times BS_{age} \times SP \times D \times CF \times 0.001/BW_{age} \tag{13.22}$$

where

D_{water} is the dose by dermal absorption via the pool water and according to the age category (mg/kg/day)
C_{water} is the concentration of insecticide in water (μg/L)
BS_{age} is the total body surface by age category (cm^2)
SP is the skin permeability coefficient (cm/h)
D is the swimming activity duration (h/day)
CF is a conversion factor (1 L = 1,000 cm^3)
0.001 is a conversion factor (mg/μg)
BW_{age} is the body weight (kg) by age category

Exposure by dermal contact with the suspended droplets in air was estimated from Eq. (13.23) [55].

$$D_{sd} = C_{air} \times T_{exp} \times Nb \times SS_{age} \times SP \times CF/BW_{age} \tag{13.23}$$

where

D_{sd} is the dose by dermal contact with the suspended droplets and according to the age category (mg/kg/day)
C_{air} is the concentration of insecticide in air (mg/m^3)
T_{exp} is the contact time (14,400 s, 4 h)
Nb is the number of contacts (one contact/day)
SS_{age} is the skin surface in contact with the insecticide, by age group: hands, feet, legs, arms, head (cm^2)
SP is the skin permeability coefficient (cm/s)
CF is a conversion factor (0.000001 m^3/cm^3)
BW_{age} is the body weight (kg) by age category

Dermal exposure to soil particles adhering to skin was estimated from Eq. (13.24) [55].

$$D_{soil} = C_{soil} \times SS_{age} \times AF \times ER \times CF/BW_{age} \quad (13.24)$$

where

D_{soil} is the dose by skin contact with soil particles by age group (mg/kg/day)
C_{soil} is the insecticide concentration in soil (mg/kg of soil)
SS_{age} is the skin surface in contact with the insecticide, by age group: hands, feet, legs, arms, head (cm^2/h)
AF is the soil to skin adherence factor (2.8 mg/cm^2) [56]
ER is the exposure rate by age category (2 h/day) [57]
CF is the conversion factor (0.000001 kg/mg)
BW_{age} is the body weight (kg) by age category

For dermal exposure to the droplets deposited on the turf, the following equation was used [55]:

$$D_{turf} = C_{ssoil} \times SS_{age} \times ER \times F_{abs} \times CF/BW_{age} \quad (13.25)$$

where

D_{turf} is the dose due to dermal contact with the droplets deposited on the ground, by age category (mg/kg/day)
C_{ssoil} is the concentration deposited on the ground (μg/cm^2)
SS_{age} is the skin surface in contact with the insecticide, by age group: hands, feet, legs, arms, head (cm^2/h)
ER is the exposure rate by age category (2 h/day) [57]
F_{abs} is the fraction of dermal absorption specific to the insecticide
CF is the conversion factor (0.001 mg/μg)
BW_{age} is the body weight (kg) by age category

The calculation of the ingestion of fruits and vegetables from the garden on which were deposited adulticide sprayed droplets on the first day requires first assessing the concentration found in the plant due to atmospheric deposition. The other ways contributing to the accumulation of contaminant in the plants (root, air–plant transfer) were not considered because of the time period assessed for acute exposure (1 day), which was too short to allow these phenomena to be meaningful. The calculation of the atmospheric deposition required first determining the daily deposition rate, which was estimated from Eq. (13.26) [55].

$$DR = [(R_g \times C_{gaz}) + (R_p \times C_p)] \times 1/6 \quad (13.26)$$

where

DR is the daily deposit rates (μg/m²/day)

R_g is the deposition velocity in gaseous form (500 m/day)

C_{gaz} is the adulticide concentration into the air, in gaseous form (μg/m³)

R_p is the deposition velocity in particulate form (600 m/day)

C_p is the adulticide concentration into the air, in particulate form (μg/m³)

1/6 represents the fraction of the day for which the adulticide is present in the air (4 h/day)

C_p is obtained from C_{gaz} by using Eq. (13.27) [55]

$$C_p = C_{gaz} \times (c \times ps)/(VP_L + (c \times ps)) \tag{13.27}$$

where

C_p is the concentration of adulticide in air in particulate form (μg/m³)

C_{gaz} is the concentration of adulticide in air in gaseous form (μg/m³)

c is a constant (1.7×10^{-4})

ps is the surface of particles supposed to be constant and equal to 1.5×10^{-6}

VP_L is the vapor pressure of the adulticide in liquid phase (At) [55]

The concentration in the vegetable resulting from the atmospheric deposition was obtained from Eq. (13.28) [55].

$$C_{veg} = ((DR \times I_{veg})/(Kt \times Pr_{veg})) \times (1 - e^{(-Kt \times Tv)}) \tag{13.28}$$

where

C_{veg} is the concentration of adulticide in the vegetable resulting from its atmospheric deposition (μg/g dry weight)

DR is the atmospheric deposition rate of the adulticide (μg/m²/day)

I_{veg} is the part intercepted by the vegetable (without unit)

Kt is the loss coefficient (day^{-1})

Pr_{veg} is the production of the eaten part of the vegetable (g/m² dry weight)

Tv is period of growth of the vegetable before its harvest (days)

The constant values have to be found or estimated for each vegetable.

Ingestion exposure is then calculated from Eq. (13.29) [55].

$$D = TI_{age} \times C_{veg} \times P_{local} / BW_{age} \tag{13.29}$$

where

D is the absorbed dose of contaminated vegetable by age category (mg/kg/day)

TI_{age} is the ingestion rate of the vegetable by age category (kg dry weight/day)

C_{veg} is the concentration of adulticide in the vegetable resulting from its atmospheric deposition (mg/kg dry weight)

P_{local} is the eaten part of the vegetable having a local origin (estimated to be 10% in urban residential areas)

The exposure by ingestion of particles is calculated by considering that 65% comes from the soil particles and 35% from the dust (Eq. 13.30).

$$D_p = ((C_{sol} \times Tipart_{age} \times 0.000001)/BW_{age}) \times (0.65 + (0.35 \times 0.5)) \quad (13.30)$$

where

D_p is the ingested dose of particles by age category (mg/kg/day)
C_{soil} is the insecticide concentration in soil surface (mg/kg dry soil)
$Tipart_{age}$ is the daily amount of particles ingested by age category (mg/day)
BW_{age} is the body weight (kg) by age category

Exposure by turf ingestion in young people can be calculated from Eq. (13.31) [55].

$$D_{turf} = C_{turf} \times 25 \times 0.67 / 16.5 \quad (13.31)$$

where

D_{turf} is the absorbed dose of adulticide by turf ingestion (mg/kg/day)
C_{turf} is the concentration of adulticide on turf surface (mg/cm^2)
25 is the turf surface ingested (cm^2/day)
0.67 is a conversion factor to obtain an amount of adulticide from a surface
16.5 is the body weight (kg)

Hand-to-mouth exposure was estimated from Eq. (13.32) [55].

$$D_{HM} = C_{turf} \times HS \times NB_C \times ER \times CF/16.5 \quad (13.32)$$

where

D_{HM} is the dose resulting from hand-to-mouth contact (mg/kg/day)
C_{turf} is the concentration of adulticide on turf surface (μg/cm^2)
HS is the hand surface in contact (430 cm^2/contact)
NB_C is contact rate (1.56 contacts/h)
ER is the exposition rate (2 h/day)
CF is a conversion factor (0.001 mg/μg)

Exposure via milk ingestion requires first estimating the intake dose received by the mother due to inhalation, dermal, and ingestion exposure. It is expressed in equivalent oral dose accounting for absorption factors for each route of exposure. This allows estimating an equivalent dose in food, which is multiplied by the bioconcentration factor of the adulticide in the human adipose tissue, which is estimated by the 1-octanol/water partition coefficient (Eq. 13.33) in order to have the concentration of adulticide in fat tissues.

$$\log BCFL = 2.54 \log Kow - 0.22\ (\log Kow)^2 - 4.56 \quad (13.33)$$

The concentration in milk was obtained from Eq. (13.34) [55].

$$C_{milk} = C_{fat} \times LF \quad (13.34)$$

where C_{milk} is the concentration of adulticide in milk (mg/kg), C_{fat} is the concentration of adulticide in fat (mg/kg), and LF is the percentage of lipid in the maternal milk (4.5%). The concentration in milligrams per kilogram of milk can be transformed into milligrams per liter from the density of the maternal milk, which is estimated to be equal to 1.03 g/mL [55].

Assuming that there are eight breastfeedings per day in 3 h intervals, the real exposure by breast milk ingestion was therefore estimated by assessing concentrations in milk at each interval. Thus, assuming that the entire exposure of the mother would take place in the first 3 h of the day, concentrations in breast milk were evaluated at times 6, 9, 12, 15, 18, 21, and 24 h, and an average concentration in breast milk at these times was evaluated. The initial concentration of adulticide in the breast milk, which was calculated from Eq. (13.34), was used for the concentration at 3 h time intervals. Multiplying the average concentration at each feeding by the average volume of milk ingested by the newborn (about 850 mL/day), exposure in this way was obtained [55].

The concentration in adulticide for the other times of the day was estimated from Eq. (13.35) [55].

$$C_T = C_{milk} \times e^{-(Kel+Kt)\times(T-3)} \quad (13.35)$$

where

- C_T is the concentration of adulticide in milk at time T = 3 h (mg/L)
- C_{milk} is the initial concentration in milk
- Kel is the elimination constant of the adulticide in the mother (h)
- Kt is the constant of elimination resulting from the breastfeeding process (h). It does not change for all chemicals for the same breastfeeding rate. It is evaluated by estimating the decrease in the concentration found in the milk from a breastfeeding to another one, based on the initial concentration assessed in Eq. (13.34), the average volume of milk in which the residual amount of chemical is distributed after a breastfeeding, and the percentage of fat in milk. It does not take into account the metabolic degradation but only the effect of the elimination by ingestion of the newborn.

The results obtained for acute exposure to the different studied adulticides are presented in Table 13.6. Over 5 years, exposure to malathion is mainly made by dermal route. Exposure of young babies in this way appears negligible. For the second age category, the dermal exposure is also important as well as the ingestion exposure, due to the hand-to-mouth behavior. Malathion is not considered lipophilic and does not tend to accumulate much in the milk. So the exposure of young babies in this way is not very significant.

TABLE 13.6
Acute Exposure Doses after Ground ULV Spraying of Malathion, Resmethrin without or with PBO, PBO Alone, and Permethrin as a Function of Age Group and Exposure Pathway

Exposure Dose (mg/kg/day)	0–6 Months	7 Months–4 Years	5–11 Years	12–19 Years	≥20 Years
		Malathion			
Ingestion	3.81E-04	1.95E-01	3.66E-05	1.95E-05	1.60E-05
Dermal	4.10E-10	1.44E-01	1.23E-01	7.48E-02	1.45E-02
Inhalation	4.59E-03	4.72E-03	3.69E-03	2.18E-03	1.32E-03
Total[a]	4.98E-03	3.43E-01	1.26E-01	7.70E-02	1.58E-02
		Resmethrin			
Ingestion	5.43E-03	6.88E-02	2.21E-05	7.21E-06	3.09E-06
Dermal	1.05E-07	6.81E-02	5.79E-02	3.49E-02	7.69E-03
Inhalation	5.45E-04	5.60E-04	4.38E-04	2.63E-04	1.55E-04
Total	5.98E-03	1.37E-01	5.83E-02	3.51E-02	7.85E-03
		Resmethrin + PBO			
Ingestion	2.41E-03	3.07E-02	9.80E-06	3.20E-06	1.37E-06
Dermal	4.66E-08	3.03E-02	2.58E-02	1.55E-02	3.42E-03
Inhalation	2.41E-04	2.48E-04	1.94E-04	1.16E-04	6.88E-05
Total	2.65E-03	6.12E-02	2.60E-02	1.56E-02	3.49E-03
		PBO			
Ingestion	1.56E-02	8.70E-02	2.05E-05	7.03E-06	3.38E-06
Dermal	3.76E-09	1.32E-01	1.13E-01	6.90E-02	1.33E-02
Inhalation	1.96E-03	2.01E-03	1.57E-03	9.44E-04	5.58E-04
Total	1.75E-02	2.22E-01	1.15E-01	6.99E-02	1.39E-02
		Permethrin			
Ingestion	5.48E-03	4.59E-02	3.84E-06	1.25E-06	5.38E-07
Dermal	7.82E-08	1.66E-01	1.41E-01	8.63E-02	1.72E-02
Inhalation	5.04E-04	5.18E-04	4.05E-04	2.43E-04	1.44E-04
Total	5.98E-03	2.12E-01	1.42E-01	8.65E-02	1.73E-02

[a] Exposure doses include background noise.

The background noise represents a negligible proportion of the total exposure doses for the malathion. Indeed, it is 3%, <0.05%, <0.1%, <0.05%, and <1% of the total exposure dose for class 1 (0–6 months) to class 5 (>20 years), respectively.

Rather similar trends are observed for resmethrin alone or with PBO, except that, for young children, the primary route of exposure is ingestion. This is also the case for PBO alone and for permethrin (Table 13.6).

For the subchronic and chronic exposures, it was considered that the air concentrations were dispersed and were negligible. Thus, inhalation and skin contact with airborne droplets were not considered. For the other exposure sources, the pathways considered were the same as those for acute exposure, with the following few differences:

- The local origin of food intake that may have bioaccumulated adulticide residues in the soil during the season through the roots was considered rather than assessing the contribution of atmospheric deposition during spraying days. In addition, accumulation by air–plant transfer was not evaluated, nor was atmospheric deposition on other days than those of spraying.
- Exposure resulting from bathing behavior was excluded because the residual concentrations in swimming pool water were so low that the corresponding exposure was considered negligible.
- Only the intake of breast milk was considered for newborns because inhalation was excluded due to a lack of adulticide concentration in air.

Estimation of the concentration in adulticide in vegetables resulting from the accumulation via the roots (CV_{root}) was calculated from the concentration found on soil (C_{surf}) by using Eq. (13.36) [55].

$$CV_{root} = C_{surf} \times 3873 \times Kow^{-0.578} \tag{13.36}$$

where C_{surf} is expressed in milligrams per kilogram of dry soil and CV_{root} in milligrams per kilogram of dry weight. Once CV_{root} is determined, its use in Eq. (13.29) allows us to evaluate the total ingestion of adulticides via plants.

All the other methods for calculating the daily exposure doses are the same as for the acute exposure. In fact, it is the different daily environmental concentrations that result in different levels of exposure. The only exception is the evaluation of the average concentration in breast milk. As the average concentration was measured over periods of 56 days (subchronic exposure) or 7 months (chronic exposure year after year), the effect of breastfeeding was not taken into account since it has been estimated that, during the whole exposure period, this effect was negligible because of the time period over which it is influential (1 day). The average concentration in the milk estimated by Eq. (13.34) for the whole period was used in this case.

The results obtained for the subchronic and chronic exposures to the different studied adulticides are presented in Table 13.7.

The results obtained for malathion show that for children under age 5, ingestion is the most important exposure pathway, due to the hand-to-mouth behavior or intake of breast milk. For the other age classes, dermal exposure is the most important. The background noise was taken into account, and it represents a negligible part of the exposure dose at the subchronic level. Indeed, its contribution is 2.2%, 0.2%, 0.3%, 0.2%, and 1.2% for class 1 (0–6 months) to class 5 (>20 years), respectively. However, at the chronic level, its contribution becomes significant, with contributions of 21%, 4%, 7%, 4%, and 18% for the class 1 to class 5, respectively.

This is explained by the fact that, at the chronic level, the additional dose is distributed over a greater period of time, resulting in lower average daily dose. In these

TABLE 13.7
Subchronic and Chronic Exposure Doses after Ground ULV Spraying of Malathion, Resmethrin without or with PBO, PBO Alone, and Permethrin as a Function of Age Group and Exposure Pathway

Exposure Dose (mg/kg/day)	0–6 Months	7 Months–4 Years	5–11 Years	12–19 Years	≥20 Years
		Malathion			
Subchronic					
Ingestion	7.34E-03	5.04E-02	3.15E-03	1.99E-03	1.59E-03
Dermal	0.00	3.40E-02	2.91E-02	1.77E-02	3.43E-03
Total[a]	7.46E-03	8.46E-02	3.23E-02	1.97E-02	5.06E-03
Chronic					
Ingestion	6.75E-04	2.37E-03	2.81E-04	1.78E-04	1.44E-04
Dermal	0.00	1.52E-03	1.30E-03	7.91E-04	1.53E-04
Total[a]	7.96E-04	4.01E-03	1.68E-03	9.89E-04	3.46E-04
		Resmethrin			
Subchronic					
Ingestion	5.26E-04	1.83E-02	3.22E-05	1.64E-05	1.15E-05
Dermal	0.00	2.36E-02	2.01E-02	1.19E-02	2.52E-03
Total	5.26E-04	4.19E-02	2.01E-02	1.19E-02	2.53E-03
Chronic					
Ingestion	2.18E-04	1.14E-03	1.11E-05	5.65E-06	3.96E-06
Dermal	0.00	4.49E-03	3.80E-03	2.19E-03	5.01E-04
Total	2.18E-04	5.63E-03	3.81E-03	2.19E-03	5.05E-04
		Resmethrin + PBO			
Subchronic					
Ingestion	2.35E-04	8.32E-03	1.43E-05	7.29E-06	5.11E-06
Dermal	0.00	1.06E-02	9.02E-03	5.34E-03	1.13E-03
Total	2.35E-04	1.89E-02	9.04E-03	5.35E-03	1.14E-03
Chronic					
Ingestion	9.65E-05	4.95E-04	4.91E-06	2.50E-06	1.75E-06
Dermal	0.00	1.98E-03	1.68E-03	9.66E-04	2.21E-04
Total	9.65E-05	2.48E-03	1.68E-03	9.68E-04	2.23E-04
		PBO			
Subchronic					
Ingestion	1.20E-02	1.86E-02	7.76E-04	4.87E-04	3.86E-04
Dermal	0.00	3.07E-02	2.62E-02	1.60E-02	3.09E-03
Total	1.20E-02	4.93E-02	2.69E-02	1.64E-02	3.48E-03

(Continued)

TABLE 13.7 (CONTINUED)
Subchronic and Chronic Exposure Doses after Ground ULV Spraying of Malathion, Resmethrin without or with PBO, PBO Alone, and Permethrin as a Function of Age Group and Exposure Pathway

Exposure Dose (mg/kg/day)	0–6 Months	7 Months–4 Years	5–11 Years	12–19 Years	≥20 Years
Chronic					
Ingestion	3.61E-03	6.21E-04	1.80E-04	1.13E-04	8.96E-05
Dermal	0.00	1.90E-03	1.62E-03	9.86E-04	1.91E-04
Total	3.61E-03	2.52E-03	1.80E-03	1.10E-03	2.81E-04
		Permethrin			
Subchronic					
Ingestion	1.44E-03	2.85E-02	4.39E-06	2.10E-06	1.40E-06
Dermal	0.00	1.02E-01	8.72E-02	5.32E-02	1.03E-02
Total	1.44E-03	1.31E-01	8.72E-02	5.32E-02	1.03E-02
Chronic					
Ingestion	5.05E-04	7.20E-03	1.63E-06	7.77E-07	5.20E-07
Dermal	0.00	2.60E-02	2.22E-02	1.35E-02	2.63E-03
Total	5.05E-04	3.32E-02	2.22E-02	1.35E-02	2.63E-03

[a] Exposure doses include background noise.

conditions, the same daily dose of background noise (which is constant, regardless of the time considered) takes on greater importance.

Regarding resmethrin without or with PBO (Table 13.7), the dermal exposure is the most important in all cases except for newborns, where only the ingestion of breast milk is the route of exposure. The doses are lower than those reported previously for malathion. This is also the case for PBO. The dermal exposure is the most important in all cases except for babies, where only the ingestion of breast milk is the route of exposure.

The results of the subchronic, and chronic exposure to permethrin resulting from barrier treatment are presented in Table 13.7. The dermal exposure is the most important in all cases except for young children, where only the ingestion of breast milk is the route of exposure.

Acute and subchronic exposures of residents to malathion after ground ULV applications were also estimated by Gosselin et al. [58] and Valcke et al. [59]. Only the oral, dermal, and inhalation routes were considered. Interesting choices were made in their scenarios that deserve to be highlighted. Thus, malaoxon, which is the metabolite of malathion, is generated in the organism and also by environmental breakdown processes. It presents a much greater intrinsic toxicity on cholinesterases than the parent molecule. As a result, the effects of both malaoxon and untransformed malathion were considered for assessing the risk of exposure of malathion via ULV treatments [58,59]. The US Environmental Protection Agency (USEPA) uses a malaoxon-to-malathion factor of 61 for the toxicity and assumes that 1%, 5%, or 10% of the deposit residues are in the form of malaoxon [60]. In terms of "malathion equivalent concentration," this means that if 1% of the deposits are present in the form of malaoxon, the amount of malathion to be considered in the risk assessment has to be 1.6 times greater (i.e., $0.99 \times 1 + 0.01 \times 61$) than the concentration predicted by modeling. Gosselin et al. [58] and Valcke et al. [59] accounted for direct exposure to malaoxon for all the population and all media. On turf and foliage, a fraction of 1% of the malathion residues was considered to be in the form of malaoxon, whereas for the other media (air, soil, and hard surfaces), a value of 5% was used. A malaoxon-to-malathion toxicity equivalent factor of 61 was also selected. Environmental concentrations were estimated from the AGDISP (version 8.13) model [29].

The maximum values obtained by AGDISP modeling for ground spraying of malathion were 159.6 μg/m^3 for the air concentration and 0.373 μg/cm^2 for the concentration on foliar residues. Assuming that the residues on the ground are distributed into the first centimeter of soil thickness and that 1 cm^3 of soil weighs 1.5 g, the malathion soil concentration was estimated to be 248.9 μg/kg after ground sprayings. These environmental concentrations were adjusted to consider the environmental transformation of malathion into malaoxon [58,59]. It is noteworthy that the environmental concentrations simulated by AGDISP are significantly different from those estimated by ISCST3 (Table 13.4) while, in both cases, 61 g/ha of malathion were used and the simulation conditions were rather similar.

The time that people spend outside influences the daily length of time during which they are directly exposed to the adulticide. Leech et al. [61] estimated that, in summer, the time spent outside by children <11 years, between 11 and 17 years of age, and by adults was 194, 233, and 148 min/24 h, respectively. These values were selected by Gosselin et al. [58] and Valcke et al. [59]. In addition, they assumed that infants spent the same time outside as adults because they are not likely to be outdoors without at least one parent present.

Maibach et al. [62] applied 4 μg/cm^2 of 14C-malathion to skin of volunteers at different anatomic sites that were not washed for 24 h following the application. These authors found that the dermal absorption fractions for the forearm, foot, and forehead were equal to 0.068, 0.068, and 0.231, respectively. These values were selected by Gosselin et al. [58] and Valcke et al. [59]. In addition, for the hands, the mean value for the hand dorsum and hand palm was taken (i.e., 0.092). Finally, the value for the forearm was taken to describe the absorption fraction for the legs [58,59].

The absorbed dose by the oral route was calculated from the ingested amounts of malathion multiplied by the gastrointestinal absorption fraction. This fraction was estimated to be 0.74, based on a kinetic analysis [63].

Daily absorbed doses, expressed in terms of "malathion equivalent doses" (mg/kg/day), following a single ground ULV spraying, as a function of the age group and exposure pathway are shown in Table 13.8. Briefly, comparison with Table 13.6 [55], which displays the same type of information, reveals significant differences in the exposure dose values, except for the last two age groups. The cumulative daily absorbed doses are in the same order of magnitude. In addition, the importance of the

TABLE 13.8
Acute Exposure Doses after Ground ULV Spraying of Malathion as a Function of Age Group and Exposure Pathway

Exposure Dose (mg/kg/day)	0–6 Months	7 Months–4 Years	5–11 Years	12–19 Years	≥20 Years
Ingestion	6.8E-07	1.33E-02	8.2E-07	8.3E-08	6.6E-08
Dermal	4.39E-02	5.18E-02	3.61E-02	3.14E-02	1.72E-02
Inhalation	7.9E-03	1.74E-02	1.24E-02	7.3E-03	5.8E-03
Total	5.18E-02	8.25E-02	4.85E-02	3.87E-02	2.30E-02

exposure dose is in the following order: dermal > inhalation > ingestion. Difference between age groups was more important in the evaluation made by INSPQ [55]. Inspection of Table 13.6 shows that the cumulative daily absorbed doses vary highly between the infants, toddlers with children, and adolescents with adults. While the dermal exposure in infants is predominant in Gosselin et al. [58] (Table 13.8), it is negligible in the evaluation made by INSPQ [55] (Table 13.6). However, both studies show that toddlers have the highest cumulative daily doses.

For subchronic exposure, malathion treatment was defined as a maximum of seven spray events distributed on days 1, 3, 8, 10, 15, 22, and 24, starting in mid-August. Based on malathion's half-lives in soil and on foliage and turf, which are, respectively, 3.0 and 1.1 days, the preceding spraying sequence does not allow complete elimination of environmental deposits from these media between the different spraying events [59]. Indeed, complete elimination from the environmental media usually requires a period of at least five to seven eliminations of half-lives. Nevertheless, accumulation in the ambient air cannot occur because the malathion droplets in the air were estimated to decrease to zero within 1 h after the spraying. Thus, a dissipation half-life of 10 min was considered for this medium [59].

In order to estimate the environmental concentrations of the residues that could be reached during the treatment period, Valcke et al. [59] assumed that each spray event generated a ground deposit equal to the one modeled with the AGDISP model following a single-spray event. Thus, the environmental concentrations encountered in any medium after a spraying consisted of the sum of the residues resulting from this spraying and the residues remaining from the previous sprayings. Therefore, after any spraying, the concentrations in the considered medium were estimated by Eq. (13.37) rooted on Eq. (13.16).

$$C^i(t) = (C^{i-1} \times e^{(-KT)} + C^1) \times e^{(-Kt)} \quad (13.37)$$

where

$C^i(t)$ is the concentration in the considered medium at time t following the *i*th spraying

t is the elapsed time since the *i*th spraying

C^{i-1} is the maximum concentration reached in the considered medium following the $(i-1)$th spraying
C^1 is the maximum concentration reached in the considered medium following the first spraying
T is the elapsed time between the $(i-1)$th and the ith sprayings
K = ln(2)/(half-life of malathion in the considered medium)

The term $C^{i-1} \times e^{(-KT)}$ corresponds to the malathion residues still present in the environment before the ith spraying, and C^1 is the contribution of the ith spraying; the exponential term describes the environmental degradation of malathion residues after the ith spraying. Eq. (13.37) allows the calculation of the concentrations in the soil, on turf, foliage, and hard surfaces, and in the ambient air reached at any moment during the treatment sequence [59].

The toxicological risks generated by ground ULV malathion treatments were quantified by calculating the risk quotients (RQs), which were obtained by dividing the cumulative equivalent daily absorbed dose (malathion and malaoxon) by the corresponding reference absorbed dose of malathion. Mean RQ values for acute and subchronic exposures and for all the age groups were inferior to 1. Monte Carlo analyses were also performed by incorporating probability density functions into the equations used to calculate the daily absorbed doses resulting from the exposure scenarios. The results showed that, for the youngest population, ground sprayings of malathion can generate acute and subchronic exposure that may exceed some levels of toxicological concern. Indeed, in the case of acute exposure following ULV ground spraying for infants, toddlers, and children, these proportions were respectively 12.5%, 24.2%, and 8.8% of the individuals, and 9.8%, 16.5%, and 7.4%. following subchronic exposure [59].

Human exposure to malathion was recently reevaluated by the EPA [64]. AGDRIFT was considered as suited only for aerial application. To calculate airborne concentrations from ULV truck fogger applications, the well-mixed box model was used (Eq. 13.38).

$$\mathrm{TWA} = (\mathrm{AR/Q} \times (1 - e^{-\mathrm{Q/V} \times \mathrm{ET}}))/\mathrm{AT} \qquad (13.38)$$

where
TWA is the time-weighted average air concentration (mg/m^3)
AR is the application rate (458 mg a.i./day)
Q, which is calculated from Eq. (13.39), represents the airflow through the treated area (5,400 m^3/h)
V is the volume of the treated space (90 m^3)
ET is the duration of exposure (1.5 h)
AT is the averaging time to match the duration of the human equivalent concentration (6 h)

$$\mathrm{Q} = \mathrm{AV} \times \mathrm{CF1} \times \mathrm{CF2} \times \mathrm{A}_{\text{cross section}} \qquad (13.39)$$

where

Q is the airflow through the treated area (m^3/h)
AV is the air velocity (0.1 m/s)
CF1 is a conversion factor (60 s/1 min)
CF2 is a conversion factor (60 min/h)
$A_{cross\ section}$ is the cross section of outdoor space treated (15 m^2)

The same approach was used for the evaluation of human exposure to chlorpyrifos for different scenarios including mosquitocide applications [65]. Interestingly, different assumptions were made. It was assumed that the aerosolized particulate that remained following application did not persist for longer than 1 h in proximity of the application source. The exposure duration was set to 1 h per day for 21 days. While the endpoint of concern was acethylcholinesterase (AChE) inhibition, which is not sex specific, the difference between the body weight of male and female adults was considered due to concerns for neurodevelopmental effects related to early exposure to chlorpyrifos [65].

Last, notice that a succinct evaluation of the potential exposure of populations to fenitrothion and deltamethrin was made in the frame of the outbreak of chikungunya in Reunion Island, which was due to *Aedes albopictus* [66]. AgDRIFT [34,35] was used for calculating the soil concentrations of insecticides after 4 × 4 pickup truck applications, while the relationships of Rautmann and Ganzelmeir [67] were employed for the knapsack sprayer applications. Air concentrations were calculated from UK-POEM [68]. These evaluations are of limited interest because only acute exposure was considered. Exposition routes were restricted to inhalation, dermal contact, and ingestion of soil dust. Only two age groups were modeled: 2–7 years and 17–60 years. The choice of the parameter values to run the models was also very often questionable [66].

13.4 CONCLUSIONS

The number of studies aiming at measuring the air concentration and/or the terrestrial deposition of insecticides after outdoor applications for the control of adult mosquitoes is rather limited. These studies generally rely on environmental concentration measurements after ultralow-volume applications. ULV spraying is one of the most effective and used techniques worldwide for fighting high-density populations of adult mosquitoes. The method, among other particularities, allows impinging mosquitoes with droplets of specific sizes and favoring the drift of these droplets contrarily to the sprayings used for the control of agricultural pests, which are designed to minimize the movements of the droplets. As a result, even if the insecticides used for the control of mosquitoes are also employed in agriculture, all the wealth of literature on the measured concentrations of insecticides in air, soil, and plants after their spraying in agriculture is of very limited interest in vector control. This means that this kind of study has to be highly encouraged, especially with experimental designs accounting for the true practice under different climatic conditions, with different human and habitat densities, and so on.

Models are useful to simulate and better understand processes and for the prediction of outcomes and behaviors in settings where measurements are not available. Nevertheless, the two main rules of thumb in modeling are, first, that a model can only be used in its domain of application and, second, the GIGO acronym, which means "garbage in, garbage out." Unfortunately, our review shows that these two principles have been violated in the analyzed papers. Indeed, the models that have been used for estimating the concentrations of adulticides in air and their deposition on the soil have not been designed for use in vector control. Their predictions, made outside their domain of application, have to be interpreted with caution, especially when default values are used to run them. This is particularly annoying because, as shown by Schleier et al. [52], air concentrations and surface depositions contribute the largest amount of variability and uncertainty to the model outputs. The MULV-Disp model [39,40] has been specifically designed for estimating the concentrations of insecticides after ground-based ULV applications of adulticides for the control of mosquitoes. Unfortunately, the model shows poor statistics and presents important methodological problems. However, because the invaluable database of experimental results used to compute the MULV-Disp model should be available [40] after data cleaning, it should be fruitful to recompute this model focusing on the number of descriptor variables, the usefulness of the interaction terms, and the selection of the best statistical tool. It should also be necessary to estimate correctly its predictive performances.

The situation is also particularly worrying with regard to estimation of the concentrations in adulticides found in plants as well as in food. Good residue concentration data are lacking as well as the models. The issue is of very high concern because adulticide treatments are very often made in suburban areas where the presence of gardens with vegetables and fruit trees is commonplace.

The characteristics of a bystander in a scenario of outdoor treatment are not well defined and hence need to be clarified, especially his or her location in relation to the functioning of the sprayer device. Those governing the resident status are a little bit better defined as, for example, there exist statistics on the different activities performed outside or on the time spent outside versus inside according to the different age groups [61]. Nevertheless, nobody has simultaneously assessed the potential outdoor and indoor exposures to adulticides, since outdoor ULV spraying operations do not exclude indoor treatments against adult mosquitoes.

ACKNOWLEDGMENTS

The financial support from the French Agency for Food, Environmental and Occupational Health & Safety (Anses) is gratefully acknowledged (CRD-2015-24). Special thanks go to Christophe Lagneau for kind permission to use the picture of ground ULV spraying application from EID Méditerranée.

REFERENCES

[1] E.J. Furtaw, *An overview of human exposure modeling activities at the USEPA's National Exposure Research Laboratory*, Toxicol. Indus. Health 17 (2001), pp. 302–314.

[2] Anonymous, *Standard Operating Procedures (SOPs) for Residential Exposure Assessments*, Residential Exposure Assessment Work Group, Office of Pesticide Programs, Health Effects Division Versar, Inc., Contract No. 68-W6-0030, 1997.

[3] M. Fryer, C.D. Collins, H. Ferrier, R.N. Colvile, and M.J. Nieuwenhuijsen, *Human exposure modelling for chemical risk assessment: A review of current approaches and research and policy implications*, Environ. Sci. Pol. 9 (2006), pp. 261–274.

[4] Anonymous, *Principles of Characterizing and Applying Human Exposure Models*, Harmonization Project Document No. 3, World Health Organization, Geneva, 2005.

[5] G.A. Mount, *A critical review of ultralow-volume aerosols of insecticide applied with vehicle-mounted generators for adult mosquito control*, J. Am. Mosq. Control Assoc. 14 (1998), pp. 305–334.

[6] P. Reiter and M.B. Nathan, *Guidelines for Assessing the Efficacy of Insecticidal Space Sprays for Control of the Dengue Vector* Aedes aegypti, World Health Organization, WHO/CDS/CPE/PVC/2001.1, 2001.

[7] Anonymous, *Space Spray Application of Insecticides for Vector and Public Health Pest Control. A Practitioner's Guide*, World Health Organization, Geneva, Communicable Disease Control, Prevention and Eradication WHO Pesticide Evaluation Scheme (WHOPES), WHO/CDS/WHOPES/GCDPP/2003.5, 2003.

[8] J.A.S. Bonds, *Ultra-low-volume space sprays in mosquito control: A critical review*, Med. Vet. Entomol. 26 (2012), pp. 121–130.

[9] G.A. Mount, T.L. Biery, and D.G. Haile, *A review of ultralow-volume aerial sprays of insecticide for mosquito control*, J. Am. Mosq. Control Assoc. 12 (1996), pp. 601–618.

[10] M.J. Perich, M.A. Tidwell, D.C. Williams, M.R. Sardelis, C.J. Pena, D. Mandeville, and L.R. Boobar, *Comparison of ground and aerial ultra-low volume applications of malathion against* Aedes aegypti *in Santo Domingo, Dominican Republic*, J. Am. Mosq. Control Assoc. 6 (1990), pp. 1–6.

[11] R.G. Knepper, E.D. Walker, S.A. Wagner, M.A. Kamrin, and M.J. Zabik, *Deposition of malathion and permethrin on sod grass after single, ultra-low volume applications in a suburban neighborhood in Michigan*, J. Am. Mosq. Control Assoc. 12 (1996), pp. 45–51.

[12] N.S. Tietze, P.G. Hester, and K.R. Shaffer, *Mass recovery of malathion in simulated open field mosquito adulticide tests*, Arch. Environ. Contam. Toxicol. 26 (1994), pp. 473–477.

[13] N.S. Tietze, P.G. Hester, K.R. Shaffer, and F.T. Wakefield, *Peridomestic deposition of ultra-low volume malathion applied as mosquito adulticide*, Bull. Environ. Contam. Toxicol. 56 (1996), pp. 210–218.

[14] J.J. Schleier III and R.K.D. Peterson, *Deposition and air concentrations of permethrin and naled used for mosquito management*, Arch. Environ. Contam. Toxicol. 58 (2010), pp. 105–111.

[15] A. Faraji, I. Unlu, T. Crepeau, S. Healy, S. Crans, G. Lizarraga, D. Fonseca, and R. Gaugler, *Droplet characterization and penetration of an ultra-low volume mosquito adulticide spray targeting the asian tiger mosquito,* Aedes albopictus, *within urban and suburban environments of northeastern USA*, PLoS ONE 11 (2016), pp. e0152069.

[16] J.C. Moore, J.C. Dukes, J.R. Clark, J. Malone, C.F. Hallmon, and P.G. Hester, *Downwind drift and deposition of malathion on human targets from ground ultra-low volume mosquito sprays*, J. Am. Mosq. Control Assoc. 9 (1993), pp. 138–142.

[17] C.J. Preftakes, J.J. Schleier III, and R.K.D. Peterson, *Bystander exposure to ultra-low-volume insecticide applications used for adult mosquito management*, Int. J. Environ. Res. Public Health 8 (2011), pp. 2142–2152.

[18] N.S. Holmes and L. Morawska, *A review of dispersion modelling and its application to the dispersion of particles: An overview of different dispersion models available*, Atmos. Environ. 40 (2006), pp. 5902–5928.

[19] EPA, *AERMOD: Description of Model Formulation*, EPA-454/R-03-004, 2004.

[20] V.V. Poosarala, A. Kumar, and A. Kadiyala, *Development of a spreadsheet for computing downwind concentrations based on the USEPA's AERMOD model*, Environ. Progr. Sustain. Energy 28 (2009), pp. 185–191.

[21] A.J. Cimorelli, S.G. Perry, A. Venkatram, J.C. Weil, R.J. Paine, R.B. Wilson, R.F. Lee, W.D. Peters, and R.W. Brode, *AERMOD: A dispersion model for industrial source applications. Part I: General model formulation and boundary layer characterization*, J. Appl. Meteor. 44 (2005), pp. 682–693.

[22] S.S. Jampana, A. Kumar, and C. Varadarajan, *Application of the United States Environmental Protection Agency's AERMOD model to an industrial area*, Environ. Prog. 23 (2004), pp. 12–18.

[23] S.G. Perry, A.J. Cimorelli, R.J. Paine, R.W. Brode, J.C. Weil, A. Venkatram, R.B. Wilson, R.F. Lee, and W.D. Peters, *AERMOD: A dispersion model for industrial source applications. Part II: Model performance against 17 field study databases*, J. Appl. Meteor. 44 (2005), pp. 694–708.

[24] P.F. Heckel and G.K. LeMasters, *The use of AERMOD air pollution dispersion models to estimate residential ambient concentrations of elemental mercury*, Water Air Soil Pollut. 219 (2011), pp. 377–388.

[25] N. Jittra, N. Pinthong, and S. Thepanondh, *Performance evaluation of AERMOD and CALPUFF air dispersion models in industrial complex area*, Air Soil Water Res. 8 (2015), pp. 87–95.

[26] EPA, *User's Guide for the Industrial Source Complex (ISC3) Dispersion Models Volume II—Description of Model Algorithms*, US Environmental Protection Agency Office of Air Quality Planning and Standards Emissions, Monitoring, and Analysis Division Research Triangle Park, NC, September 1995, EPA-454/B-95-003b, 1995.

[27] S. Diez, E. Barra, F. Crespo, and J. Britch, *Uncertainty propagation of meteorological and emission data in modeling pollutant dispersion in the atmosphere*, Ingen. Invest. 34 (2014), pp. 44–48.

[28] S.R. Hanna, B.A. Egan, J. Purdum, and J. Wagler, *Comparison of AERMOD, ISC3, and ADMS model performance with five field data sets*, Int. J. Environ. Poll. 16 (2005), pp. 301–314.

[29] M.E. Teske, T.B. Curbishley, and H.W. Thistle, *AGDISP Version 8.25 User Manual*, C.D.I. Report No. 09-27C, 2011.

[30] W.W. Wilkes, C.N. Lewis, J.R. Brown, R.A. Allen, and M.V. Meisch, *Optimized aerial applications of two resmethrin formulations against caged* Anopheles quadrimaculatus, J. Am. Mosq. Cont. Assoc. 25 (2009), pp. 194–198.

[31] A. Chaskopoulou, M.D. Latham, R.M. Pereira, R. Connelly, J.A.S. Bonds, and P.G. Koehler, *Efficacy of aerial ultra-low volume applications of two novel water-based formulations of unsynergized pyrethroids against riceland mosquitoes in Greece*, J. Am. Mosq. Cont. Assoc. 27 (2011), pp. 414–422.

[32] S.J.R. Woodward, R. Connell, J. Zabkiewicz, K. Steele, and J. Praat, *Evaluation of the AGDISP ground boom spray drift model*, New Zealand Plant Protect. 61 (2008), pp. 164–168.

[33] S.A. Nsibande, J.M. Dabrowski, E. van der Walt, A. Venter, and P.B.C. Forbes, *Validation of the AGDISP model for predicting airborne atrazine spray drift: A South African ground application case study*, Chemosphere 138 (2015), pp. 454–461.

[34] M.E. Teske, S.L. Bird, D.M. Esterly, T.B. Curbishley, S.L. Ray, and S.G. Perry, *AgDRIFT: A model for estimating near-field spray drift from aerial applications*, Environ. Toxicol. Chem. 21 (2002), pp. 659–671.

[35] M.E. Teske, S.L. Bird, D.M. Esterly, S.L. Ray, S.G. Perry, and D.R. Johnson, *A User's Guide for AgDRIFT® 2.0.07: A Tiered Approach for the Assessment of Spray Drift of Pesticides*, C.D.I. Report No. 01-02, 2003.

[36] M.E. Teske, H.W. Thistle, and J.A.S. Bonds, *A technical review of MULV-Disp, a recent mosquito ultra-low volume pesticide spray dispersion model*, J. Am. Mosq. Cont. Assoc. 31 (2015), pp. 262–270.

[37] R.E. Mickle, O. Samuel, L. St-Laurent, P. Dumas, and G. Rousseau, *Comparaison des Dépôts de Malathion Générés par les Applications UBV Terrestres et Aériennes avec les Estimations d'AGDISP*, Centre de Toxicologie Direction des Risques Biologiques, Environnementaux et Occupationnels, Institut National de Santé Publique du Québec, Canada, 2006.

[38] J.J. Schleier III, R.K.D. Peterson, K.M. Irvine, L.M. Marshall, D.K. Weaver, and C.J. Preftakes, *Environmental fate model for ultra-low-volume insecticide applications used for adult mosquito management*, Sci. Total Environ. 438 (2012), pp. 72–79.

[39] J.J. Schleier III and R.K.D. Peterson, *The mosquito ultra-low volume dispersion model for estimating environmental concentrations of insecticides used for adult mosquito management*. J. Am. Mosq. Control Assoc. 30 (2014), pp. 223–227.

[40] J.J. Schleier III, *Development of an Environmental Fate Model for Risk Assessment of Ultra-Low-Volume Insecticides*, thesis, Montana State University, Bozeman, 2012.

[41] M.E. Teske, H.W. Thistle, and J.A.S. Bonds, *Rebuttal to letter to the editor concerning Teske et al. (2015)*, J. Am. Mosq. Control Assoc. 32 (2016), pp. 66–67.

[42] J.J. Schleier III and R.K.D. Peterson, *Letter to the editor concerning Teske et al. (2015)*, J. Am. Mosq. Control Assoc. 32 (2016), pp. 63–65.

[43] S. Wold, M. Sjöström, and L. Eriksson, *PLS-regression: A basic tool of chemometrics*, Chemom. Intell. Lab. Syst. 58 (2001), pp. 109–130.

[44] J. Devillers, *Neural Networks in QSAR and Drug Design*, Academic Press, London, 1996.

[45] J. Devillers, *Artificial neural network modeling of the environmental fate and ecotoxicity of chemicals*, in *Ecotoxicology Modeling*, J. Devillers, ed., Springer, New York, 2009, pp. 1–28.

[46] L.M. Barraj, B.J. Petersen, J.R. Tomerlin, and A.S. Daniel, *Background Document for the Sessions: Dietary Exposure Evaluation Model (DEEM™) and DEEM™ Decompositing Procedure and Software*, United States Environmental Protection Agency (US EPA) Office of Pesticide Programs, Washington, DC, 2000. https://www.epa.gov/pesticide-science-and-assessing-pesticide-risks/deem-fcidcalendex-software-installer#info

[47] R.K.D. Peterson, P.A. Macedo, and R.S. Davis, *A human-health risk assessment for West Nile virus and insecticides used in mosquito management*, Environ. Health Perspect. 114 (2006), pp. 366–372.

[48] J.S. Fletcher, J.E. Nellessen, and T.G. Pfleeger, *Literature review and evaluation of the EPA food-chain (Kenaga) nomogram, an instrument for estimating pesticide residues on plants*, Environ. Toxicol. Chem. 13 (1994), pp. 1383–1391.

[49] T.G. Pfleeger, A. Fong, R. Hayes, H. Ratsch, and C. Wickliff, *Field evaluation of the EPA (Kenaga) nomogram, a method for estimating wildlife exposure to pesticide residues on plants*, Environ. Toxicol. Chem. 15 (1996), pp. 535–543.

[50] Anonymous, *Generic Risk Assessment Model for Indoor and Outdoor Space Spraying of Insecticides*, WHO Pesticide Evaluation Scheme Department of Control of Neglected Tropical Diseases Cluster of HIV/AIDS, Tuberculosis, Malaria and Neglected Tropical Diseases and Chemical Safety Department of Public Health and Environment Cluster of Health Security and Environment, WHO Press, Geneva, Switzerland, 2011.

[51] J. Xue, V. Zartarian, J. Moya, N. Freeman, P. Beamer, K. Black, N. Tulve, and S. Shalat, *A meta-analysis of children's hand-to-mouth frequency data for estimating nondietary ingestion exposure*, Risk Anal. 27 (2007), pp. 411–420.

[52] J.J. Schleier III, P.A. Macedo, R.S. Davis, L.M. Shama, and R.K.D. Peterson, *A two-dimensional probabilistic acute human-health risk assessment of insecticide exposure after adult mosquito management*, Stoch. Environ. Res. Risk Assess. 23 (2009), pp. 555–563.

[53] P.A. Macedo, R.K.D. Peterson, and R.S. Davis, *Risk assessments for exposure of deployed military personnel to insecticides and personal protective measures used for disease-vector management*, J. Toxicol. Environ. Health Part A 90 (2007), pp. 1758–1771.

[54] US Environmental Protection Agency, *Exposure Factors Handbook*, Vol. I. EPA/600 /P-95/002Ba. Washington, DC, National Center for Environmental Assessment, 1997.

[55] INSPQ (Institut National de Santé Publique du Québec), *Evaluation des Risques Toxicologiques Associés à l'Utilisation d'Adulticides dans le Cadre d'un Programme de Lutte Vectorielle contre la Transmission du Virus du Nil Occidental*, ISBN 2-550-39331-7 Québec, QC, Canada, 2002. Available at http://www.inspq.qc.ca/pdf/publications /091 AdulticidesLutteVectoVNO.pdf

[56] D.J. Pausterbach, *The practice of exposure assessment: A state-of-the-art review*, J. Toxicol. Environ. Health, part B, 3 (2000), pp. 179–291.

[57] US Environmental Protection Agency, *Malathion: Revisions to the Preliminary Risk Assessment for the Registration Eligibility Decision (RED) Document*, chemical no. 057701, case no. 0248, barcode D265482, Office of Prevention, Pesticides and Toxic Substances, 2000, 83 pp., http://www.epa.gov/pesticides/op/malathion.htm

[58] N.H. Gosselin, M. Valcke, D. Belleville, and O. Samuel, *Human exposure to malathion during a possible vector-control intervention against West Nile virus. I: Methodological framework for exposure assessment*, Hum. Ecol. Risk. Assess. 14 (2008), pp. 1118–1137.

[59] M. Valcke, N.H. Gosselin, and D. Belleville, *Human exposure to malathion during a possible vector-control intervention against West Nile virus. II: Evaluation of the toxicological risks using a probabilistic approach*, Hum. Ecol. Risk. Assess. 14 (2008), pp. 1138–1158.

[60] US Environmental Protection Agency, *Reregistration Eligibility Decision (RED) for Malathion*, EPA/738/R/06/030, case no. 0248, Office of Prevention, Pesticides and Toxic Substances, Washington, DC, 2006.

[61] J.A. Leech, W.C. Nelson, R.T. Burnett, S. Aaron, and M.E. Raizenne, *It's about time: A comparison of Canadian and American time–activity patterns*, J. Expo. Anal. Environ. Epidemiol. 12 (2002), pp. 427–432.

[62] H.I. Maibach, R.J. Feldmann, T.H. Milby, and W.F. Serat, *Regional variation in percutaneous penetration in man: Pesticides*, Arch. Environ. Health 23 (1971), pp. 208–211.

[63] M. Bouchard, N.H. Gosselin, R.C. Brunet, O. Samuel, M.J. Dumoulin, and G. Carrier, *A toxicokinetic model of malathion and its metabolites as a tool to assess human exposure and risk through measurements of urinary biomarkers*, Toxicol. Sci. 73 (2003), pp. 182–194.

[64] *Malathion: Human Health Draft Risk Assessment for Registration Review*, US Environmental Protection Agency, Washington, DC, 2016.

[65] Federal Register, Environmental Protection Agency, *Chlorpyrifos: Tolerance, Revocations, Proposed Rule*, 40 CFR 180, 2015.

[66] AFSSET, *La Lutte Antivectorielle dans le Cadre de l'Épidémie de Chikungunya sur l'Île de la Réunion, Évaluation des Risques et de l'Efficacité des Produits Adulticides*, AFSSET, Paris, 2007.

[67] H. Ganzelmeier and D. Rautmann, *Drift, drift-reducing sprayers, and sprayer testing*, Aspects Appl. Biol. 57 (2000), pp. 1–10.

[68] UK POEM, *Predictive Operator Exposure Model*, http://www.hse.gov.uk/pesticides /topics/pesticide-approvals/pesticides-registration/data-requirements-handbook/operator -exposure.htm

14 Occupational Exposure Scenarios and Modeling in Vector Control

James Devillers

CONTENTS

ABSTRACT

The use of insecticides for controlling adult mosquitoes introduces occupational hazards and health risks to workers involved in handling and application tasks. The main hazard to operator health is direct chemical exposure, especially during mixing and loading. Training, personal protective equipment, and use of appropriate material are of paramount importance for reducing the exposure of workers to pesticides. The aim of this chapter is to review the treatment scenarios and biological monitoring data dealing with the exposure of applicators to insecticides during adulticide treatments. Occupational exposure modeling focused on adult mosquito control has also been analyzed. Our review shows that it is absolutely necessary to set precise scenarios corresponding to the different practices in vector control and to develop specific occupational exposure models for workers in mosquito control.

KEYWORDS

- DDT
- Pyrethroids
- Organophosphorus insecticides
- Operator
- Occupational exposure

14.1 INTRODUCTION

Operator exposure to insecticides during vector control operations has become a major issue in occupational health. Indeed, the operators are required to work in close proximity to spray emissions, thereby increasing the risk of contamination during spraying. The potential for exposure will largely be determined by the type of sprayer and the spray characteristics. Thus, it is assumed that indoor space spraying using manually carried thermal fogging equipment induces higher operator exposures than both outdoor spraying vehicle-mounted sprayers and the application of cold fogs [1]. Spray characteristics such as spray droplet spectrum, volume application rates, and spray concentrations influence the efficacy of the sprayed insecticides as well as their capacity to be potentially in contact with the operators. The intrinsic physicochemical properties of the insecticides and the formulations used for the spraying operations are also of first importance. The number and frequency of the treatments in normal situations and during outbreaks must be considered. The professional experience of the operators and the number of years working in vector control are also crucial. The personal behavior of the operators intervenes directly in their likelihood to be exposed to insecticides during vector control operations. Use of personal protective equipment (PPE) (e.g., gloves, respirators, face shields, boots, or overalls), personal work habits and hygiene (e.g., changing into clean clothes/washing hands or taking a bath/shower after spraying, frequency of healthcare visits), and being a smoker or not widely affect the potential exposure level of the operators to insecticides used in vector control.

The main tasks associated with exposure of operators to adulticides are opening containers, mixing and loading spraying solutions, spraying insecticide products with hand-carried or vehicle-mounted equipment, washing and maintaining spray equipment, and disposal of empty containers (Figures 14.1 through 14.3). These tasks can be handled by the same person or by different workers, occasionally or permanently. The

FIGURE 14.1 Mixing and loading a spraying solution in a knapsack hand-operated compression sprayer.

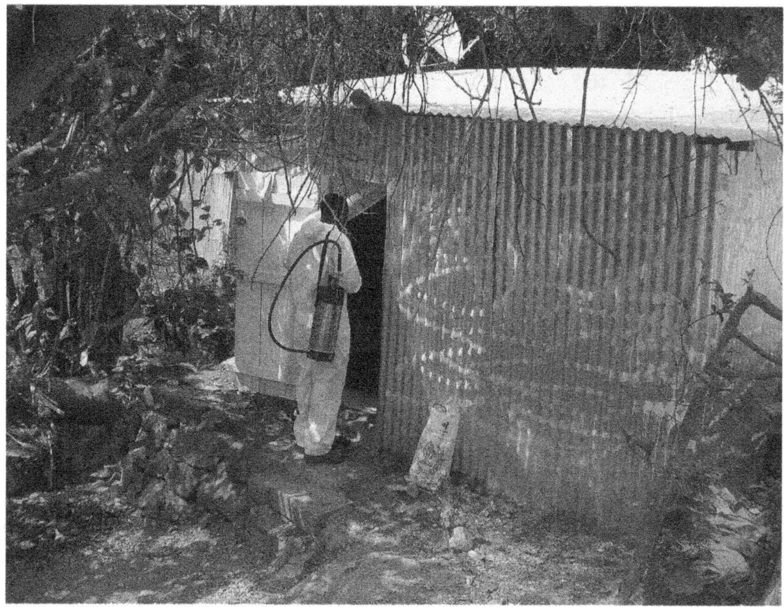

FIGURE 14.2 Knapsack spraying operation in Mayotte, France.

FIGURE 14.3 Ground vehicle-mounted ULV space spraying operation in Martinique, France.

activities of a vector control unit (VCU) are rather similar, but substantial variations in job duties can be observed depending on the country in which the VCU is located, the regulation in place in the country, the size of the VCU, and so on. As a result, differences of exposure exist among the individual members of a VCU as well as between the operators of different VCUs even if they are located in the same country and use the same protocols. This means that any attempt to generalize in the assessment of the potential exposure of operators to insecticides should be made with caution.

Different methods are used to measure the acute, subchronic, and/or chronic exposure of operators to adulticides. These techniques may be divided conveniently into direct and indirect methods of measurement.

The direct methods involve the use of some apparatus to trap the insecticide as it comes in contact with the operator during a treatment. The amount of insecticide, as determined by chemical analysis or another method, is then a direct measure of the acute exposure. The indirect methods involve the measurement of some physiological effect resulting from the exposure of the operator to the insecticide. This can be a biomarker, which can be related to present and/or past exposure to the adulticide. It is important to note that questionnaires covering the work and medical history of the operators as well as their personal activities leading to potential exposure during spraying also offer invaluable information.

Chemical analyses and biological monitoring studies are undoubtedly the most suitable approaches to estimate the occupational exposure of an operator to an adulticide. However, depending on the nature of the endpoints investigated, their number, and the number of samples collected, such studies can be very costly. Moreover, these approaches can be difficult to implement due to technical problems (e.g., number of persons sufficient to perform blood sampling, conditions for good sample preservation), as well as human ones (e.g., reluctance to participate in the study), and so on.

Modeling approaches allow us to overcome these problems. To estimate correctly the acute, subchronic and/or chronic exposure of operators to adulticides, the models have to account for the key variables that intervene in the occupational exposures in vector control. Exposure modeling is prompt to favor comparison of chemicals based on the same scenarios for selecting chemicals with the best compromise between high mosquitocidal activity and low toxicity. Modeling also allows us to save time and money in the exposure assessment process.

The aim of this review is to analyze the scenarios and biological monitoring studies dealing with actual exposures of applicators to insecticides during adulticide treatments. Occupational exposure modeling focused on mosquito treatments is also examined. Ultimately, the goal of this study is to establish an inventory on the existing information to see what can be used or adapted and what has to be investigated to fill the gaps.

14.2 OCCUPATIONAL EXPOSURE MEASUREMENT IN VECTOR CONTROL

14.2.1 DDT

Rivero-Rodriguez et al. [2] assessed the occupational history of DDT exposure in operators dedicated to spraying houses for controlling malaria in Veracruz, Mexico.

Exposure was estimated directly by analysis of DDT metabolites in adipose tissue samples of 40 operators and indirectly via a questionnaire distributed to 331 operators. The age of the participants ranged from 20 to 70 years, with a mean of 45 years. Eighty percent of the operators had been employed in the treatment campaigns for at least 20 years. The geometric mean of total DDT = 104.48 μg/g fat, *p,p′*-DDT = 31 μg/g fat, *o,p′*-DDT = 2.1 μg/g fat, *p,p′*-DDE = 60.98 μg/g, and *p,p′*-DDD = 0.95 μg/g fat [2]. It is noteworthy that the body burdens reported for the general population in Veracruz were 15.65 mg/kg for total DDT and 14.10 mg/kg for *p,p′*-DDE, which could be related to direct exposure due to vector control campaigns and also to the consumption of animal products contaminated with DDT [2]. Half of the operators referred to the use of PPE (41% used masks for respiratory protection, 37% gloves, 15% plastic goggles). Eighty-one percent of them reported eating during the day in a local house, and 42% reported eating occasionally in the field from a packed lunch. Upon finishing their spraying activities, 57% reported changing clothes, 60% reported bathing, and 91% reported washing hands before eating [2]. The reproductive history of 2,033 male malaria control operators was also analyzed in Mexico [3]. The ratio of boys to girls was similar for pregnancies before and after the father's exposure to DDT. However, the number of birth defects was higher in the latter case (i.e., 0.77% vs. 0.17%). This was also the case for spontaneous abortions (i.e., 5.1 vs. 2.46) [3]. The same kind of study was performed in Limpopo Province, South Africa [4–7]. Tests including a questionnaire, a physical examination of the reproductive system, semen analysis, and measurements of DDT and its isomers were performed in 60 malaria vector-control operators [5]. The mean number of years to work as an operator was 15.8 ± 7.8 years, and mean serum DDT was 94.3 ± 57.1 μg/g of lipid. Self-perceived current problems with sexual function ranged between 10.2% (inability to ejaculate) and 20.3% (difficulties in full erection). The most prevalent genital abnormality was abnormal testis disposition at 71.1%. There were few significant associations between DDT exposure measures and reproductive outcomes. *p,p′*-DDT was negatively associated with semen count but even if semen quality was less than normal, no strong evidence for a DDT effect was found [5]. Hormone analyses at −15, 0, 30, 90, and 120 min after the administration of gonadotropin-releasing hormone in a stimulation test for disruption in the hypothalamic–pituitary–gonadal axis were performed [6]. These analyses included luteinizing hormone, follicle-stimulating hormone, testosterone, sex hormone-binding globulin, estradiol, and inhibin. Associations between DDT exposure measures (years of work as operator and DDT isomers) and hormonal outcomes were weak and inconsistent. The results therefore do not suggest an explicit antiandrogenic or estrogenic effect of long-term DDT exposure on hormone levels, but correlations do exist in a manner that is not understood [6]. Sixty-eight operators for DDT treatments were screened for their exposure to this organochlorine insecticide during a spraying period from September to December in Zimbabwe [8]. Venous blood samples were obtained from them and the serum was used for the analysis of vitamin A. A questionnaire was also provided to the operators to obtain information on their health and their level of use of protective clothing. There was a relationship between the number of operators with vitamin A levels above the control range and with detectable DDE levels, especially among the smokers. Of respondents, 68% had partial knowledge of the kind of clothing required to spray DDT and 59% had no knowledge of the probable health effects of DDT [8].

Ferreira et al. [9] analyzed DDT exposure in 119 workers who had been involved in malaria control activities in Pará state, Brazil. Most of them were either sprayers (89.1%) or drivers (7.6%) who handled and transported DDT to endemic areas. The estimated duration of exposure to DDT ranged from 11 to 340 months (mean = 84.3 months). The recommended dose of technical DDT for spraying the walls was 2 g/m^2. This means that the average amount of DDT used per house in the Amazon region was 500 g (i.e., DDT load), while one worker sprayed five houses per day. Thus, the occupational dose per sprayer was 2,500 g of technical DDT per working day. In these conditions, the mean cumulative dose (i.e., number of working days × the average amount of DDT sprayed per working day) was equal to 4,309 kg. The mean time between the last DDT spray and medical investigation was 94 days. Blood serum levels of DDT and metabolites were measured in 1997, 1998, and 2001. The mean levels of Σ-DDT and *p,p′*-DDE were respectively equal to 231.5 and 156.9 μg/L in 1997, 126.4 and 83 μg/L in 1998, and 50.43 and 39.4 μg/L in 2001. The mean half-life of *p,p′*-DDE was estimated to be 29.5 months [9].

Carvalho and Berbet [10] measured serum DDT and HCH levels in occupationally exposed vector control workers in the state of Bahia (Brasilia). They reported high absorption levels, exceeding the limits established by occupational health legislation. This excessive absorption occurred partially due to the lack of use of personal protective equipment, workers' habits, and lack of knowledge concerning proper handling measures and techniques for pesticide application. Mean serum concentration in a control group was 3.23 μg/L for total HCH and 8.55 μg/L for total DDT. In this group, *o,p′*-DDT and *p,p′*-DDD were not detected. *p,p′*-DDT was found in few samples, while *p,p′*-DDE appeared in all samples, with a mean concentration of 8.32 μg/L. Among the three exposed groups that were investigated—namely, one with recent but not past exposure to DDT; another one with past but not recent exposure; and the last one with both past and recent exposure—the latter showed the highest mean serum concentration of 732.50 μg/L. From an occupational health point of view, all the exposed groups presented total serum DDT levels above the values considered normal at that time in Brazilian legislation (i.e., 30 μg/L) [10].

Blood samples (collected in June 1999 and October 2000) were analyzed from 41 sprayers, 20 drivers, and 14 unexposed workers in two Brazilian regions in Mato Grosso state: Sinop and Cáceres [11]. Sprayers and drivers were occupationally exposed, and no significant differences were found in serum DDT levels between these two groups in either region. No significant differences were found in *p,p′*-DDE and total DDT levels between Cáceres and Sinop. However, *p,p′*-DDT levels were higher in Sinop due to a more recent intensive use of this insecticide. The two regions together showed the following results: total DDT ranging from 7.50 to 875.5 μg/L (median = 135.5 μg/L) for sprayers; from 34.5 to 562.3 μg/L (median = 147.7 μg/L) for drivers; and from undetected to 94.8 μg/L (median = 22.5 μg/L) for unexposed workers [11].

Minelli and Ribeiro [12] determined serum insecticide residues in 26 sprayers who had handled HCH and DDT in malaria vector control campaigns during 22 to 35 years in São José do Rio Preto, São Paulo state, Brazil. The control group of unexposed workers consisted of 16 individuals with 8 to 38 years of work. All the serum samples of the exposed workers were contaminated. Residues of α-HCH, β-HCH, γ-HCH, *p,p′*-DDE, *o,p′*-DDT, or *p,p′*-DDT were found in the workers. Residues of *p,p′*-DDT were detected in the 26 exposed workers. The concentrations ranged from 1.6 to 62.9 μg/L. *p,p′*-DDE was also found in all the samples with concentrations ranging from 5.9 to 405.9 μg/L. β-HCH was found in all the samples with concentrations ranging from 3.4 to 129 μg/L. *o,p′*-DDT and γ-HCH were detected at low concentrations in 62% and 92% of the samples, respectively. All blood samples were contaminated with α-HCH at concentrations ranging from <0.2 (limit of detection, LOD) to 0.9 μg/L. All the serum samples of the unexposed workers were contaminated by *p,p′*-DDE with concentrations ranging from 5.1 to 31.6 μg/L. *o,p′*-DDT was not detected in the 16 serum samples analyzed. The other residues were detected very often at low concentrations [12].

14.2.2 Pyrethroids

Icon 10 WP, a wettable powder formulation containing 10% of the pyrethroid insecticide lambda-cyhalothrin, was evaluated for possible adverse effects on three teams of workers, each including one supervisor, one mixer-loader, and five spraymen [13]. Another sample of 15 male villagers was monitored by analyzing their urine during three consecutive days following spraying of their houses. A suspension of 0.75 g of lambda-cyhalothrin per liter was applied to the walls and ceilings of the houses at the rate of 30 mg of active ingredient per square meter using a handheld compression sprayer. Each spray pump discharged 8.3 L (i.e., 6.25 g of lambda-cyhalothrin). Urine outputs and blood serum samples were collected for analysis of the three major metabolites of lambda-cyhalothrin—namely, 3-(2-chloro-3,3,3-trifluoroprop-1-enyl)-2,2-dimethylcyclopropanecarboxylic acid (CTFMVA), 3-phenoxybenzoic acid (3-PBA), and 3-(4-hydroxyphenoxy)benzoic acid (4-OH-3-PBA). Creatinine levels in urine were also determined. No traces of lambda-cyhalothrin metabolites were found in the serum samples analyzed. The average total amounts of metabolites excreted in 15 days by the spraymen, mixer-loaders, and supervisors were 0.81, 0.24, and 0.23 mg, respectively. This means that the average amount of active ingredient absorbed by the spraymen was 54 μg/day. Assuming that the average amount of lambda-cyhalothrin handled by them was 65 g per day, the average amount absorbed per day was about 8.3×10^{-5}% [13]. Lambda-cyhalothrin metabolites were detected in a limited number of urine samples collected from villagers whose homes had been sprayed. Only one villager excreted more than 0.1 mg of such metabolites in his urine samples [13].

Dermal exposure to malathion and bifenthrin, analysis of urinary metabolites, and evaluation of health symptoms in mosquito control sprayers were assessed by Kongtip et al. [14]. The 54 subjects were interviewed with a questionnaire consisting of general characteristics (sex, age, marital status, smoking, alcohol drinking, and personal diseases), working and exposure characteristics (working duration, frequency, experiences, use of PPE), health effects (respiratory and other health symptoms) from application of pesticides. The urine samples were collected before and after the working period. Cotton patches were placed on the skin of upper legs of subjects before working. They were removed immediately after ending the work for analysis. All the subjects were males with an average age of 38.2 years old (<20 to >50 years old). They sprayed insecticides for 3–4 h/day, mostly twice a week, and their average working experience was 11.2 years (<1 to 30 years). They used personal protective devices except gloves. The spraying characteristic in the mosquito control application was of 100% smoke type. The spraying was performed with a portable sprayer (Super Hawk Model 2605, USA) at a pressure of 10 to 15 bars (Figure 14.4). The highest power of the sprayer was 30 hp/h; the spraying flow rate of liquid solution ranged from 0 to 40 L/h and a diameter of pesticide aerosols ranging from 0.5 to 50 μm. The concentrations of malathion and bifenthrin were 0.5% and 0.03% W/V, respectively. The flow rate of spraying mixed pesticides was recommended at 1.02 mL/m^2 [14]. Health symptoms were observed in 59.3% of the subjects after 1–3 h of spraying. They were skin and upper respiratory irritation (75%), dizziness/nausea (59.4%), headache (37.5%), shortness of breath (18.8%), chest tightness (12.5%), hands and face numbness (3.1%), and tinnitus (6.3%). Analysis of the cotton patches showed that the operators were exposed to average concentrations of malathion and bifenthrin of 0.2 ± 0.1 μg/cm^2 (0.1 to 0.6 μg/cm^2) and 0.3 ± 0.2 μg/cm^2 (0.1 to 1.2 μg/cm^2), respectively [14]. Most of the CTFMVA concentrations before the spraying period were below the limit of detection (LOD = 0.23 μg/mL); whereas the average CTFMVA concentration of the sprayers after the working period was 39.2 (0.6 to 261.2) mg/g creatinine [14].

Pyrethroids are widely used for indoor pest control due to their high toxicity against insects, whereas their mammalian toxicity is considerably lower compared with other pesticides. As a result, investigations of potential exposure of operators deal not only with mosquito control but also with any kind of pest insect. Thus, for example, Wang et al. [15] investigated the seasonal variation of urinary concentration of 3-PBA among Japanese pest control operators occupationally exposed

FIGURE 14.4 Portable fogger.

to pyrethroids (phenothrin, etofenprox, cyphenothrin, and mainly permethrin). Respectively, 78 and 66 operators took part in the winter and summer checkups, which consisted in urinary analyses and an interview via a health questionnaire asking about the names of the pesticides used and their spraying frequencies. Thus, 14.1% and 36.4% of the operators sprayed pesticides within 2 days before the survey in winter and summer, respectively. The average hours of spraying were 3.4 h (1 to 8) and 7.9 h (1 to 14), respectively. In addition, 37.1% and 95.4% of them had sprayed pesticides within a month before the survey in winter and summer, respectively. This represented 12.9 h (3 to 40) and 20.1 h (4 to 52) of spraying, respectively. The geometric means of urinary 3-PBA in winter and summer for workers spraying in both seasons were 3.9 and 12.2 μg/g creatinine, respectively, evidencing a statistically significant difference between the two investigated seasons [15]. The urinary 3-PBA levels of the operators who sprayed or not within 2 days before the survey in winter were equal to 5.4 and 0.9 μg/g creatinine, respectively. In summer, the concentrations were respectively equal to 12.3 and 8.7 μg/g creatinine, showing no statistically significant differences between the two groups. Leng et al. [16] demonstrated that pyrethroid metabolites were detectable in urine for a period of elimination up to 3.5 days after exposure to cyfluthrin. They also showed that cyfluthrin metabolites were eliminated following first-order kinetics ($t_{1/2}$ = 6.4 h) [17].

14.3 OCCUPATIONAL EXPOSURE MODELING IN VECTOR CONTROL

14.3.1 DDT

In Ethiopia, DDT has been widely used for indoor residual spraying since the 1950s for controlling malaria. The coverage of DDT-sprayed households has increased from 20% in 2006 to 65% in 2009. Due to a total lack of knowledge about operator exposure to DDT, 57 operators were monitored, using a patch sampling method for dermal exposure and a personal air sampler for inhalation exposure [18]. A 75% water dispersible powder (WDP) DDT formulation of 540 g active ingredient (a.i.)/kg was used. The spray solution was prepared by mixing 535 g DDT formulation, in one unit-dose package, with 8 L of water in a bucket, and then poured into the hand-operated compression sprayer. The spray tank was pumped until the pressure gauge showed 3.8 bar, which was supposed to give the specified nozzle discharge rate of 760 mL spray per min. The final concentration was 0.05 g total DDT (0.036 g a.i.)/mL spray solution. This resulted in a target application dose rate of 2 g total DDT (1.44 g a.i.)/m^2 [18]. The application rate was 40 mL/m^2 with a nozzle type of 8002E flat fan with an 80° spray angle. The spray was applied in vertical swathes; each was 75 cm wide and the swathes overlapped by 5 cm. The applicators sprayed from roof to floor, using a downward motion, and then stepped sideways and sprayed upward from floor to roof, although they did not cover the full size of the roof, which was sprayed last. Operators were in the middle of the house and sprayed the highest parts of thatched-roofed (covered with grass) and corrugated iron sheet-roofed houses. The length of the lances was about 60 to 70 cm. During spraying, operators tried to keep the tip of the lance (nozzle) about 45 cm away from the wall. All the applicators sprayed 6 days per week for 6 weeks [18]. The operators wore 100% cotton coveralls over their normal clothing. They used only one coverall for the 36 days of the spraying campaign, and they did not use gloves or hats. They also did not take showers throughout the spraying campaign, and they did not wash their hands thoroughly before eating food or chewing khat leaves, which was done by all the operators [18]. The actual dermal exposure (ADE) was calculated as the sum of the amount of DDT deposited on the head patch and cotton gloves, and the amount deposited on the clothing (potential dermal exposure) multiplied by the penetration factors. A protection factor of 0.01 was used, except for the head and hands, because the operators were not supposed to use a hat or gloves during spraying. Dermal absorption rates of 1.8%, 10%, and 25% were selected.

Total DDT amounts detected in milligrams per patch were added together and extrapolated to the defined surface areas (square centimeters) of the body regions where the patches were attached. Surface areas of 1,160; 3,150; 3,150; 1,730; 720; 4,046; and 1,554 cm^2 were used for the head and face, the chest, the back, both upper arms, both lower arms, both upper legs, and both lower legs, respectively. In the inhalation exposure assessment, the amount of DDT (milligrams per cotton segment) was equal to the amount of available DDT in the total volume of air (0.0289 m^3) sucked in by the personal air sampler at the flow rate of 2.41 L/min in 12 min. Therefore, this has to be extrapolated to the inhalation rate of the workers, estimated to be 32.9 m^3/day for a person with a body weight of 60 kg. This means that the volume of air inhaled by an applicator, in 12 min, was 0.274 m^3. The amount of DDT calculated in 0.274 m^3 was expressed as milligrams per person.

The internal dose was estimated by combining the ADE with data on the dermal absorption rate (Eq. 14.1).

$$\text{ADD} = (\text{PDE} \times \text{CPF1} \times \text{CPF2} \times \text{DA}) + (\text{ADE} \times \text{DA}) \tag{14.1}$$

where

ADD is the absorbed dermal dose
PDE is the potential dermal exposure
CPF1 and CPF2 are the clothing penetration factors (default 10%) for coverall and clothing under the coverall, respectively
DA is the dermal absorption rate (1.8%, 10%, and 25%)
ADE is the actual dermal exposure for the hands and head

The absorbed inhalation dose (AID) was estimated from the inhalation exposure multiplied by the inhalation absorption rate, which was estimated to be 100%. The total absorbed dose (TAD) was obtained from the summation of ADD and AID. ConsExpo 5.0 (beta version 01) [19] and spraying models 1 and 10 [20] were used for estimating the dermal and inhalation exposures.

The dermal route of exposure to DDT was much more important than the inhalation route of exposure. The average potential dermal exposure for 25 min during spraying of one house was 779.60 mg with a minimum of 79.01 mg and a maximum of 5,041.08 mg. The average potential inhalation exposure for 12 min during spraying of one house was 0.56 mg with a minimum of 0.02 mg and a maximum of 2.26 mg. The deposition of DDT on the upper body parts (hands, chest, back, and head) was greater than on the lower body parts (legs). The hands and head of applicators were more exposed than the rest of the body parts and significantly contributed to the dermal exposure results.

The rate of ADE (milligrams per minute and milligrams per cubic meter volume of a house) was higher in thatched-roofed houses than in corrugated iron sheet-roofed houses. The mean TAD during spraying a house per 25 min was estimated as 0.14, 0.76, and 1.88 mg/kg of body weight (BW)/day with dermal absorption rates of 1.8%, 10%, and 25%, respectively. The mean TAD during the spraying of 15 houses in 1 day was equal to 2.16, 11.36, or 28.20 mg/kg BW/day, depending on the use of a dermal absorption rate of 1.8%, 10%, and 25%, respectively. The acceptable operator exposure level (AOEL) and the acute oral minimal risk level (MRL) were estimated as 0.01 and 0.0005 mg/kg BW/day, respectively [18]. As a result, all the ratios of the internal exposure of applicators to these end points were >1, leading to an unacceptable risk.

The internal exposure obtained with the ConsExpo 5.0 (version beta 01) model was situated between the median and the 75th percentile of the experimental data for the acute and chronic dermal exposure. The chronic internal dose was calculated as ADE (mg/kg BW) × dermal absorption factor × 15 (number of houses sprayed per day) × 36 (total number of spraying days/year))/365 days/year [18]. Regarding inhalation, the predictions made by the ConsExpo model were lower than the mean acute and chronic inhalation absorbed doses. Spraying models 1 and 10 overestimated exposure of applicators by a factor of about 2 to 10. These three models appeared not suited to assessing the potential exposure of operators during DDT spraying when their outputs were compared to actual values [18].

14.3.2 Organophosphorus and Pyrethroid Insecticides

Occupational exposure of workers in vector control was estimated during ground ultralow-volume (ULV) spraying of malathion (95%), resmethrin without (40%) or with (18%, 54%) PBO and fence spraying of permethrin (50%) [21]. Dermal and inhalation exposures were estimated from Eqs. (14.2) and (14.3), respectively.

$$D_{cut} = (Conc \times SC \times dur \times cont \times Kp \times fc \times F_{abscut})/BW \quad (14.2)$$

where

D_{cut} = worker dermal dose (mg/kg/day)

Conc = concentration of the handled liquid formulation (mg/L)

SC = body surface contact with the liquid adulticides (assumed to be the arms and the hands for the handlers mixing and loading the spraying solutions and the arms, hands, and head for the sprayers, cm^2)

dur = contact time (hours/contact)

cont = number of contacts (assumed to be 1 h/day for the handlers and 7 h/day for the sprayers)

Kp = skin permeability coefficient (cm/h); for malathion, resmethrin, PBO, and permethrin, Kp equaled 8.71E-04, 1.84, 1.83E-02, and 1.47 cm/h, respectively [21]

fc = conversion factor (0.001 L/cm^3)

F_{abscut} = dermal absorption factor, assumed to be 10% for malathion and resmethrin, 20% for PBO, and 50% for permethrin [21]

BW = body weight (kg)

$$D_{inhal} = Conc_{air} \times inhal \times dur \times cont/BW \tag{14.3}$$

where

D_{inhal} = worker inhalation dose (mg/kg/day)

$Conc_{air}$ = concentration in air (mg/m^3, value calculated from the industrial source complex model version 3, short-term mode, ISCST3 [22])

inhal = inhalation rate (m^3/day)

dur = contact time (7/24[e] of day/contact)

cont = number of contacts (one per day)

BW = body weight (kg)

The calculated acute exposure doses for malathion, resmethrin (with and without PBO), PBO, and permethrin are listed in Table 14.1. Inspection of this table shows that people mixing and loading insecticides are the most exposed to insecticides, especially via the dermal route. As a result, INSPQ [21] only considered this route of exposure for handlers. Moreover, in cases in which the two types of activities were performed by the same person, only the dermal exposure of handlers was considered [21].

A database was developed by the US EPA and Health Canada including results from field studies of workers to standardize assessments of their potential exposure to pesticides. In this database, named the Pesticide Handlers Exposure Database (PHED) [23], the results of these studies are extrapolated, classified by type of activity involving the handling of pesticides (applicator, mixer, etc.), the type of application (ULV aerial or ground, barrier treatment, etc.), and the target (mosquitoes, orchards, fields). Thus, with regard to malathion, the ULV ground application against mosquitoes involves the use of 60.56 L of malathion formulation with an application time estimated at 4 h. A working day was assumed to have a duration of 8 h. In these conditions, exposure to malathion for handlers and sprayers was estimated to the values shown in Table 14.2, when personal protections are considered or not. Because these exposure data result from the work of a single day, they are considered acute exposure doses.

The subchronic and chronic doses of malathion for handlers, extrapolated from PHED [23], were estimated as 1.36 and 0.05 mg/kg/day with the basic protection equipment and 0.011 and 0.001 mg/kg/day with PPE, respectively.

In the same way, the subchronic and chronic doses of malathion for sprayers, extrapolated from PHED [23], were estimated to be 0.17 and 0.01 mg/kg/day with basic protection equipment and 0.103 and 0.004 mg/kg/day with PPE, respectively.

TABLE 14.1
Acute Exposure Doses of Malathion, Resmethrin, PBO, and Permethrin[a] for Workers with Basic Protection[b]

Insecticide/Worker	Route of Exposure	Acute Dose
	Malathion	
Handler	Dermal	4.8E+01
	Inhalation	6.5E-04
Sprayer	Dermal	1.8E-06
	Inhalation	4.5E-03
	Resmethrin	
Handler	Dermal	3.5E+03
	Inhalation	1.6E-03
Sprayer	Dermal	1.3E-03
	Inhalation	1.5E-03
	Resmethrin + PBO	
Handler	Dermal	3.5E+03
	Inhalation	7.4E-04
Sprayer	Dermal	5.9E-04
	Inhalation	7.2E-04
	PBO	
Handler	Dermal	1.1E+02
	Inhalation	2.2E-03
Sprayer	Dermal	1.8E-05
	Inhalation	2.2E-03
	Permethrin	
Handler	Dermal	7.1E+02
	Inhalation	8.4E-04
Sprayer	Dermal	6.7E-04
	Inhalation	8.3E-04

[a] Milligrams per kilogram per day.
[b] The worker wore long pants and a long-sleeved jacket, but no gloves and no mask.

Potential exposure of operators to fenitrothion, naled, deltamethrin, or pyrethrum was estimated in the frame of the outbreak of chikungunya fever in Reunion Island in the Indian Ocean [24]. Spraying scenarios included the use of a knapsack sprayer and a 4 × 4 sprayer pickup truck. A solution of 530 g/L of fenitrothion was considered for hand spraying. The concentration in the aerosol was 2.5–2.65 g/L. Spraying occurred 4 h/day and 5 days/week. The spraying surface was of 5 ha/day, and 200 g/ha (or 0.377 L/ha) were used. Inhalation and dermal exposures were estimated from spraying model 1 (1 to 3 bar pressure), spraying model 2 (4 to 7 bar pressure), and UK POEM (predictive operator exposure model) [20,25]. With spraying model 1, an

TABLE 14.2
Acute Exposure Doses of Malathion[a] for Workers with Different Levels of Protection according to PHED

Level of Protection	Route of Exposure	Acute Dose
	Basic[b]	
Handler	Dermal	6.6
	Inhalation	0.003
Sprayer	Dermal	0.81
	Inhalation	0.01
	PPE[c]	
Handler	Dermal	0.052
	Inhalation	0.003
Sprayer	Dermal	0.50
	Inhalation	0.002

Source: PHED, *Pesticide Handler Exposure Database*, US Environmental Protection Agency, Office of Pesticide Programs, Occupational Pesticide Handler Unit, Exposure Surrogate Reference Table, September 2015 https://www.epa.gov/sites/production/files/2015-09/documents/handler-exposure-table-2015.pdf.

[a] Milligrams per kilogram per day.

[b] The worker wore long pants and a long-sleeved jacket, but no gloves and no mask.

[c] The worker wore long pants, a long-sleeved jacket, and PPE consisting of gloves, an extra layer of clothing (not waterproof), and a mask.

atmospheric concentration of 130 mg/m^3, corresponding to the 75th percentile [20], was selected. This led to an atmospheric concentration in fenitrothion of 0.325 mg a.i./m^3 (i.e., 130 × 0.0025). The inhalation dose was estimated to be 1.14×10^{-3} mg a.i./kg/day when PPE was used. The dose reached 0.023 mg a.i./kg/day in the absence of PPE (except gloves). Considering that the dermal exposure was 6% with the use of PPE, the dermal dose was estimated to be 0.47 mg a.i./kg/day (95th percentile) or 0.14 mg a.i./kg/day (75th percentile) when PPE was used. Without PPE (100% exposure), dermal dose was equal to 2.5 mg a.i./kg/day (95th percentile) or 0.88 mg a.i./kg/day (75th percentile). Atmospheric concentrations of 198 and 76 mg/m^3, corresponding respectively to the 95th and 75th percentiles, were selected with spraying model 2. This led to atmospheric concentrations in fenitrothion of 0.5 and 0.19 mg a.i./m^3, respectively. The inhalation dose was estimated as 1.7×10^{-3} mg a.i./kg/day (95th percentile) or 0.67×10^{-3} mg a.i./kg/day (75th percentile) when PPE was used. Without PPE, the values were equal to 0.035 mg a.i./kg/day (95th percentile) or 0.013 mg a.i./kg/day (75th percentile). The dermal dose with PPE was equal to 4.94 mg a.i./kg/day (95th percentile) or 0.23 mg a.i./kg/day (75th percentile). Without PPE, spraying model 2 led to dermal doses of 21.86 mg a.i./kg/day (95th percentile) and 2.01 mg a.i./kg/day (75th percentile). When UK-POEM [25] was used with a hand spraying scenario, an atmospheric concentration in fenitrothion of 0.042 mg a.i./m^3 was found. The inhalation dose for a sprayer weighing 70 kg and protected by PPE was estimated to be 0.15 × 10^{-3} mg a.i./kg/day. The dermal exposure to fenitrothion was estimated as 377.58 mg a.i./day and the corresponding dermal dose with an absorption rate of 100% was estimated to be 5.39 mg a.i./kg day. Regarding the ground spraying application scenario, the formulation of fenitrothion used was at 500 g/L. UK-POEM was used with the application corresponding to tractor-mounted/ trailed boom sprayer and rotary atomizers [25]. An atmospheric concentration of 0.01 mg a.i./m^3 was estimated. This led to an inhalation dose of 0.035 × 10^{-3} mg a.i./ kg/day for an operator of 70 kg with PPE. If we consider that neurotoxic effect may be observed in the operators when they use fenitrothion at 0.013 mg/kg/day [24], the risk is unacceptable with both scenarios.

Naled is also an organophosphorus insecticide used in vector control. The DIBROM® emulsion formulation was applied at 22.4 g/L, during 4 h by an operator of 60 kg using a knapsack sprayer [24]. With spraying models 1 and 2, dermal exposure was considered to be 100% and 6% without and with PPE, respectively. The inhalation absorption of naled was 100% without PPE and reduced to 5% with PPE. With spraying model 1, the inhalation doses (75th percentile) were estimated to 0.02 mg a.i./ kg/day without PPE and 0.001 mg a.i./kg/day with PPE. In the same way, the dermal dose equaled 0.316 mg a.i./kg/day without PPE and 0.05 mg a.i./kg/day with PPE. If we consider the 95th percentile, a dermal dose of 0.894 mg a.i./kg/day was calculated without PPE and 0.168 mg a.i./kg/day with PPE. With spraying model 2, the inhalation doses (75th/95th percentiles) were estimated to be 0.0139/0.0363 mg a.i./ kg/day without PPE and 0.0007/0.0018 mg a.i./kg/day with PPE. The dermal dose calculated for an operator weighing 70 kg equaled 0.822/6.61 mg a.i./kg/day without PPE and 0.179/0.526 mg a.i./kg/day with PPE (75th/95th percentiles). UK POEM [25] was also used in the knapsack sprayer scenario. The container size was 1 L, the application dose was 0.1 L solution/ha during mixing and loading and 10 L/ha during spraying. Time of spraying was 4 h for the treatment of 5 ha/day. Transmission to skin during mixing and loading was 100% without PPE and with only gloves. With PPE, the transmission was considered to be 10%. Hand, trunk, and leg penetration during spraying was respectively 100%, 15%, and 20% without PPE, 10%, 15%, and 20% with only gloves, and 10%, 5%, and 5% with PPE. The total operator (weighing 60 kg) exposure dose was estimated as 1.68, 0.513, and 0.248 mg a.i./kg/day in absence of PPE, with only gloves, and with PPE, respectively. To estimate the potential operator exposure during ground application with a 4 × 4 pickup truck, the UK POEM was also used. During mixing and loading, a transmission of 100% of naled to skin was assumed without PPE and with gloves. Hand, trunk, and leg penetration during spraying was respectively of 75%, 15%, and 10% without PPE or with gloves. The total operator exposure dose was estimated to be 0.13 and 0.015 mg a.i./kg/day in absence of PPE and with only gloves, respectively. With a selected NOEL (no observed effect level) of 0.2 mg/kg/day (safety factor of 100) and an RfD equal to 0.039 (safety factor of 200 for inhalation exposure), all the scenarios and modeling conditions led to unacceptable risks for the operators [24]. It is worth noting that a model specifically designed for contamination in gardens and called "modèle expo jardin v3_4" was also tested, but because the inputs showed mistakes, the output results are not presented.

Operator exposure to Aqua K Othrine, including 20 g/L of deltamethrin, was also estimated [24]. Spraying occurred 4 h/day and 5 days/week. The spraying surface was 5 ha/day and 1 g/ha (0.05 L/ha). With spraying model 1, an atmospheric concentration of 130 mg/m^3 was selected, corresponding to 75% percentile. The inhalation dose was estimated as 0.18 × 10^{-3} and 0.0093 × 10^{-3} mg a.i./kg/day without and with PPE, respectively. In the former case, dermal exposure and inhalation absorption equaled 100%. In the latter option, the dermal exposure was estimated as 6% and the inhalation exposure was considered to be 5%. The dermal doses (75th/95th percentiles) were estimated to be 0.7 × 10^{-3}/2 × 10^{-3} mg a.i./kg/day without PPE and 0.11 × 10^{-3}/0.37 × 10^{-3} mg a.i./kg/day with PPE. Use of UK POEM [25] with the knapsack sprayer scenario led to an inhalation dose of 0.0011 × 10^{-3} mg a.i./kg/day (atmospheric concentration = 0.00032 mg/m^3) and a dermal dose of 12.5 × 10^{-3} mg a.i./kg/day with PPE. If the ground application scenario is considered, UK POEM calculation led to an inhalation dose of 0.14 × 10^{-3} mg a.i./kg/day (atmospheric concentration = 0.04 mg/m^3) and a dermal dose of 0.011 mg a.i./kg/day with PPE. With an RfD equal to 0.01 mg kg/day, all the selected scenarios and modeling conditions led to acceptable risks for the operators [24].

Potential exposure of operators to pyrethrum (Pynet® 5% EC formulation) was estimated with spraying models 1 and 2. With PPE, the dermal exposure was estimated as 6% and the inhalation exposure was considered to be 5%. Both routes of exposure were equal to 100% without PPE. With an applied dose of 1% (10 g/L), use of spraying model 1 led to total doses of exposure (75th/95th percentiles) of 4.22/11.72 mg a.i./kg/day without PPE and 0.654/2.18 mg a.i./kg/day with PPE. With an applied dose of 0.05%, the total doses of exposure (75th/95th percentiles) were equal to 0.21/0.59 mg a.i./kg/day without PPE and 0.05/0.109 mg a.i./kg/day with PPE. Use of spraying model 2 with an applied dose of 1% led to total doses of exposure (75th/95th percentiles) of 10.74/85.96 mg a.i./kg/day without PPE and 2.33/6.84 mg a.i./kg/day with PPE. With an applied dose of 0.05%, the total doses of exposure (75th/95th percentiles) were equal to 0.537/4.30 mg a.i./kg/day without PPE and 0.116/0.342 mg a.i./kg/day with PPE. When the UK-POEM was used with an application volume of 0.5 L/ha, the hand spraying scenario led to an operator exposure estimated as 8.27, 5.26, and 1.84 mg a.i./kg/day, without PPE, with only gloves, and with PPE, respectively. With an application volume of 10 L/ha, the operator exposure was estimated to be 0.58, 0.28, and 0.109 mg a.i./kg/day, without PPE, with only gloves, and with PPE, respectively. "Modèle expo jardin v3_4" led to a total exposure dose of 0.235 mg a.i./kg/day. With the ground application scenario, UK POEM calculation led to total exposure doses of 1.028 mg a.i./kg/day without PPE and 0.121 mg a.i./kg/day with PPE for an application volume of 0.5 L/ha. Total exposure doses of 0.059 mg a.i./kg/day without PPE and 0.007 mg a.i./kg/day with PPE were obtained with an application volume of 10 L/ha. During the use of a knapsack sprayer, the risk was always considered as unacceptable except in the case of the use of "modèle expo jardin v3_4," but this model is far from vector control practice. With the knapsack sprayer scenario, use of the models led to an unacceptable risk, except with "modèle expo jardin v3_4" when the operator uses gloves for mixing and loading pyrethrum before application. With the ground application scenario, use of UK POEM led to an unacceptable risk [24].

14.4 CONCLUSIONS

Experimental studies aimed at estimating operator exposure to the insecticides used in vector control are very limited in number. These are often *a posteriori* studies showing residual levels of contamination in workers living in countries where hydrophobic chlorinated insecticides have a long history of use. This is the case of DTT, studies of which are the most numerous. Even if the organophosphorus and pyrethroid insecticides are widely used in vector control worldwide, the number of studies focusing on the potential exposure of these chemicals to workers in mosquito control is much lower than the works dealing with the exposition potential of the aquatic and terrestrial ecosystems to these chemicals and their effects on the biota. In fact, although the use of professional people to fight mosquitoes with insecticides started nearly 60 years ago, we are amazed by the few number of studies aimed at measuring their potential level of contamination. Moreover, due to the paucity of the studies on actual exposure measurements on workers during vector control, the existing data have to be considered with care.

Nevertheless, the available studies highlight important points:

- They show that operators are exposed during vector control operations and therefore such studies are necessary and must imperatively be encouraged. The presence and persistence of insecticides in operators' tissues and biological fluids depend on the physicochemical properties of molecules. The exposure concentrations found vary also greatly from one individual to another. This variation is even more important when there is no PPE. These results highlight the need to avoid working on mean values that are very often not representative due to the high variability in the results, expressed by large standard deviation values.
- These studies reveal gaps in the experimental protocols. Insecticide air concentrations are almost never measured. A typical day of work for an operator is never described, although it is often different from one vector control center to another. The studies focus on a single treatment operation; they do not try to follow the same operator or the exposure during a normal season and during outbreak events—and, even more, over several years.

To address the problem of lack of experimental data, exposure models have been used but none of them is suitable for vector control. To estimate the concentrations of adulticides in air, other models are used. Unfortunately, they are not any better suited to be used in vector control than the previous ones. This increases the uncertainty of the results, leading to very doubtful conclusions.

Our study shows that, very often, the models are used with a "button press" mode without prior knowledge of their characteristics and without carrying out sensitivity analysis to detect the most influential parameters.

It is absolutely necessary to establish precise scenarios corresponding to the different practices in vector control and to perform inhalation and dermal exposure measurements during vector control operations under controlled conditions. It is also crucial to develop models specifically dedicated to the evaluation of the exposure of workers during mixing, loading, and spraying insecticides to control mosquito populations in various scenarios corresponding to the main types of indoor and outdoor environments.

ACKNOWLEDGMENTS

The financial support from the French Agency for Food, Environmental and Occupational Health & Safety (Anses) is gratefully acknowledged (CRD-2015-24). Special thanks go to Christophe Lagneau for kind permission to use the pictures of EID Méditerranée.

REFERENCES

[1] Anonymous, *Generic Risk Assessment Model for Indoor and Outdoor Space Spraying of Insecticides*, WHO Pesticide Evaluation Scheme Department of Control of Neglected Tropical Diseases Cluster of HIV/AIDS, Tuberculosis, Malaria and Neglected Tropical Diseases and Chemical Safety Department of Public Health and Environment Cluster of Health Security and Environment, WHO Press, Geneva, Switzerland, 2011.

[2] L. Rivero-Rodriguez, V.H. Borja-Aburto, C. Santos-Burgoa, S. Waliszewskiy, C. Rios, and V. Cruz, *Exposure assessment for workers applying DDT to control malaria in Veracruz, Mexico*, Environ. Health Perspect. 105 (1997), pp. 98–101.

[3] F. Salazar-Garcia, E. Gallardo-Diaz, P. Cerón-Mireles, D. Loomis, and V.H. Borja-Aburto, *Reproductive effects of occupational DDT exposure among male malaria control workers*, Environ. Health Perspect. 112 (2004), pp. 542–547.

[4] M.A. Dalvie, *The Reproductive Health Effects of Long-Term DDT Exposure on Malaria Vector Control Workers In Limpopo Province, South Africa*, doctoral thesis, Department of Public Health and Primary Health Care, University of Cape Town, Cape Town, 2002.

[5] M.A. Dalvie, J.E. Myers, M.L. Thompson, T.G. Robins, S. Dyer, J. Riebow, J. Molekwa, M. Jeebhay, R. Millar, and P. Kruger, *The long-term effects of DDT exposure on semen, fertility, and sexual function of malaria vector-control workers in Limpopo Province, South Africa*, Environ. Res. 96 (2004), pp. 1–8.

[6] M.A. Dalvie, J.E. Myers, M.L. Thompson, S. Dyer, T.G. Robins, S. Omar, J. Riebow, J. Molekwa, P. Kruger, and R. Millar, *The hormonal effects of long-term DDT exposure on malaria vector-control workers in Limpopo Province, South Africa*, Environ. Res. 96 (2004), pp. 9–19.

[7] M.A. Dalvie, J.E. Myers, M.L. Thompson, T.G. Robins, S. Omar, and J. Riebow, *Exploration of different methods for measuring DDT exposure among malaria vector-control workers in Limpopo Province, South Africa*, Environ. Res. 96 (2004), pp. 20–27.

[8] C.F.B. Nhachi and O.J. Kasilo, *Occupational exposure to DDT among mosquito control sprayers*, Bull. Environ. Contam. Toxicol. 45 (1990), pp. 189–192.

[9] C.P. Ferreira, A. Cecilia, A.X. de Oliveira, and F.J.R. Paumgartten, *Serum concentrations of DDT and DDE among malaria control workers in the Amazon region*, J. Occup. Health 53 (2011), pp. 115–122.

[10] W.A. Carvalho and P.R. Berbet, *Resíduos de inseticidas organoclorados em sangue de indivíduos ocupacionalmente expostos em Campanhas de Saúde Pública no estado da Bahia*, in *XI Encontro Nacional de Analistas de Resíduos de Pesticidas, Relatório*, São Paulo: Instituto Adolfo Lutz, 1987, pp. 175–192.

[11] E.F.G. de Carvalho Dores, L. Carbo, and A.B. Gonçalves de Abreu, *Serum DDT in malaria vector control sprayers in Mato Grosso state, Brazil*, Cad. Saúde Pública, Rio de Janeiro, 19 (2003), pp. 429–437.

[12] E.V. Minelli and M.L. Ribeiro, *DDT and HCH residues in the blood serum of malaria control sprayers*, Bull. Environ. Contam. Toxicol. 57 (1996), pp. 691–696.
[13] G. Chester, N.N. Sabapathy, and B.H. Woollen, *Exposure and health assessment during application of lambda-cyhalothrin for malaria vector control in Pakistan*, Bull. World Health Org. 70 (1992), pp. 615–619.
[14] P. Kongtip, S. Sasrisuk, S. Preklang, W. Yoosook, and D. Sujirarat, *Assessment of occupational exposure to malathion and bifenthrin in mosquito control sprayers through dermal contact*, J. Med. Assoc. Thai 96 Suppl. 5 (2013), pp. S82–S91.
[15] D. Wang, M. Kamijima, R. Imai, T. Suzuki, Y. Kameda, K. Asai, A. Okamura, H. Naito, J. Ueyama, I. Saito, T. Nakajima, M. Goto, E. Shibata, T. Kondo, K. Takagi, K. Takagi, and S. Wakusawa, *Biological monitoring of pyrethroid exposure of pest control workers in Japan*, J. Occup. Health 49 (2007), pp. 509–514.
[16] G. Leng, K.H. Kühn, and H. Idel, *Biological monitoring of pyrethroid metabolites in urine of pest control operators*, Toxicol. Lett. 88 (1996), pp. 215–220.
[17] G. Leng, K.H. Kühn, and H. Idel, *Biological monitoring of pyrethroids in blood and pyrethroid metabolites in urine: Applications and limitations*, Sci. Total Environ. 199 (1997), pp. 173–181.
[18] F. Wassie, P. Spanoghe, D.A. Tessema, and W. Steurbaut, *Exposure and health risk assessment of applicators to DDT during indoor residual spraying in malaria vector control program*, J. Expos. Sci. Environ. Epidemiol. 22 (2012), pp. 549–558.
[19] H.J. Bremmer, L.C.H. Prud'homme de Lodder, and J.G.M. van Engelen, *General fact sheet, limiting conditions and reliability, ventilation, room size, body surface area*, updated version for ConsExpo 4, RIVM report 320104002/2006.
[20] EC (European Commission), *Technical Notes for Guidance Human Exposure to Biocidal Products Guidance on Exposure Estimation, B4-3040/2000/291079/MAR/E2 for the European Commission, DG Environment*, June 2002.
[21] INSPQ (Institut national de santé publique du Québec), *Evaluation des Risques Toxicologiques Associés à l'Utilisation d'Adulticides dans le Cadre d'un Programme de Lutte Vectorielle contre la Transmission du Virus du Nil Occidental*, ISBN 2-550-39331-7 Québec, QC, Canada, 2002. Available at http://www.inspq.qc.ca/pdf/publications/091 AdulticidesLutteVectoVNO.pdf
[22] EPA, *User's Guide for the Industrial Source Complex (ISC3) Dispersion Models Volume II—Description of Model Algorithms*, US Environmental Protection Agency Office of Air Quality Planning and Standards Emissions, Monitoring, and Analysis Division Research Triangle Park, North Carolina 27711, September 1995, EPA-454/B-95-003b, 1995.
[23] PHED, *Pesticide Handler Exposure Database*, US Environmental Protection Agency, Office of Pesticide Programs, Occupational Pesticide Handler Unit, Exposure Surrogate Reference Table, September 2015 https://www.epa.gov/sites/production/files/2015-09/documents/handler-exposure-table-2015.pdf
[24] AFSSET, *La Lutte Antivectorielle dans le Cadre de l'Épidémie de Chikungunya sur l'Île de la Réunion, Évaluation des Risques et de l'Efficacité des Produits Adulticides*, AFSSET, Paris, 2007.
[25] UK POEM, *Predictive Operator Exposure Model*, http://www.hse.gov.uk/pesticides/topics/pesticide-approvals/pesticides-registration/data-requirements-handbook/operator-exposure.htm

15 Use of Insecticides Indoors for the Control of Mosquitoes

Exposure Scenarios and Modeling

James Devillers

CONTENTS

ABSTRACT

The various indoor insecticide diffusion devices used for controlling the adverse effects of adult mosquitoes produce surface deposition and air distribution patterns that can vary sharply in fate, concentration, and kinetics. The influential parameters explaining this variability include the type of device, the time during which it is used, and whether it is employed by trained or untrained individuals. The physicochemical properties of the released insecticide or product mixture are also of first importance. In this chapter, the exposure scenarios and models related to indoor residual spraying, spot spraying in different types of dwellings or in aircraft, as well as following the use of electric vaporizers, insecticide-treated bed nets, mosquito coils, and impregnated clothes are reviewed. Gaps in the scenarios and in the modeling processes are identified.

KEYWORDS

- Indoor exposure
- Sprayer
- Treated bed net
- Electric vaporizers
- Pyrethroids
- Fugacity model

15.1 INTRODUCTION

Assessment of indoor contamination by xenobiotics, including insecticides used in vector control, becomes a growing issue of concern, due to people spending more and more time in indoor environments. This is particularly true for those living in industrialized countries. Thus, for example, Leech et al. [1] estimated that the mean time spent indoors at home per 24 h in six major locations in the United States was, in winter and in summer, 971 and 898 min for adults, 893 and 906 min for adolescents (i.e., 11 to 17 years old), and 1,067 and 989 min for children <11 years, respectively. In Canada, the mean time spent indoors at home by adults, adolescents, and children, in winter and summer, was of 999 and 837 min/24 h, 1,004 and 979 min/24 h, 1,137 and 931 min/24 h, respectively [1]. It is noteworthy that, depending on the age group, other indoor environments have to be accounted for (e.g., schools, bars/restaurants, vehicles). In the same way, indoor time–environment–activity patterns were investigated by Schweizer et al. [2] in cities in seven European countries. They showed that the time spent indoors at home ranged from 56% to 66% per day. Interestingly, more than 90% of the variance in indoor time–activity patterns originated from differences between and within subjects rather than between cities [2].

In this context, the increasing use of insecticides indoors, to protect people against adult mosquitoes, is raising questions due to their potential adverse effects since the exposure level to these insecticides and the duration of exposure are likely to be significant. This is even more important because the indoor environment is already contaminated by a huge number of xenobiotics [3,4]. Exposure to adulticides can occur from the use of sprayers, electro vaporizers, treated bed nets, or mosquito coils. For scenarios involving such diffusion devices, risk assessment needs to address the adverse effects that are of potential concern after a single, multiple, or continuous exposure. Depending on the type of insecticide diffusion technique, the active substances may be different. Nevertheless, most of them belong to the pyrethroid family due to their low toxicity to mammals compared to the organochlorine and organophosphorus insecticides. Generally, the persistence of these insecticides is longer indoors than in agricultural environments due to diminished or filtered sunlight, reduced moisture and air movement, surface area, and lack of soil microorganisms [5]. It is important to note that the high persistence of an insecticide, which is intrinsically governed by its physicochemical properties, can be a criterion of selection for indoor treatments. Indeed, it was shown that spraying with residual insecticide was particularly important in control of the mosquito vectors of malaria, which often rest on walls before and after feeding [6,7]. This is the case of DDT (4,4′-dichloro-diphenyl-trichloroethane) which is used by way of derogation to control malaria in more than 10 countries [7,8], although this organochlorine insecticide is listed as a persistent organic pollutant (POP) under the Stockholm Convention. As a POP, this means that once released, it persists in the environment, undergoes long-range environmental transport, and causes adverse environmental and/or health effects [9]. Nevertheless, sprayed on walls and ceilings in houses, DDT is able to kill mosquitoes for several months and it shows noncontact spatial repellent properties [10].

This study was intended to investigate the fate of insecticides after their application indoors for the control of mosquitoes. The air concentration and deposition of the different adulticides were characterized according to different experimental conditions. The goal was to examine the different indoor exposure scenarios, the diverse dispersion techniques, and different insecticides commonly used indoors for the control of adult mosquitoes. It is obvious that some studies do not make the distinction between mosquitoes and other flying insects that can be found in residences. We claim that such studies are also interesting to consider. Ultimately, the study was intended to review the modeling approaches applied to these different indoor scenarios. A distinction was made between the generic and specific models. In the former situation, the objective was mainly to evaluate whether their use was justified or not, while in the latter situation other questions, such as reliability, domain of application, and availability, were necessary to address.

15.2 SPRAYING SCENARIOS

15.2.1 Indoor Residual Spraying

The main studies dealing with the assessment of exposure of residents to DDT and metabolites after indoor residual spraying (IRS) have been compiled and analyzed by the WHO's International Program on Chemical Safety (IPCS) [11]. An example of IRS is given in Figure 15.1. Simple generic models for ingestion and dermal exposure during IRS were also proposed. Residential postapplication ingestion exposure due to contaminated foodstuff was estimated from Eq. (15.1).

$$\text{Predicted dose} = \text{translodgeable residue} \times \text{surface area of food/body weight (BW)} \tag{15.1}$$

To calculate the translodgeable residue, it was assumed that the amount available for transfer from contaminated surfaces was 50% of the active ingredient on the contact surfaces, which were estimated to be covered with 30% of the target concentration of the treated walls (2,000 mg/m^2 75% WP DDT). The concentration decreased with a default half-life of 60 days. The exposure time was 365 days/year. The translodgeable residue was therefore equal to $600 \times 0.42 \times 0.5 = 126$ mg/m^2.

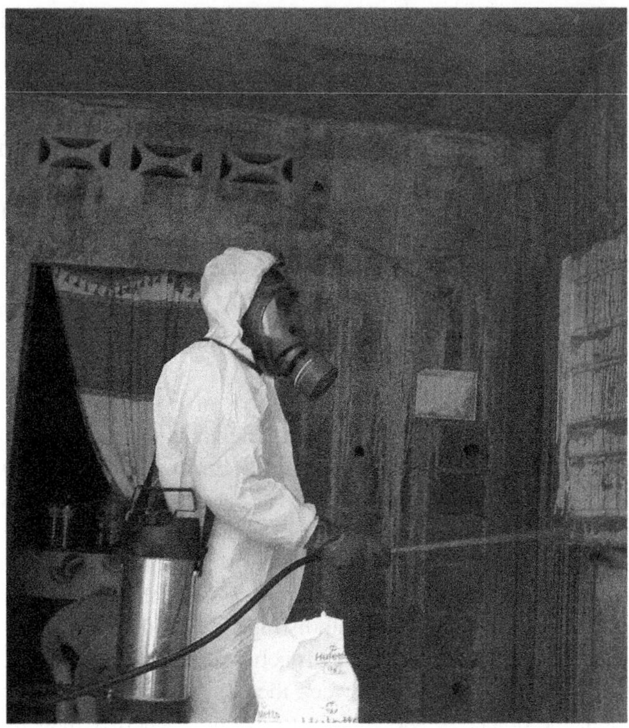

FIGURE 15.1 **(See color insert.)** Indoor residual spraying in Mayotte, France.

The surface area of food was calculated from Eq. (15.2).

$$\text{Surface area of food} = [(\text{volume of food eaten})^{1/3}]^2 \tag{15.2}$$

The volume of the food eaten per day was estimated to be 1.1 dm^3 for adults (BW = 60 kg), 0.85 dm^3 for children (BW = 40 kg), and 0.55 dm^3 for toddlers (BW = 13 kg). Using Eq. (15.2), the surface area available for contamination was 1.07 dm^2 for adults, 0.9 dm^2 for children, and 0.7 dm^2 for toddlers.

The introduction of the preceding calculated values in Eq. (15.1) led to residential postapplication ingestion exposure values of 0.022, 0.028, and 0.068 mg/kg for adults, children, and toddlers, respectively.

The residential postapplication ingestion exposure via breast milk was very roughly estimated by considering that the median concentration of total DDT in milk from direct IRS exposure was 312 μg/L, or 0.000000 312 kg/L, which has to be multiplied by the ingestion rate of milk (0.95 kg/day). This led to a predicted median dose of 0.0988 mg active ingredient (a.i.)/kg BW/day (BW = 3 kg).

The residential postapplication ingestion exposure of toddlers via hand-to-mouth behavior was also crudely evaluated [11]. It was estimated that the concentration of DDT in dust was approximately 1 mg/kg (1 ng/mg) after IRS at the WHO recommended dose rate of 2 g/m^2. The 95th percentile of dust eaten was estimated to be 587 mg/day, so the daily dose for a 13 kg child was estimated to be 1 ng/mg × 587 mg/day/13 kg = 45 ng/kg BW per day [11].

The residential postapplication dermal exposure due to touching of contaminated surfaces (walls, floors, furniture) was estimated from Eq. (15.3).

$$\text{Predicted dose} = (0.2 \times \text{AV} \times \text{Conc.} \times \text{P} \times \text{ESA} \times \text{A})/\text{BW} \qquad (15.3)$$

where

A is the fraction absorbed, which is defaulted as 0.1 (10%)

AV is the average proportion of the sprayed DDT of the original concentration (0.42)

P is the proportion translodged onto the skin (50%)

Conc. is the target concentration of DDT on the wall (2 g/m^2)

The predicted doses due to touching of contaminated surfaces were equal to 0.028 mg/kg/day, 0.042 mg/kg/day, and 0.27 mg/kg/day for adults, children, and toddlers, respectively.

For toddlers, hand-to-mouth transfer was assumed to be 10% of the amount on the hands (surface area, 0.032 m^2), and the gastrointestinal absorption to be 100%. As a result, an absorbed dose of 0.021 mg/kg/day was calculated.

The annual estimated exposure of residents following IRS with DDT was estimated to be 0.05, 0.07, 0.359, and 0.0988 mg/kg/day for adults, children, toddlers, and breastfed infants, respectively [11].

Ritter et al. [7] performed a meta-analysis from the peer-reviewed literature to compare the human exposure to DDT in four populations: a population living in houses treated by indoor residual spraying (IRS) with DDT, a general population from tropical regions, a population from Greenland consuming marine mammals in their diet, and a general population from northern regions. They showed that DDT concentrations varied by a factor of about 60 between people living in IRS-treated houses and those living in northern regions. Inhalation exposure explained most of the difference in concentration between the highly exposed and the general population in the Tropics [7]. Booij et al. [12] have confirmed that inhalation exposure was a relevant exposure pathway, but they also showed the importance of dust as a route of exposure. Although DDT was only sprayed indoors, they revealed that the insecticide and its metabolites were also found in outdoor soil. This can lead to the contamination of water and crops commonly planted in close proximity to the dwellings [13].

Other insecticides have also been recommended for IRS [14]. They are listed in Table 15.1 with some of their characteristics.

TABLE 15.1
Insecticides Currently Recommended for IRS

Name	Formulation	Dosage (g/m^2)	Residual Effect in Mud Surfaces (Months)	Nb Spraying (per 6 Months)	Total Dosage (per 6 Months, g/m^2)	Amount of Formulated Product per House and per 6 Months, (kg)
DDT	75% WP	2	6	1	2	0.5
Deltamethrin	2.5% WP	0.025	3	2	0.05	0.4
Malathion	50% WP	2	2	3	6	2.4
Lambda-cyhalothrin	10% WP	0.03	3	2	0.06	0.1
Bendiocarb	80% WP	0.4	2	3	1.2	0.3
Fenitrothion	40% WP	2	3	2	4	2
Propoxur	80% WP	2	3	2	4	1

Source: S. Sadasivaiah, Y. Tozan, and J.G. Breman, Am. J. Trop. Med. Hyg. 77 (Sup. 6), 2007, pp. 249–263.

15.2.2 Spot Spraying

Spot applications involve insecticides sprayed either directly onto the mosquitoes or in places where residents have seen or would likely find them. Spray time is generally 10 s and the insecticide discharge rate is 1.5 g/s [15]. Class and Kintrup [16] performed indoor application experiments during the summer in a study room (20 m^2, 50 m^3, 20°C) with wallpaper, carpet, bookshelves, a desk and a sofa, a wooden closet, and some plants. Two insect sprays were tested. IS1, based on an aqueous formulation, included pyrethrins (1 mg/g), tetramethrin (0.5 mg/g), cyphenothrin (1 mg/g), and piperonyl butoxide (PBO) (2.8 mg/g). IS2, based on oily formulation with little matrix and gaseous propellant, included tetramethrin (9 mg/g), cyfluthrin (2.7 mg/g), and PBO (60 mg/g). Spraying was performed in the center of the room by hand directing the spray in all directions and moving the spray can up and down. Spraying was done in the evening with the first 12 h of deposition occurring in the dark followed by about 12 h of diffuse daylight. During spraying (IS1 = 15 s, IS2 = 10 s), the window of the room was closed; later it was opened to allow air exchange. Just after spraying, pyrethrins and tetramethrin in IS1 were present in air at concentrations of 125 and 55 μg/m^3, respectively. Half an hour after spraying, the concentrations were respectively 48 and 22 μg/m^3, while after 1 h, nothing was detected. Right after spraying, tetramethrin and cyfluthrin in IS2 were found in air at concentrations of 300 and 40 μg/m^3, respectively. The air concentrations were respectively 160 and 40 μg/m^3 30 min after spraying and 45 and 10 μg/m^3 after 1 h [16].

Surfaces placed in the room were covered with about 400 μg/m^2 of tetramethrin and 900 μg/m^2 of pyrethrins 15 or 30 min after spraying IS1. A strong reduction to about 10% of the initial concentration was observed after 60 h. Surface depositions for IS2 were about 200 μg/m^2 for cyfluthrin and 900 μg/m^2 for tetramethrin, 15–30 min after spraying.

If a drastic decrease in concentration of tetramethrin was observed after 24 h, cyfluthrin persisted for 48 h without any losses and, after 60 h, only a slight reduction was observed [16].

A characterization of exposure scenarios during self-use of five commercially available insect sprays was made by Berger-Preiβ et al. [17]. For the sake of simplicity, commercial sprays were termed CS1 (*d*-tetramethrin, 0.04 g/100 g; *d*-phenothrin, 0.06 g/100 g), CS2 (cyfluthrin, 0.04 g/100 g; tetramethrin, 0.2 g/100 g; PBO, 1 g/100 g), CS3 (*d*-tetramethrin, 0.15 g/100 g; *d*-phenothrin, 0.15 g/100 g), CS4 (pyrethrum extract, 0.25 g/100 g; PBO, 1 g/100 g), and CS5 (chlorpyrifos, 0.5 g/100 g; bioallethrin, 0.15 g/100 g; PBO, na). Each spray was applied in a model room of 16 m^2 (40 m^3) including basic living equipment. Walls were covered with woodchip wallpaper and floors with carpets. Air circulation was simulated with a ventilator. Spray products were applied in the middle of the room between the door and the window. CS1 was spayed for 20 s and CS2-CS5 for 10 s, following manufacturer instructions. Then the rooms were kept closed for 20 min and, after that, ventilated for 30 min. To simulate worst-case conditions, in another experiment, the products were sprayed for 2 min and the rooms were not ventilated. Inhalation exposure of the spray users was evaluated with a personal Respicon™ 3F (Hund, Wetzlar, Germany), which allows the simultaneous sampling and monitoring of the three health-relevant particle size fractions (respirable, thoracic, inhalable). Inhalation exposure was recorded during the spraying operation and for up to 2–3 min thereafter with the normal scenario and 5 min under the worst-case conditions. A measurement was also made over a period of 60 min. Pads of filter paper were positioned on the head, back, chest, upper left and right arms, left and right forearms, left and right thighs, and left and right shins of the sprayer representing about 8% of the total body surface. After spraying, the pads were removed for analysis following the different scenarios. The particle diameter of the droplets was measured by laser diffraction to characterize the sprays. The median particle diameter (X50) of the spray droplet distributions of CS1 to CS5 was equal to 27.94, 28.41, 18.14, 33.02, and 12.07 μm, respectively [17]. Metabolite concentrations in the urine of the spray user and of a bystander were determined after 10 s of spraying and staying in the room for 5 min. Twenty-four-hour urine samples were collected. The metabolites (E)-*cis/trans*-chrysanthemum dicarboxylic acid ((E)-*cis/trans*-CDCA), *cis/trans*-3-(2,2-dichlorovinyl) -2,2-dimethylcyclopropane carboxylic acid (*cis/trans*-DCCA), and 4-fluoro-3-phenoxybenzoic acid (4-F-3-PBA) were analyzed.

Exposure modeling was made with SprayExpo, a simple box model freely available as a Microsoft Excel® worksheet [18]. This short-term model calculates the airborne concentrations of respirable, thoracic, inhalable, or any other size fractions of aerosols of interest generated during spraying. It is assumed that the sprayed product is composed of a nonvolatile active substance dissolved in a solvent with known volatility. Regarding dermal exposure, the model can only take into account the sedimentation flow of the airborne droplets, but not accidentally occurring splashes [18].

The exposure concentrations in micrograms of active substance per cubic meter of inhaled air, the inhalable doses (respiratory minute volume of 12 L/min), and the dermal dose related to the duration of exposure for the sprays CS1 to CS5 are listed in Table 15.2. Inspection of this table shows that, during correct spraying, average exposure concentrations of the pyrethroids were between about 70 and 590 µg/m^3. Calculated inhalable active substance doses ranged from 2 to 16 µg. The exposure concentration of chlorpyrifos was about 850 µg/m^3 with a calculated inhalable dose of about 20 µg. Highest concentrations were measured for PBO (1,100–2,100 µg/m^3), leading to an inhalation intake of 30–60 µg. Much higher exposure concentrations and inhalable doses were very often found with the worst-case scenarios.

Moreover, summarizing all results (not shown), it was shown that between 19% and 48% of the active substance-carrying particles were in the respirable fraction and 59%–80% in the thoracic fraction [17].

During correct use of the sprays, between 80 and 1,086 µg of active substances/application were deposited on the body of the sprayers. Again, significant highest levels were obtained for PBO. Active substance concentrations, which vary considerably in the different spray formulations, had an important influence on the contamination levels. The potential total dermal exposure was far higher when spraying was performed under worst-case conditions.

In most cases, the calculated potential dermal doses of single active substances were about five to seven times higher than during correct application (except CS1, three times). Upper body parts were contaminated more strongly than lower ones.

After exposure to CS1 to CS4, concentrations of (E)-*trans*-CDCA in urine samples were between 1.7 and 7.1 µg/L. (E)-*cis*-CDCA was not found in any of the urine samples. Concentrations in the urine of the user were slightly lower than in the urine of bystanders (their location in the room was not given). The metabolites *cis*/*trans*-DCCA and 4-F-3-PBA were below the limit of detection [17].

Use of SprayExpo to calculate the airborne concentrations of the respirable, the thoracic, and the inhalable fractions of active substances released following the different scenarios gave acceptable results except for the products having the highest median particle diameter of the spray droplet distribution (X50) [17].

A fugacity model level IV called SPRAY-MOM (spraying model by Matoba, Ohnishi, and Matsuo) was developed by Matoba et al. [19] to simulate the dynamics of aerosol droplets and amount of active ingredient in a defined Japanese indoor environment. Fugacity modeling was developed by Mackay [20]. Briefly, fugacity, which means escaping or fleeing tendency, has units of pressure (F in pascals) and can be viewed as the partial pressure that a chemical exerts as it attempts to escape from one phase and migrate to another. With this formalism, chemical concentrations in various media are converted to a common currency of fugacity, and calculations are performed using fugacity as the descriptor of the presence of the studied chemical. The advantages of this approach are that the algebraic expressions are more easily described and manipulated, the calculations can be extended to any number of compartments, and interpretation of the results is easier. Level I estimates the equilibrium partitioning of a quantity of organic chemical between the homogeneous environmental media with defined characteristics. Level II is similar to the level I, but is a steady-state model with a constant input rate, rather than a single dose of chemical. Level III, which is the most widely used, does not assume an equilibrium state, but only steady state. Level IV allows the description of an

TABLE 15.2
Exposure Concentrations[a], Calculated Doses for Inhalable Fraction of Active Substance[b], and Dermal Dose[c] with the Five Sprays and Three Different Scenarios

Spray	T1[d]	T2[e]	Active Substance	EC	ID	DD
CS1	20	3	*d*-Tetramethrin	67.7	2.4	80
			d-Phenothrin	85.6	3.1	100
	120	5	*d*-Tetramethrin	231.2	13.9	267
			d-Phenothrin	306.2	18.4	360
	120	60	*d*-Tetramethrin	114.1	82.2	NA[f]
			d-Phenothrin	176.8	127.3	NA
CS2	10	2.2	Cyfluthrin	120.9	3.2	114
			Tetramethrin	587.1	15.5	334
			PBO	2,122	56	1,138
	120	5	Cyfluthrin	424.4	25.5	712
			Tetramethrin	3,223	193.4	1,973
			PBO	8,014	480.8	6,328
	120	60	Cyfluthrin	267.4	192.5	NA
			Tetramethrin	1,992	1,434	NA
			PBO	4,518	3,253	NA
CS3	10	3	*d*-Phenothrin	144.5	5.2	722
			d-Tetramethrin	147.4	5.3	752
	120	5	*d*-Phenothrin	1,416	84.9	4,017
			d-Tetramethrin	1,935	116.1	3,931
	120	60	*d*-Phenothrin	538.3	387.6	NA
			d-Tetramethrin	695.6	500.8	NA
CS4	10	3	Pyrethrins	344.7	12.4	1,086
			PBO	1,467	52.8	2,969
	120	5	Pyrethrins	2,616	156.9	4,736
			PBO	5,920	355.2	20,895
	120	60	Pyrethrins	753.3	542.4	NA
			PBO	2,466	1,775	NA
CS5	10	2.2	Chlorpyrifos	845.5	22.3	249
			Bioallethrin	426.9	11.3	124
			PBO	1,167	30.8	264
	120	5	Chlorpyrifos	3,444	206.6	1,817
			Bioallethrin	1,565	93.9	571
			PBO	4,831	289.9	1,792
	120	60	Chlorpyrifos	2,036	1,466	NA
			Bioallethrin	888.6	639.8	NA
			PBO	3,578	2,576	NA

[a] EC in micrograms per cubic meter.
[b] ID in micrograms.
[c] DD in micrograms.
[d] Spraying time (seconds).
[e] Duration of exposure (minutes).
[f] No measurements were made.

unsteady-state situation but needs the use of differential equations for the different compartments [20–22].

SPRAY-MOM [19] includes five compartments: aerosol droplet, air, floor, ceiling, and wall compartment. The aerosol droplet compartment was assigned to individual particles and classified into large, medium, and small particle-diameter compartment. When spraying, the solvent of droplets evaporates and the volume of the compartments decreases with time. The volume of the air compartment was equal to that of the sprayed room. The movement of the active ingredient in the air compartment is caused by ventilation and transference with aerosol droplet, floor, ceiling, and wall compartments as well as photodegradation, which was described by a first-order rate constant. The floor, wall, and ceiling compartments increased in volume with time. A spraying experiment was performed with an oil-based formulation (active ingredient unknown) during the summer in a typical Japanese apartment room. The floor area, which consisted of six mats, was 9.72 m^2 and the volume of the room was 23.3 m^3. The room temperature was 25°C. The biocide was released all at once in the center of the room during 15 s by omnidirectional spraying. It was shown that the time-dependent aerial concentrations agreed with the measured ones. The model showed that the air concentration on the first day was highly influenced by the behavior of small droplets and ventilation rate [19]. However, nothing can be verified in this study.

SPRAY-MOM was modified [23] to describe the time-dependent aerial concentration and amounts of pesticides on floor, wall, and ceiling under various temperature and humidity conditions, and it was included in the InPest model [24–26] with other fugacity models developed by these authors. An aerosol canister containing a mixture of 0.45 g of *d*-tetramethrin and 0.06 g of *d*-resmethrin in a 300 mL product was applied in a Japanese room of 3.6 m (L), 2.7 m (W), and 2.4 m (H) [26]. The walls and ceiling were covered with polyvinyl chloride wallpaper. There were two windows of 1.2 × 1.2 m each. Aerosol was sprayed for a total of 10 s through four small windows 1.6 m above the floor. Five experiments were done (El to E5). The air conditioner was always off, except in E4. Windows were closed except in E3, where they were open 5 min or 2 h. Air exchange was about 1.55 in all experiments except in E2 (0.5 to 4.1) and E3 (1.5 to 12.7). Illumination equaled 830 lux (El), 0–100 lux or 0–650 lux (E3), and 0 lux in E2, E4, and E5. In El (T = 24°C, H = 60%), spraying was done twice a day at 10 a.m. and 6 p.m. for 30 days and air concentration in insecticide was measured for 113 days. Average air concentrations of *d*-tetramethrin and *d*-resmethrin just after spraying were 148 and 20.5 μg/m^3, respectively. The half-life in air was 23 min for *d*-tetramethrin and 16 min for *d*-resmethrin. The profile of residues during the 30-day period was quite comparable to the first-day profile, and from the 31st to 113th days, no insecticide was detected (detection limit: 0.05 μg/m^3). Average depositions on the walls during the 30 days of spraying ranged from 89 to 167 μg/m^2 for *d*-tetramethrin. Its half-life was 18 days on the floor and 30 days on the walls or ceiling. No *d*-resmethrin deposition was recorded (<10 μg/m^2) and the half-life was equal to 3 days [26]. In E2, air concentrations of *d*-tetramethrin equaled 8.3, 11.6, and 3.33 μg/m^3, with an air exchange rate of 1.58, 0.5, and 4.1/h, respectively. In E3, the average daily concentrations were 1.28 and 1.05 μg/m^3 when the windows were open 5 min and 2 h, respectively. In E4, the average daily air concentration in *d*-tetramethrin was 7.12 μg/m^3. Depending on the method used, dislodgeable rates of *d*-tetramethrin were 0.38%–0.27% and 0.11%–0.09% at 6 and 24 h postapplication in E5, respectively. For *d*-resmethrin the rate was 0.12% at 6 h postapplication [26]. Measured values for *d*-tetramethrin were compared to the simulations obtained with inPest when the E1 conditions were used. Good agreement was obtained between the actual and calculated data [26]. It is important to note that the inPest model is not available and hence no external validation was made. Moreover, from the information provided in different publications, it is impossible to recompute the model.

It is noteworthy that Matoba et al. [27] also developed a more specific unsteady-state fugacity model for broadcast spraying to a room carpet in order to control harmful household pests. The model was included in the inPest model as well as another component dealing with crack and crevice treatments [28].

15.2.3 Spraying in Aircraft

Insecticides are applied in aircraft for the following four reasons: (1) to comply with foreign quarantine regulations applicable to certain international flights, (2) to control insects in the cargo holds where food and food waste are stored, (3) to avoid contamination of ecosystems by foreign insect species, and (4) to respond to insect sightings reported by passengers or crew [29]. In-flight spraying is applied by flight attendants using aerosolized canisters of a pyrethroid insecticide formulation at "blocks away" or "top of descend." The former refers to the time at which an aircraft is loaded, closed, and ready to depart from the gate. The latter refers to that segment of a flight when the aircraft is ready to descend to its destination [30]. Different exposure scenarios were investigated by Berger-Preiß et al. [31] using the in-flight spraying method in parked Airbus A310 aircraft of 244 m^3. The aerosol spray consisted of 1.25% pyrethrum containing 25% pyrethrins, 2.6% PBO, and the propellants butane and propane (d_{50} = 23.4 μm). The amounts of spray applied in the six different experiments (E1 to E6) were equal to 107, 200, 168, 176, 204, and 170 g, respectively. Four spray containers of 100 mL were used except for E1 (50 mL). Fresh air exchange rate was 22.2 h^{-1} except for E3, where the air conditioning system failed 5 min after spray application. E4 was performed with heated spray containers. The time of application varied from 25 to 30 s. Inhalation exposure was evaluated with a personal Respicon 3F (Hund, Wetzlar, Germany), which allows simultaneous sampling and monitoring of the three health-relevant particle size fractions (respirable, thoracic, inhalable). Potential dermal exposure was made via pads (10 × 10 cm) covering 8% of the total body surface.

The pyrethrin concentrations (mean values of each experiment) varied from 11 to 65 μg/m^3 in the air of the passenger cabin, which was collected during the spray operation and 40 min after. For PBO, the mean concentrations in the air ranged from 200 to 485 μg/m^3. In some experiments, measurements were made 1–2 h after spraying. On average, the pyrethrin and PBO concentrations were equal to 0.05 and 0.7 μg/m^3, respectively. In the cockpit, the pyrethrin and PBO concentrations were equal to 0.8–1 and 6–10 μg/m^3, respectively. The concentrations of pyrethrins and PBO on the different surfaces are listed in Table 15.3.

In galley areas, concentrations ranged from <1 to 4 ng/cm^2 for pyrethrins and 6 to 24 ng/cm^2 for PBO, while in the cockpit the maximum concentrations were

TABLE 15.3
Median Concentrations in Pyrethrins/PBO[a] in the Different Experiments (E1 to E6)

Surface	E1	E2	E3	E4	E5	E6
Folding table vertical	<1	<1	<1	<1	<1	<1
	<1	<1	5	<1	9	12.5
Under the seats	14.5	31.5	35.5	8.5	25.5	28.5
	279.5	150.5	194.5	44	88	152.5
On the seats	46	38.5	55.5	12	45.5	41.5
	1,162.5	227	324	53.5	179	225.5
Headrests	21	37.5	52	20	34	35
	533.5	179.5	302.5	71	109.5	138
Overhead bins (closed)	<1	2	<1	<1	1	<1
	<1	14	5	<1	10	17

Note: Data in rows 2, 4, 6, 8, and 10 are PBO concentrations.
[a] Nanograms per square centimeter.

respectively 2 and 15 ng/cm^2. In settled dust, only the concentrations of PBO were determined. They varied mainly between 43 and 70 mg/kg in dust collected from the seats, while those collected from the floor yielded 34 to 54 mg/kg [31].

During spraying of 100 g of spray, the calculated inhaled doses for the sprayers were 3–12 µg pyrethrins (25–69 µg PBO, except E1: 200 µg). The total potential dermal doses (without the hands) were 200 to 830 µg/person for pyrethrins and 2,140 to 8,840 µg/person for PBO. For persons sitting in the passenger cabin during spraying, the calculated inhaled doses varied from 4 to 17 µg for pyrethrins and 50 to 103 µg for PBO. Their potential dermal exposure was 120 to 300 µg/person for pyrethrins and 1,080 to 4,040 µg/person for PBO [31].

Other scenarios with *d*-phenothrin-containing (2%) aerosol sprays were investigated by Berger-Preiß et al. [32]. The amounts of spray applied in the five different experiments (El to E5) were equal to 173, 144, 202, 113, and 91 g, respectively. Sprays used in El and E4 contained 90% water, while in E2, E3, and E5, sprays contained 40%–60% aliphatic hydrocarbons. In the former case, the aerosol particles in the spray showed a d_{50} equal to 48.53 µm, while in the latter case, d_{50} equaled 29.56 µm. The pre-embarkation method was used in El to E3, while the top-of-descent method was simulated in E4 and E5. E3 and carried out in a Boeing 747-400 (cabin volume: 887m^3; two-thirds of the passenger cabin sprayed). The other experiments were made in an Airbus A310 (244 m^3, passenger cabin sprayed). The air conditioning system was operating (exchange rate 20–22 h^{-1}) during spraying in E3–E5. In El and E2, the air conditioning packs and recirculation fans were turned off during and for 20 min after spraying. Four spray containers were used, except in E3 where eight were employed. Each had 250 mL (E2, E3, E5) or 50 mL (El and E4). Persons sprayed 0.7 g (El), 0.6 g (E2), 0.35 g (E3), 0.47 g (E4), and 0.38 g (E5) of formulation per m^3 cabin volume. Inhalation and dermal exposures were measured as in Berger-Preiß et al. [31]. Twenty-four hour urine samples from persons staying in

TABLE 15.4
Median Concentrations in *d*-Phenothrin[a] in the Different Experiments (E1 to E5)

Surface	E1	E2	E3	E4	E5
Folding table vertical	0.45	0.75	0.5	3	0.25
Floor under seats	732.5	1,159	386.9	261.5	219
On seats	376	622	115.5	714.5	545
Headrests	243.5	376	96	1,005	425.5
Overhead bins (closed)	0.9	2.2	9.95	2	0.35

[a] Nanograms per square centimeter.

the cabins for about 1 h after spraying were collected for analysis. Concentrations of *d*-phenothrin in air were equal to 182 μg/m^3 (El), 1,203 μg/m^3 (E2), 168 μg/m^3 (E3), 133 μg/m^3 (E4), and 224 μg/m^3 (E5). During spraying of 100 g of spray, the calculated inhaled doses for the sprayers were 39 μg *d*-phenothrin (10% in the respirable fraction, RF) in El, 235 μg *d*-phenothrin (30% RF) in E2, 29 μg *d*-phenothrin in E4, and 87 μg *d*-phenothrin in E5. For persons present in the passenger cabins, the highest calculated inhaled dose was 249 μg *d*-phenothrin in E2 for 100 g of spray applied.

The amounts of *d*-phenothrin on interior surfaces at the 8–12 seat positions are shown in Table 15.4.

Concentrations of *d*-phenothrin in settled dust, collected from the floor of the passenger cabin, varied between 122 and 262 mg/kg using the pre-embarkation technique (El–E3). With the top-of-descent method (E4–E5), concentrations were of 45 and 75 mg *d*-phenothrin per kilogram of dust. For sprayers, on the whole body surface (without hands) about 1,700–2,800 μg *d*-phenothrin were calculated per 100 g spray applied with the pre-embarkation method (El and E2, no data were recorded with E3). The potential dermal doses for sprayers were considerably higher during top-of-descent spraying with about 27,700 μg *d*-phenothrin (E4) and 4,100 μg *d*-phenothrin (E5) calculated per 100 g spray applied. During top-of-descent spraying (E5), a potential total dermal exposure of about 6,000 μg *d*-phenothrin was calculated for passengers per 100 g spray applied.

The metabolite e-*trans*-chrysanthemumdicarboxylic acid was detected in concentrations between 0.62 and 1.21 μg/L for persons entering the passenger cabin immediately after spraying. The metabolite concentration was lower (0.11 μg/L) in urine of a person entering the main cabin 10 min after spraying (e-*cis*-form was not detected) [32].

15.3 ELECTRIC VAPORIZER

The FLUENT model of fluid dynamics (version 4.11), developed by Fluent Inc. (Lebanon, NH), was used by Matoba et al. [33] to simulate the release of an active ingredient by an electric vaporizer located in a Japanese room in summer (9.72 m^2, 23.3 m^3, 298 K). The electric vaporizer was located in the center of the floor and at a distance of 0.6 m from the inlet-sided front wall. The air inlet (0.21 × 0.30 m) was on the wall at a height of 0.15 m from the floor. The air outlet (0.25 × 0.25 m) was on the back wall 2.05 m from the floor. The windows and doors were closed and thus the air inside the room moved from the air inlet to the outlet and the air exchange rate was 0.57 h^{-1}. Model simulations showed that condensed droplets were dispersing in the air with the rising flow just after evaporating from the vaporizer and that most droplets then settled on the ceiling even if some droplets diffused into the indoor air. Sensitivity analysis showed that the behavior of the evaporated active ingredient from the vaporizer was related to the air exchange rate, the location of the air inlet and vaporizer, and room temperature [33]. Results of fluid dynamics obtained with the FLUENT model and the parameterized Japanese room described were used for designing an unsteady-state fugacity model called VAPOR-MOM (vaporizer model by Matoba, Ohnishi, and Matsuo) [34] including five principal compartments: condensed droplets, air, floor, wall, and ceiling. The condensed droplets were divided into three compartments by generation and disappearance times. The air was classified into vapor, droplet-supplying, and breathing air compartments. The ceiling was divided into three compartments: the first compartment absorbing the droplets, the second connecting the droplet-supplying air compartment, and the third covering the previous two. The authors claim that the model is able to predict the behavior of an active ingredient in the different compartments in varying conditions of air exchange rate, photodegradation and oxidation rate, transfer rate, and physicochemical properties of the active ingredient. However, the model is not available and the information provided is not sufficient to confirm these claims or to recreate the model. The VAPOR-MOM model was further improved to include room temperature and humidity influence [23]. VAPOR-MOM was included as a scenario of the inPest model [24–26].

The behavior of prallethrin evaporated from an electric vaporizer was investigated by Matoba et al. [35] under different conditions. In the first experiment, an electric vaporizer was placed in the center of the floor of a Japanese room, about 60 cm from the inlet. It dispersed 0.87% (w/w) liquid formulation containing prallethrin for 12 h. The average air exchange rate, temperature, and relative humidity were 1.58/h, 26°C, and 29%, respectively. No natural sunlight or other illumination was provided except incidentally when the sheets for residue analysis were collected. The total amount of residue on all of the sheets, which covered 27% of the floor area, was 64.9 μg and hence the average amount was 24.5 μg/m^2. The amount of prallethrin evaporated for 12 h was 8.79 mg, or 1.01 g of the 0.87% (w/w) liquid formulation containing prallethrin. Thus, the portion of prallethrin present on the floor after evaporation was 2.7% [35]. In the second experiment, time-dependent floor residue levels were measured over a 30-day evaporation period in summer, in a tatami-floored room with the walls and ceiling covered with polyvinyl chloride. An electric vaporizer placed in the center of the floor, about 60 cm from the inlet, dispersed a 0.91% (w/w) liquid formulation containing prallethrin for 12 h each day (6 p.m. to 6 a.m.). The room was lit by an electric lamp from 7 to 10 a.m. and. 6 to 11 p.m. daily, and by natural sunlight through two windows. The average air exchange rate, room temperature, and relative humidity were 1.52/h, 25°C, and 70%, respectively. The strongest illumination equaled 920 lux due to sunlight. During the 30 days, a total of 31.8 g of formulation was evaporated, at an average rate of 9.65 mg of prallethrin per day. The amount of prallethrin increased with each 12 h vaporization, reaching a plateau at around the sixth day. The time–weight average amount on the floor was 114 μg/m^2 during the 30-day period [35]. The third experiment was performed in a small chamber of 1.2 m (L), 1.2 m (W), and 2.4 m (H). Floor material included tatami, carpet, and wood. The electric vaporizer was placed on the floor at the center of the chamber and operated for a 12 h period each day; wipe tests, designed to simulate contacts by a walking or crawling infant, were conducted immediately after each day's evaporation, over periods of 1 to 7 days. Initial transfer efficiencies measured immediately after 12 h were 1.51% for tatami, 0.485% for carpet, and 0.371% for wood. Results from multiple wipes of similarly treated areas showed that the transfer efficiency decreased with time according to a function of time $t^{-0.5}$. Indirect dermal exposures to prallethrin evaporated 12 h per day for 90 days from tatami/carpet/wood flooring were equal to 0.311/0.129/0.117, 0.254/0.105/0.1, 0.196/0.081/0.074, and 0.189/0.08/0.07 μg/kg/day for children of 1 year old, 6 years old, female adults, and male adults, respectively. The direct exposure was estimated as 0.321, 0.253, 0.183, and 0.171 μg/kg/day for children of 1 year old, 6 years old, female adults, and male adults, respectively [35].

Berger-Preiβ et al. [17] tried to characterize inhalation and dermal exposure resulting from the use of two electric vaporizers that were named EV1 and EV2 for sake of simplicity. EV1 contained 20 mg pyrethrins and 40 mg PBO per pad, while EV2 included 38 mg of *d*-allethrin and 38 mg of PBO per pad. Both electric vaporizers were applied 6 h per day in a room of 40 m^3. Use of EV1 led to indoor air concentrations ranging from 0.4 to 5 μg pyrethrins/m^3 and 1 to 7 μg PBO/m^3. With EV2, the indoor air concentrations ranged from 5 to 12 μg *d*-allethrin/m^3 and 0.5 to 2 μg PBO/m^3. A 2 h stay in the room close to the vaporizer led to potential total dermal exposure of about 80 μg of pyrethrins and 190 μg of PBO with EV1, and of about 450 μg of *d*-allethrin and 50 μg of PBO with EV2 [17].

Vesin et al. [36] investigated the potential respiratory exposure to transfluthrin resulting from the use of electric vaporizers. Two electric vaporizers, named EVA1 and EVA2, included 13.4% w/w of transfluthrin in a solid formulation form. Another electric vaporizer (EVA3) included 0.88% w/w of transfluthrin in a liquid formulation form. The vaporizers were plugged in at 1 m above the floor in an empty room of 46.70 m^3 with a window. The air exchange rates were kept constant, at 0.35 h^{-1} for the experiments with EVA1 and EVA3 and at 0.14 h^{-1} for the experiment with EVA2. The concentration of transfluthrin was monitored 1 h before the beginning of the spreading period, during the application and once the vaporizer was unplugged 8 h later, until the concentration level was stable and close to the initial background level. A high-sensitivity proton-transfer-reaction mass spectrometer was used to measure the gaseous transfluthrin emitted by the electric vaporizer refills [37]. A scanning mobility particle sizer (SMPS) device was used to detect eventual particle formation. Transfluthrin is likely to be adsorbed on airborne particles present in the room.

Transfluthrin equilibrium partitioning in the air compartment between the gas phase and airborne particles was evaluated from a model initially developed by Weschler and Nazaroff [38]. Briefly, the model assumes that most indoor surfaces have organic films at their interfaces with room air. The partitioning coefficients specific to the different sorptive surfaces present in indoor atmospheres can be described as a function of an octanol–air partition coefficient (K_{OA}) The gas-phase concentration in transfluthrin was also modeled via the ConsExpo 4.0 software [39]. Exposure was supposed to be 5 months per year of 8 h per day. The mass of commercial product emitted during the 8 h application was equal to 18.44, 18.56, and 463.06 mg for EVA1, EVA2, and EVA3, respectively. The average transfluthrin peak concentrations in the gaseous phase were equal to 4.9, 8.5, and 5.6 μg/m^3, for EVA1, EVA2, and EVA3, respectively. The concentrations calculated with ConsExpo were respectively equal to 25.6, 45.7, and 42.2 μg/m^3. According to Vesin et al. [36], the differences between the actual and modeled concentrations were due to the fact that a large proportion of emitted transfluthrin is directly adsorbed on the different surfaces. No significant formation of particles was observed with the SMPS device during the application of the electric vaporizers. However, the background concentration of PM1 airborne particles was detected at around 10 μg/m^3. As a result, these particles can serve as a support for adsorption of transfluthrin. However, because the transfluthrin proportion being absorbed on PM1 airborne particles was around 0.11% relative to the quantity present in the gas phase, it was assumed that nearly the totality of transfluthrin present in the air compartment was found in the gas phase [36]. The air exchange rate widely influences the concentration in insecticide. Thus, the concentrations in transfluthrin after 8 h emission were 8.1–11, 4.5–6.1, 3.3–4.5, and 1.7–2.3 μg/m^3 for air exchange rates of 0.14, 0.35, 0.5, and 1 h^{-1}, respectively. After 24 h of emission, the concentrations were equal to 11.5–15.7, 4.8–6.5, 3.3–4.6, and 1.7–2.3 μg/m^3 for air exchange rates of 0.14, 0.35, 0.5, and 1 h^{-1}, respectively. Average inhaled concentrations of transfluthrin over 1 h, 1 week, and 5 months were estimated to be 8.3, 1.8, and 1.8 μg/m^3, respectively, during the use 8 h per day of an electric vaporizer with an air exchange rate of 0.14 h^{-1}. A no-observed-adverse-effect level (NOAEL) of 11 mg/m^3 was selected for exposures ranging from 1 h to 1 week, while a NOAEL of 19 mg/m^3 was selected for the 5-month exposure duration. No chronic toxicological studies were found. Margins of exposure of 1,300, 6,100, and 10,500 were obtained for 1 h, 1 week, and 5 months of exposure, respectively. This indicated that adverse effects were not likely to occur [36]. Considering the ingestion route, on the basis of a log K_{OA} for transfluthrin equal to 8.43 and its concentration in the gas phase of 1.8 μg/m^3 (8 h inhaled concentration), the mass fraction of transfluthrin in dust can be estimated to be 48.4 μg/g [40]. Using a mean dust intake rate of 60 mg/day for a child, the mean intake via dust ingestion would be equal to 2.9 μg/day. Breathing 8.9 m^3/day (child of 2–3 years old) of air containing 1.8 μg/m^3 of transfluthrin leads to an inhalation intake of 16 μg/day. Therefore, inhalation intake is expected to be more than five times higher than ingestion. Dermal intake was roughly estimated to be in the same order of magnitude as inhalation. Thus, dermal pathway is expected to double inhalation exposure, whereas ingestion is expected to add 20% [36].

It is worth noting that Nazimek et al. [41] also have measured the air concentration in transfluthrin following the use of electric vaporizers. The insecticide was at 37.5% in the form of gel or liquid in the electric vaporizers. During the application of the insecticide in the form of a gel, the mean concentration of transfluthrin in the air of a room of 46.70 m^3 varied from 1.295 to 2.422 μg/m^3, on different days. It varied from 3.817 to 5.227 μg/m^3 with the liquid form. Due to a lack of information in the publication, the results cannot be analyzed more thoroughly.

15.4 INSECTICIDE-TREATED BED NETS

The efficacy and cost effectiveness of insecticide-treated bed nets in reducing malaria-related morbidity and mortality is recognized and in recent years has led to massive efforts to distribute millions of free bed nets impregnated with insecticides to vulnerable populations, mainly in Africa [42,43]. Pyrethroids are used to impregnate bed nets because of their high reactivity against mosquitoes and low toxicity for humans. Barlow et al. [44] assessed the risks of bed net treatments by deltamethrin and those resulting from the use of these bed nets.

15.4.1 Treatments of Bed Nets

Treatment was made by a tablet of K-O TAB® [45] including 0.4 g of deltamethrin. One tablet is needed for a net of 15 m^2. The volumes of suspension for a synthetic net and cotton net are 0.5 and 2 L, respectively. The maximum time spent manually dispersing a tablet in water was fixed to 60 s. A total contact time of 15 min was assumed for the entire process [44]. When dipping was made by individuals, two scenarios were considered depending on whether or not all was made to avoid contamination. With the safer scenario, three different ways were estimated. By volume of suspension on the skin, it was assumed that a maximum of 12 mL of dipping suspension contaminated the hands (6 mL/hand) via drips, splashes, and so on [46] and that 400 mg of deltamethrin was evenly distributed in 0.5 L of dipping suspension, yielding a concentration of 0.8 mg/mL. Consequently, the amount deposited on the skin was estimated to be 9.6 mg (i.e., 12 × 0.8), equivalent to 0.16 mg/kg BW for an adult of 60 kg and twice as much for a child of 30 kg. By amount of active ingredient on the skin, it was assumed that for product densities of 1 g/mL, the 95th percentile rate of exposure inside protective gloves was 72 mg/min of diluted product (median 1.4 mg/min) [46]. The potential contact time was estimated as 15 min. The calculated amount deposited on the skin was equal to 0.864 mg (i.e., 72 × 15 × 0.8/1,000), which was equivalent to 0.014 and 0.028 mg/kg BW for an adult and a child, respectively.

The third scenario consisted in estimating the systemic absorption of deltamethrin following its deposition onto the skin. Using the default dermal penetration of 10% proposed in UK POEM [47] and the value of 9.6 mg deposited on the skin obtained with the first scenario, a calculated amount of 0.96 mg was obtained, which was equivalent to 0.016 and 0.032 mg/kg BW for an adult and a child, respectively.

In the unsafe scenario, no gloves were worn, the face might be touched by contaminated hands, splashing occurred, and hands as well as contaminated clothing were not washed when dipping was finished [44]. It was assumed that 12 mL was in contact with the hands and lower arms and the same volume was spilled onto the legs and feet. In that case, the amount of deltamethrin deposited on the skin was estimated to be 28.8 mg (i.e., 36 × 0.8), which was equivalent to 0.48 and 0.96 mg/kg BW for an adult and a child, respectively. By using the alternative approach with 10% of dermal penetration, the amount absorbed was estimated as 2.88 mg (i.e., 28.8 × 0.1), equivalent to 0.048 and 0.096 mg/kg BW for an adult and a child, respectively.

For adults' dipping bed nets on a commercial basis, it was assumed that gloves and perhaps other protective clothing were worn and were changed when necessary, hands were washed, and so on. Contact time during the working day was estimated to be 6 h. The amount of deltamethrin deposited on the skin was estimated as 0.40 mg/day (i.e., 1.4 × 60 × 6 × 0.8/1,000) in a safe situation (0.007 mg/kg BW/day) and 21 mg/day (i.e., 72 × 60 × 6 × 0.8/1,000) in the worst case (0.35 mg/kg BW/day). With the alternative approach, if 21 mg/day of deltamethrin were deposited on the skin and 10% of dermal penetration was considered, the amount absorbed became 2.1 mg/day (i.e., 21 × 0.1), leading to 0.035 mg/kg BW/day for a 60 kg adult [44].

The influence of washing treated nets on exposure was also evaluated by Barlow et al. [44]. An interval of 4–6 weeks was assumed as a reasonable frequency of washing. Each washing event was considered as a new acute exposure. However, the amount of exposure was estimated as being only one-third of that estimated for the original treatment process.

15.4.2 Sleeping under a Treated Net

Potential routes of exposure taken into account by Barlow et al. [44] were inhalation, dislodgeable residues deposited on the skin, and sucking for young people.

Barlow et al. [44] assumed that a concentration of 0.055 μg/m^3 of deltamethrin was found in the breathing zone of people sleeping under treated nets (the value was extrapolated from an experiment with cyfluthrin). It is worth noting that from the same result obtained on cyfluthrin, a value twice as high was selected by Afsset [48]. If the time spent under the net is 8 h for an adult (60 kg, respiratory volume [RV] = 18.5 m^3/24 h), 12 h for a child (10 kg, RV = 4 m^3/24 h), and 12 h for a newborn (3 kg, RV = 0.9 m^3/24 h), the amount of deltamethrin inhaled was equal to 0.0056, 0.011, and 0.0086 μg/kg BW/day, respectively [44]. This corresponded to 0.34, 0.11, and 0.026 μg/day, respectively.

The surface areas potentially in contact with the treated net are given in Table 15.5.

With a target dose of 25 mg/m^2 and a transfer coefficient of 2.5%, the amount dislodgeable was estimated to be 0.0000625 mg/cm^2. The potential daily dermal exposure was therefore estimated as 0.003 mg/kg BW/day (0.16 mg/day) for the adults, 0.004 mg/kg BW/day (0.04 mg/day) for the children, and 0.006 mg/kg BW/day (0.02 mg/day) for the newborns [44].

TABLE 15.5
Skin Surface Areas[a] Potentially in Contact with the Net

Body Part	Adult	Child	Newborn
Hands	250	60	30
Arms	1,240	310	155
Lower legs	700	175	85
Feet	400	100	50
Total	2,590	645	320

Source: S.M. Barlow, F.M. Sullivan, and J. Lines, Food Chem. Toxicol. 39 (2001), pp. 407–422.

[a] Square centimeters.

Estimated exposures by oral exposure from hand-to-mouth transfer and from chewing and sucking of nets were also made by Barlow et al. [44]. It was assumed that 10% of the dislodgeable residues present on the hand were transferred to the mouth and swallowed. The authors also considered a worst-case situation in which an area of 50 cm^2 was chewed and sucked overnight and 30% of the deltamethrin present in the area was absorbed, leading to an amount absorbed of 0.0375 mg/day. The total oral exposure from these two routes was 0.04 mg/day, equivalent to 0.004 mg/kg BW/day for a child and 0.013 mg/kg BW/day for a newborn infant. Notice that Afsset [48] considered that an area of 10 cm^2 and a transfer rate of 3% were more realistic.

Barlow et al. [44] showed that in all the situations, the dermal exposure was low, 1/10 or less of the AEL dermal (10 mg/kg BW/day) and the margins of safety for systemic exposure derived from oral data were acceptable, ranging from 10 to 3,300.

Macedo et al. [49] also assessed the risk associated with the use of impregnated bed nets. It was assumed that the nets were impregnated with permethrin (60.33 or 500 mg/m^2), deltamethrin (25 mg/m^2), lambda-cyhalothrin (20 mg/m^2), α-cypermethrin (40 mg/m^2), or cyfluthrin (50 mg/m^2) and that the size of the bed net was 15 m^2. Military personnel were assumed to spend 8 h/night under the bed net; acute (1 day), subchronic (180 days), chronic (250 days for 10 years), and cancer (250 days each year for 10 years in a lifetime) risks were assessed. Inhalation exposure was estimated by considering the air concentration in insecticide under the net equal to 0.55 $\mu g/m^3$ and the respiratory rate equal to 0.4 m^3/h.

For a bed net impregnated with 1,000 mg/m^2 of permethrin, Afsset [48] assumed that the inhalation exposure was 1.1 $\mu g/m^3$.

Dermal exposure was estimated by considering that the surface area potentially in contact with the net was 50% of the total area of head, trunk, arms, legs, hands, and feet. A transfer coefficient of 2.5% was used [49].

A deeper risk assessment was performed by Peterson et al. [50] on 17 different age groups (0 to 70 years) of Africans. Exposure and risk were estimated for both genders but only results on females were presented. Five impregnated bed nets were considered: BN1 with 30 mg/m^2 of permethrin, BN2 with 63 mg/m^2 of deltamethrin, BN3 with 55 mg/m^2 of deltamethrin, BN4 with 80 mg/m^2 of deltamethrin and 261 mg/m^2 of α-cypermethrin, and BN5 with 200 mg/m^2 of α-cypermethrin.

The inhalation exposure was estimated from Eq. (15.4).

$$PE_{inh} = (AC \times RR \times T \times (T/24) \times CF)/BW \tag{15.4}$$

where

- PE_{inh} is the potential exposure by inhalation (mg/kg BW/day)
- AC is the air concentration of the insecticide ($\mu g/m^3$) under the boundaries of the net
- RR is the respiratory rate (m^3/h)
- T is the sleep time under the net (h)
- T/24 is the correction factor for the proportion of day exposed
- CF is the conversion factor from micrograms to milligrams
- BW is the body weight (kg)

The air concentrations of the insecticide were assumed to be uniformly distributed, with a minimum of 0.02 μg/m^3 and a maximum of 0.06 μg/m^3. Sleep durations for infants and toddlers (0–3 years), children (3.5–10.4 years), youth (11–14 years), and adults (15–70 years) were equal to 10–12, 8–10, 6–10, and 6–10 h, respectively. The respiration rates for children (3.5–10.4 years), youth (11–18 years), adults 1 (19–30 years), adults 2 (31–60 years), and adults 3 (≥61 years) were equal to 0.24, 0.35, 0.33, 0.32, and 0.30 m^3/h, respectively.

The dermal exposure was estimated from Eq. (15.5).

$$PE_{derm} = (TD \times SA \times TC \times AbR)/BW \tag{15.5}$$

where

- PE_{derm} is the potential dermal exposure (mg/kg BW/day)
- TD is the dose of insecticide impregnated into the bed net (mg/m^2)
- SA is the surface area of the person's body in contact with the net (Eq. 15.6)
- TC is the transfer coefficient of the amount of insecticide transferred to the skin while the person is in contact with the net
- AbR is the dermal absorption rate
- BW is the body weight (kg)

$$SA = (4 \times BW + 7)/(BW + 90) \tag{15.6}$$

where SA is the surface area (m^2) and BW is the body weight (kg).

The percentage of unclothed body surface area in contact with the net each night was selected from a uniform distribution ranging from 10% to 60%. The transfer coefficient rate was assumed to be 0.49% for permethrin and was assumed to follow a triangular distribution for the other insecticides. The dermal absorption rates for α-cypermethrin, deltamethrin, and permethrin were equal to 2.5%, 6.7%, and 15%, respectively [50].

Sucking and hand-to-mouth behaviors for children till 10 years old were estimated from Eqs. (15.7) and (15.8), respectively.

$$PE_{suck} = (NA \times TD \times WTC \times SEF)/BW \tag{15.7}$$

where

- PE_{suck} is the potential exposure (mg/kg BW/day)
- NA is the net area sucked on per night (0.15% of 15 m^2 or 225 cm^2)
- TD is the dose of insecticide impregnated into the bed net (mg/m^2)
- WTC is the wet-transfer coefficient, which is the percentage of insecticide transferred from the net to the child (20%)
- SEF is the saliva extraction factor ranged from 25% to 75%
- BW is body weight (kg)

$$PE_{hm} = (TD \times TC \times SA_h \times DR)/BW \quad (15.8)$$

where

PE_{hm} is the potential exposure (mg/kg BW/day)
TD is the dose of insecticide impregnated into the bed net (mg/m^2)
TC is the transfer coefficient
SA_h is the surface area of the child's hand touching the net each night (30% of the total surface area of the hand)
DR is the dislodgeable residue of the insecticide assumed to be dislodged from the hand, transferred to the mouth, and swallowed (10%)

Potential exposures from inhalation, dermal, and oral routes were summed to estimate potential total systemic exposures, and it was assumed that individuals in each age group slept under the treated nets each night for extended periods of time; hence, the exposures were considered chronic. Total exposures were compared to chronic, noncancer endpoints in a risk quotient (RQ) as a ratio of exposure to toxic threshold, which was the acceptable daily intake (ADI) because exposures were standardized by body weight. The ADI values used were 0.05 mg/kg/day for permethrin, 0.01 mg/kg/day for deltamethrin, and 0.02 mg/kg/day for α-cypermethrin. It was shown that RQs at the 50th and 90th percentiles for noncancer risks were <1.0 for lifetime adjusted risk and all youth and adult age groups. RQs for infants and toddlers (0–3 years) and child groups from 3 to 10 years were ≥1.0 for specific treated bed nets (e.g., BN5).

The cancer oral slope quotient (Q*) multiplied by the total potential exposures were used to calculate the population cancer risk rate. Q* was only available for permethrin and was equal to 9.567×10^{-3} mg/kg BW/day. The mean, 50th, and 90th percentile cancer risk for BN1 equaled 8.7×10^{-6}, 8.6×10^{-6}, and 10.9×10^{-6}, respectively, for the whole population.

A sensitivity analysis allowed Peterson et al. [50] to identify three input variables that could be refined to meaningfully improve future risk assessments. They were the saliva extraction factor, the body surface area in contact with the bed net, and the transfer coefficient from the dry bed net to the skin.

It is worth noting that generic equations have been proposed by the World Health Organization (WHO) [51] for the risk assessment of exposure to insecticides of individuals sleeping under insecticide-treated nets, during washing of nets, and during the conventional treatments of nets with insecticides. One scenario, which was not discussed in the preceding models on bed nets, is the potential exposure via breast milk. Indeed, pregnant and lactating women may not only sleep under treated bed nets but also wash and dip them with insecticides. Infants may therefore also be exposed through breast milk, especially for chemicals with a long half-time in the mother's body.

The predicted daily systemic dose ($SysD_O$) from mother's milk was estimated from Eq. (15.9) [51].

$$\text{SysD}_o = \text{Abs}_o \times \text{IngR} \times \text{C}_{\text{milk}}/\text{BW} \tag{15.9}$$

where

Abs_o is the gastrointestinal absorption (default value = 100%)
IngR is the ingestion rate of milk (default value = 950 mL/day)
C_{milk} is the concentration of the insecticide in the mother's milk, which is calculated from Eq. (15.10)
BW is the body weight, assumed to be 4.8 kg for a newborn

$$C_{\text{milk}} = 0.361 \times \text{DoseM} \times T^{1/2} \tag{15.10}$$

where

DoseM is the total exposure of the mother resulting from sleeping under the bed net and the operation of treatments on the bed net (0.61 μg/kg BW/day)
$T^{1/2}$ is the half-time in blood (assumed to be 5.5 h = 0.3 day)
$C_{milk} = 0.361 \times 0.61 \times 0.3 = 0.066$ μ/L

With an Abs_o equal to 75%, a SysD_O equal to 0.01 μg/kg BW/day was obtained [51].

Because pyrethroid resistance is now widespread among malaria vector populations, attempts have been made to use at least two molecules having different modes of action to impregnate the same bed net. Thus, the combination consists of bifenthrin (pyrethroid) with carbosulfan (carbamate) [52] or chlorpyrifos methyl (organophosphorus) [53], deltamethrin (pyrethroid) with PBO and dinotefuran [54], or acetamiprid, nitenpyram, thiamethoxam, clothianidin, imidacloprid, and thiacloprid [55], which all belong to the neonicotinoids, α-cypermethrin (pyrethroid) with chlorfenapyr (pyrrole) [56], and permethrin (pyrethroid) with pyriproxyfen (juvenile hormone analog) [57].

In the same way, insecticide-treated nets and indoor residual spraying are often used together in the same households with the aim of achieving greater impact on mosquitoes than either method alone [58–61].

Nevertheless, to our knowledge, no study has tried to evaluate the cumulative potential exposure to these insecticides used together.

15.5 MOSQUITO COILS

Mosquito coils are licensed as repellent and should be used either outdoors or in rooms where there is enough ventilation to prevent smoke inhalation. Ramesh and Vijayalakshmi [62] investigated the extent of deposition of residues of *d-trans*-allethrin on various surfaces of a room resulting from the use of a mosquito coil for a period of 30 days. Experiments were performed in a 26.3 m^3 (3 × 3.5 × 2.5 m) room. The ceiling, floor, and four sides were covered with 25 × 25 cm papers to measure the residual deposition and dissipation of the insecticide in the room. The mosquito coil presented a diameter of 12 cm, a length of 83 cm and a weight of 14 g. The *d-trans*-allethrin (0.2% w/w) was allowed to burn for 8 h. The experiments were conducted during 30 days by igniting one mosquito coil every day at 8 p.m. Two different experiments were conducted: one keeping the room fully closed and the other keeping the doors and windows (number not given) open during daytime for a period of 14 h. Samples were collected for analysis on days 1, 3, 5, 7, 10, 15, 20, 25, 30, 35, 40, and 45 [62]. After the initial use of 8 h in the closed room, the residues of *d-trans*-allethrin deposited on the ceiling, sides, and floor were equal to 6.34, 4.68, and 20 μg/m^2, respectively. The amounts gradually increased and reached a maximum at 30 days of 148.63, 170.72, and 184.52 μg/m^2 for the ceiling, sides, and floor, respectively. After the termination of the use of the mosquito coil at day 30, the concentrations gradually decreased to reach 58.12, 76.44, and 90.25 μg/m^2 at day 45 on the ceiling, sides, and floor, respectively. After the initial use of 8 h in the open room, the residues of *d-trans*-allethrin deposited on the ceiling, sides, and floor were equal to 1.65, 2.69, and 4.46 μg/m^2, respectively. Again, an increase was recorded till the 30th day, where the concentrations were equal to 13.88, 17.08, and 18.56 μg/m^2 on the ceiling, sides, and floor, respectively. After 45 days, the concentrations were all <0.1 μg/m^2 [62]. Other attempts have been made to evaluate the potential exposure to pyrethroids resulting from the use of mosquito coils [16,63,64]. However, due to some gaps in the protocols used they have been excluded from this review.

15.6 IMPREGNATED CLOTHES

Macedo et al. [49] estimated the exposures via permethrin-impregnated battle-dress uniforms (BDUs). It was assumed that BDUs were used 18 h/day and 250 days/year with no wash-off or degradation of permethrin on the BDUs over time (worst-case scenario). Inhalation exposure was considered negligible due to the low vapor pressure of permethrin and this particular use.

The acute and subchronic dermal exposures from BDUs were estimated from Eq. (15.11).

$$\mathrm{PE} = (\mathrm{CR} \times \mathrm{SA} \times \mathrm{TF} \times \mathrm{AbR} \times \mathrm{EF})/\mathrm{BW} \tag{15.11}$$

where

- PE is the potential exposure (mg/kg BW)
- CR is the clothing residue (mg a.i./cm^2), which is assumed to be 0.125 mg permethrin/cm^2
- SA is the surface area (cm^2)
- TF is the transfer factor from the clothing to the skin, estimated to be 0.49%
- AbR is the dermal absorption rate
- EF is exposure frequency (h/day)

It is noteworthy that a TF of 2.5% was assumed by Afsset [48]. Dermal exposure was calculated for both adults (60 kg) and children (10 kg) with clothes impregnated with permethrin at 1,250 mg/m^2. The dermal doses for adults and children were estimated as 36.57 and 12.91 mg/day, respectively [48].

Macedo et al. [49] estimated the chronic dermal exposure using BDUs from Eq. (15.11) by multiplying BW by (D × Y)/AT, where D is the exposure duration (d), Y is number of years, and AT is averaging time (365 days × 10 years = 3,650) [49].

Afsset [48] also assumed that the inhalation exposure was negligible but a potential oral exposure was considered for children by hand-to-mouth transfer and from chewing/sucking. In the former case, it was considered that 10% of the insecticide on the hand was transferred to the mouth, leading to an exposure potential of 0.028 mg/day. In the latter case, the surface in contact with the mouth was equal to 0.005 m^2 and an exposure potential of 0.1875 mg/day was calculated. By considering an oral absorption of 100% and a dermal absorption of 2%, total doses of 0.012 and 0.047 mg/kg/day were obtained. The ADI of permethrin being equal to 0.05 mg/kg/day, this led to margin of safety (MOS) values of 4.2 and 1.1 for adults and children, respectively.

Interestingly, Afsset [48] also considered the scenario where impregnated clothes were re-impregnated by a spray can at 4% of permethrin. A spray can of 100 mL, including 4 g of permethrin, allows us to impregnate 5.1 m^2 (0.78 g of permethrin/m^2 of cloth). With a dermal contamination of 1% (i.e., 40 mg) and a dermal absorption of 2% (i.e., 0.8 mg), a dose of 0.013 mg/kg/day was obtained for an adult of 60 kg. With the ADI of permethrin equal to 0.05 mg/kg/day, this led to a MOS of 3.85 [48].

15.7 CONCLUSIONS

Applying insecticides indoors for the control of mosquitoes raises several important considerations concerning exposure and safety.

Indoor insecticide diffusion devices produce surface deposition and air distribution patterns that can vary sharply in fate, concentration, and kinetics. Thus, for example, insecticides contained in vaporizers are released slowly near the floor in a room and have a particular potential for long-term and low-level exposure (about 6 h per day). Conversely, insecticides in spray cans are dispersed directionally at human height and are characterized by a short-term (around 10 s) but high-level exposure potential. The extent and duration of the exposure are therefore highly product specific but also person dependent. The age of the person in relation to the type of product is of paramount importance. Thus, for example, young children are potentially more exposed than adults in impregnated bed nets. Crawling children exhibiting hand-to-mouth behavior are more exposed to contaminated indoor dust than adults. The diversity of the diffusion devices and the exposure scenarios they imply explain the rather high number of experimental results available in the literature for the different age groups.

However, studies dealing with the deposition and distribution of insecticides indoors to control mosquitoes are made in controlled environments and by respecting the recommendations of use of the devices. The obtained results obligatorily differ from those found in occupied residential settings. The normal functioning of a multiroom house with its occupants of different ages is much more complex than that of an experimental room without residents or at best with mannequins equipped with samplers [15]. Users do not always comply with the recommendations of use of these insecticide dispersion devices because it is commonly believed that increasing the application time is beneficial, due to phobias, and so on. In a more normal way, it is evident that depending on the country or region of life, the use of these devices may be occasional or almost continuous throughout the year. This is the case in regions with outbreaks of dengue, chikungunya, or Zika.

Insecticide diffusion devices are always studied separately, whereas in practice they are often used together such as in indoor residual spraying and insecticide-treated nets or mosquito coils and spot spraying. Moreover, the indoor environment is not free from pollutants, especially insecticides, and this contamination should be taken into account when assessing exposure to the insecticides used for control of mosquitoes. Finally, there is a continuum between the indoor and outdoor environment that, excluding outdoor contamination, especially when considering mosquito control programs, minimizes the levels of contamination and, as a result, those of potential exposures.

Faced with the diversity of exposure scenarios to be taken into account in indoor environments, it is obvious that *in silico* tools provide invaluable help for simulating any kind of real or hypothetical situations of human exposure to vector control insecticides. Their ability to fit best with reality is generally directly linked to their level of complexity. Thus, for example, SprayExpo [18] is a freely available simple box model aimed at calculating the human inhaled and dermal doses following the release of an active substance by a spray process. It requires a limited number of input data; the processes are described by few equations but the output results are obligatorily far from reality, although they remain interesting, especially for comparison purposes. Matoba and co-workers [19,24–28,34,35] have computed different fugacity models allowing simulation of the dispersion of biocides within a typical Japanese room. These models could be of interest to simulate indoor transport and fate of insecticides applied by space spraying, electric vaporizing, and residual spraying. However, they are not available and the information provided by the authors is too limited to recompute these models. Nevertheless, others [65,66] have designed more documented fugacity models to simulate the transport and fate of chemicals in indoor environments. Fugacity-based models can include any kinds of indoor compartments and subcompartments, simulate indoor intermedia transport and removal processes, and account for various physicochemical processes. In the same way, there exist more complex dermal [67] and inhalation [68] exposure models, but their use requires input data that are not always available.

ACKNOWLEDGMENTS

The financial support from the French Agency for Food, Environmental and Occupational Health & Safety (Anses) is gratefully acknowledged (CRD-2015-24). Special thanks go to Christophe Lagneau for kind permission to use the picture of IRS from EID Méditerranée.

REFERENCES

[1] J.A. Leech, W.C. Nelson, R.T. Burnett, S. Aaron, and M.E. Raizenne, *It's about time: A comparison of Canadian and American time–activity patterns*, J. Expo. Anal. Environ. Epidemiol. 12 (2002), pp. 427–432.

[2] C. Schweizer, R.D. Edwards, L. Bayer-Oglesby, W. J. Gauderman, V. Ilacqua, M. J. Jantunen, H.K. Lai, M. Nieuwenhuijsen, and N. Künzli, *Indoor time–microenvironment–activity patterns in seven regions of Europe*, J. Expo. Sci. Environ. Epidemiol. 17 (2007), pp. 170–181.

[3] R. Barro, J. Regueiro, M. Llompart, and C. Garcia-Jares, *Analysis of industrial contaminants in indoor air: Part 1. Volatile organic compounds, carbonyl compounds, polycyclic aromatic hydrocarbons and polychlorinated biphenyls*, J. Chromatogr. 1216 (2009), pp. 540–566.

[4] C. Garcia-Jares, J. Regueiro, R. Barro, T. Dagnac, and M. Llompart, *Analysis of industrial contaminants in indoor air. Part 2. Emergent contaminants and pesticides*, J. Chromatogr. 1216 (2009), pp. 567–597.

[5] R.I. Krieger, C.E. Bernard, T.M. Dinoff, J.H. Ross, and R.L. Williams, *Biomonitoring of persons exposed to insecticides used in residences*, Ann. Occup. Hyg. 45 (2001), pp. S143–S153.

[6] M.N. Manaca, J.O. Grimalt, M. Gari, J. Sacarlal, J. Sunyer, R. Gonzalez, C. Dobaño, C. Menendez, and P.L. Alonso, *Assessment of exposure to DDT and metabolites after indoor residual spraying through the analysis of thatch material from rural African dwellings*, Environ. Sci. Pollut. Res. 19 (2012), pp. 756–762.

[7] R. Ritter, M. Scheringer, M. MacLeod, and K. Hungerbühler, *Assessment of nonoccupational exposure to DDT in the tropics and the north: Relevance of uptake via inhalation from indoor residual spraying*, Environ. Health Perspect. 119 (2011), pp. 707–712.

[8] H. van den Berg, *Global status of DDT and its alternatives for use in vector control to prevent disease*, Environ. Health Perspect. 117 (2009), pp. 1656–1663.

[9] Stockholm Convention, *Protecting Human Health and the Environment from Persistent Organic Pollutants*, http://chm.pops.int/Home/tabid/2121/Default.aspx

[10] F.W. Gaspar, J. Chevrier, R. Bornman, M. Crause, M. Obida, D.B. Barr, A. Bradman, H. Bouwman, and B. Eskenazi, *Undisturbed dust as a metric of long-term indoor insecticide exposure: Residential DDT contamination from indoor residual spraying and its association with serum levels in the VHEMBE cohort*, Environ. Intern. 85 (2015), pp. 163–167.

[11] IPCS, *DDT in Indoor Residual Spraying: Human Health Aspects*, Environmental Health Criteria 241, World Health Organization, Geneva, Switzerland, 2011.

[12] P. Booij, I. Holoubek, J. Klánová, J. Kohoutek, A. Dvorská, K. Magulová, S. Al-Zadjali, and P. Čupr, *Current implications of past DDT indoor spraying in Oman*, Sci. Total Environ. 550 (2016), pp. 231–240.

[13] J.C. van Dyk, H. Bouwman, I.E.J. Barnhoorn, and M.S. Bornman, *DDT contamination from indoor residual spraying for malaria control*, Sci. Total Environ. 408 (2010), pp. 2745–2752.

[14] S. Sadasivaiah, Y. Tozan, and J.G. Breman, *Dichlorodiphenyltrichloroethane (DDT) for indoor residual spraying in Africa: How can it be used for malaria control*? Am. J. Trop. Med. Hyg. 77 (Sup. 6), 2007, pp. 249–263.

[15] W. Steiling, M. Bascompta, P. Carthew, G. Catalano, N. Corea, A. D'Haese, P. Jackson, L. Kromidas, P. Meurice, H. Rothe, and M. Singal, *Principal considerations for the risk assessment of sprayed consumer products*, Toxicol. Lett. 227 (2014), pp. 41–49.

[16] T.J. Class and J. Kintrup, *Pyrethroids as household insecticides: Analysis, indoor exposure and persistence*, Fresenius J. Anal. Chem. 340 (1991), pp. 446–453.

[17] E. Berger-Preiß, W. Koch, S. Gerling, H. Kock, and K.E. Appel, *Use of biocidal products (insect sprays and electro-vaporizer) in indoor areas—Exposure scenarios and exposure modeling*, Int. J. Hyg. Environ. Health 212 (2009), pp. 505–518.

[18] SprayExpo 2.2, *A Computer Based Model for the Assessment of Inhalation and Dermal Exposure During Spray Application*, http://www.baua.de/en/Topics-from-A-to-Z/Hazardous-Substances/SprayExpo.html

[19] Y. Matoba, J.I. Ohnishi, and M. Matsuo, *A simulation of insecticides in indoor aerosol space spraying*, Chemosphere 26 (1993), pp. 1167–1186.

[20] D. Mackay, *Multimedia Environmental Models. The Fugacity Approach*, Lewis Publishers, Mi, 1991.

[21] D. Mackay and S. Paterson, *Fugacity models*, in *Practical Applications of Quantitative Structure–Activity Relationships (QSAR) in Environmental Chemistry and Toxicology*, W. Karcher and J. Devillers, eds., Kluwer Academic Publishers, Dordrecht, the Netherlands, 1991, pp. 433–460.

[22] J. Devillers, S. Bintein, and W. Karcher, *CHEMFRANCE: A regional level III fugacity model applied to France*, Chemosphere 30 (1995), pp. 457–476.

[23] Y. Matoba, J.I. Ohnishi, and M. Matsuo, *Temperature and humidity dependency of pesticide behavior in indoor simulation*, Chemosphere 30 (1995), pp. 933–952.

[24] Y. Matoba, J. Yoshimura, J.I. Ohnishi, N. Mikami, and M. Matsuo, *Development of the simulation model InPest for prediction of the indoor behavior of pesticides*, J. Air Waste Manag. Assoc. 48 (1998), pp. 969–978.

[25] Y. Matoba and M.P. van Veen, *Predictive residential models*, in *Occupational and Residential Exposure Assessment for Pesticides*, C.A. Franklin and J.P. Worgan, eds., Wiley & Sons, Chichester, England, 2005, pp. 209–242.

[26] Y. Matoba, Y. Takimoto, and T. Kato, *Indoor behavior and risk assessment following space spraying of d-tetramethrin and d-resmethrin*, Am. Ind. Hyg. Assoc. J. 59 (1998), pp. 181–190.

[27] Y. Matoba, J.I. Ohnishi, and M. Matsuo, *Indoor simulation of insecticides in broadcast spraying*, Chemosphere 30 (1995), pp. 345–365.

[28] Y. Matoba, Y. Takimoto, and T. Kato, *Indoor behavior and risk assessment following residual spraying of d-phenothrin and d-tetramethrin*, Am. Ind. Hyg. Assoc. J. 59 (1998), pp. 191–199.

[29] J. Murawski, *Insecticide use in occupied areas of aircraft*, in *Handbook of Environmental Chemistry*, Springer–Verlag, Berlin, vol. 4, part H (2005), pp. 169–190.

[30] C. van Netten, *Analysis and implications of aircraft disinsectants*, Sci. Total Environ. 293 (2002), pp. 257–262.

[31] E. Berger-Preiβ, W. Koch, W. Behnke, S. Gerling, H. Kock, L. Elflein, and K.E. Appel, *In-flight spraying in aircraft: Determination of the exposure scenario*, Int. J. Hyg. Environ. Health 207 (2004), pp. 419–430.

[32] E. Berger-Preiβ, W. Koch, S. Gerling, H. Kock, J. Klasen, G. Hoffmann, and K.E. Appel, *Aircraft disinsection: Exposure assessment and evaluation of a new pre-embarkation method*, Int. J. Hyg. Environ. Health 209 (2006), pp. 41–56.

[33] Y. Matoba, T. Hirota, J.I. Ohnishi, N. Murai, and M. Matsuo, *An indoor simulation of the behavior of insecticides supplied by an electric vaporizer*, Chemosphere 28 (1994), pp. 435–451.

[34] Y. Matoba, J.I. Ohnishi, and M. Matsuo, *Indoor simulation of insecticides supplied with an electric vaporizer by the fugacity model*, Chemosphere 28 (1994), pp. 767–786.

[35] Y. Matoba, A. Inoue, and Y. Takimoto, *Clarifying behavior of prallethrin evaporated from an electric vaporizer on the floor and estimating associated dermal exposure*, J. Pestic. Sci. 29 (2004), pp. 313–321.

[36] A. Vesin, P. Glorennec, B. Le Bot, H. Wortham, N. Bonvallot, and E. Quivet, *Transfluthrin indoor air concentration and inhalation exposure during application of electric vaporizers*, Environ. Intern. 60 (2013), pp. 1–6.

[37] A. Vesin, E. Quivet, B. Temime-Roussel, and H. Wortham, *Indoor transfluthrin concentration levels during and after the application of electric vaporizers using a proton-transfer-reaction mass spectrometer*, Atmos. Environ. 65 (2013), pp. 123–128.

[38] C.J. Weschler and W.W. Nazaroff, *Semivolatile organic compounds in indoor environments*, Atmos. Environ. 42 (2008), pp. 9018–9040.

[39] J.E. Delmaar, M.V.D.Z. Park, and J.G.M. van Engelen, *ConsExpo 4.0 Consumer Exposure and Uptake Models Program Manual*, RIVM report 320104004/2005, http://www.rivm.nl/en/Topics/C/ConsExpo

[40] C.J. Weschler and W.W. Nazaroff, *SVOC partitioning between the gas phase and settled dust indoors*, Atmos. Environ. 44 (2010), pp. 3609–3620.

[41] T. Nazimek, M. Wasak, W. Zgrajka, and W.A. Tursk, *Content of transfluthrin in indoor air during the use of electro-vaporizers*, Ann. Agric. Environ. Med. 18 (2011), pp. 85–88.

[42] B.G. Blackburn, A. Eigege, H. Gotau, G. Gerlong, E. Miri, W.A. Hawley, E. Mathieu, and F. Richards, *Successful integration of insecticide-treated bed net distribution with mass drug administration in central Nigeria*, Am. J. Trop. Med. Hyg. 75 (2006), pp. 650–655.

[43] J.L Vanden Eng, J. Thwing, A. Wolkon, M.A Kulkarni, A. Manya, M. Erskine, A. Hightower, and L. Slutsker, *Assessing bed net use and non-use after long-lasting insecticidal net distribution: A simple framework to guide programmatic strategies*, Malaria J. 9 (2010), pp. 133.

[44] S.M. Barlow, F.M. Sullivan, and J. Lines, *Risk assessment of the use of deltamethrin on bednets for the prevention of malaria*, Food Chem. Toxicol. 39 (2001), pp. 407–422.

[45] K-O TAB, *Deltamethrin-based tablet for bed net impregnation*, http://www.vectorcontrol.bayer.com/~/media/Products/Links/Brochure-K-OTab.ashx

[46] A.N.I. Garrod, A.M. Philipps, and J.A. Pemberton, *Potential exposure of hands inside protective gloves. A summary of data from non-agricultural pesticide surveys*, Ann. Occup. Hygien. 45 (2001), pp. 55–60.

[47] UK POEM, *Predictive Operator Exposure Model*, http://www.hse.gov.uk/pesticides/topics/pesticide-approvals/pesticides-registration/data-requirements-handbook/operator-exposure.htm

[48] AFSSET, *La Lutte Antivectorielle dans le Cadre de l'Epidémie de Chikungunya sur l'Île de la Réunion, Évaluation des Risques Liés à l'Utilisation des Produits Insecticides d'Imprégnation des Moustiquaires et des Vêtements*, Afsset, Maisons-Alfort, France, 2007.

[49] P.A. Macedo, R.K.D. Peterson, and R.S. Davis, *Risk assessments for exposure of deployed military personnel to insecticides and personal protective measures used for disease-vector management*, J. Toxicol. Environ. Health Part A 90 (2007), pp. 1758–1771.

[50] R.K.D. Peterson, L.M. Barber, and J.J. Schleier III, *Net risk: A risk assessment of long-lasting insecticide bed nets used for malaria management*, Am. J. Trop. Med. Hyg. 84 (2011), pp. 951–956.

[51] WHO, *A Generic Risk Assessment Model for Insecticide-treated Nets*, rev. ed., World Health Organization (WHO), 2012.

[52] V. Corbel, F. Darriet, F. Chandre, and J.M. Hougard, *Insecticides mixtures for mosquito net impregnation against malaria vector*, Parasite 9 (2002), pp. 255–259.

[53] F. Darriet, V. Corbel, and J. M. Hougard, *Efficacy of mosquito nets treated with a pyrethroid-organophosphorous mixture against Kdr- and Kdr+ malaria vectors* (Anopheles gambiae), Parasite 10 (2003), pp. 359–362.

[54] F. Darriet and F. Chandre, *Combining piperonyl butoxide and dinotefuran restores the efficacy of deltamethrin mosquito nets against resistant* Anopheles gambiae *(Diptera: Culicidae)*, J. Med. Entomol. 48 (2011), pp. 952–955.

[55] F. Darriet and F. Chandre, *Efficacy of six neonicotinoid insecticides alone and in combination with deltamethrin and piperonyl butoxide against pyrethroid-resistant* Aedes aegypti *and* Anopheles gambiae *(Diptera: Culicidae)*, Pest. Manag. Sci. 69 (2013), pp. 905–910.

[56] R. N'Guessan, C. Ngufor, A.A. Kudom, P. Boko, A. Odjo, D. Malone, and M. Rowland, *Mosquito nets treated with a mixture of chlorfenapyr and alphacypermethrin control pyrethroid-resistant* Anopheles gambiae and Culex quinquefasciatus *mosquitoes in West Africa*, PLoS ONE 9 (2) (2014), pp. e87710.

[57] A.A. Koffi, L.P. Ahoua Alou, A. Djenontin, J.P.K. Kabran, Y. Dosso, A. Kone, N. Moiroux, and C. Pennetier, *Efficacy of Olyset®Duo, a permethrin and pyriproxyfen mixture net against wild pyrethroid-resistant* Anopheles gambiae s.s. *from Côte d'Ivoire: An experimental hut trial*, Parasite, 22, 28, (2015), pp. 1–8.

[58] M.J. Hamel, P. Otieno, N. Bayoh, S. Kariuki, V. Were, D. Marwanga, K.F. Laserson, J. Williamson, L. Slutsker, and J. Gimnig, *The combination of indoor residual spraying and insecticide-treated nets provides added protection against malaria compared with insecticide-treated nets alone*, Am. J. Trop. Med. Hyg. 85 (2011), pp. 1080–1086.

[59] C. Ngufor, R. N'Guessan, P. Boko, A. Odjo, E. Vigninou, A.N. Asidi, M. Akogbeto, and M. Rowland, *Combining indoor residual spraying with chlorfenapyr and long lasting insecticidal bednets for improved control of pyrethroid-resistant* Anopheles gambiae: *An experimental hut trial in Benin*, Malaria J. 10 (2011), pp. 343.

[60] F.O Okumu, S.S. Kiwarel, S.J. Moore, and G.F. Killeen, *Mathematical evaluation of community level impact of combining bed nets and indoor residual spraying upon malaria transmission in areas where the main vectors are* Anopheles arabiensis *mosquitoes*, Paras. Vect. 6 (2013), pp. 17.

[61] M. Pinder, M. Jawara, L.B.S. Jarju, K. Salami, D. Jeffries, M. Adiamoh, K. Bojang, S. Correa, B. Kandeh, H. Kaur, D.J. Conway, U. D'Alessandro, S.W. Lindsay, *Efficacy of indoor residual spraying with dichlorodiphenyltrichloroethane against malaria in Gambian communities with high usage of long-lasting insecticidal mosquito nets: A cluster-randomised controlled trial*, Lancet 385 (2015), pp. 1436–1446.

[62] A. Ramesh and A. Vijayalakshmi, *Impact of long-term exposure to mosquito coils: Residual deposition and dissipation of d-trans-allethrin in a room*, J. Environ. Monit. 4 (2002), pp. 202–204.

[63] A. Ramesh and A. Vijayalakshmi, *Monitoring of allethrin, deltamethrin, esbiothrin, prallethrin and transfluthrin in air during the use of household mosquito repellents*, J. Environ. Monit. 3 (2001), pp. 191–193.

[64] H. Li, M.J. Lydy, and J. You, *Pyrethroids in indoor air during application of various mosquito repellents: Occurrence, dissipation and potential exposure risk*, Chemosphere 144 (2016), pp. 2427–2435.

[65] D.H. Bennett and E.J. Furtaw, *Fugacity-based indoor residential pesticide fate model*, Environ. Sci. Technol. 38 (2004), pp. 2142–2152.

[66] H.-M. Shin, T.E. McKone, and D.H. Bennett, *Intake fraction for the indoor environment: A tool for prioritizing indoor chemical sources*, Environ. Sci. Technol. 46 (2012), pp. 10063–10072.

[67] T. Schneider, R. Vermeulen, D.H. Brouwer, J.W. Cherrie, H. Kromhout, and C.L. Fogh, *Conceptual model for assessment of dermal exposure*, Occup. Environ. Med. 56 (1999), pp. 765–773.

[68] M.A. Jayjock, C.F. Chaisson, S. Arnold, and E.J. Dederick, *Modeling framework for human exposure assessment*, J. Expos. Sci. Environ. Epidemiol. 17 (2007), pp. S81–S89.

Index

A

B

N

P

Q

R

S

T

U

V

W

Y

Z